Sustainability Sciences in Asia and Africa

Sustainable Agriculture and Food Security

Series Editor

This book series support the global efforts towards sustainability by providing timely coverage of the progress, opportunities, and challenges of sustainable food production and consumption in Asia and Africa. The series narrates the success stories and research endeavors from the regions of Africa and Asia on issues relating to SDG 2: Zero hunger. It fosters the research in transdisciplinary academic fields spanning across sustainable agriculture systems and practices, post- harvest and food supply chains. It will also focus on breeding programs for resilient crops, efficiency in crop cycle, various factors of food security, as well as improving nutrition and curbing hunger and malnutrition. The focus of the series is to provide a comprehensive publication platform and act as a knowledge engine in the growth of sustainability sciences with a special focus on developing nations. The series will publish mainly edited volumes but some authored volumes. These volumes will have chapters from eminent personalities in their area of research from different parts of the world.

Ram Swaroop Meena • Sandeep Kumar
Gourisankar Pradhan • Manoj Kumar Jhariya
Cherukumalli Srinivasarao
Editors

Carbon-Negative Agriculture

Springer

Editors
Ram Swaroop Meena
Department of Agronomy
Banaras Hindu University
Varanasi, Uttar Pradesh, India

Gourisankar Pradhan
Department of Agriculture and Farmer's Empowerment
Government of Odisha
Bhubaneswar, Odisha, India

Cherukumalli Srinivasarao
ICAR - Indian Agricultural Research Institute
New Delhi, India

Sandeep Kumar
ICAR - Indian Agricultural Research Institute, Regional Station
Karnal, Haryana, India

Manoj Kumar Jhariya
Department of Farm Forestry
Sant Gahira Guru Vishwavidyalaya
Ambikapur, Chhattisgarh, India

ISSN 2730-6771 ISSN 2730-678X (electronic)
Sustainability Sciences in Asia and Africa
ISSN 2730-6798 ISSN 2730-6801 (electronic)
Sustainable Agriculture and Food Security
ISBN 978-981-95-1482-3 ISBN 978-981-95-1483-0 (eBook)
https://doi.org/10.1007/978-981-95-1483-0

This Springer imprint is published by the registered company Springer Nature Singapore Pte Ltd.
The registered company address is: 152 Beach Road, #21-01/04 Gateway East, Singapore 189721, Singapore

Preface

In the face of accelerating climate change, the global agricultural sector is at a crossroads. Agriculture, while being a key driver of economic growth and food security, is also a significant contributor to greenhouse gas emissions. However, within this challenge lies a profound opportunity—transforming agriculture into a solution for climate mitigation through carbon-negative farm practices.

This edited book, *Carbon-Negative Agriculture*, presents a timely and comprehensive exploration of emerging strategies, technologies, and policy frameworks aimed at transitioning agriculture from a carbon-emitting to a carbon-sequestering enterprise. The concept of carbon-negative agriculture—going beyond neutrality to actively remove and store atmospheric carbon—offers a visionary yet achievable path toward sustainability.

The book brings together 17 meticulously curated chapters contributed by leading scientists, experts, and policymakers. Each chapter addresses a critical dimension of carbon-negative agriculture, ranging from the potential of carbon credits, regenerative and circular farming systems, and agroforestry practices to the integration of digital technologies, renewable energy solutions, and policy instruments. The book also illuminates the institutional and economic enablers for large-scale adoption.

The book is primarily intended for postgraduate students, research scholars, scientists, and academicians across disciplines such as agronomy, environmental sciences, plant biology, and ecology. It also serves as a vital resource for land managers, food scientists, and emerging researchers who are contributing to innovative solutions in climate-smart agriculture. Furthermore, this book will benefit policymakers, program planners, development practitioners, and stakeholders in both the public and private sectors engaged in climate and agriculture policy formulation, implementation, and monitoring.

By combining scientific rigor with policy relevance, this book aspires to create greater awareness and build capacity around the pressing agenda of achieving net-zero and carbon-negative goals within agroecosystems. We hope that the insights, methodologies, and technologies shared herein will inspire collaborative action and innovation, ultimately contributing to a more resilient, regenerative, and climate-responsible global food system.

We express our sincere gratitude to all contributing authors, reviewers, and collaborators whose commitment and expertise have shaped the quality of this work. We believe that *Carbon-Negative Agriculture* will serve as both a reference and a roadmap for students, professionals, and changemakers striving to turn agricultural systems into powerful allies in the fight against climate change.

Varanasi, Uttar Pradesh, India — Ram Swaroop Meena
Karnal, Haryana, India — Sandeep Kumar
Bhubaneswar, Odisha, India — Gourisankar Pradhan
Ambikapur, Chhattisgarh, India — Manoj Kumar Jhariya
New Delhi, India — Cherukumalli Srinivasarao

Preface to the Book Series

The UN's second sustainable goal (SDG) aims to eliminate world hunger, achieve food security and improved nutrition, and promote sustainable agriculture within the next seven years. Yet despite the tireless efforts of the global scientific community, national agricultural research systems, and nongovernment organizations the world over, more than 800 million people still suffer from undernourishment. To address this food security gap, it's estimated that global food production systems will need to expand their capacities and increase production by 70% to feed the world's burgeoning population.

It is a momentous challenge that is further exacerbated by more frequent extreme weather events, unpredictable rainfall patterns, and pest-pathogen dynamics associated with global climate change. With time, these shifts will have a profoundly negative impact on agricultural productivity and farm incomes, particularly in the developing world. To address these challenges, we need sustainable innovations in our agri-farming systems that lead to improved food production in resource-constrained agricultural settings, and more equitable, gender-diverse farming practices that encourage the participation of women in agriculture.

This book series, Sustainable Agriculture and Food Security, has been planned to highlight and support global efforts toward sustainability by providing timely coverage of the progress, opportunities, and challenges of sustainable food production and consumption in Asia and Africa. The series highlights the success stories and research endeavors from the two regions on issues relating to SDG 2: Zero Hunger, outlining new approaches and technological innovations that are enhancing agricultural productivity. It fosters research in transdisciplinary academic fields spanning sustainable agriculture systems and practices, post-harvest, and food supply chains.

The focus of the series is to provide a comprehensive publication platform to stakeholders across the agricultural value chain, and to act as a knowledge engine in the growth of sustainability sciences. So far in this series, we've highlighted innovative research approaches, nascent technologies, and new farming practices.

One such shift is the niche, yet growing practice of Carbon Negative Agriculture. Our food systems contribute to an estimated 30% of global emissions. Major sources of these emissions include livestock, rice cultivation, and the use of nitrogen fertilizers. Therefore, practices that reverse agricultural emissions would be highly beneficial, resulting in better soil health and water retention, an increased resilience to drought and extreme weather, and, in the long term, better productivity and profitability for farmers.

While the benefits of a carbon-negative agriculture sector are clear, there are significant roadblocks and challenges to address before it can become a mainstream approach. This includes the impact to yields, training, and management of carbon-neutral practices, particularly in developing nations, and the complexity of accurately measuring and verifying carbon-negative farming practices.

In view of this, the present book, *Carbon-Negative Agriculture*, edited by Ram Swaroop Meena, Sandeep Kumar, Gourisankar Pradhan, Manoj Kumar Jhariya, and Cherukumalli Srinivasarao, provides an important platform to discuss the potential of carbon-negative agriculture approaches in delivering sustainable global food systems. The book presents 16 chapters written by multidisciplinary experts and practitioners, each containing important insights into the topic. I wholeheartedly congratulate the Editors and all contributing authors for their valuable insights. I have no doubts that this book will serve as an essential resource for students, researchers, farmers, educators, policymakers, and other stakeholders engaged in transforming agriculture around the world.

I wish to extend my sincere thanks and gratitude to the Springer staff, particularly Sanchi Bhimrajka, Associate Editor (Books, Medicine and Life Sciences), and Nick Melchior, Editorial Director (Medicine and Life Sciences Books, Asia), for their constant support for the accomplishment of this compendium. The cooperation received from my supervisors and research leaders such as Prof. Andrew Deeks (Vice Chancellor), Prof. Peter Eastwood (Deputy Vice Chancellor—Research), Prof. Peter Davies (Pro Vice Chancellor), and my laboratory colleagues—Rhys Copeland, Anu Chitikineni, and Vanika Garg from Murdoch University (Australia)—is also gratefully acknowledged. I would like to thank my family members—Monika Varshney, Prakhar Varshney, and Preksha Varshney—for their love and support in discharging my duties as Series Editor.

Perth, WA, Australia Rajeev K. Varshney

Contents

Editors and Contributors

About the Editors

Ram Swaroop Meena born into a farmer's family, is an Associate Professor (L-13A) in the Department of Agronomy at the Institute of Agricultural Sciences, BHU, Varanasi (UP). Dr. Meena has secured first division in all his classes, with triple NET, JRF, and SRF qualifications from ICAR, and is an RGNF Awardee from UGC, GOI. He is a Fellow of the National Academy of Agricultural Sciences (ICAR-FNAAS), Fellow of the National Academy of Sciences, India (NASI) and the National Academy of Biological Sciences (FNABS), as well as a Fellow of the Society for Rapeseed-Mustard Research (ICAR-SRMR). He served as a visiting scientist at IRRI-ISARC under an INSA fellowship. Dr. Meena has received the Raman Research Fellowship awarded by the Ministry of Education, GOI. He has been listed in the World's Top 2% Scientists by Stanford University, USA, and the Elsevier report. He is also listed in 8th position out of 100 scientists from India in the field of Agriculture and Forestry Sciences by the AD Scientific Index at the global level ranking. Dr. Meena did his post-doctoral fellowship on soil carbon sequestration at the CFAES Rattan Lal Center for Carbon Management and Sequestration, The Ohio State University, Columbus, OH, United States. Dr. Meena has more than 15 years of research and teaching experience at the under/postgraduate/and PhD levels and has contributed significantly to soil sustainability, crop productivity, resource use efficiency, and climate-resilient agriculture with 10 externally funded projects, including MOE, ICAR, and DST (GOI). He has to his credit more than 130 national and international research and review papers with an H-index of 75, an i-10 index of 218, and citations of over 16,500. He has published 28 books, including 24 books at the international level (Springer, Elsevier, *etc.*), and 85 book chapters. Dr. Meena is credited with a patent for developing a novel, affordable biochar separation technique for integrated nutrient management. Dr. Meena has contributed to the agricultural extension activities in the three adopted villages under the DST project and is active as an associate coordinator in trainings, meetings, workshops, and farmers' fairs.

Sandeep Kumar was born into a farming family in the village Jaswantpura, Rajgarh (Churu), Rajasthan, India. He joined the Agricultural Research Services (ARS) in 2020 at the ICAR-Indian Institute of Pulses Research (IIPR), Regional Centre, Bhopal. Currently, he is working as a Scientist (Agronomy) at the ICAR-Indian Agricultural Research Institute (IARI), Regional Station, Karnal. Dr. Kumar completed his schooling in his native village and earned his Bachelor's degree in Agriculture (Hons.) in 2013 from the College of Agriculture, Lalsot (SKRAU, Bikaner, Rajasthan). He completed his Master's degree from Banaras Hindu University, Varanasi, and his PhD in Agronomy from Chaudhary Charan Singh Haryana Agricultural University, Hisar, with a specialization in nutrient management in grain legumes. Throughout his academic career, he secured first division and qualified for the ICAR-NET three times, in addition to receiving the prestigious SRF fellowships from the Indian Council of Agricultural Research. His areas of expertise include improving resource-use efficiency in conservation agriculture-based rice–wheat systems, cropping systems diversification, carbon sequestration, and the carbon credits for sustainable crop production. As an editor and author, Dr. Kumar has published ten books with reputed publishers such as Academic Press Elsevier, Springer, and leading national publishers. He has authored over 60 research and review papers in high-impact, peer-reviewed national and international journals (with NAAS ratings up to 20.0) and contributed to more than 60 book chapters with publishers like Elsevier and Springer. Dr. Kumar has received several accolades, including the Land 2020 Best Paper Award from MDPI and Wiley's Top Cited Article Award for 2022–2023. He was also honored with the Best Young Scientist Award (Regional Station) for 2024–2025 by ICAR-IARI, New Delhi. With over 6000 citations, an h-index above 45, and an i10-index exceeding 80, Dr. Kumar is actively engaged in teaching postgraduate and PhD students at the IARI–IIWBR Mega Hub, Karnal. He also plays a key role in institutional development and resource generation initiatives. Currently, he is guiding both PhD and Master's research scholars. His future research goals focus on soil carbon sequestration, climate change adaptation, carbon credit generation, regenerative agriculture, and low-emission technologies in a systems-based approach.

Gourisankar Pradhan is working as an Assistant Agriculture Officer in the Department of Agriculture and Farmers' Empowerment (DAFE), Government of Odisha, India. Dr. Pradhan did his schooling at Kisinda High School, intermediate at Veer Surendra Sai College, Sambalpur, and graduated from agriculture in 2016 from the College of Agriculture, Bhawanipatna, OUAT, Odisha. He was awarded a gold medal for his Master's degree from Visva-Bharati University, West Bengal, India, in 2018. He got his PhD degree from Banaras Hindu University, UP, in 2022. He secured first division throughout his learning career with double NET (ICAR-ASRB), ICAR-NTS, and SRF from the Indian Council of Agricultural Research. He also got the best MSc (Agronomy) thesis award and the best oral presentation award. His research interests have focused on soil carbon management, sustainable crop production, soil health, nutrient management, etc. He has published 15 research papers and 3 review papers in leading national and international peer-reviewed journals with impact factors and has also contributed 10 book chapters to Elsevier and Springer, and more than 8 popular articles. He also attended and presented more than 30 research papers and posters at international and national congresses, seminars, symposiums, conferences, workshops, training programs, etc. He has over 1117 citations with an h-index of 17 in Research Gate and more than 987 citations with an i10 index of 18 in Google Scholar. He is a life member of the Indian Society of Agronomy, the National Academy of Biological Sciences, the *Journal of Agrometeorology*, and the Agricultural Technology Development Society.

Manoj Kumar Jhariya is working as Senior Assistant Professor and HOD in the Department of Farm Forestry at Sant Gahira Guru Vishwavidyalaya (Formerly, Sarguja University), Sarguja, Ambikapur, Chhattisgarh, India. He is working on forest ecology, natural resource management and sustainability, carbon sequestration, and agroforestry. Dr. Jhariya is listed as one among the World's Top 2% Scientists by Stanford University (USA) and Elsevier Data in 2024, and has over 7580 citations with an *h*-index of 48 and an i10-index of 123. Dr. Jhariya has published more than 105 international research and review papers in peer-reviewed, reputed journals. He has published 21 books (Springer, Elsevier, Wiley, Apple Academic Press, Taylor and Francis, CRC, etc.), 120 book chapters, and several extension articles. Dr. Jhariya acquired BSc (Agriculture), MSc, and PhD (both in Forestry) degrees from Indira Gandhi Krishi Vishwavidyalaya, Raipur, Chhattisgarh, India. He received the University Gold Medal, Chhattisgarh Young Scientist Award, UGC-GoI Fellowship, Best Paper Award of MDPI, Switzerland, Sir Dietrich Brandis Award, Excellence of Research Award-2024, Young Achiever Award-2024, and Paryavaran Padam Puraskar in 2025. He is editorial board member of several journals and life member of The Indian Science Congress Association, Applied and Natural Science Foundation, Society for Advancement of Human and Nature, Medicinal and Aromatic Plants Association of India, and ISDS Society, *International Journal of Development and Sustainability*, The Asia Society of Researchers, The Agriculture Forum for Technical Education of Farming Society, Society of

Green Development, India, National Academy of Biological Sciences, and The Society for Science of Climate Change and Sustainable Environment. He has supervised more than 121 MSc (Farm Forestry) students and is involved in teaching and research.

Cherukumalli Srinivasarao has been serving as Director, ICAR-Indian Agricultural Research Institute (IARI) since December 26, 2024. Dr. Rao, born in Anigandlapadu village, Krishna District of Andhra Pradesh, did his BSc and MSc in Agriculture from Agricultural College, Bapatla (ANGRAU); PhD from IARI, and post-doctoral fellowship from Tel-Aviv University, Israel. Before joining as Director, ICAR-IARI, he worked as Director, ICAR-National Academy of Agricultural Research Management (NAARM), Hyderabad from April 28, 2017 to December 26, 2024; Director, ICAR-Central Research Institute for Dryland Agriculture (CRIDA) during 2014-17 and as Project Coordinator, All India Coordinated Research Project on Dryland Agriculture (AICRPDA), Hyderabad during 2013–14 and several other positions. He had 33 years of rich experience in research management. He is the recipient of 48 National and International awards from ICAR, NAAS, ISCA, ICRISAT, IARI, DST, FAI, including the top ICAR Award of Rafi Ahmed Kidwai for Outstanding Research in Agricultural Sciences. Dr. Rao received twice prestigious Sardar Patel Outstanding ICAR Institution Awards from the Hon'ble Prime Minister for CRIDA (2015) and the Hon'ble Union Agriculture Minister for NAARM (2021) under his leadership. He is recipient of the Top ICAR Award of Rafi Ahmad Kidwai. Dr. Rao was invited recently by the Royal Society of London to the workshop and made a presentation on the biodiversity-water nexus with Food Security. Recently, he was awarded the Coromandel Plant Nutrition Award by the FAI and was selected as the Dr. N.S. Randhawa Memorial of NAAS, received during the XVII Agricultural Science Congress held on February 22nd, 2025. He is a Fellow of several reputed professional bodies such as INSA, NASI, NAAS, Indian Society of Soil Science, A.P. Academy of Sciences, and Telangana Academy of Sciences, which speaks of his caliber and achievements in the field of agricultural sciences, climate change, etc. He is the President of the Biology Section of NASI for 2024. Dr. Rao is credited with 361 Research papers, authored/edited 56 books, 109 national and international book chapters, 69 bulletins, 37 policy papers, 38 popular articles, and obtained 18 copyrights. He served as Member/Chairman of various National and International Committees—International Dryland Development Commission (IDDC), Cairo, Egypt; Asian Long Term Experimental Network for Agriculture (ALTENA), Japan; Expert Committee of National Biodiversity Authority; Chairman of Capacity Building Unit of DARE/ICAR. He visited about 20 countries for various scientific meetings. Dr. Rao also held key responsibilities in National Missions such as National Mission for Sustainable Agriculture (NMSA) as its Technical Chairman and ICAR flagship program, NICRA, as National Coordinator. Dr. Rao represented India in UNFCCC CoP Meetings held in France, Germany, and Spain as Global Climate Change Negotiator and in IPCC Meeting in Japan and APN Meeting in Bangkok, Thailand. His major research interests are: Climate Change, Soil Carbon Sequestration, Integrated Genetic Natural Resource Management (IGNRM), Water Stressed Agro Ecosystems, Contingency Crop Planning, Agriculture Research Management, Human Resource Development, Building Agri Startups, Agri Business Management, and Research Policy Development in Agriculture and Allied Sectors. Dr. Rao has accumulated over 12,900 citations and holds an h-index exceeding 55.

Contributors

Ashna Acharya Department of Floriculture and Landscaping, Punjab Agriculture University, Ludhiana, India

Newlove A. Afoakwah Department of Food Science and Technology, Faculty of Agriculture, Food and Consumer Sciences, Nyankpala Campus, University for Development Studies, Tamale, Ghana

A. T. Ahamed Akram Faculty of Technology, Department of Biosystems Technology, South Eastern University of Sri Lanka, Oluvil, Sri Lanka

K. R. Anusha Department of Vegetable Science, Punjab Agricultural University, Ludhiana, Punjab, India

P. Ashoka Department of Agronomy, College of Agriculture (University of Agricultural Sciences, Dharwad), Haveri, Karnataka, India

Tanmaya Kumar Bhoi ICFRE-Arid Forest Research Institute, Jodhpur, Rajasthan, India

Peter Osei Boamah Department of Ecological Agriculture, Bolgatanga Technical University, Bolgatanga, Ghana

Sharanjit Kaur Brar Department of Soil Science, Punjab Agricultural University, Ludhiana, Punjab, India

P. Chaitra Department of Soil Science, VC Farm, University of Agricultural Sciences, Bangalore, Karnataka, India

Dixit Chaudhary Department of Floriculture and Landscape Architecture, Dr. Y S Parmar University of Horticulture and Forestry, Nauni, Solan, Himachal Pradesh, India

Amit Kumar Dash ICAR-National Institute of Seed Science and Technology, Mau, Uttar Pradesh, India

Saikat Dey Division of Agronomy, School of Agriculture and Rural Development, Ramakrishna Mission Vivekananda Educational and Research Institute, Narendrapur, West Bengal, India

Anannya Dhar Division of Agronomy, School of Agriculture and Rural Development, Ramakrishna Mission Vivekananda Educational and Research Institute, Narendrapur, West Bengal, India

Saheed Garnaik ICAR-National Academy of Agricultural Research Management, Hyderabad, Telangana, India

Shambhunath Ghosh ICAR-Indian Institute of Wheat and Barley Research, IARI Karnal Hub, Karnal, Haryana, India

Debasis Golui ICAR-Indian Agricultural Research Institute, New Delhi, India

Dennis Agyemanh Nana Gookyi Institute for Scientific and Technological Information, Council for Scientific and Industrial Research, New Delhi, Delhi, India

Ram Gopal North Eastern Hill University, Tura, Meghalaya, India

N. J. Gowthamchand Department of Soil Science, Dr. Rajendra Prasad Central Agricultural University, Pusa, Bihar, India

Alekhya Gunturi Division of Agronomy, ICAR-Indian Agricultural Research Institute, New Delhi, India

Soura Shuvra Gupta ICAR-Indian Agricultural Research Institute, New Delhi, India

Raymond Gyaang Department of Electrical and Electronics Engineering, Bolgatanga Technical University, Bolgatanga, Ghana

Heena Department of Chemistry, Baba Farid College of Engineering and Technology, Bathinda, Punjab, India

M. B. Fathima Jemziya Department of Biosystems Technology, Faculty of Technology, South Eastern University of Sri Lanka, Oluvil, Sri Lanka

Manoj Kumar Jhariya Department of Farm Forestry, Sant Gahira Guru Vishwavidyalaya, Ambikapur, Chhattisgarh, India

Anamika Kandel Department of Agricultural and Environmental Biology, The University of Tokyo, Tokyo, Japan

M. Karthika Amrita Vishwa Vidyapeetham, Coimbatore, Tamil Nadu, India

Abhay Kumar Faculty of Forestry, Department of Silviculture and Agroforestry, Birsa Agricultural University, Ranchi, Jharkhand, India

Sandeep Kumar ICAR-Indian Agricultural Research Institute, Regional Station, Karnal, Haryana, India

Sanjeev Kumar Institute of Agricultural Sciences, Banaras Hindu University, Varanasi, Uttar Pradesh, India

Sunita Kumari Faculty of Forestry, Department of FPU, Birsa Agricultural University, Ranchi, Jharkhand, India

Kerobim Lakra Department of Agricultural Economics, Ranchi Agriculture College, Birsa Agricultural University, Ranchi, Jharkhand, India

Ritika Lamichhane Nebraska Extension, University of Nebraska-Lincoln, Lincoln, NE, USA

Dinesh Mahto KVK, Jehanabad, Bihar Agricultural University, Bhagalpur, Bihar, India

Lalit Kumar Mainway Division of Agronomy, ICAR-Indian Agricultural Research Institute, New Delhi, India

M. S. Malik Faculty of Forestry, Department of Silviculture and Agroforestry, Birsa Agricultural University, Ranchi, Jharkhand, India

Priyanka Meena ICAR-National Institute of Biotic Stress Management, Baronda, Raipur, Chhattisgarh, India

Ram Swaroop Meena Department of Agronomy, Institute of Agricultural Sciences, Banaras Hindu University, Varanasi, Uttar Pradesh, India

Krishna Kant Mina Institute of Agricultural Sciences, Banaras Hindu University, Varanasi, Uttar Pradesh, India

Hemant Kumar Mina Institute of Agricultural Sciences, Banaras Hindu University, Varanasi, Uttar Pradesh, India

Muneeb M. Musthafa Faculty of Technology, Department of Biosystems Technology, South Eastern University of Sri Lanka, Oluvil, Sri Lanka

B. Mahesh Naik Malla Reddy University, Hyderabad, Telangana, India

M. M. M. Najim Faculty of Agriculture, Sultan Sharif Ali Islamic University, Bandar Seri Begawan, Brunei

M. N. F. Nashath Department of Biosystems Technology, Faculty of Technology, South Eastern University of Sri Lanka, Oluvil, Sri Lanka

Jacqueline Onumah Department of Ecological Agriculture, Bolgatanga Technical University, Bolgatanga, Ghana

P. R. Oraon Faculty of Forestry, Department of Silviculture and Agroforestry, Birsa Agricultural University, Ranchi, Jharkhand, India

Harshita Pant Department of Environmental Studies, University of Delhi, Delhi, India

C. M. Parihar ICAR-Indian Agricultural Research Institute, New Delhi, India

Vikas Patidar ICAR-Indian Institute of Wheat and Barley Research, IARI Karnal Hub, Karnal, Haryana, India

Kiranmoy Patra ICAR-Indian Agricultural Research Institute, New Delhi, India

Panneerselvam Peramaiyan International Rice Research Institute (IRRI), South Asia Regional Centre, Varanasi, Uttar Pradesh, India

Gourisankar Pradhan Institute of Agricultural Sciences, Banaras Hindu University, Varanasi, Uttar Pradesh, India

Sunil Kumar Prajapati Division of Agronomy, ICAR-Indian Agricultural Research Institute, New Delhi, India

Muneshwar Prasad KVK, Jehanabad, Bihar Agricultural University, Bhagalpur, Bihar, India

Satya Priya Division of Agricultural Extension, ICAR-Indian Agricultural Research Institute, New Delhi, India

Sanskriti Rai Institute of Agricultural Sciences, Banaras Hindu University, Varanasi, Uttar Pradesh, India

Md Basit Raza ICAR-Directorate of Floricultural Research, Pune, Maharashtra, India

Gongalla Sreeja Reddy ICAR-Indian Agricultural Research Institute, New Delhi, India
Division of Agricultural Extension, ICAR-Indian Agricultural Research Institute, New Delhi, India

K. Srikanth Reddy Division of Agronomy, ICAR-Indian Agricultural Research Institute, New Delhi, India
International Rice Research Institute (IRRI), South Asia Regional Centre, Varanasi, Uttar Pradesh, India

K. Sreenivasa Reddy Division of Agronomy, ICAR-Indian Agricultural Research Institute, New Delhi, India
International Rice Research Institute (IRRI), South Asia Regional Centre, Varanasi, Uttar Pradesh, India

Suddamalla Manoj Kumar Reddy Division of Agricultural Extension, ICAR-Indian Agricultural Research Institute, New Delhi, India

Arkaprava Roy Division of Agronomy, ICAR-Indian Agricultural Research Institute, New Delhi, India
ICAR-National Institute of Biotic Stress Management, Baronda, Raipur, Chhattisgarh, India

Swati Sabnam School of Agriculture, Rural and Tribal Development, Ramakrishna Mission Ashrama, Ranchi, Jharkhand, India

Anusmita Saha ICAR-Indian Institute of Wheat and Barley Research, IARI Karnal Hub, Karnal, Haryana, India

Priyanka Saha Division of Agronomy, ICAR-Indian Agricultural Research Institute, New Delhi, India
ICAR-Krishi Vigyan Kendra, West Garo Hills, Tura, Meghalaya, India

Nilutpal Saikia Division of Agronomy, ICAR-Indian Agricultural Research Institute, New Delhi, India

Tayari Salifu Department of Agricultural Engineering, Bolgatanga Technical University, Bolgatanga, Ghana

Ayan Sarkar ICAR-Indian Agricultural Research Institute, New Delhi, India

Sukamal Sarkar Division of Agronomy, School of Agriculture and Rural Development, Ramakrishna Mission Vivekananda Educational and Research Institute, Narendrapur, West Bengal, India

Suchismita Satapathy Institute of Agricultural Sciences, Banaras Hindu University, Varanasi, Uttar Pradesh, India

Priyanka Sharma NICE School of Business Studies, Shobhit University, Meerut, India

Tarun Sharma Division of Agronomy, ICAR-Indian Agricultural Research Institute, New Delhi, India

Ankita Shukla South Pole Group UK Limited, London, UK

Ayushi Singh Department of Soil Science, Punjab Agricultural University, Ludhiana, Punjab, India

Alok Sinha Division of Agronomy, ICAR-Indian Agricultural Research Institute, New Delhi, India

Neha Thakur University School of Agriculture Sciences, Rayat Bahra University, Mohali, Punjab, India

M. G. Mohamed Thariq Faculty of Technology, Department of Biosystems Technology, South Eastern University of Sri Lanka, Oluvil, Sri Lanka

Ajad Singh Verma Institute of Agricultural Sciences, Banaras Hindu University, Varanasi, Uttar Pradesh, India

Samuel Dodobatia Wetajega Department of Electrical and Electronics Engineering, Bolgatanga Technical University, Bolgatanga, Ghana

Fortunatus Aabangbio Wulnye Department of Telecommunication Engineering, Kwame Nkrumah University of Science and Technology, Kumasi, Ghana

Amit Kumar Yadav Institute of Agricultural Sciences, Banaras Hindu University, Varanasi, Uttar Pradesh, India

Sufyan Yakubu Department of Electrical and Electronics Engineering, Bolgatanga Technical University, Bolgatanga, Ghana

Chapter 1
Potential of Carbon Credits and Solutions Through Carbon Farming to Achieve Carbon-Negative Agriculture

Ankita Shukla and Harshita Pant

Abstract Rapid climate change has left the world struggling with its impact on human and natural systems. Agriculture is one such bridging sector serving as both the cause and victim of climate extremes if not managed well. Contrarily, agricultural land carries enough potential to sequester carbon due to its sheer size. Hence, it is crucial to switch conventional/traditional agricultural practices to sustainable regenerative practices that not only sequester soil organic carbon (SOC) but also improve the overall soil and crop health for long-term viability and profitability. Currently, most farmers raising crops and managing agricultural lands in developing and least developed countries are small-scale and belong to marginalized sections in financial terms. One of the most credible solutions to support agriculture and/or green technology projects is through carbon finance in the form of carbon credits led by carbon platforms like the Voluntary Carbon Market (VCM). Recently, there has been an upsurge of carbon farming projects willing to change to regenerative practices such as no tillage, cover cropping, and low chemical utilization to be able to register and seek financial support through carbon credits. Currently, there are 23 projects sequestering carbon and 14 projects reducing emissions in agriculture, already issuing 13.35 million and 4.65 million carbon credits, respectively. However, VCM is an evolving carbon market that presents its unique challenges, particularly in relation to agricultural project types, including additionality, monitoring, reporting, and verification (MRV) costs, permanence, and social acceptance, among others. Nonetheless, it still stands as one of the great platforms to serve financial incentives to farmers for implementing sustainable practices. Monetization of carbon credits motivates farmers to switch to non-conventional practices and equips them with the tools to implement such practices for the long run until they observe the improved soil health and crop yields themselves.

A. Shukla (✉)
South Pole Group UK Limited, London, UK

H. Pant
Department of Environmental Studies, University of Delhi, Delhi, India

R. S. Meena et al. (eds.), *Carbon-Negative Agriculture*, Sustainability Sciences in Asia and Africa, https://doi.org/10.1007/978-981-95-1483-0_1

Keywords Carbon credits · Carbon farming · Emissions trading · Leakage · Permanence · Voluntary carbon market

Abbreviations

ACCUs	Australian Carbon Credit Units
ACoGS	Avoided Conversion of Grasslands and Shrublands
ACR	American Carbon Registry
AFOLU	Agriculture, Forestry, and Other Land Uses
ALM	Agricultural Land Management
ARR	Afforestation, Reforestation and Revegetation
CAP	Carbon Agriculture Policy
CDM	Carbon Development Mechanism
CH_4	Methane
CHSP	California's Healthy Soils Program
CO_2	Carbon Dioxide
COP	Conference of the Parties
CORSIA	Carbon Offsetting and Reduction Scheme for International Aviation
ETS	Emission Trading Scheme
EU	European Union
GAEC	Good Agri-Environmental Conditions
GHGs	Greenhouse gases
GLAS	Green Low-Carbon Agri-Environmental Scheme
GS	Gold Standard
IET	International Emissions Trading
IFM	Improved Forest Management
IPCC	International Panel on Climate Change
ITMOs	Internationally Transferred Mitigation Outcomes
JI	Joint Implementation
MRV	Monitoring, Reporting, and Verification
N_2O	Nitrous Oxide
NBS	Nature-Based Sources
NDCs	Nationally Determined Contributions
NPRT	Non-Permanence Risk Tool
REDD	Reduced Emissions from Deforestation and Degradation
SDG	Sustainable Development Goals
SOC	Soil Organic Carbon
TIST	The International Small Group and Tree Planting Program
UNFCCC	United Nations Framework Convention on Climate Change
VCM	Voluntary Carbon Market
VCS	Verified Carbon Standard
WRC	Wetlands Restoration and Conservation

1.1 Introduction

Agriculture, Forestry, and Other Land Uses (AFOLU), if managed properly, can yield enough produce to sustain the population while also bringing co-benefits to humans and the natural systems of the Earth. With the increasing climate adversities, the agriculture sector has recently gained attention for its significant mitigation potential in carbon reduction and avoidance of greenhouse gases (GHGs). The agriculture sector, on average, accounted for 13–20% of anthropogenic GHG emissions from 2010 to 2019 (Nabuurs et al. 2023). On the contrary, the managed and natural ecosystems of the Earth acted as carbon sinks for one-third of the anthropogenic carbon emissions during the same time (Nabuurs et al. 2023). Considering the high emission and mitigation potential of agricultural and forestry land, it is important to manage such areas to act as carbon sinks rather than sources of carbon emissions.

The agriculture sector holds the largest potential to either reduce/or remove significant volumes of GHGs into the atmosphere due to the vast land use changes that have occurred over the years. To achieve global food security in the future, it is essential to not only improve soil health for better yields but also effectively reduce and/or remove GHG emissions, helping to combat climate change. With the rapid increase in population, the agricultural land use sector has grown rapidly with unsustainable and conventional practices aimed at achieving the highest yield. However, with the rapidly changing climate, farmlands are becoming vulnerable to withstand extreme conditions, leading to soil erosion, deteriorated soil health, and low crop yields.

To combat the situation of trembling agri-food systems and rapid global warming, it is crucial to motivate farmers across the globe to switch to sustainable agricultural practices. Conservative and sustainable agricultural practices are proven to be the best land-based management solution for carbon sequestration (Kumara et al. 2023). However, implementing such activities at all farm levels requires funding, regulations, and a well-established market/system to inspect the execution and successful continuation of mitigation activities over the years.

Monetizing each ton of soil organic carbon (SOC) sequestered through sustainable agricultural practices in the form of carbon credits is seen as a pivotal financial instrument to achieve this transition. Following specific rules and regulations set by a carbon standard body, a project is developed under the voluntary carbon market (VCM) to accurately measure and monitor the SOC flux that shall be accounted for as the carbon credit. Once the project is successfully registered, monitored, and verified under a carbon standard body, carbon credits are then issued to the project owner to be further traded with buyers. Although the carbon market is an evolving market that comes with its own set of challenges and opportunities, it is still considered one of the vital climate finance instruments to boost carbon farming toward sustainable agricultural practices.

1.2 Concept of Carbon Credit in the Context of Climate Change

1.2.1 Climate Change Trends

Climate change has been an inevitable phenomenon since the time of Earth and life formations. However, the pace and severity of climate change have been altered since the Industrial Revolution. Before the industrial era, atmospheric carbon dioxide (CO_2) remained almost constant at around 280 ppm. Since then, each decade post-1990 has been warmer than all previous decades, and the CO_2 gas increased by 50%, reaching 422.33 ppm in the year 2024 (CO_2 Earth 2024).

Anthropogenic activities such as the burning of fossil fuel, deforestation, and changes in land use led to the increase of three major GHGs—CO_2, methane (CH_4), and nitrous oxide (N_2O). An increase in the atmospheric mixing ratio of these three gases has led to recurrent, extreme, and irregular natural calamities such as land and marine heat waves, cyclones, and droughts. World Economic Forum 2023 confirms climate change and environmental stress to be among the top four major global risks in the long term, based on the severity. The risks are directly associated with extreme weather events, critical changes to Earth systems, biodiversity loss, ecosystem collapse, and natural resource shortages (World Economic Forum 2023). These risks indirectly lead to other devastating impacts, particularly on food security and human mobility. As a result of such climate impacts, nations are becoming vulnerable to climate resilience, narrowing down the opportunity for global sustainable development goals (SDGs). So, it is critical to limit the global temperature to 1.5° C and well below 2° C as urged by organizations like the International Panel on Climate Change (IPCC) and the United Nations Framework Convention on Climate Change (UNFCCC).

1.2.2 Trends in Agriculture

Each decade since 1990, 55% of carbon emissions have been absorbed by the oceans and sinks on land. This reflects a constant absorption rate, despite the decadal growth of GHG emissions (World Meteorological Organization 2023). However, land in general serves as a net source of carbon emissions due to unprecedented anthropogenic activities. Agriculture and forestry contribute the highest to CH_4 gas emissions, one of the major GHGs responsible for global warming. A classic example of this is rice cultivation, which is the major source of CH_4 emissions, yet an important staple food for most of the Asian continent. Additionally, N_2O emissions are also greatly influenced by agricultural activities, by using excess fertilizers on the field. In any case, agriculture/farming is an important land use change to ensure global food security, therefore essential to be well managed to maximize the mitigation opportunities while delivering food and other renewable resources.

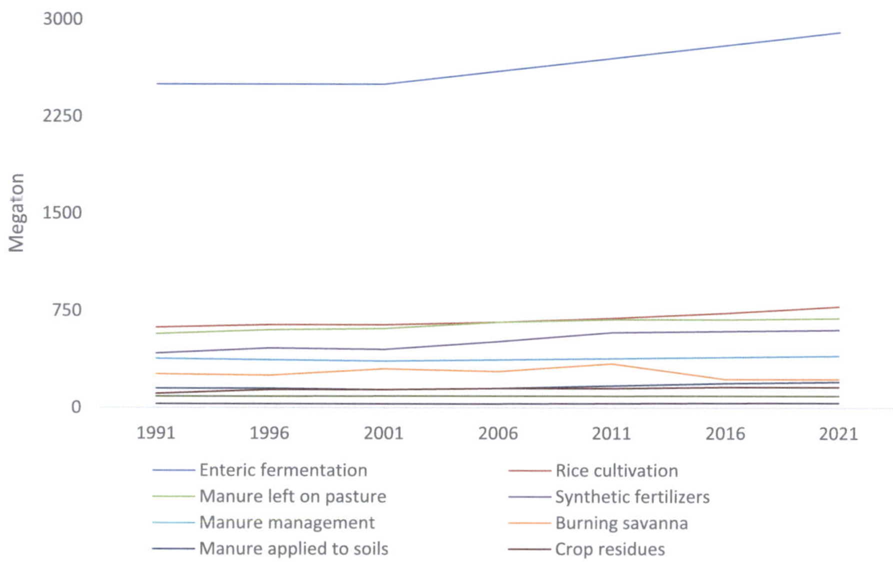

Fig. 1.1 Chart demonstrating the emissions under different sub-sectors of agriculture for the period 1991–2021. (*Data Source*: Climate Watch 2024)

Figure 1.1 below demonstrates the major drivers of historical emissions in the agricultural sector across the years. Large amounts of CH_4 and N_2O are emitted through activities such as enteric fermentation, manure management, rice cultivation, and biomass burning. Most emissions from rice cultivation, enteric fermentation, and manure management originate from the Asian and African continents. In all global regions, agriculture is the dominant emissions source, thereby gaining attention in recent times (Nabuurs et al. 2023). All these activities are heavily responsible for the cumulative increase in GHG emissions to the atmosphere. However, if these practices are managed well, they can severely reduce and/or remove the GHGs from the atmosphere.

Agriculture, despite being the dominant reason for CH_4 and N_2O emissions, has only seen a steady increase in cropland area over the decades. Figure 1.2 represents the ever-increasing global cropland up to the year 2023. The land use change to agriculture and farming is increasing at a constant rate, despite the high GHG emissions it causes, as observed from both Figs. 1.1 and 1.2. This is due to the increasing pressure on agri-food systems to sustain the basic livelihood needs of the ever-growing population. This can be achieved not only by increasing the cropland area but also by improving the yields per hectare. However, with the extremes of climate change, the agriculture sector has become vulnerable to producing high yields while being resilient. Sustainable agricultural practices are the only plausible solution to build strong, yielding croplands with reduced GHG emissions and to tackle both the current global issues of food scarcity and the impact of climate change simultaneously.

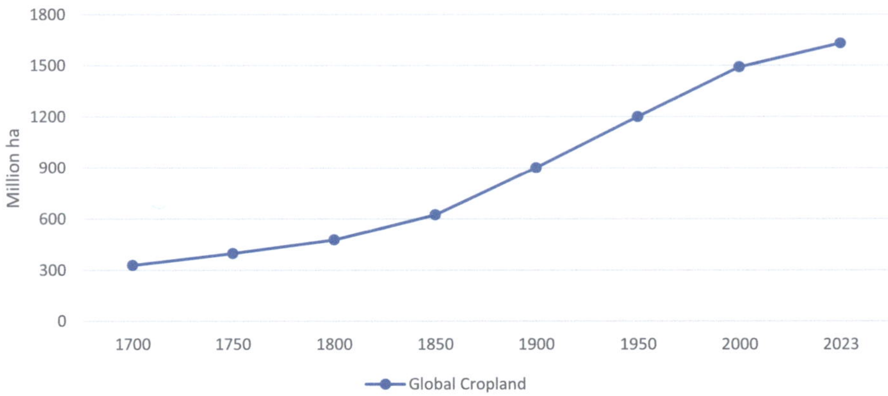

Fig. 1.2 Increase in global cropland from the period 1700 to 2023. (*Data Source*: Taylor & Rising (2021); UN World Population Prospects (2021) – processed by Our World in Data)

1.2.3 Concept of Carbon Credit and Voluntary Carbon Market

To ensure that agricultural systems become climate-resilient and adopt sustainable practices, it is essential to have government support programs and adequate funding. As realized under the Paris Agreement, countries are required to submit their nationally determined contributions (NDCs) to limit the temperature increase to 1.5 °C above pre-industrial levels. In recent times, the agriculture and forestry sectors have gained attention for their large-scale carbon removal potential through afforestation/reforestation, SOC management, and other sustainable practices (Nabuurs et al. 2023). Still, the ground policies and funding required to achieve this net carbon removal through the agriculture sector remain highly unregulated and fragile. One of the most credible solutions to support agriculture and/or green technology projects is through carbon finance led by carbon platforms like the VCM. As the name suggests, VCM is a voluntary market established to help reduce global carbon emissions by incentivizing climate action. VCM allows all levels of entities/organizations, such as governments, non-governmental organizations (NGOs), corporations, and individuals, to offset their carbon emissions and trade them in the form of carbon credits. The history of VCM lies in the hands of the Kyoto Protocol in the year 1997, which led to the emergence of compliance carbon market mechanisms such as the Carbon Development Mechanism (CDM), International Emissions Trading (IET), and Joint Implementation (JI). CDM soon became one of the popular instruments to allow Annex I countries to financially support the carbon projects being developed in the non-Annexe I countries (developing) in the form of certified emissions reductions (CERs). There was a spike of interest observed from both government and corporate sectors that further led to the development of several other voluntary carbon offset programs and registries, such as Gold Standard (GS) and VERRA, in the early 2000s. Subsequently, CDM lost

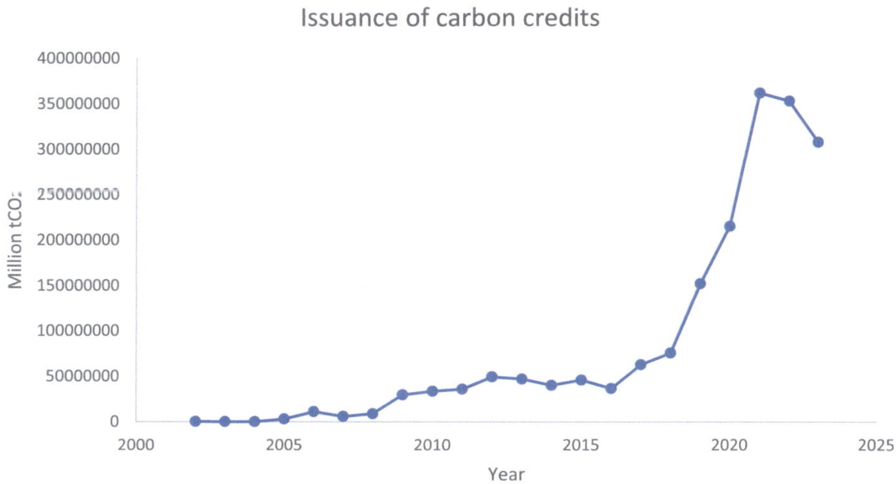

Fig. 1.3 Trajectory of VCM through the years 2000–2023, including all registries of the time. (*Data Source*: Bravo and Mikolajczyk 2022)

relevance due to the unpredictable future of Kyoto Protocol commitments, thereby leaving VERRA and GS to be the dominant carbon standards/registries in the VCM alongside the existing compliance market (regulated by the government). Since then, the VCM has been considered one of the important instruments for achieving net-zero commitments and later NDCs for countries under the Paris Agreement (Spilker and Nugent 2022). Figure 1.3 below describes the trajectory of carbon credit issuance across the years from 2000 onward. Post-2016, there was an upsurge of carbon credits in the VCM, seeking attention from all actors of the market, including government officials, brokers, financial institutions, scientists, and project developers.

The VCM is run by independent carbon standards/registries like Verified Carbon Standard (VCS), GS, and American Carbon Registry (ACR), to name a few. These registries are responsible for laying out the rules, methodological approaches, and monitoring, reporting, and verification (MRV) requirements for each project type capable of carbon emission reduction, removal, and/or avoidance. Once the project qualifies the registry's criteria, it is then registered under the relevant standard. The registered project further executes an annual cycle of MRV, which, when verified, leads to the issuance of carbon credits each year. A detailed description of the typical carbon project development phases is provided below in Fig. 1.4.

One carbon credit is equivalent to one ton of carbon emissions reduced, removed, and/or avoided from the environment. Carbon credits are issued periodically till the end of the project crediting period as set under the registry. Major share of carbon credits issued so far under the four largest carbon standards is described in Fig. 1.5. Once the carbon credits are issued, these credits/units/certificates are capable of being traded to buyers seeking to offset their emissions and reach carbon-neutral targets for their organizations (Spilker and Nugent 2022).

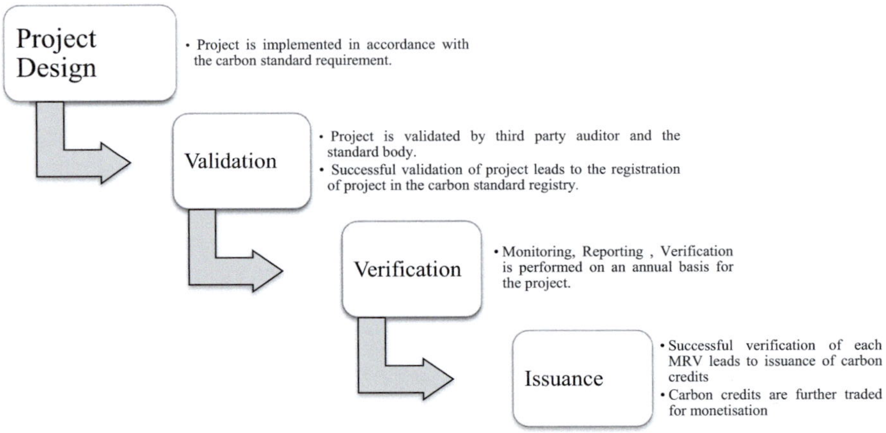

Fig. 1.4 Process of carbon project development in a voluntary carbon market

Fig. 1.5 Four largest carbon standard bodies with their respective percentage of total carbon credits issued until the year 2023. (*Data Source:* Bravo and Mikolajczyk 2022)

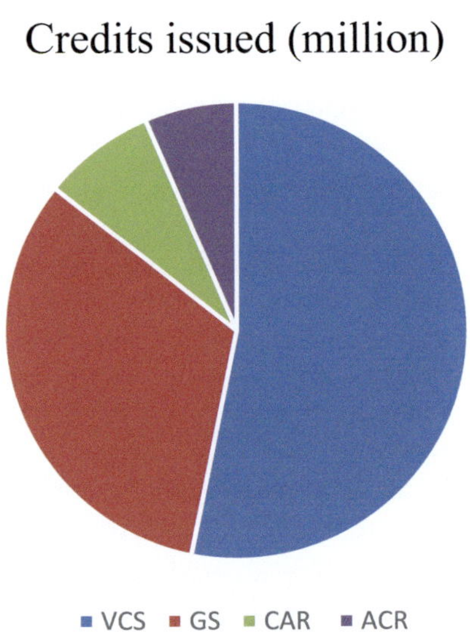

1.2.4 Linking Agriculture with a Climate Mitigation Strategy Under the Carbon Credit

Carbon farming can therefore serve as a great tool in the management of the agriculture and forestry sectors. "Carbon farming refers to sequestering and storing carbon and/or reducing GHG emissions at the farm level" (McDonald et al. 2021). Improved agricultural practices are recognized as an important mitigation strategy to combat climate change under the Paris Agreement. VERRA, supposedly the largest carbon registry under VCM, promotes carbon farming to be developed and registered under the VM00042 methodology. There are several other methodologies, such as the Soil Organic Carbon Framework methodology by GS and the methodology for regenerative land management by Social Carbon, to name a few. Following the example case of VM00042, farmers are allowed to develop carbon farming projects if they follow one or more of the following regenerative practices on their farms: Improve fertilizer (organic or inorganic) management; Improve water management/irrigation; Reduce tillage/improve residue management; Improve crop planting and harvesting (e.g., improved agroforestry, crop rotations, cover crops); and/or Improve grazing practices (VERRA 2024a). Most of the barriers and opportunities for the implementation of such practices lie in the non-Annex countries that are still underfunded by the government at large. So far, the most promising financial instruments to cover the carbon emission reduction, removal, and/or avoidance project activities are the VCM and carbon credits.

1.3 Status, Challenges, and Opportunities of Carbon Farming in Terms of Carbon Credits

1.3.1 Market Trend/Current Status of Carbon Farming in the VCM Market

Voluntary Carbon Market issued the maximum amount of carbon credits in the year 2021, post which, the issuance was dropped by 22%, accounting for 475 million credits issued in the year 2022 (World Bank 2023). The demand side of the VCM market is primarily derived from corporate commitments to achieve net-zero targets. Oil and gas firms, food and beverage, and tourism companies bought the highest number of credits in the year 2021 (World Bank 2022).

Figure 1.6 represents the total issuance across years for different project technologies in the voluntary carbon market. As can be observed, the market has been dominated by renewable energy activities; however, nature-based sources (NBS), that is, agriculture and forestry, are subsequently gaining attention from both the supply and demand side. Till the year 2022, renewable energy projects accounted for 45% of the total registered projects under the VCM (World Bank 2023). Nonetheless, the drastic decline in the cost of renewable energy project implementations, affordability, and accessibility for this project type has now raised concerns about the additionality criterion. Therefore, a shift of focus and a large opportunity for other project types as agriculture and NBS, are witnessed on the supply side predominantly. Agriculture-based projects not only claim carbon emission reductions or removals but also serve many other co-benefits, generating additional value.

In addition to the demand–supply ratio, carbon prices tend to vary based on factors like project type, geography, age/vintage of credits, and co-benefits. There is also a variation observed under different trading platforms for different project types. Removals and avoidance credits typically generated from the NBS and agriculture

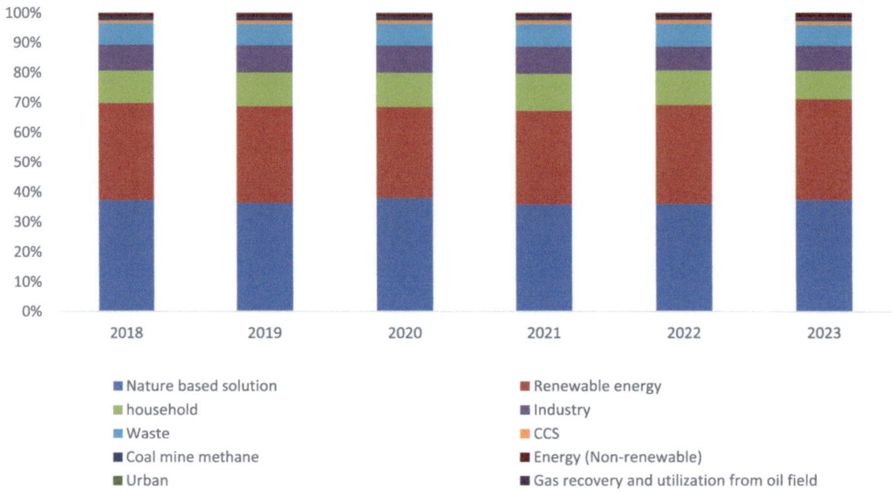

Fig. 1.6 Total issuance for different project technologies for the period 2018–2023. (*Data Source*: Bravo and Mikolajczyk 2022)

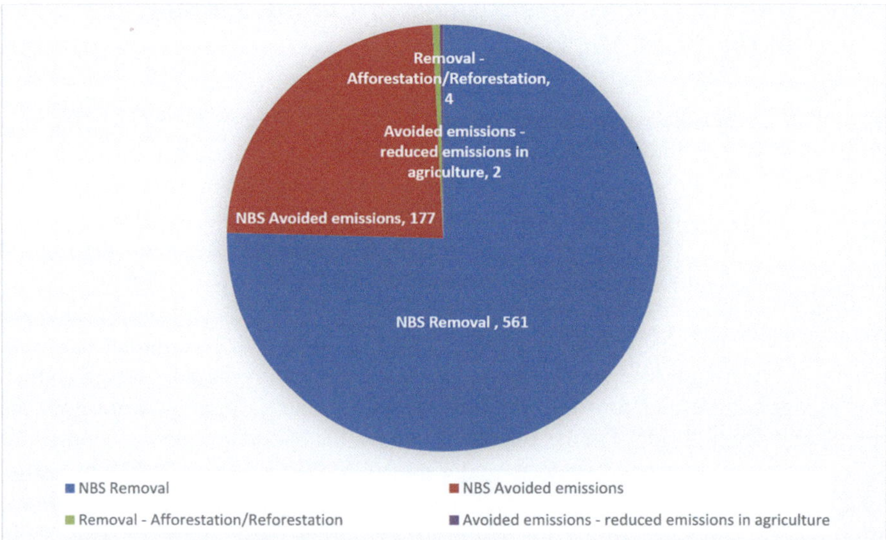

Fig. 1.7 Overview of nature-based solution projects. (*Data Source*: Bravo and Mikolajczyk 2022)

sectors are currently placed on the high-price end. Presumably, the project complexity and shortage on the supply side can be the reason for higher prices in the sector (World Bank 2022). As depicted in Fig. 1.7, agriculture sector projects for reduced emissions are less than 1% in the NBS sector. However, carbon removals obtained through carbon farming, including regenerative practices and agroforestry, still account for a higher share in the NBS Removal category already.

1.3.2 Agriculture Sector Under the VCM

For the year 2023, nature-based solution projects were mainly dominated by the forestry sector as compared to agriculture. Most projects were registered under afforestation/reforestation, Improved Forest Management (IFM), and avoided deforestation activities, as compared to only 14 registered projects under reduced emissions from agriculture and 23 registered carbon sequestration agricultural activities (Bravo and Mikolajczyk 2022). Those 23 projects sequestering carbon and 14 reduced emission projects in agriculture are already issuing carbon credits for 13.35 million and 4.65 million, respectively (Bravo and Mikolajczyk 2022). Although most projects were based on carbon removals, the issuance of credits was heavily inclined toward the avoidance credits. Presumably, the complexity and uncertainties around carbon removal technologies are the prime reason for the lower demand for removal credits. Also, a longer period and robust measurement techniques are required to develop an optimum removal project. On the other side, avoidance credits, although selling on higher demand, can only pool a higher number of credits if large-scale conservation projects are being undertaken (Dibattista et al. 2023).

1.3.3 Challenges in the VCM Market for Carbon Credit

As witnessed above in Sect. 1.3.2, VCM experienced a large upsurge in carbon credit issuance ever since it was established. Entities and organizations from various sectors became seemingly interested and invested in the VCM market, both from the supply and demand sides. The year 2021 recorded the highest issuance volume of around 362 million credits from the VCM, consolidating all registries' data (Bravo and Mikolajczyk 2022), post which, VCM started facing scrutiny and criticism on the credibility and viability of carbon credits.

1.3.3.1 Governance Challenge

As discussed in Sect. 1.3.3, the corporate sector drove the demand for carbon credits to offset their carbon emissions and reach net-zero targets. However, to integrate such offsets in the wider emission reduction contributions for the countries and to avoid double accounting, a legal architecture was developed under the Paris Agreement. VCM currently lacks a viable business model to align with the NDCs as laid out under the Paris Agreement, ensuring the environmental integrity and credibility of carbon credits (Kreibich and Hermwille 2021). Currently, there is a lot of commotion about the uncertainties of how the voluntary carbon credit is aligned with the NDCs. Further, countries and governments are yet to clearly define the types of voluntary credits that will be considered for NDCs, disabling the VCM from basing its emission reduction models.

To conform to environmental integrity, corresponding adjustments are required from both the demand and supply sides to ensure the avoidance of double claiming. If the mitigation outcome is reflected in the NDCs of the buyer country, it should be deducted from the NDCs of the host country. This concept is referred to as a corresponding adjustment that requires parties to apply robust accounting when transferring Internationally Transferred Mitigation Outcomes (ITMOs) (Kreibich and Hermwille 2021).

To establish robust voluntary credits that align with the NDCs criteria, it is desirable for projects to seek additional certifications or labels presenting co-benefits. This would certainly challenge the infrastructure and monitoring capacity of such labels. However, once achieved, it can certainly leverage the reputational benefits and credibility of carbon offsets (Kreibich and Hermwille 2021).

1.3.3.2 Market Heterogeneity and Fragmentation

The VCM carries a US$50 billion market potential by the year 2030 (Chen et al. 2021). A market with this monetary potential struggles with its heterogeneous business model, accommodating a wide range of project types. Each project type is unique on the grounds of geography, technology, scale, methodology, and standards/registries under which it is developed. Each standard/registry differs in the rules and requirements, methodological approach, MRV procedures, and costing for projects, leading to severe heterogeneity both at the project and standard levels (Chen et al. 2021). Each standard under the VCM is focused on different aspects of the project and aims to work toward environmental integrity through its unique approach. A classic example is the GS registry, which tends to focus on the overall SDGs that the project is contributing to, in addition to reducing carbon emissions. Furthermore, due to the constantly evolving nature of this sector, standards are always upgrading to new taxonomies and requirements. Lack of standardization among registries and regulations thus leads to issues like double accounting, transparency, and integrity for the VCM, whereas a lack of complete knowledge about the markets leads to suspicion about their credibility.

1.3.3.3 Demand-Supply Imbalance

The VCM has always suffered from a demand–supply imbalance due to its loose structure and regulation on this criterion. Several actors at the local level are responsible for creating supply and distorted prices that interfere with the effectiveness of the market (Miltenberger et al. 2021). Recently, the VCM has suffered

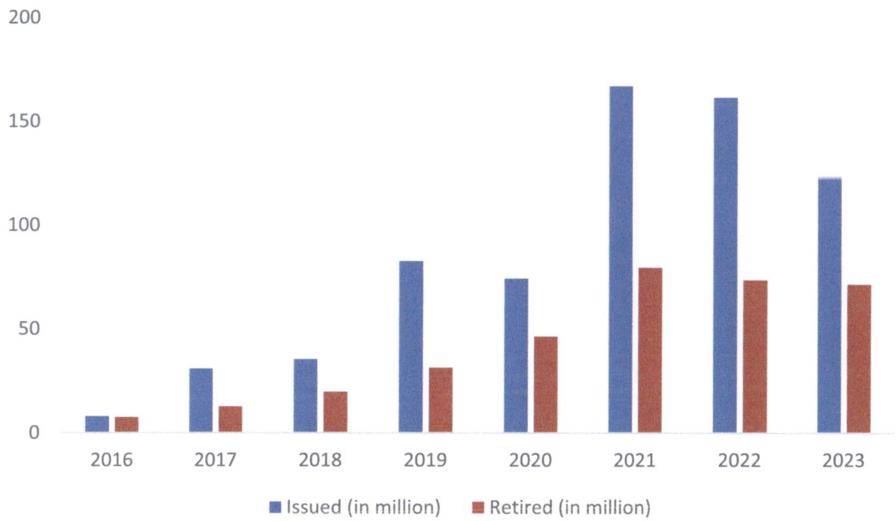

Fig. 1.8 Ratio of issuance and retirement volume of credits for the NBS sector. (*Data Source*: Bravo and Mikolajczyk 2022)

upheaval in its credibility and viability; the market needs to focus on the kind of credit supplies it is generating. This is another reason why the shift in demand from renewables to NBS projects has been observed lately (World Bank 2023). The supply–demand imbalance is also observed in the retirement ratio of credits from the VCM, as described in Fig. 1.8 (specifically dedicated toward the NBS sector). The issuance volume for a project reflected in the registry account determines the volume that is now available to be traded. As soon as the credit is traded, it must simultaneously be "retired" from the registry and accounted for in the buyer's targets and obligations. The carbon credit cannot be retired until the removal, reduction, and/or avoidance mitigation outcome is complete. Therefore, there is always a lag between the issuance of the credits and the actual number of credits retired and traded in the market. The entire process of credit to complete its mitigation activity takes a longer duration during which it may suffer from price fluctuations, regulation changes, and/or the viability of the credits due to the market's nascency (Dibattista et al. 2023).

1.3.3.4 Quality and Integrity of the VCM Market

The quality of credits is highly dependent on the monitoring and measurement approach of the project type. Depending on the project type and its suited methodology, the project may be required to undertake direct measurements or modeling approaches that tend to be an expensive procedure. This is especially the case with agriculture project types and removals, where the most relevant measurement approach with a cost-effective mechanism for robust project development is still not clearly defined. Since farming project types are most susceptible to baseline miscalculations and permanence issues, there is a high chance that credit quality and integrity will be sacrificed (World Bank 2023). Such project types suffer higher risks of emission reversals due to interfering anthropogenic and/or natural phenomena. Specific risks associated with this sector and its respective mitigation strategy are defined in Sects. 1.6 and 1.7.

1.3.4 Opportunities in the VCM Market and Carbon-Credits Ecosystem

1.3.4.1 Carbon Market Governance

The current market governance challenge to align carbon credits with the NDCs under the Paris Agreement also brings in a huge opportunity for the VCM market. So far, the VCM demand has mainly been driven through the corporate sector as discussed in the previous section. To expand the scope of carbon credits from corporates to sovereign/compliance, airlines, and other government-related buyers, VCM needs to convert Verified Emission Reductions (VERs) to ITMOs. Since the Paris Rulebook demands bookkeeping of all mitigation claims being made, it further helps reduce and/or eliminate the risk of double accounting.

Article 6 of the Paris Rulebook is responsible for designing the transfer of ITMOs and bookkeeping criteria for the countries to ensure environmental integrity. One of the major criteria and differentiating factors between the VCM work strategy and Article 6 is the authorization of the host country required to accept the mitigation outcome and report it to the UNFCCC by both the host and buyer countries (Greiner et al. 2019). Countries meet every year at the Conference of the Parties (COP) to discuss the status of their NDCs and progress in the Paris Agreement framework. From COP 28, Articles 6.2 and 6.4 are a descriptive rulebook that is still a work in progress and shall soon lay down its complete framework for countries to issue ITMOs. Presently, Article 6.2 allows countries to sign bilateral agreements for ITMOs and record it under UNFCCC, whereas Article 6.4 provides a broad framework and regulatory development for CDM and other projects to develop projects under Article 6 (UNFCCC 2023).

1.3.4.2 Consolidation of the Market

Most of the projects are developed under four registries, namely Verra (VCS), GS, ACR, and Climate Action Reserve (CAR). A database developed by the Berkeley Carbon Trading Project contains all the information for offset projects developed under these registries. This database allows transparency in the carbon market by allowing comparisons between projects on different parameters such as location and type (So et al. 2023). It was observed that projects following standards' developed methodologies allowed higher transparency and comparability as compared to projects developing projects with its own methodology. Hence, relying on credible methodologies and procedures developed by major standards allows the market to be less fragmented and comparable for further credit sales in the market.

1.3.4.3 Striking the Demand–Supply Balance

Reporting the carbon credit demands by standardizing the accounting scales and systems between corporate net-zero targets and NDCs under the Paris Agreement can present a transparent status of credit demand and supply (Miltenberger et al. 2021). Addressing interoperability through environmental bookkeeping under Article 6 can provide robust data on the supply–demand and project-type needs in the country.

1.3.4.4 Improving the Quality and Integrity of the VCM

Due to varied project types in the VCM, there has always been a disparity in credit price and quality checks. To optimize the credit price based on project credibility and increase transparency, several exchange platforms have been developed to trade credits under the VCM. CBL Xpansiv is the largest spot exchange for offset credits. Carbon-offset volume transacted on its platform exceeded 120 million metric tons of CO_2 equivalents (Mt CO_2e), a 288% increase from 2020 levels (Spilker 2022). Several other platforms based on the growing demands of project types are being developed, such as CBL Nature-based Global Emissions Offset (N-GEO) to support the trading for VERRA-based AFOLU projects (Spilker and Nugent 2022).

1.4 Voluntary Carbon Market

1.4.1 Evolution and Progress in the Context of Agriculture

The evolution of agricultural project types, especially agriculture in the VCM, is fairly recent. Most NBS projects for the agriculture and forestry sector are registered under the removals category of afforestation/reforestation, IFM, and avoided emissions under the avoided deforestation category. So far, more than 50 projects are registered for the agricultural sector under both reduction and removal of GHG emissions, as depicted in Fig. 1.9 below.

Figure 1.9 also provides an overview of different mitigation projects registered under the VCM for both NBS removals and avoided emissions. Agriculture itself comprises 4% of carbon sequestration projects and 7% of emissions reduction projects reflecting the huge potential still untapped for this sector (Fig. 1.9). However, there has been a spike in carbon farming projects seeking registration under

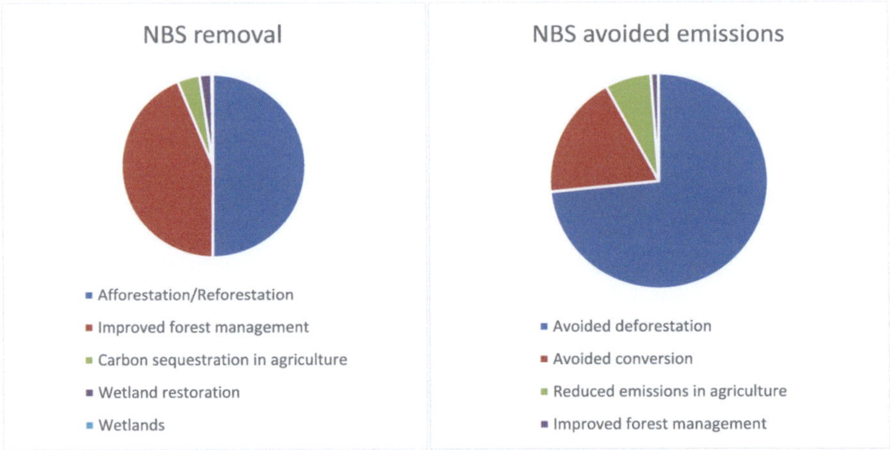

Fig. 1.9 Overview of sub-sector of (**a**) NBS removal and (**b**) NBS avoided emissions. (*Data Source*: Bravo and Mikolajczyk 2022)

VERRA's VM0042 methodology for improved agricultural land management (ALM). So far, 182 projects are either under development or under validation phases to seek registration for their projects, out of which India is already emerging as a leader with 50 carbon farming projects seeking registration that would account for 16.5 million hectares of agricultural land (VERRA 2024b; https://registry.verra.org/app/search/VCS), although so far only three carbon farming projects have been successfully able to register their projects under VERRA, which shall now be able to issue credits after each verification cycle (VERRA 2024b; https://registry.verra.org/app/search/VCS). The ratio of projects seeking registration and finally registered is severely high due to the stringent rules and criteria set by the standard and methodology to ensure a credible carbon credit for carbon farming project types.

1.4.2 Carbon Farming and the VCM

Carbon farming, as described in Sect. 1.3.2, refers to sequestering and storing carbon and/or reducing GHG emissions at the farm level. This can be done through a single or a combination of mitigation activities, most suitable for farming operations at different topographies. There is no set definition of carbon farming; however, it certainly reflects an additional business model to encourage farmers to mitigate agriculture-based emissions through financial incentives.

With the growing urgency to tackle climate change, agriculture and forestry have gained attention across several industries and funding from both the public and private sectors. In the case of VCM, registries like VERRA and GS shifted focus and broadened the portfolio of agriculture and forestry-based projects. AFOLU projects at large are defined in the following categories as per VERRA Afforestation,

Reforestation and Revegetation (ARR), Agricultural Land Management (ALM), IFM, Reduced Emissions from Deforestation and Degradation (REDD), Avoided Conversion of Grasslands and Shrublands (ACoGS), and Wetlands Restoration and Conservation (WRC) (Bravo and Mikolajczyk 2022), out of which, carbon farming projects generating carbon credits can be developed in the ALM category. There are also agroforestry projects aiming to increase the sequestration by combining carbon farming with forestry.

Based on the carbon farming interventions, the mitigation approach can be further classified into peatland rewetting and restoration, agroforestry system establishment and maintenance, maintenance and enhancement of SOC on mineral soils, livestock, and manure management, and nutrient management on croplands and grasslands. Each of the mitigation approaches has significant potential if implemented in the right way, as per the biogeographical makeup of the region. Agriculture activities not only help reduce/avoid emissions but also bring in co-benefits to the farmers and the ecosystem. Farmers may reach a higher yield, diversification in crops, less input costs for fertilizers and energy, and an overall improved soil health and productivity, whereas the ecosystem or societal benefits include biodiversity, water retention, and quality, nutrient retention, to name a few (McDonald et al. 2021).

Carbon farming under the VCM follows a similar ecosystem of project owners, intermediaries, buyers, and the governance body (Fig. 1.10). Project owners in agriculture project types are the farmers who shall be responsible for carrying out the mitigation activities on farmland. Classic examples of agricultural land management could include activities but are not limited to crop management, for example, high input carbon practices such as improved crop varieties, crop rotation, use of cover crops, perennial cropping systems, integrated production systems, crop diversification, agricultural biotechnology; nutrient management including fertilization with organic amendments/green manures; reduced tillage intensity and residue retention; improved water management, including drainage of waterlogged mineral soils and irrigation of crops in arid/semi-arid conditions; and improved rice management and biochar application (Nabuurs et al. 2023). However, to include such activities in the regular farm operations requires additional efforts and resources to be continued for the entire project cycle. So, farmers need to gain confidence in the carbon markets, their incentive schemes, and the rightful yield to be generated following land management practices. This is a voluntary activity, and so the farmers must confirm to follow and apply the practices for the entire project cycle.

Intermediaries can be government bodies, private entities, consultants, and/or NGOs that take responsibility for aligning the agricultural land management project as per the VCM registries for successful registration and issuance of credits for farmers. It is responsible for conducting the measurements as per the monitoring and verification requirements of the registries under VCM. Finally, once the project is registered and available to issue credits, intermediaries also source relevant buyers with premium credit pricing based on the project type and quality (Larson 2024).

Buyers can include all private and public entities trying to reach their net-zero target or NDCs for the country. The credit prices and buyers' inclination for AFOLU projects are highly dependent on the quality and credibility of the project that is

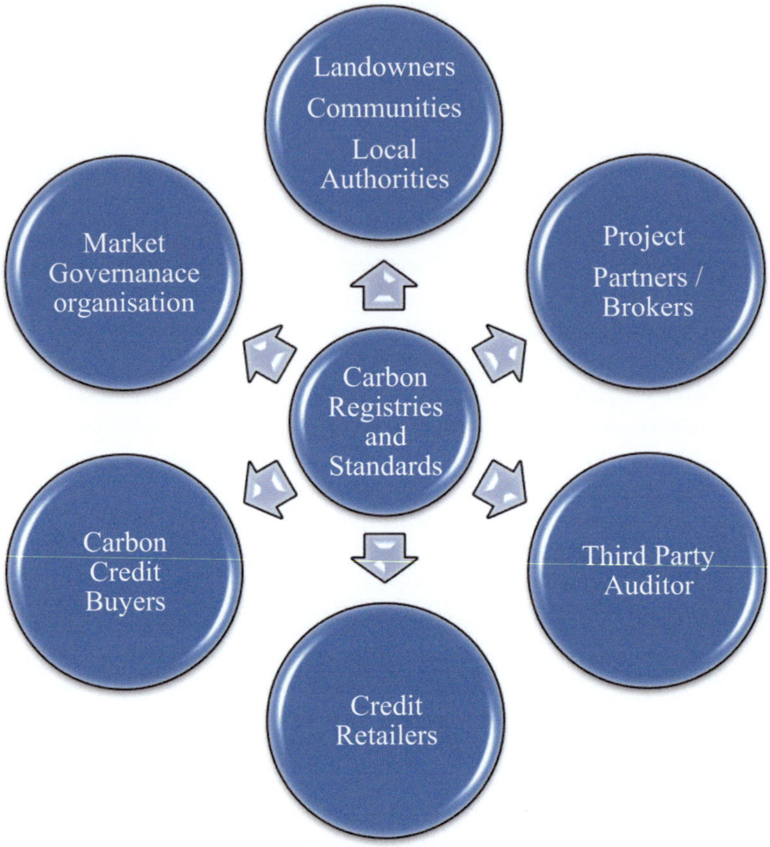

Fig. 1.10 Typical actors involved in a VCM ecosystem

developed under the registries. The governance body plays an important role in laying down the MRV criteria for projects, carefully assessing and validating project alignment as per the registry's requirements to issue only high-quality credits for projects (Larson 2024).

1.4.3 Types of Carbon Farming and Its Impact

Carbon farming can introduce or implement one or more new changes to pre-existing ALM practices—such as Improve fertilizer (organic or inorganic) management; Improve water management/irrigation; Reduce tillage/Improve residue management; Improve crop planting and harvesting (e.g., improved agroforestry, crop rotations, cover crops); Improve grazing practices; etc.—to be able to seek carbon credits. Farmers are advised through the intermediaries responsible for knowledge sharing, data collection, and SOC analysis via soil sampling and soil stratification of the project region. Baseline surveys and data collection through measurements and modeling are performed in the targeted project areas for carbon farming to understand the already existing SOC levels in the soil. So, regenerative practices to be implemented in the project should not have existed in the historical baseline to understand the exact SOC levels in the baseline with no regenerative practice that will be included in the project to seek carbon credits. Based on these results, farmers are expected to change to one or multiple regenerative practices. Mostly after 3–5 years of adoption of regenerative practices, project-level measurements and modeling are re-performed based on a sampling basis to determine the change in SOC levels from the baseline pre-historic levels of carbon in the soil.

To measure the SOC levels both in the baseline (pre-project year) and the project implementation years, soil sampling is conducted by the intermediaries or responsible parties on a random basis. Selected samples undergo the process of soil stratification based on several factors as elevation, slope, and soil properties and an appropriate layer depth is extracted from each point to further extrapolate the data. In most cases, 0–30 cm depth represents the top soil characteristics, whereas 30–60 cm depth represents the bulk density of soil to analyze the full profile of soil and its potential SOC levels through direct measurements and later model it for the entire project/farm area. Soil sampling and stratification must be conducted at each verification cycle using the same approach throughout for most standards as VERRA and GS to ensure the credibility of the carbon credit issued. Therefore, farmers are required to maintain their farm areas intact with continuous use of regenerative practices to sequester high SOC levels, generating higher carbon credits each year.

There are several studies and baseline data to suggest the best regenerative practice, or, in combination, to suggest to the farmers. It is observed that soil texture and soil type, coupled with environmental conditions, are highly variable and influential factors responsible for bringing significant changes in the SOC levels irrespective of the practices that are followed by the farmers (Rosinger et al. 2023). Hence, farmers farming in different regions can lead to different sequestration levels resulting in different carbon credits generated for the same regenerative practices. Additionally, studies have shown that cover cropping also dependent on the type of cover crops, can lead to significant carbon sequestration in the soil. Further addition of organic amendments and low mechanical intensity for tillage are considered great practices to adopt to sequester significant levels of SOC (Dupla et al. 2022). Most practices are expected to sequester higher SOC levels in the top layer of soil if the practices are kept intact with no tillage, cover cropping, and rotational cropping throughout the project cycle of at least 100 years (Slepetiene et al. 2024). In addition to these common yet effective regenerative practices, based on the topography of the project region, all other activities undertaken by farmers to improve soil health, such as the addition of organic amendments, a decrease in fertilizer use, improved nitrogen-fixation process in the fields leading to increased SOC sequestration levels, are eligible for carbon credits.

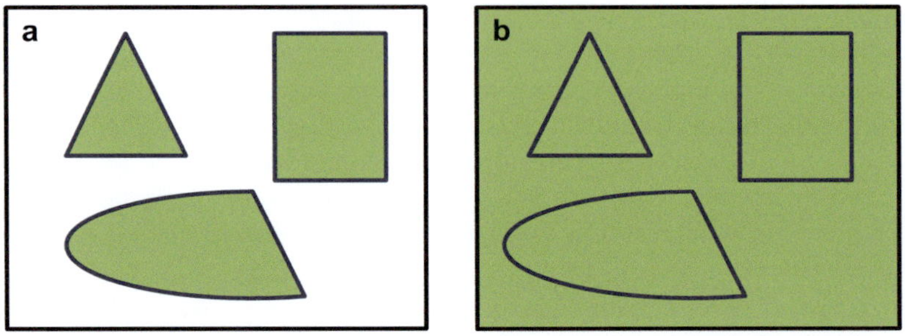

Fig. 1.11 Demonstration of project and jurisdictional approach in carbon farming. The project boundary in the image (**a**) only includes the farm units included in the project. However, the project boundary in the image (**b**) includes the entire jurisdictional boundary of the project area, which is considered the scope of the project

However, adoption and continuation of a new practice to generate carbon credits is seen as a major challenge in this sector, especially in cases where there is a decline in productivity observed in the initial years of the practice change. In addition to this, several other challenges as permanence, leakage, MRV costs, etc. are explained in Sect. 1.6 to understand major hurdles preventing farmers from adopting sustainable practices leading to carbon credit generation. Nonetheless, potential solutions to overcome such issues are explained in Sect. 1.8 to help widen the scope of carbon farming across countries. Furthermore, agriculture project types are categorized based on the mitigation activity executed, as explained in Sect. 1.4.2.

All projects in general can be defined into three categories based on their ways of combating GHG emissions, that is, either through emission reductions, removals, or avoidance. Projects developed in carbon farming for emission reductions, removals, and/or avoidance require measuring SOC to account for carbon flux between land and atmosphere. Change in SOC levels post the application of mitigation activities in a project then quantifies the GHG emissions that have been reduced, removed, or avoided. Under agriculture projects, measuring and re-measuring SOC of each project is technically not feasible (Bradford et al. 2023). Hence, the project area is grouped under two major types: project-based and jurisdictional groupings.

As demonstrated in Fig. 1.11, the project-level approach involves only farmers with farm lands that have enrolled in the project activity. Net emissions from the project area are then estimated against the business-as-usual (BAU) scenario. The project-level approach can suffer from an overestimation of carbon while determining the BAU scenario. This approach focuses only on the project boundary, with less stakeholder engagement.

The jurisdictional approach takes a wider approach that uses government administrative boundaries at the national and/or sub-national level to define the scope of project activity, further involving broader beneficiaries and strong stakeholder engagement. Emissions from all farms under the jurisdiction, irrespective of their involvement in the project activity, are accounted for to establish the baseline level against which the progress is measured. The jurisdictional approach allows for negating the effects of the leakage of carbon. It is empirically measured and does not require assessment and demonstration of additionality. Further, the risk of non-permanence is also tackled via the jurisdictional authority as well as the stakeholders (Larson 2024).

1.4.4 Payments for Carbon Farming Projects

Carbon farming is largely influenced by the socio-economic dynamics of the respective project's country. The agriculture sector at large struggles with adequate financial support in most parts of the world, especially in the Asian and African regions. Further, farmers are socio-culturally inclined to perform agricultural practices in a certain way for ages. Shifting those practices to enhance carbon sequestration and soil health requires entrusting and incentivizing farmers to encourage them toward new mitigation approaches. Three major ways exist so far to incentivize carbon farming at different stages of implementation, including result-based payments, action-based payments, and hybrid payments.

1.4.4.1 Result-Based Payments

It is a widely adopted approach under VCM to pay farmers after the implementation, validation, and verification of the project's mitigation activity is completed. As the name suggests, it allows the payment to be made to farmers once the project activity has been successfully validated and its carbon emission reductions or removals have been verified through a certain registry under VCM. Payment is based on the amounts of certified emission reductions or removals that are achieved by the project activity. One carbon credit is equivalent to one ton of reduction or removal achieved and verified against which the payment is made (Davide 2021). This requires an upfront cost for farmers to measure the carbon stock through a certain set of methodologies and standards that the project has chosen to follow. The final deliverable is the carbon credit unit and/or the labels that can be sold in the market. To produce premium prices for the deliverables in the form of carbon credit and/or labels, the project activity must follow stringent MRV requirements to ensure the quality and credibility of carbon farming (McDonald et al. 2021). Payments for result-based farming activities are generally implemented through three ways, which include auctions, voluntary tradeable emission reduction certificates, and payment for products/labels.

The auction of carbon reductions/removals is done through a bidding process. Each bid is set by a single seller, in this case, a farmer selling at a fixed price for their carbon mitigation activity. Multiple buyers, such as the government and other private entities, participate in a competitive auction to ensure enough funds for farmers to execute the farming activity and maximize the result for carbon abatement per hectare per dollar. A classic example of this reverse auction was led by the

Australian government buying Australian Carbon Credit Units (ACCUs) for bids with the cheapest price. However, farmers also have the flexibility to sell ACCUs to private entities offering better prices (Salas Castelo 2017).

Payments in voluntary tradable emission reduction certificates are made through verified carbon credits that are sold in the market either directly or through intermediaries. The carbon price in the VCM for carbon farming is highly variable depending on factors such as the quality and credibility of credits, standards, and MRV requirements used to quantify the change in carbon stock, project location, etc. Recently, some intermediaries have grown to specialize in carbon farming activities to help align the projects with the MRV and standard requirements to ensure credibility, environmental integrity, and a premium price in the market. Classic examples of such intermediaries are Nori, Indigo Ag, and Truterra, which invest with the farmers during the initial stages to align and bridge the gap between farmers and VCM requirements, further setting a minimum carbon price already for the activity that shall be conducted on the ground (Havens 2021).

Payment for products is done when carbon abatement is achieved through certificates and/or labels claiming carbon neutrality or carbon-negative emissions through estimations of their complete supply chain and stringent MRV requirements. Example cases of such products are the U.S. Department of Agriculture (USDA), Certified Organic labeling, and Beyond California's Healthy Soils Program (CHSP) (Cruz et al. 2020; Biggs et al. 2021; Raina et al. 2024).

1.4.4.2 Action-Based Payments

Such payments are focused on the delivery of specific actions as defined by local, regional, or national programs and/or policies. Farmers are required to carry out carbon mitigation activities capable of carbon sequestration or reduction to bring environmental co-benefits in the region. Such activities require a low level of monitoring with a minimum standard of practice to be followed as per the relevant program or policy. Action-based payment can be carried out in two ways: upfront and post-implementation.

An example case of upfront payment for action-based practices is executed under the Carbon Agriculture Policy (CAP) in Europe. European Union (EU) CAP allows farmers to follow farming practices that can comply with good agri-environmental conditions (GAEC), contributing to climate mitigation either directly or indirectly. An eco-scheme was introduced recently under the CAP to pay farmers per hectare of land dedicated toward climate action. This upfront payment allows the farmers to cover the cost required to transition to agricultural practices as required under the CAP (European Commission 2021).

Payments post-implementation is done only after the carbon action is implemented and incorporated by the farmer into their farming approach. Several schemes, such as the Green Low-Carbon Agri-Environmental Scheme (GLAS), CHSP Incentives Program, and REDD+ schemes, work on incentivizing farmers to adopt carbon mitigation activity that leads to improved soil health, carbon removal/reduction, and additional co-benefits in the region (Raina et al. 2024).

1.4.4.3 Hybrid Payments

Hybrid payments can be considered the most practical solution for incentivizing farmers to adopt carbon mitigation practices. This is because carbon farming and AFOLU in general are exposed to a higher risk of delivery and permanence of carbon sequestration as opposed to other project types. Further, quantifying the flux of SOC post-carbon farming practices is an additional upfront cost for the farmers. Therefore, hybrid payments allow farmers to be incentivized at both the pre- and post-adoption stages of carbon farming practices. The International Small Group and Tree Planting Program (TIST) spread across Africa and Asia provides strong incentives to farmers to plant trees and trains farmers to consolidate tree growth data to be later quantified under the VCM project for carbon credits (Buxton et al. 2021).

1.5 Potential of Carbon Credit Through Farming-Based Modules

As described in Sect. 1.3.2, sustainable agriculture can serve as a major tool to help combat climate change due to its carbon sequestration capacity. It is highly essential to do so since agriculture is being massively hit by the impact of climate change. Despite the growing land use and land change to croplands, the agriculture sector still struggles to maintain food security globally. This is primarily due to extreme weather conditions like heavy rainfall, droughts, and salinity that severely impact the soil health by turning it alkaline and low resistant to climate and pests. This has overall led to a decline in agricultural productivity and ecosystem health.

Most Asian countries, like India, comprise "smallholder farmers producing one-third of the world's food farming on 12% of agricultural land" (Ahuja 2023). The rural livelihood to raise crops is heavily dependent on the natural systems of land and water; this is because smallholder farmers belong to the economically weaker section of society, mostly raising their crops on as little as 1 ha of land (Ahuja 2023). Thus, introducing mitigation activities on farmlands requires adequate funding and a regulatory market to establish and encourage farmers to implement such practices. Amid this, carbon credits have emerged as a promising financial instrument to close the gap between farmers and the resources required to adapt to sustainable practices.

Although VCM comes with its own set of challenges and opportunities, there are several other platforms inspired by VCM that have been established to finance the value of carbon credits for agricultural projects. "Over 50 agricultural projects are currently active in India, claiming to contribute to an annual reduction of 37 million tons of CO_2 equivalent (CO_2e) emissions" (Cariappa et al. 2024). In addition to this, nations like India and Thailand have also set up their national carbon market, namely "Indian Carbon Market" and "Thailand Voluntary Emission Reduction Program," respectively. Under the Carbon Registry India, special emphasis is laid on "Sustainable Agriculture green credits" to promote conservation and sustainable agriculture management practices.

Sustainable and/or conservation agricultural practices have proved to be significantly effective in improving soil health through no-tillage practice, cover cropping through rotation, residue retention, intercropping, nutrient management, and biochar. Minimum soil disturbance and soil cover are reportedly the best land use management solutions to sequester carbon in the soil for a longer period (Krishna et al. 2022; Kumara et al. 2023).

1.6 Challenges for Carbon Farming in the VCM

In any financial instrument, there are always constant challenges and difficulties, allowing the scope of improvement and upgrade with time, and VCM is no exception. There are certain market challenges, such as additionality, permanence, time, and costs, that are in general for the VCM. Additionally, there are specific challenges to agriculture and AFOLU projects that may jeopardize the credibility of carbon credits if not executed well. Some of these challenges are related to leakage, temporal fluctuations, timing and duration, non-compliance of farmers through the project cycle, extended practice shifts, etc.

1.6.1 Market-Related Challenges

1.6.1.1 Additionality

Additionality is one of the prime tools under VCM to mark the eligibility criterion for projects. To develop a project of any technology, it is required to prove that the project would not have occurred or completed its activity in a business-as-usual scenario. UNFCCC CDM Tool 01 and 02 well define the criteria to prove additionality, with largely fewer than four parameters: regulatory analysis, investment analysis, barrier analysis, and common practice analysis (UNFCCC, 2021). Regulatory analysis should prove that there are no mandatory regulations set by the local or national government for the project activity to be conducted in the region. So, this project is a voluntary activity proving its additionality to the standard/registry. To understand this better, let us take a real case example of an agricultural practice, that is, the burning of crop residues in the states of Punjab and Haryana in

India, which is banned by the government of India. Hence, this activity is non-additional and ineligible to be registered under the VCM (Cariappa et al. 2024).

Other additionality parameters, such as investment analysis and barrier analysis, are performed to prove that the project suffers from either financial or other technological constraints, limiting the execution of the project activity. Thus, project types such as carbon farming are required to prove that the farming activity is not led in the BAU scenario due to a lack of financial support for the implementation of such mitigation activities.

Lastly, the common practice analysis is carried out to understand the adoption rate of the project activity in the defined geographical project boundary. This is crucial to understand and complement the analysis done for investment and barrier analysis, too. In general, a less than 20% adoption rate is considered additional for agriculture project activity under VCM (Cariappa et al. 2024).

1.6.1.2 Permanence

Projects issuing carbon credits under VCM are expected to produce a lifetime reduction or removal effect. In most project types, such as sustainable technologies, this is the case. However, for carbon farming project types, carbon sequestration poses a high risk of leakage across a lifetime, raising questions about its permanence. This risk is primarily due to the chemical composition of soil carbon, susceptibility toward natural calamities such as drought, forest fires, etc., and changes in the activity regime by the farmers. Several agricultural practices may release carbon back into the atmosphere at some point once the farmers stop implementing the project mitigation activities, thus undermining the environmental integrity of such credits. Therefore, it is highly important to tap the time and expected duration of the mitigation activity in the agriculture sector to understand the permanence level of the mitigation activity that is carried out (Larson 2024).

1.6.2 Project-Related Challenges

1.6.2.1 Leakage

The concept of leakage is similar to permanence. Permanence refers to the longevity of carbon that is permanently sequestered from a project activity, whereas leakage refers to the release of carbon during the project activity cycle due to direct or indirect causes. In the case of agricultural GHG mitigation activities, leakage might occur if there is a change in the farmer's yield. Farmers may prefer to increase their production land size to match the yield or decreased yield may lead to increased production by adjacent farm lands to compensate for the demand. Either case can lead to carbon emissions and thus impact the overall mitigation cause of the project (Larson 2024).

1.6.2.2 Temporal Fluctuation

Temporal fluctuations are the major concerns for permanence and leakage in an agricultural project activity. The agriculture sector at large is susceptible to more challenges and changes over time, accounting for that risk while setting the baseline. Weather shocks, changing economic conditions, and changes in farming practices by farmers lead to a change in the net GHG flux from soil carbon. There have been reported issues in the past that were subject to AFOLU project scrutiny under VCM for overestimating the carbon sequestered as compared to the actual baseline scenario.

1.6.2.3 Monitoring, Reporting, and Verification Costs

Once the project is validated and registered under VCM, it is required to monitor its carbon reduction or removals regularly to achieve carbon credits. Typically, an annual cycle of monitoring the carbon emission reductions and/or removals is followed, which is reported under VCM and later verified by a third-party auditor and standard bodies to issue credits. A robust MRV procedure allows the credibility and quality of carbon credits in the market, ensuring premium prices in the market. However, the MRV procedure in the agriculture sector is a relatively complex process as it involves the estimation of soil carbon flux over time. This can be achieved using two ways—direct measurement and modeling. Direct measurement involves on-site measurements of carbon stored in the soil at every verification cycle. Although this is a relatively accurate method to measure the soil carbon flux, it is prohibitively expensive for farmers and other intermediaries to perform such a method. Another method, that is, modeling, involves measuring the soil carbon flux based on a combination of proxy data and already existing scientific relationships for the project region (McDonald et al. 2021).

An agricultural SOC project, where humus is built up through regenerative practices, especially involves large upfront sampling costs. This includes a baseline sampling to determine the initial carbon content in the soil, and regular verification sampling, on average done every 3–5 years, to determine the amount of sequestered carbon over time. Since farmers usually do not own technological equipment for soil sampling, these services are outsourced to service providers and need to be paid accordingly. The problem with these large upfront sampling costs for agricultural producers is that carbon stocks first need to be composited in the soil. Financial payback through selling carbon credits to third parties thus only happens after the verification sampling, meaning 3–4 years from the baseline sampling. When deciding whether to transition to regenerative agriculture, these costs depict a major investment and thus not only an entry barrier but also a long-term project risk, as other internal and external risks could also result in project failures without a return on investment. An initial conviction of project success and financial profitability can thus prove to be difficult for farmers, who rely on financial resources in case of unforeseen challenges like crop shortages or failures. To ensure a robust MRV model for the AFOLU sector, it is essential for different entities or intermediaries to either fund or devise solutions to reduce the costs of this process.

1.6.2.4 Cultural and Social Acceptance by Farmers

Agriculture is often associated with strong cultural and social norms passed along for several generations. Thus, switching native agricultural practices to better mitigation activities requires more effort to successfully implement the practices across the entire project cycle. Farmers may under- or over-implement the new mitigation activity during the project cycle or even afterward, leading to an imbalance in the estimated amount of carbon sequestration from the project (Nabuurs et al. 2023). According to both the literature and experiences shared by project owners, one of the major barriers for farmers to transition to regenerative agriculture is their conservativeness or willingness to change (Nguyen and Drakou 2021). This can be directly linked to internal factors, such as perceptions of own capabilities related to existing knowledge and training not being sufficient, or external factors, like the lack of availability of time and financial resources, and subsequent dependency on others (Peden et al. 2018). As described already in the previous section, farmers often feel trapped by their current practices but are also afraid of drastic changes. When farmers are performing well financially, they are not inclined to change as involved uncertainties might disrupt their success. On the other hand, when performing poorly financially, they lack the financial resources to try something different in the first place. This often results in farmers maximizing their current efforts to their highest potential.

1.7 Potential Solutions for Addressing Challenges to Carbon Farming in the VCM

Since the agriculture sector has recently gained attention and spiked in the project development under VCM, there have been considerable challenges or barriers confronted by this sector. However, standard bodies and stakeholder entities in the VCM are already devising potential solutions to overcome the challenge of carbon farming.

1.7.1 Additionality

To ensure the credibility of any project type, it is essential to prove additionality under VCM. Carbon farming certainly poses a higher challenge in proving the mitigation activity as an additional and voluntary practice against the business-as-usual scenario, especially when there is no historical or literature evidence present in the region. Therefore, it is essential to transparently present all the relevant laws applicable to the project activity in the region. In addition to this, it is important to mention any additional schemes and incentives that are already applicable to the mitigation activity or agricultural practice of the region. This ensures that the farmers are not over-compensated by different schemes for the same agricultural practice, nor are they compensated for real voluntary mitigation practice in the region.

1.7.2 Permanence

The permanence of carbon reduced or removed from the atmosphere is the primary objective of developing a GHG mitigation project under VCM. Carbon reduced or removed from the atmosphere is expected to stay such for at least 100 years. This is for most agriculture project cycles having a crediting period of around 20 years to be renewed four times under most registry requirements. Thus, most farm practices are based on the continuous payment cycles for farmers, and MRV can only be conducted for the limited time frame of the project cycle. Hence, it is necessary to draft a short-term permanence contract for the agriculture project, alongside managing the risk of intentional or non-intentional release of carbon through a risk buffer system. Based on the type of agriculture mitigation practices and the potential risk associated with the project activity, a buffer of 5–60% can be set aside to compensate for carbon release during the project activity.

There are several ways to manage the risks associated with the release of carbon back into the atmosphere. Setting a buffer pool is one of them, where a proportion of carbon credits is set aside during the project cycle duration to compensate for the risk. Secondly, setting a separate short-term and long-term contract with farmers depending on their mitigation practices and their impact on carbon sequestration and permanence. Thirdly, separate result-based schemes for farmers on a long-term contract under the AFOLU sector. This is essential for such a project type, as the real-time carbon sequestration impact occurs after a long time, such as in the planting of trees (Radley et al. 2021).

1.7.3 Leakage

One of the major reasons for leakage, as discussed above, is a change in yield size due to the adoption of mitigation practices. Such cases are best handled when the geographical boundary for the agriculture projects is not specific to the project land but to the entire jurisdictional area. Any change in carbon emissions of the area can then be adjusted at the jurisdictional level, and leakage can be well managed in such a case (Larson 2024). Other potential reasons for leakage on carbon farming land are changes in the livestock management of farms, import of organic soil amendments, and changes in the yield productivity (VERRA 2024a). All leakage is then accounted for and deducted from the final GHG reductions/removals of the project activity.

1.7.4 Temporal Fluctuations

As described above for the permanence issue, the buffer system is currently considered the best tool to mitigate any temporal fluctuation risks from the project, too. Standard bodies as VERRA, have devised tools such as the AFOLU Non-Permanence Risk Tool that allow project proponents to take into consideration all the possible risks associated with the project (VERRA 2024b).

1.7.5 MRV Costs and Financial Support

To develop a credible system for carbon credits under VCM, it is highly essential to develop cost-effective, robust, and reliable MRV tools. For agricultural projects in general, the measurement methods for carbon sequestration tend to differ based on differences in standard requirements, geographical variability, farm area specificities, etc. In any case, the MRV should aim for robust procedures to ensure premium carbon prices in the VCM. As explained in Sect. 1.6, MRV methods can be of two types—direct measurements and modeling. While direct measurements provide nearly accurate results, it is an expensive procedure and is logically expected to be conducted through sampling. However, soil sampling also requires a large dataset to bring results on close estimations of carbon sequestered across a period. This is because the soil carbon levels are highly variable within a field. Hence, it is quite a challenge to be addressed through statistical and modeling studies to understand the feasible amount of samples to be conducted in a specific farm size area, which can compensate for the within-field variation (Bradford et al. 2023). To further refine the results for measuring the change in SOC, sampling can then be coupled with modeling that allows extrapolating the data for the entire project area based on other factors such as topography and mitigation activities implemented. This recent approach of coupling the sample and models for the farm area is seen as a positive MRV development for the agriculture sector, capable of managing the cost at some levels.

In addition to improving MRV tools with cost-effective methods, it is highly essential to provide financial support at the ground level that can motivate the farmers to implement mitigation activities.

1.8 Future Research Toward Monetizing Carbon Credits Through Agricultural Land Use

There is certainly a large amount of scrutiny and challenges around the VCM as it has grown rapidly through the years to compensate for the emissions caused by big corporations. However, in the current times, net-zero commitments and strategies for corporations are not well considered in the national environmental bookkeeping of their country, raising issues on the demand side of the credits.

On the contrary, the compliance market surges the importance and monetization of carbon credits under national/sub-national carbon pricing schemes such as Emission Trading Scheme (ETS) and international targets such as the Carbon Offsetting and Reduction Scheme for International Aviation (CORSIA) and Article 6. The compliance market is the market driven by the government to meet its NDCs through various ways as carbon prices, schemes, allowances, and carbon credits. The large demand for carbon credits is expected to be driven by the CORSIA sector, which is a baseline and offset system for the international aviation sector that is required to buy carbon credits whenever they are above the baseline limit. Second, is the development of the Article 6 mechanism under the UNFCCC Paris Agreement. Article 6 is the current ongoing international market-based mechanism that is described in Sect. 1.3.3, where the COP is held each year to bring together all countries and work on the rulebook for Article 6. In the recent COP 29 held at Baku, Article 6 has finally reached the operational framework to enable global collaboration on market and non-market approaches to tackle climate change.

Several other national mechanisms, such as the European Union's Carbon Removal Certification Framework (EU CRCF), are currently under development to produce a certification framework for carbon removals, carbon farming, and carbon storage. Understanding the shortcomings of carbon farming in the VCM, new structures are under development aiming to use the available digital methodology as electronic databases and GIS, remote sensing, novel on-site carbon quantification systems, artificial intelligence, and machine learning to decrease the cost of establishing standardized baselines and ensure the robustness of the monitoring of the activities (European Union 2024). Meanwhile, the standard bodies under the VCM are also working on digitalizing the AFOLU Non-Permanence Risk Assessment Calculator, which automates the generation of non-permanence risk reports to achieve accuracy (VERRA 2024a, b). Since carbon removals are considered one of the requisite tools to limit the global temperature to 1.5° C and well below 2° C, there is extensive research and work under development to improve the baseline, MRV, and leakage requirements and calculations for carbon farming and the AFOLU sector in general.

1.9 Plan and Policy on Carbon Farming-Based Carbon Credit

Formulating plans and policies under the carbon market for both voluntary and compliance sectors is a different approach to the government's policy and regulations set toward carbon farming or practices. Almost all carbon frameworks and markets developed to generate carbon credits require the project to be additional. As explained in Sect. 1.6, additionality can be attested through four phases—regulatory analysis, barrier analysis, investment analysis, and common practice analysis. So, any policy and regulation set up by the government for carbon farming is considered non-additional under VCM. For example, burning crop residue is banned in northern states of India like Punjab and Haryana. Since this is a regulation set by the government, farmers following this practice are considered non-additional, even if the rate of practice adoption on the ground is much lower. Only if the farmers adopt practices that are non-regulated, that is, no policies mandate the farmer to adopt such agri-sustainable practices, and the adoption rate of such a practice is also at least less than 20%, only then the project is additional and eligible for carbon credits (Cariappa et al. 2024).

As briefly mentioned in Sect. 1.8, several plans, policies, and frameworks at both national and international levels are under development to create a robust model for carbon removals, carbon farming, and carbon storage in general.

Carbon farming credits generally account for removals, avoidance, and reduction and can be utilized at several platforms if developed robustly. Table 1.1 describes the different levels of policies and frameworks for the development of credits and further utilization at premium prices.

The carbon market is an ever-evolving space that continuously works on establishing frameworks and policies to improve the methodologies, MRVs, additionality, and leakage criteria for different project types. There are several rating agencies such as in the market that proofread the quality and credibility of projects developed under the carbon standards to achieve legitimate pricing of credits. Further, several nations are entering the VCM through their domestic market, such as Thailand and India. India recently launched its carbon registry, known as Carbon Registry India. The registry is developed by the Network for Certification and Conservation of Forests (NCCF), aiming to achieve its NDCs and combat climate change through global environmental markets. The standard body largely focuses on forest management, sustainable agricultural practices, and the land use sector (Carbon Standard Version 1 2021).

Table 1.1 Current carbon platforms for carbon credit development and utilization

Carbon development platforms			
Voluntary carbon market (VCM)	Compliance carbon market		
Offsetting, compensation, mitigation, contribution	National/domestic scheme	International sectoral targets (CORSIA)	International target (Article 6)
Big corporations and individuals purchase carbon credits through VCM to achieve their set net-zero targets	Governments on a national level propose carbon pricing schemes such as carbon tax and EU-ETS to stay within the set limits of carbon emissions for the country; carbon credits can further be utilized to reduce or compensate one's tax liability	Sectors such as the CORSIA are compliant with working on a baseline and offset system; in case airlines emit emissions over the baseline, this needs to be compensated through carbon credits	Governments have agreed on their NDCs under the Paris Agreement to develop an international carbon mechanism; carbon credits are expected to be utilized in the form of ITMOs, where corresponding adjustments will be made to the NDC contributions of the countries involved in ITMO transactions

1.10 Recommendation

To limit the ever-increasing global temperature below 1.5° C, it is essential to not only drastically reduce carbon emissions but also implement negative emission technologies that help remove and/or avoid emissions from the atmosphere. Carbon farming is one such vital tool that, if implemented well, can serve as a major tool for negative emissions. This is not because agricultural land holds the maximum capacity in the AFOLU sector, but instead because of the sheer size of agricultural land that has been converted globally to sustain the growing population. In recent times, there has already been an increase in farmland practicing regenerative agriculture to be able to generate carbon credits and leverage the financial incentive. However, some major issues/challenges are faced by carbon farming sectors, which can be improved through the recommendations below at both the project level and organizational level:

- Pre-funding: Providing incentives to farmers to undertake sustainable agricultural practice that was not part of traditional practices can certainly motivate farmers to make this switch. Intermediaries such as Nori and Indigo are already pioneering ways to incentivize farmers to initiate and adapt to sustainable agricultural practices as required under the VCM.
- Methodology and tool improvisation: Carbon standard bodies and/or carbon removal frameworks are required to improvise on the methodology and calculations to precisely calculate the flux between baseline and project SOC emissions.

A unanimously accepted methodology and carbon credit calculation measures can significantly eliminate heterogeneity between standards, thus reflecting a similar price range for credits based on the project type.

- Monitoring, Reporting, and Verification costs: MRV costs and procedures for estimating the carbon flux between baseline and project year are currently carried out through different methods of measurement and modeling. Project owners and intermediaries involved in carbon project development should aim for the most accurate method of measuring the carbon flux to ensure the credibility of carbon credits, while bearing the costs on themselves. Carbon standard bodies and different national frameworks as the EU CRCF, are constantly working on improving and automating the MRV and Non-Permanence Risk Tool (NPRT) to achieve accuracy while still remaining cost-effective.
- Intermediaries' role: The carbon market is a relatively less-known topic for stakeholders such as farmers. The intermediaries and/or brokers need to play an important role during the implementation of the entire carbon project cycle. They should be responsible for managing all the documentation and MRV implementation while maintaining a transparent exchange of information and benefits with the farmer. Further, it should be the intermediary's responsibility to make sure farmers implement and continue the regenerative agriculture practices throughout the project cycle to ensure carbon credit reliability and credibility.

1.11 Conclusion

Agricultural lands are becoming vulnerable to climate extremes, leading to poor soil health, low yields, and low crop health, further worsening farm profitability. Carbon farming through carbon credits is among the most plausible solutions to educate, motivate, and incentivize farmers to adopt sustainable agricultural practices such as conservation tillage, cover cropping, and integrated nutrient management. There is no major barrier that can disrupt carbon credit generation for farmers if the project is aligned with additionality and standard/methodology requirements. It is highly essential to incentivize farmers to switch to non-conventional sustainable practices, which may, in some cases, bring a lower yield at the initial stages without the start of revenue generation from carbon credits, which can only happen after the complete validation and verification of the project. Additionally, to motivate farmers and compensate for all the additional costs and technology required under carbon project development, as discussed in the chapter, for a successful implementation of agricultural practices capable of sequestering carbon in the long run, farmers should continue executing such practices for the longest time possible. Therefore, different actors such as intermediaries, standard bodies, the government, and farmers together need to act efficiently and seriously for the successful and continued implementation of sustainable agricultural practices over the years. Farmers struggling with crop yields due to conventional practices are expected to be motivated to adopt mitigation practices on their land. However, this can only be achieved if the carbon mechanisms provide enough funds to farmers, enabling them to implement new measures. Although VCM and carbon markets come with certain limitations and challenges, as discussed in this chapter, they still hold several opportunities and scope for advances that can build this market even stronger in the coming times. At the project level, in the course of implementation years, farmers are soon to realize the advantages of sustainable agricultural practices with improved soil health and crop yields.

References

Ahuja R (2023) Carbon offsetting and smallholder Indian agriculture. In: State of India's livelihoods. ACCESS Development Services, New Delhi, p 73

Biggs NB, Hafner J, Mashiri FE, Huntsinger L, Lambin EF (2021) Payments for ecosystem services within the hybrid governance model: evaluating policy alignment and complementarity on California rangelands. Ecol Soc 26(1)

Bradford MA, Eash L, Polussa A, Jevon FV, Kuebbing SE, Hammac WA, Oldfield EE (2023) Testing the feasibility of quantifying change in agricultural soil carbon stocks through empirical sampling. Geoderma 440:116719

Bravo F, Mikolajczyk S (2022) The voluntary carbon market dashboard [Data set]. Climate Focus. https://climatefocus.com/initiatives/voluntary-carbon-market-dashboard/

Buxton J, Powell T, Ambler J, Boulton C, Nicholson A, Arthur R et al (2021) Community-driven tree planting greens the neighboring landscape. Sci Rep 11(1):18239

Carbon standard Version 1 (2021) Retrieved on September 6, 2024, from https://cri.nccf.in/home/standards/

Cariappa AA, Konath NC, Sapkota TB, Krishna VV (2024) Evaluating the potential and eligibility of conservation agriculture practices for carbon credits. Sci Rep 14(1):9193

Chen S, Marbouh D, Moore S, Stern K (2021) Voluntary carbon offsets: an empirical market study. SSRN. https://doi.org/10.2139/ssrn.3981914

Climate Watch (2024) Climate Watch. Washington, DC: World Resources Institute

CO_2 Earth (2024) CO_2 Earth: Current and historic atmospheric CO_2 data. https://www.co2.earth/

Cruz A, Foote D, Janosy A, Poirier K, VanVranken A (2020) Carbon markets: assessing opportunities, risks, and challenges for Minnesota agriculture. ASU Library, Phoenix

Davide M (2021) Technical guidance handbook-setting up and implementing result-based carbon farming mechanisms in the EU. Ecologic Institute, Berlin

Dibattista I, Manini Tiwari M, Mazzarano M, Borghesi S (2023) Carbon farming and voluntary carbon markets in the EU: an updated guide. In: Transforming our world: interdisciplinary insights on the sustainable development goals. SDSN, New York, pp 70–102

Dupla X, Lemaître T, Grand S, Gondret K, Charles R, Verrecchia E, Boivin P (2022) On-farm relationships between agricultural practices and annual changes in organic carbon content at a regional scale. Front Environ Sci 10:834055

European Commission (2021) Commission staff working document sustainable carbon cycles—carbon farming accompanying the communication from the Commission to the European Parliament and the Council on sustainable carbon cycles. EC, Brussels

European Union (2024) Regulation of the European Parliament and the Council. EU, Brussels. Retrieved on Novemeber 20, 2024. https://data.consilium.europa.eu/doc/document/PE-92-2024-INIT/en/pdf%22

Greiner S, Krämer N, Michaelowa A, Espelage A (2019) Article 6 corresponding adjustments: key accounting challenges for Article 6 transfers of mitigation outcomes. Climate Focus BV, Amsterdam

Havens D (2021) Agricultural carbon markets: how might they work? Presentations, working papers, and grey literature: agricultural economics, p 51

Taylor & Rising (2021); United Nations, World Population Prospects (2021) Dataset processed by Our World in Data. https://ourworldindata.org/ https://ourworldindata.org/grapher/global-cropland

Kreibich N, Hermwille L (2021) Caught in between: credibility and feasibility of the voluntary carbon market post-2020. Clim Pol 21(7):939–957

Krishna VV, Keil A, Jain M, Zhou W, Jose M, Surendran-Padmaja S, Erenstein O (2022) Conservation agriculture benefits Indian farmers, but technology targeting needed for greater impacts. Front Agron 4:772732

Kumara K, Pal S, Chand P, Kandpal A (2023) Carbon sequestration potential of sustainable agricultural practices to mitigate climate change in Indian agriculture: a meta-analysis. Sustain Prod Consum 35:697–708

Larson K (2024) Design concerns for agricultural soil carbon markets: an economic research agenda. Environmental defense fund economics discussion paper series, EDF EDP, 24-01

McDonald H, Frelih-Larsen A, Keenleyside C, Lóránt A, Duin L, Andersen SP, Hiller N (2021) Carbon farming: making agriculture fit for 2030. European Parliament, Luxembourg

Miltenberger O, Jospe C, Pittman J (2021) The good is never perfect: why the current flaws of voluntary carbon markets are services, not barriers to successful climate change action. Front Clim 3:686516

Nabuurs GJ, Mrabet R, Hatab AA, Bustamante M, Clark H, Havlík P, Steinfeld JP (2023) Agriculture, forestry and other land uses (AFOLU). In: Climate change 2022: mitigation of climate change. Cambridge University Press, Cambridge, pp 747–860

Nguyen N, Drakou EG (2021) Farmers intention to adopt sustainable agriculture hinges on climate awareness: the case of Vietnamese coffee. J Clean Prod 303:126828

Peden RS, Akaichi F, Camerlink I, Boyle LA, Turner SP (2018) Factors influencing farmer willingness to reduce aggression between pigs. Animals 9(1):6

Radley G, Keenleyside C, Frelih-Larsen A, McDonald H, Pyndt Andersen S, Qwist-Hoffmann H et al (2021) Setting up and implementing result-based carbon farming mechanisms in the EU: technical guidance handbook. Publications Office of the European Union, Brussels

Raina N, Zavalloni M, Viaggi D (2024) Incentive mechanisms of carbon farming contracts: a systematic mapping study. J Environ Manag 352:120126

Rosinger C, Keiblinger K, Bieber M, Bernardini LG, Huber S, Mentler A, Bodner G (2023) On-farm soil organic carbon sequestration potentials are dominated by site effects, not by management practices. Geoderma 433:116466

Salas Castelo EM (2017) The role of factors that influence the adoption of the Australian carbon farming initiative-emissions reduction fund: a mixed methods study. Doctoral dissertation, James Cook University, Douglas, Australia

Slepetiene A, Kadziene G, Suproniene S, Skersiene A, Auskalniene O (2024) The content and stratification of SOC and its humified fractions using different soil tillage and inter-cropping. Sustainability 16(3):953. https://doi.org/10.3390/su16030953

So IS, Haya BK, Elias M (2023) Voluntary registry offsets database, Version 8. Berkeley Carbon Trading Project. University of California, Berkeley

Spilker G (2022) The future(s) of emissions trading—COP26 and the market for voluntary carbon offsets. Open Markets. Retrieved on September 6, 2024. https://www.cmegroup.com/openmarkets/commodities/2022/The-Future-of-Emissions-Trading-COP26-and-the-Market-for-Voluntary-Carbon-Offsets.html

Spilker G, Nugent N (2022) Voluntary carbon market derivatives: growth, innovation & usage. Borsa Istanbul Rev 22:S109–S118

UNFCCC (2021) Methodological tool – tool for the demonstration and assessment of additionality. UNFCCC, Rio de Janeiro

UNFCCC (2023) Article 6 – cooperative implementation. Retrieved on December 10, 2024 https://unfccc.int/process/the-paris-agreement/cooperative-implementation#Intersessional-and-in-session-work-first-half-of-2023
Verra (2024a) VCS methodology-VM0042 improved agricultural land management. Verra, Washington, DC
Verra (2024b) Verra registry. Verified carbon standard. https://registry.verra.org/app/search/VCS
World Bank (2022) State and trends of carbon pricing 2022. State and Trends of Carbon Pricing. ©. World Bank, Washington, DC. http://hdl.handle.net/10986/37455
World Bank (2023) State and Trends of Carbon Pricing 2023. © http://hdl.handle.net/10986/39796
World Economic Forum (2023) The global risks report 2023. Retrieved on October 6, 2024 from https://www3.weforum.org/docs/WEF_Global_Risks_Report_2023.pdf
World Meteorological organization (2023) Rate and impact of climate change surges dramatically in 2011–2020. Retrieved on September 8, 2024 from https://wmo.int/news/media-centre/rate-and-impact-of-climate-change-surges-dramatically-2011-2020#:~:text=Since%20then%2C%20CO2%20has%20increased,it%20rose%20to%20402.0%20ppm

Chapter 2
Prospects, Challenges and Policies for Carbon-Negative Circular Agriculture

M. G. Mohamed Thariq, A. T. Ahamed Akram, M. M. M. Najim, and Muneeb M. Musthafa

Abstract Global conventional agricultural, livestock production and related land uses contribute a large proportion to the total global greenhouse gas (GHG) emissions. On the other hand, circular agriculture reduces carbon emissions and has the potential to be a carbon sink within the agricultural systems. However, the successful adoption of circular agricultural practices requires a national-level policy framework. The key challenges for adopting circular agriculture practices are: financial obstacles to acquiring technology, limited access to affordable credits and subsidies, technical incompetence of farmers and lack of institutional support, agribusiness objectives and environmental attitudes. This chapter identifies key prospects of circular agriculture, including climate change mitigation through carbon capture, utilisation and storage (CCUS) technologies, reducing atmospheric carbon dioxide (CO_2) levels by recycling carbon into valuable products, and enhancing carbon sequestration in forest soils and oceans. The integration of circular economy principles fosters economic opportunities, supports energy security and accelerates decarbonisation, ecosystem restoration and nature-based solutions. Policy challenges include fragmented governance, weak policy enforcement and limited institutional coordination. The policies to adopt circular agriculture by farmers need to focus on investment in circular agricultural research and development, institutional support, rewarding with carbon credits, incorporating circular agriculture models in regional development plans, development of technical standards and monitoring and evaluation capacity for the circular agriculture model, reforming agricultural subsidies and introducing innovative policy support, development of circular agricultural value chains, enhancement of private sector investment in circular agriculture, the establishment of more intersectoral circular bioeconomy models for sustainable and low-carbon rural development, and enhancement of international cooperation. The

M. G. M. Thariq (✉) · A. T. A. Akram · M. M. Musthafa
Faculty of Technology, Department of Biosystems Technology, South Eastern University of Sri Lanka, Oluvil, Sri Lanka
e-mail: mgmthariq@seu.ac.lk; akram.at@seu.ac.lk; muneeb@seu.ac.lk

M. M. M. Najim
Faculty of Agriculture, Sultan Sharif Ali Islamic University, Bandar Seri Begawan, Brunei
e-mail: najim.mujithaba@unissa.edu.bn

R. S. Meena et al. (eds.), *Carbon-Negative Agriculture*, Sustainability Sciences in Asia and Africa, https://doi.org/10.1007/978-981-95-1483-0_2

circular agriculture contributes to achieving maximum benefits from natural processes within the ecosystems through the application of environmentally friendly technologies with minimum input dependency, and efficient recycling of nutrients, energy and water. These processes transform waste into valuable food products with minimum costs and food losses by using the principles of reduce, reuse and recycle in the agri-food production and supply chain while contributing to tackling climate change.

Keywords GHG emissions · Reuse · Recycle · Ecosystems · Carbon sink · Circular economy

2.1 Introduction

The climate crisis is one of the most pressing issues of our time. Implementing long-term carbon sink measures is necessary to confront climate change and achieve the 1.5 °C target of the Conference of the Parties (COP 28) (Singh et al. 2024; Meena et al. 2025). As the Earth's system undergoes rapid transformation, fundamental life-supporting processes are undermined, which can lead to very serious consequences for society to tipping points that can irreversibly destabilise the Earth (Rockström et al. 2023). Global agricultural and related land use emissions in 2018 were 9.3 billion gigatonnes (Gt) of carbon dioxide (CO_2) equivalent (Gt CO_2eq.) (FAO 2020). About half (5.3 Gt CO_2eq.) of this total came from crop and livestock activities within the farm gate, whereas the land use and land use change activities are responsible for nearly 4 Gt CO_2eq. (Tubiello et al. 2014; Meena et al. 2024c).

Conventional agricultural practices, comprising crop cultivation, livestock management, fertiliser usage, waste disposal and energy consumption, significantly contribute to direct and indirect carbon emissions and impact climate change (Singh et al. 2024). Soil tillage and rice cultivation release CO_2 and methane (CH_4) (Fernández-Ortega et al. 2024), whereas livestock farming (ruminants) release CH_4 by ruminating and manure (Faizan 2024; Sikiru et al. 2024). Excessive usage of synthetic fertiliser and inappropriate application procedures increase emissions of ammonia (NH_3) and nitrous oxide (N_2O) (Sarkar et al. 2024; Thirunagari et al. 2024). Open burning of crop leftovers and organic waste, like poor waste management practices, increases CO_2 and CH_4 emissions (Khan et al. 2024; Kumar et al. 2024; Meena et al. 2024a), as well as carbon footprint from the use of fossil fuel in agricultural machinery and transportation (Qin et al. 2024; Wang et al. 2024). All these activities further deteriorate soil health (El-Ramady et al. 2024; Meena et al. 2024b), biodiversity (Selvakumar 2024) and ecosystem resilience while worsening climate change. In light of these emissions, an urgently needed sustainable and climate-resilient agricultural system is required. Circular agriculture is considered a sustainable solution to minimise carbon emissions.

Circular agriculture reduces carbon emissions and enhances the circulation of carbon while emphasising the reuse and recycling of resources in the agricultural ecosystem (Yang et al. 2022). It not only reduces carbon emissions but also supports carbon sequestration and retention within agricultural systems, functioning as a potential carbon sink (Singh et al. 2024). Though various circular agricultural practices exist and are implemented in different parts of the world, governmental policies are lacking in those countries to make circular agriculture a national need. The key prospects of circular agriculture include climate change mitigation through carbon capture, utilisation and storage (CCUS) technologies (Roy et al. 2023; Tebbiche et al. 2021; Meena et al. 2023), the restoration of degraded ecosystems (Priyadarshini and Abhilash 2020) and the integration of circular economy principles for economic gain and improved energy and food security (Selvan et al. 2023; Sharma et al. 2021; Kumar et al. 2025). These systems maximise resource efficiency and minimise environmental foodprints.

However, significant challenges remain in circular agriculture, such as high initial investment costs for technology adoption, lack of technical skills among farmers, limited access to credit and subsidies, inadequate policy support and fragmented institutional frameworks (Jaroenkietkajorn et al. 2024; Sharma et al. 2021; Meena and Pradhan 2023). To overcome these barriers, effective policy interventions are crucial. This chapter attempts to identify key aspects, such as challenges associated with traditional agriculture regarding carbon emissions, the prospects of various circular agricultural practices and their contribution to minimising carbon emissions, and policies to establish and regulate circular agricultural practices to make agricultural food production a sustainable and low-carbon emission system.

2.2 Global Carbon Emissions from Agriculture

The global distribution of agricultural emissions is not consistent and is influenced by resource availability, climate and farming practices in different regions of the world (Lakhouit et al. 2025; Li et al. 2025). In Asia, CH_4 emissions originating from rice paddies are predominant. Many countries like India, China and Indonesia mainly depend on the flooded rice culture, which accounts for roughly 20% of atmospheric CH_4 emissions (Darikandeh et al. 2025; Li et al. 2021; Pradhan and Meena 2023b). However, livestock farming is the most significant source of emissions in Europe and North America, with large amounts produced through dairy and beef farming (O'Mara 2011). These regions also depend on synthetic fertilisers, which produce N_2O. The manufacturing of synthetic nitrogen fertilisers constitutes around 2% of global energy consumption (Walling and Vaneeckhaute 2020). Africa contributes a relatively small proportion (2–3%) of global agriculture emissions (IPCC 2018). Predictions indicate that Africa's lowest anticipated proportion of world emissions will approach 10% by 2050 (Mostefaoui et al. 2024).

Because of continuous human-induced releases of CO_2 and other greenhouse gases (GHGs) into the atmosphere, the Earth's climate is changing quickly (IPCC 2018). Among various GHGs (e.g. N_2O and CH_4), CO_2 stands out as the primary driver of global climate change due to its substantial increase from pre-industrial times to the present (Table 2.1). Specifically, atmospheric CO_2 concentrations have

Table 2.1 Global warming potential of greenhouse gases

Greenhouse gas	Main source	Average lifetime in the atmosphere	100-year GWP
CO_2	Fossil fuels; land use change	Carbon lifetime cannot be represented with a single value	1
CH_4	Fossil fuels, rice and livestock	12	28
N_2O	Fertilisers	121	265
CHF_3	Refrigerants	222	12,400

Source: Alsarhan et al. (2021), GWP - Global Warming Potential

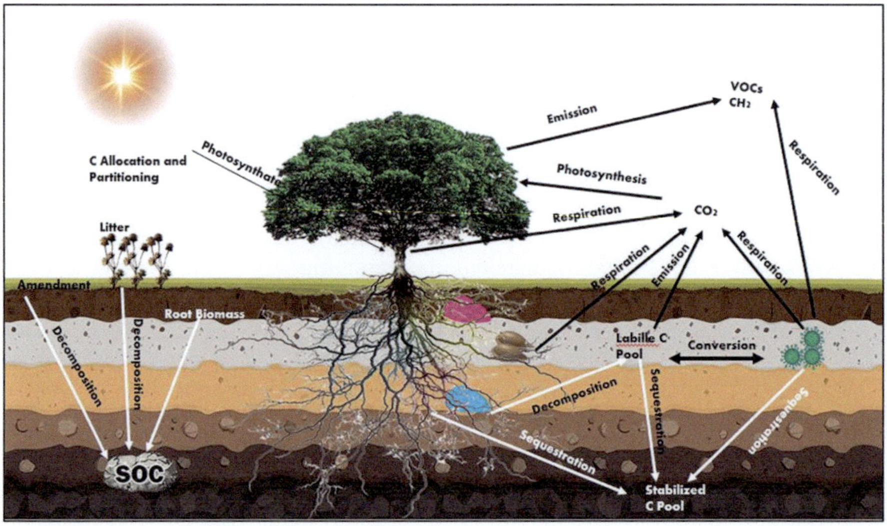

Fig. 2.1 Carbon circulation in the environment

surged from approximately 300 ppm before 1850 to 420 ppm in 2023 (NOAA 2022; Richardson et al. 2023; WMO 2021). Predictions estimate that if existing trends continue, CO_2 levels may attain 550 ppm by 2050 and 731 ppm by 2130 (Pacheco-Torgal 2024). Most of this overall rise in atmospheric CO_2 is attributed to the burning of fossil fuels, accounting for about two-thirds of the increase (IPCC 2018). The remaining portion of the increase is linked to the loss of soil organic carbon (SOC) due to changes in land use (Singh et al. 2024), such as deforestation (Amoakwah et al. 2022) and crop cultivation (Fig. 2.1) (Nazir et al. 2024). Soil contains a significant portion of the Earth's terrestrial carbon, with a total carbon content of approximately 3170 Gt (Jansson et al. 2021). Precisely, 2500 Gt of carbon are in the top 3 m of soil. Of this, 1550 Gt are SOC and 950 Gt are soil-inorganic carbon (Singh et al. 2024; Friedlingstein et al. 2023). This soil carbon reservoir is 3.3 times greater than the air carbon reservoir, which holds 760 Gt (Lal 2004). The conversion of natural ecosystems into farmland has resulted in a decline in SOC levels. This alteration has led to the emission of 50–100 Gt of carbon from the soil into the atmosphere since the onset of the Industrial Revolution in 1950 (IPCC 2018).

2.3 Carbon Emissions by Various Traditional Agriculture

2.3.1 Crop Farming

Traditional crop farming systems, which rely heavily on monocultures and synthetic fertilisers, play a major role in GHG emissions (Basheer et al. 2024; Xing and Wang 2024; Pradhan and Meena 2023a). Fertilisation with nitrogen-based fertilisers also releases N_2O, a very potent GHG with a global warming potential of 300 times that of CO_2 (Caillol 2011; Punia et al. 2020). Inefficient irrigation practices aggravate this problem, as wet fields produce anaerobic conditions that allow CH_4 emissions, mainly from rice cultivation (Darikandeh et al. 2025; Li et al. 2021; Jangir et al. 2024). Burning crop residues, particularly in countries like India and China, also emits millions of tonnes of CO_2 each year, helping to cause air pollution and harming public health (Li et al. 2021). Most crop farming systems using linear production tend to maximise productivity in the short term at the cost of soil health and biodiversity (Kumar et al. 2022; Chung et al. 2024), where the comparisons between linear and circular agriculture are listed in Table 2.2. Intensive tillage methods change soil structure and promote microbial degradation of soil organic matter,

Table 2.2 The comparisons between linear and circular agriculture

Criteria	Linear agriculture	Circular agriculture
Development history	Widely practised by industrial societies	Widely practised in pre-industrial society
Business model	Large-scale, specialised agricultural firms	Small-scale, integrated agriculture smallholders
Farming practices	Modern farming, large-scale, monoculture and inorganic practices	Indigenous, small-scale, polyculture, bio-organic practices
Input use	More capital-intensive with off-farm inputs	More labour-intensive with on-farm inputs
Focus	Maximising profit over the protection of the environment	Balancing economic, environmental and social aspects
Reuse	Downcycling; low-grade recycling	Upcycling, cascading and high-grade recycling
Products that have reached the end of their life	Invaluable and waste	Valuable next-use resources
Impacts	Increased resource requirements and waste; less sustainable	Reduced resource requirements and waste; more sustainable

Source: Chung at al. (2024)

resulting in soil CO_2 release (Nunes et al. 2020). In turn, they deplete the soil's organic carbon and reduce its ability to sequester carbon, aggravating climate change. Monocultures reduce ecosystem richness, destabilise these systems' resilience and increase their vulnerability to pests, diseases and climatic variability (Khatri et al. 2024; Huss et al. 2022).

Transitioning to circular farming systems is a way to reduce emissions and promote sustainability. The approach is recycling the nutrients within the agricultural ecosystem (Valve et al. 2020). Moreover, organic substitutes for synthetic inputs (compost and animal manure) promote soil health, mitigate N_2O emissions (Lazcano et al. 2021; Adegbeye et al. 2020) and decrease energy consumption in synthetic fertiliser production (Vaneeckhaute et al. 2013; Tuomisto et al. 2012; Pradhan et al. 2023). Rotational practices and intercropping practices also help to diversify root systems, maintain continual ground cover, and help soil fertility for carbon sequestration (Eddy and Yang 2022). Global Positioning System (GPS) guided equipment and soil sensors are used to optimise the usage of water, fertiliser and pesticides to reduce waste and emissions (Vellingiri et al. 2025; Gautam et al. 2023). Such tactics can help balance production and environmental responsibility on farms.

2.3.2 *Livestock Farming*

In ruminants (cattle, sheep and goats), enteric fermentation produces 90 million tonnes of CH_4 per year, approximately 40% of all agricultural GHG emissions (FAO 2013). This makes livestock responsible for about 14.5% of total global GHG emissions, with CH_4 being accountable for the majority of these (FAO 2013). CH_4 is a GHG that has the potential to cause 25–30 times more global warming than CO_2 over 100 years (Arif 2024; Jackson et al. 2024; Sobanaa et al. 2024). Beef and milk production from cattle make about 3.8 Gt of CO_2 equivalent every year, which is 62% of all the emissions from animals. Fourteen percent of the emissions come from pigs, 9% from chickens, 8% from buffaloes, and 7% from small ruminants (Džermeikaitė et al. 2024). Manure management techniques also impact climate change, especially in large-scale operations that emit CH_4 and N_2O (Ouatahar et al. 2024; Yan et al. 2024). Traditional livestock systems rarely operate efficiently enough to minimise emissions. Poor-quality feed causes animals to need more feed to meet energy needs, increasing the amount of CH_4 emitted per unit of meat or milk produced (Nedelkov et al. 2024; Waghorn and Hegarty 2011). Using a balanced ration feeding routine, Garg et al. (2013) observed a significant increase in animal productivity in lactating cows and buffaloes in India. Integrating livestock into the agricultural system modifies interactions among soil, vegetation and the atmosphere. Animals increase biodiversity and nutrient and water flow channels, improving nutrient recycling, food production efficiency and fertiliser and resource use (Fig. 2.2) (Bhagat et al. 2024; Paramesh et al. 2020).

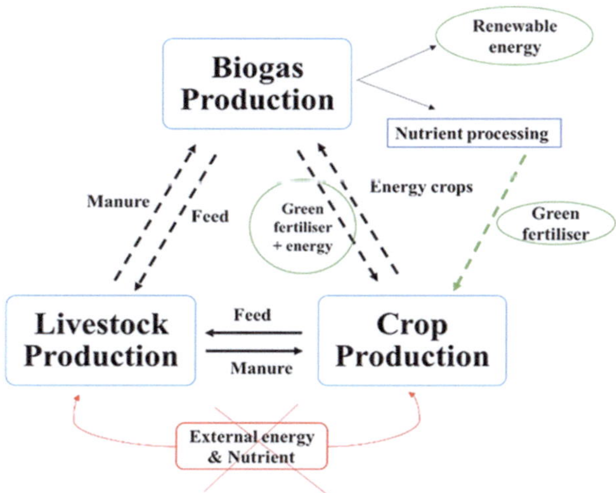

Fig. 2.2 Livestock and crop integration

2.3.3 Agricultural Land Use and Land Use Change

Changes in agricultural land use, such as deforestation and wetland conversion, are primary sources of GHG emissions (Hergoualc'h and Verchot 2014; Kreileman and Bouwman 1994). Significant deforestation is documented in tropical countries, particularly in South America and Southeast Asia, for palm oil production (Chiriacò et al. 2024; Vijay et al. 2016; Ravindranath et al. 2009). However, these land use changes release substantial quantities of carbon formerly sequestered deep into the atmosphere, accounting for around 12.5% in 2010 (Houghton et al. 2012) to 25% in 2017 (Hong et al. 2021) of anthropogenic GHG emissions. Additionally, they reduce ecosystems' capacity to absorb CO_2, which is the primary cause of climate issues. Conventional agricultural systems often fail to overlook the long-term environmental impacts of land use changes. Wetland conversion to agriculture causes CO_2 emissions (Tan et al. 2020) and loss of important ecosystem services such as water filtration and animal habitat (Kingsford et al. 2016; Verhoeven and Setter 2010). These changes make the ecosystems less resilient and less varied. That makes them more susceptible to climate change and extreme weather.

Circular agriculture is a mechanism to achieve economic productivity while conserving the environment. Agroforestry is an important technique for enhancing carbon sequestration and generating economic benefits from timber, crop, animal shade and fruit production (Keprate et al. 2024; Montagnini 2024; Jhariya et al. 2022). Soil organic carbon (SOC) stores can be restored through the use of organic amendments (compost, biochar) to restore water retention and increase microbial activity levels (Fahad et al. 2022). In addition, these techniques help make farming systems more resilient and effective and also lower emissions.

2.4 Circular Agriculture as a Key to Sustainable and Low-Carbon Food Production

2.4.1 *Circular Agricultural Practices*

Circular agriculture is based on recycling resources within the agricultural environment, which turns waste into useful inputs (Hoogstra et al. 2024; Chung et al. 2024; Rodino et al. 2023). Another way to include them is to turn food waste into animal feed and animal dung into biofertilisers. By closing the nutrient cycle and eliminating reliance on artificial inputs, this method keeps waste in the system to a minimum and maximises resource usage (Pal et al. 2024). Circular agriculture exemplifies agroforestry and integrated farming systems. Agroforestry integrates trees with agricultural and livestock cultivation to provide advantages such as carbon sequestration, enhanced soil fertility and increased revenue via timber and crop production (Keprate et al. 2024; Montagnini 2024). Integrated farming systems integrate crop and animal production, facilitating nutrient recycling and minimising waste (Kumar et al. 2023a; Bhagat et al. 2024). Manure that fertilises crops and agricultural wastes that nourish animals result in a symbiotic connection that enhances production and environmental health.

Precision agriculture technologies enhance the circularity of the agricultural system (Sanyaolu and Sadowski 2024). Utilising sensors and GPS-guided equipment, farmers may accurately administer water, fertilisers and pesticides at the appropriate times and locations (Vellingiri et al. 2025; Gautam et al. 2023). This not only conserves resources but also reduces emissions with minimal impact on yields or maybe enhances them. In addition, circular practices, such as composting and applying biochar, turn organic waste into fertiliser (Gunasekara et al. 2024). Biochar is a carbon-rich substance derived from biomass, which enhance soil structure and fertility while sequestering carbon, contributing to climate change mitigation (Gyawali 2024).

Circular agriculture approaches are not limited to farms but extend to the post-harvest process. Cold storage facilities driven by renewable energy mitigate food loss and decrease the carbon footprint of storage activities. Farmers may utilise solar-powered refrigeration units in remote locations to keep perishables for extended periods, reducing spoiling and improving marketing availability (Rami and Allouhi 2024). Adopting these comprehensive methods will enable the farms to become sustainable systems that satisfy food security, economic resilience and global climate goals.

2.4.2 *Circular Carbon Economy*

The economic model proposed to find the balance between the economy and the environment is termed the Circular Carbon Economy (Fig. 2.3) (Marku et al. 2024;

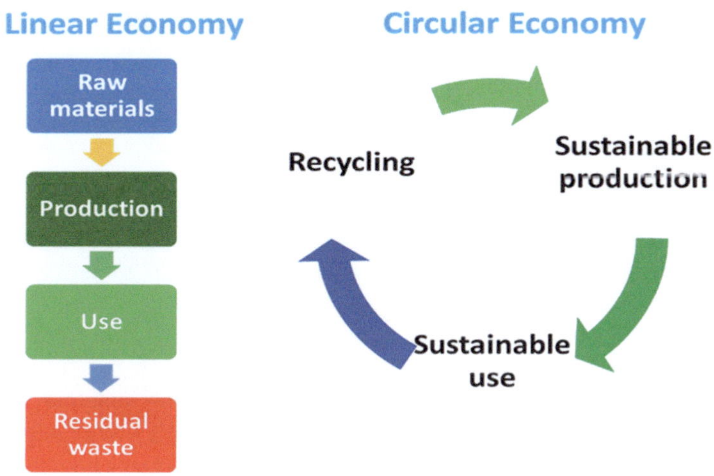

Fig. 2.3 Linear economy versus circular economy

Minh Khue et al. 2024; Yang et al. 2023), which is based on the 3Rs concept: reduce, reuse and recycle (Zoboli et al. 2016). From this perspective, circular agriculture emerged, focusing on preserving agroecosystems, reducing waste and optimising resource use and energy use (Rauw et al. 2023; Silvius et al. 2023). One of the main goals of circular agriculture is to reduce the reliance on external inputs by applying a closed nutrient loop system within the agricultural models (Minh Khue et al. 2024).

The organic carbon levels increase when soil is enriched by cover cropping (keeping ground cover year-round) and reduced tillage (less soil disturbance). They also improve soil fertility, prevent erosion and provide water retention (Koushika et al. 2024; Leharwan et al. 2023). The other key element of the circular carbon economy is converting agricultural waste into bioenergy (Yrjälä et al. 2022). Biogas facilities convert manure and agricultural byproducts into renewable energy and nutrient-rich digestate. These organic fertilisers are crucial for decreasing dependence on synthetic fertilisers and promoting nutrient recycling (Islam et al. 2021). Renewable energy is integrated into agricultural processes to advance the circular carbon economy further. A solar-powered irrigation system is a cost-effective, emission-efficient substitute for diesel pumps. Wind turbines and small hydropower systems empower farms to generate their clean energy. These renewable sources enable farms to minimise their carbon footprint while preserving their resources (Velasco-Muñoz et al. 2021).

2.4.3 Challenges for Carbon-Negative Circular Agriculture

Circular agriculture offers many benefits, but there are also significant barriers to broader adoption. Financial obstacles are the most important. Anaerobic digesters, precision farming equipment and renewable energy technology all have prohibitive upfront costs for most smallholders. Moreover, the limited access to affordable credits and subsidies further complicates the challenge by restricting farmers from adopting sustainable practices (Singh et al. 2024; Chung et al. 2024).

Furthermore, there are several technological limitations. Farmers in developing countries may encounter challenges using advanced technologies such as composting units, biogas plants and digital monitoring systems. Although these technologies are available, farmers do not use them due to a lack of technical skills or training. For example, smallholder farmers lack the tools to perform precision agriculture. Such tools may be unaffordable to them, or they may have trouble maintaining them, which will have little impact on their agricultural practices (Velasco-Muñoz et al. 2021).

Implementing collaborative strategies or capacity-building initiatives is essential to tackling these challenges. Multi-stakeholder partnerships bridge resources and expertise, guiding farmers to adopt circular practices. Circular agriculture has the potential to become a reality and contribute to the maintenance of global food systems.

2.5 Guiding Policies for Carbon-Negative Circular Agriculture

2.5.1 Enhancing Producer and Consumer Awareness

Raising awareness among farmers and consumers about the benefits of transitioning to circular agriculture is essential. Farmers generally neglect the long-term sustainability and cost-saving potential of circular processes. Implementing successful educational campaigns can boost trust and demonstrate the economic advantages of circular systems. Customers also play a significant role in creating demand for food produced through sustainable practices. Eco-labelling and public awareness initiatives might educate customers about the environmental effects of the foods they choose. Labelling such as carbon footprints or circular production processes can influence purchase decisions and create market incentives for farmers to use sustainable practices (Velasco-Muñoz et al. 2021).

Governmental and non-governmental organisations (NGOs) can collaborate to establish knowledge-sharing platforms, develop digital tools and conduct workshops. However, these platforms must solve the practical challenges of circular practices like composting, renewable energy integration and waste recycling to materialise them. Empowering farmers and consumers with some information makes switching to a sustainable food system more feasible.

2.5.2 *Investment in Circular Agricultural Research and Development*

Advances in the technology and practical framework of circular agriculture require research and development (R&D). Investment is needed to fund innovative solutions such as bio-based fertilisers, carbon capture technology and precision farming instruments. Scaling this up will require both private and public agencies to join forces and give smallholder farmers access to these technologies (Haque et al. 2023; Kucher et al. 2022). Furthermore, public–private partnerships can hasten research and development. Universities and agribusiness joint ventures are a perfect example of how they could work together to develop climate-smart solutions for their region. Such a collaboration fills the gap between theory and real application and makes the ideas reach farmers (Osumba et al. 2021). Field-based studies also replicate findings. The demonstration farms are real-life examples of the benefits of circular practice. These farms can be used as demonstration farms so that farmers can learn about and get comfortable with alternative approaches.

2.5.3 *Institutional Reinforcement*

Strong institutional support is essential for the successful implementation of circular agriculture. However, extension services must be improved to enable farmers to access resources, training and technical support. Customised solutions are needed to address region-specific concerns, such as water shortages and soil degradation (Velasco-Muñoz et al. 2021). Another important factor is policy consistency. Frequent revisions of policies create confusion and slow implementation (Zarb and Taylor 2023). Moreover, this requires collaboration among stakeholders. Multi-stakeholder platforms can be used to share ideas and resources with farmers, researchers and policymakers. They can help to reduce systemic hurdles and enable an easier transition to circular agriculture (Seifu et al. 2022).

2.5.4 *Rewarding with Carbon Credits*

Carbon credit programmes compensate farmers for lowering emissions and increasing their carbon sequestration. In contrast, farmers practising agroforestry or no-till enter crediting for the carbon stored in their soils or trees (Raina et al. 2024; Singh et al. 2024; Tamba et al. 2021). Governments and international organisations can create specific access to carbon markets for small-scale farmers. It is easier to sign up for and accumulate points, so even resource-poor people can join (Tamba et al. 2021). Moreover, associations with financial institutions can enable the farmers to borrow or invest in their agricultural infrastructure using credit. They can be made more attractive through co-benefits like the protection of biodiversity or the water cycle. These systems provide incentives for reduction in emissions, as well as other environmental and social benefits (Raina et al. 2024).

2.5.5 *Incorporate Circular Agriculture Models in Regional Development Plans*

Integrating circular agriculture within regional development plans brings its principles into the social and economic strategies. Agroforestry and rainwater collection should be part of the plans for areas that are prone to soil erosion or a lack of water. Regional planning should also orient towards the growth of the value chain (Velasco-Muñoz et al. 2021). This can lead to job opportunities by setting up processing facilities for biogas or biofertilisers and implementing circularity. These actions align with supporting rural development goals by building economic resilience and environmental sustainability. This allows governments to push municipalities, agricultural institutions and local communities to work on circular objectives together. However, joint approaches can potentially optimise resource use and group work among stakeholders (Chung et al. 2024; Velasco-Muñoz et al. 2021).

2.5.6 *Develop Technical Standards and Monitoring and Evaluation Capacity for the Circular Agriculture Model*

Measures that can standardise the assessment of circular agriculture performance will be necessary to ensure consistency and the potential for scaling (Soto et al. 2021). These metrics need to account for carbon sequestration, nutrient recycling efficiency and optimisation of water usage. Monitoring frameworks must have rigour and be friendly to the farmer. Apps and satellite images can allow farmers to provide information on the health of their soil, their emissions, and their resource use. Farmers can use these tools to receive real-time feedback and adapt their practices to optimise their sustainability and efficiency. Circular agricultural initiatives must be reported and audited annually for accountability and transparency. These requirements will be facilitated by government and non-governmental organisations (NGOs) to support farmers and institutions in effectively meeting these challenges. This also allows for capacity building (Velasco-Muñoz et al. 2021).

2.5.7 Reform Agricultural Subsidies and Introduce Innovative Policy Support

Agricultural subsidies frequently support resource-exhausting methods under a linear, extractive model of agriculture. Reforming these subsidies for embedded sustainable and circular economies could accelerate the transition (Sreekumar et al. 2024). Subsidies for the use of synthetic fertilisers could support precision farming technology or the development of biofertiliser. Financial uncertainty with circular agriculture can be mitigated by innovative policies, including risk-sharing arrangements and insurance plans. New tax incentives for businesses that invest in circular technology may also help governments ensure the involvement of the private sector (Glauber et al. 2021). Future policy frameworks may also be furthered through public–private collaborations. The collaboration of both sectors for common goals might drive investment and promote innovation (Yin et al. 2024).

2.5.8 Develop Circular Agricultural Value Chains

Circular value chains begin and end with recycling and reuse across manufacturing, processing and distribution. Waste generated from food processing can be converted into bioenergy or animal feed, which both minimises waste and elevates the value of byproducts. Collaboration along the value chain is critical to maximise the effort to produce the product (Santiago et al. 2024). Closing the material loop requires farmers, processors, distributors and retailers to align their processes. Farmers can sign contracts with biogas plants to deliver a constant flow of organic waste and get renewable energy in return. Conversely, circular value chains create opportunities for development in rural areas. Moreover, it reduces transportation emissions and creates jobs.

2.5.9 Enhance Private Sector Investment in Circular Agriculture

Expanding circular agriculture necessitates the involvement of the private sector. Investing in technologies like anaerobic digesters and green energy systems can help ensure that these trends continue to spread broadly. These expensive technologies can be adopted in resource-constrained places and have a lower risk in public–private partnerships. Co-financing solutions enable smallholder farmers to access innovative technologies and infrastructure (Velasco-Muñoz et al. 2021). Circular agriculture may potentially get advantages from corporate social responsibility (CSR) initiatives (Santiago et al. 2024). Companies may assist with sustainable packaging and integrating renewable energy into farming. These initiatives simultaneously fulfil the goals of companies and the environment.

2.5.10 *Establish More Intersectoral Circular Bioeconomy Models for Sustainable and Low-Carbon Rural Development*

Intersectoral bioeconomy models integrate agriculture, energy, waste management and forestry sectors to enhance resource utilisation (Taron and Gebrezgabher 2024). Agricultural waste can be converted into bioplastics or bioenergy, allowing us to move away from fossil fuels. These approaches require strong collaboration among stakeholders. Policies and incentives can nurture synergies across sectors, enabling governments to work together more successfully. Support for renewable energy programmes that use agricultural waste serves both the energy sector and agriculture itself (Kumar et al. 2023b). Bioeconomy models also offer income diversification and job opportunities besides enhancing rural development. These models must be prioritised in national development strategies to optimise their potential.

2.5.11 *Enhance International Cooperation*

Climate change and food security are the current global concerns that require a collective response. Knowledge-sharing platforms facilitate the dissemination of best practices and technology to promote circular agriculture. Circular agricultural initiatives in underdeveloped countries will receive money from international sources like the Green Climate Fund. The allocated funds may be utilised for the development of infrastructure, as well as for training and capacity-building initiatives. Research programmes can be expedited by coordinated efforts across multiple nations (Velasco-Muñoz et al. 2021).

2.6 Conclusion

The conventional agricultural sector is facing several challenges, leading to unsustainable development and environmental degradation. Circular agriculture is a potential alternative, and governments around the world may integrate it into their agriculture and food security policy objectives to foster sustainable agricultural development. Sustainable agricultural practices include agroforestry, cover cropping, rotational grazing, reduced chemical fertiliser use and minimal tillage. The application of environmentally friendly technologies in a rational way can achieve maximum benefits from natural processes with minimum input dependency, efficient recycling of nutrients, energy and water within ecosystems and transforming waste into valuable food production inputs, minimising costs and food losses using the principles of reduce, reuse and recycle in the agri-food value chain. Implementing circular agriculture practices requires national-level policies and institutional mechanisms to deliver relevant extension, and technological and financial services to farmers in a timely and efficient manner.

References

Adegbeye MJ, Reddy PRK, Obaisi AI, Elghandour MMMY, Oyebamiji KJ, Salem AZM, Camacho-Díaz LM (2020) Sustainable agriculture options for production, greenhouse gases and pollution alleviation, and nutrient recycling in emerging and transitional nations – an overview. J Clean Prod 242:118319

Alsarhan LM, Alayyar AS, Alqahtani NB, Khdary NH (2021) Circular carbon economy (CCE): a way to invest CO_2 and protect the environment, a review. Sustainability 13(21):11625

Amoakwah E, Lucas ST, Didenko NA, Rahman MA, Islam KR (2022) Impact of deforestation and temporal land-use change on soil organic carbon storage, quality, and lability. PLoS One 17(8): e0263205

Arif NA (2024) Characteristics of the distribution of greenhouse gases into the atmosphere. In: The 1st international scientific and practical conference "advanced technologies for the implementation of new ideas (January 9–12, 2024), Brussels, Belgium. International Science Group. 349p (p 24)

Basheer S, Wang X, Farooque AA, Nawaz RA, Pang T, Neokye EO (2024) A review of greenhouse gas emissions from agricultural soil. Sustainability 16(11):4789

Bhagat R, Walia SS, Sharma K, Singh R, Singh G, Hossain A (2024) The integrated farming system is an environmentally friendly and cost-effective approach to the sustainability of agri-food systems in the modern era of the changing climate: a comprehensive review. Food Energy Secur 13(1):e534

Caillol S (2011) Fighting global warming: the potential of photocatalysis against CO_2, CH_4, N_2O, CFCs, tropospheric O_3, BC and other major contributors to climate change. J Photochem Photobiol C: Photochem Rev 12(1):1–19

Chiriacò MV, Galli N, Santini M, Rulli MC (2024) Deforestation and greenhouse gas emissions could arise when palm oil is replaced with other vegetable oils. Sci Total Environ 914:169486

Darikandeh D, Shahnazari A, Khoshravesh M, Yousefian M, Porter CH, Hoogenboom G (2025) Optimizing rice management to reduce methane emissions and maintain yield with the CSM-CERES-rice model. Agric Syst 224:104248

Džermeikaitė K, Krištolaitytė J, Antanaitis R (2024) Relationship between dairy cow health and intensity of greenhouse gas emissions. Animals 14(6):829

Eddy WC, Yang WH (2022) Improvements in soil health and soil carbon sequestration by an agroforestry for food production system. Agric Ecosyst Environ 333:107945

El-Ramady H, Brevik EC, Abowaly M, Ali R, Saad Moghanm F, Gharib MS, Mansour H, Fawzy ZF, Prokisch J (2024) Soil degradation under a changing climate: management from traditional to nano-approaches. Egypt J Soil Sci 64(1):287–298

Fahad S, Chavan SB, Chichaghare AR, Uthappa AR, Kumar M, Kakade V, Pradhan A, Jinger D, Rawale G, Yadav DK, Kumar V, Poczai P (2022) Agroforestry systems for soil health improvement and maintenance. Sustainability 14(22):14877

Faizan M (2024) Enteric methane production in ruminants: its effect on global warming and mitigation strategies – a review. Pak J Sci 76(01):16–38

Fernández-Ortega J, Álvaro-Fuentes J, Cantero-Martínez C (2024) Double-cropping, tillage and nitrogen fertilisation effects on soil CO_2 and CH_4 emissions. Agric Ecosyst Environ 359:108758

Food and Agriculture Organization of the United Nations (FAO) (2013) Greenhouse gas emissions from ruminant supply chains—a global life cycle assessment. FAO, Rome

Food and Agriculture Organization of the United Nations (FAO) (2020) Global Forest Resources Assessment 2020: Main report. Rome

Friedlingstein P, O'Sullivan M, Jones MW, Andrew RM, Bakker DC, Hauck J, Landschützer P, Le Quéré C, Luijkx IT, Peters GP, Peters W (2023) Global carbon budget 2023. Earth Syst Sci Data 15(12):5301–5369

Garg MR, Sherasia PL, Bhanderi BM, Phondba BT, Shelke SK, Makkar HPS (2013) Effects of feeding nutritionally balanced rations on animal productivity, feed conversion efficiency, feed nitrogen use efficiency, rumen microbial protein supply, parasitic load, immunity and enteric methane emissions of milking animals under field conditions. Anim Feed Sci Technol 179(1–4): 24–35

Gautam K, Tripathi KS, Pandey K (2023) Concepts of remote sensing, GIS & GPS. In: Modern agronomy-theory and practices. Navi Publications, Gwalior, p 49

Glauber J et al (2021) Design principles for agricultural risk management policies, OECD food, agriculture and fisheries papers, no. 157. OECD Publishing, Paris. https://doi.org/10.1787/1048819f-en

Gunasekara NS, Ariyawansha RTK, Basnayake BFA (2024) Effective organic fertiliser through a novel approach of biocatalyst activated from biochar. Int J Environ Waste Manag 35(3): 323–337

Gyawali A (2024) Biochar for soil health and climate change mitigation: a sustainable approach. Ecofem Clim Change 5(2):94–103. https://doi.org/10.26480/efcc.02.2024.94.103

Haque F, Fan C, Lee YY (2023) From waste to value: addressing the relevance of waste recovery to agricultural sector in line with circular economy. J Clean Prod 415:137873

Hergoualc'h K, Verchot LV (2014) Greenhouse gas emission factors for land use and land-use change in southeast Asian peatlands. Mitig Adapt Strateg Glob Chang 19:789–807

Hong C, Burney JA, Pongratz J, Nabel JE, Mueller ND, Jackson RB, Davis SJ (2021) Global and regional drivers of land-use emissions in 1961–2017. Nature 589(7843):554–561

Hoogstra AG, Silvius J, de Olde EM, Candel JJL, Termeer CJAM, van Ittersum MK, de Boer IJM (2024) The transformative potential of circular agriculture initiatives in the north of The Netherlands. Agric Syst 214:103833

Houghton RA, House JI, Pongratz J, Van Der Werf GR, Defries RS, Hansen MC, Le Quéré C, Ramankutty N (2012) Carbon emissions from land use and land-cover change. Biogeosciences 9(12):5125–5142

Huss CP, Holmes KD, Blubaugh CK (2022) Benefits and risks of intercropping for crop resilience and pest management. J Econ Entomol 115(5):1350–1362

Intergovernmental Panel on Climate Change (IPCC) (2018) Global warming of 1.5° C: an IPCC special report on the impacts of global warming of 1.5° C above pre-industrial levels and related global greenhouse gas emission pathways, in the context of strengthening the global response to the threat of climate change, sustainable development, and efforts to eradicate poverty. Intergovernmental Panel on Climate Change, Geneva

Islam KN, Sarker T, Taghizadeh-Hesary F, Atri AC, Alam MS (2021) Renewable energy generation from livestock waste for a sustainable circular economy in Bangladesh. Renew Sust Energ Rev 139:110695

Jackson RB, Saunois M, Martinez A, Canadell JG, Yu X, Li M et al (2024) Human activities now fuel two-thirds of global methane emissions. Environ Res Lett 19(10):101002

Jangir CK, Sangwan PS, Panghaal D, Kumar S, Meena RS, Bharti, Jat RD, Singh N (2024) Spatial variability and statistical analysis of soil properties in the rice wheat-based systems of northwest India. Commun Soil Sci Plant Anal 55(8):1205–1223

Jansson C, Faiola C, Wingler A, Zhu XG, Kravchenko A, De Graaff MA, Ogden AJ, Handakumbura PP, Werner C, Beckles DM (2021) Crops for carbon farming. Front Plant Sci 12:636709

Jaroenkietkajorn U, Gheewala SH, Mungkung R, Jakrawatana N, Silalertruksa T, Lecksiwilai N, Prasara AJ, Nilsalab P (2024) Challenges and opportunities of bio-circular-green economy for agriculture. Circ Econ Sustain 4(3):1729–1750

Jhariya MK, Raj A, Banerjee A, Meena RS, Bargali SS, Kumar S, Nema S, Poonam, Oraon PR (2022) Plan and policies for soil organic carbon management under agroforestry system. In: Plans and policies for soil organic carbon management in agriculture. Springer, Singapore, pp 191–219

Keprate A, Bhardwaj DR, Sharma P, Verma K, Abbas G, Sharma V, Sharma K, Janju S (2024) Climate resilient agroforestry systems for sustainable land use and livelihood. In: Transforming agricultural management for a sustainable future: climate change and machine learning perspectives. Springer, Cham, pp 141–161

Khan MN, Sial TA, Ali A, Wahid F (2024) Impact of agricultural wastes on environment and possible management strategies. In: Frontier studies in soil science. Springer, Cham, pp 79–108

Khatri P, Kumar P, Shakya KS, Kirlas MC, Tiwari KK (2024) Understanding the intertwined nature of rising multiple risks in modern agriculture and food system. Environ Dev Sustain 26(9): 24107–24150

Kingsford RT, Basset A, Jackson L (2016) Wetlands: conservation's poor cousins. Aquat Conserv Mar Freshwat Ecosyst 26(5):892–916

Koushika SP, Krishnaveni A, Pazhanivelan S, Bharani A, Arunkumar V, Devaki P, Muthukrishnan N (2024). Carbon economics of different agricultural practices for farming soil. arXiv preprint arXiv:2403.07530

Kreileman GJJ, Bouwman AF (1994) Computing land use emissions of greenhouse gases. Water Air Soil Pollut 76:231–258

Kucher L, Kucher A, Morozova H, Pashchenko Y (2022) Development of circular agricultural economy: potential sources of financing innovative projects. Agric Resour Econ Int Sci E-J 8(2): 206–227

Kumar S, Sharma SK, Thakral SK, Bhardwaj KK, Jhariya MK, Meena RS, Jangir CK, Bedwal S, Jat RD, Gaber A, Atta AA (2022) Integrated nutrient management improves the productivity and nutrient use efficiency of *Lens culinaris* Medik. Sustainability 14(3):1284

Kumar S, Sharma SK, Dhaka AK, Bedwal S, Sheoran S, Meena RS, Jangir CK, Kumar D, Kumar R, Jat RD, Meena AK (2023a) Efficient nutrient management for enhancing crop productivity, quality and nutrient dynamics in lentil (*Lens culinaris* Medik.) in the semi-arid region of northern India. PLoS One 18(2):e0280636

Kumar SP, Subudhi S, Bhatia L, Saha K, Mudgil D, Prasad Shadangi K, Srivastava RK, Pattnaik B, Arya RK (2023b) Utilization of agricultural waste biomass and recycling toward circular bioeconomy. Environ Sci Pollut Res 30(4):8526–8539

Kumar R, Kaur A, Sharma S, Bharti H, Kumar R (2024) Advancements and challenges in agriculture waste management: a comprehensive. Educ Adm Theory Pract 30(5):7253–7273

Kumar S, Meena RS, Kumar S, Pradhan G, Jangir CK, Ghosh S, Punia H, Sheoran P, Meena R, Ahmad MA, Goyal SK (2025) Designing a diversified Indian mustard production system for energy-carbon-cum-heat use efficiency and sowing dates assessment. GCB Bioenergy 17(6): e70044

Lakhouit A, Rashed WSA, Abbas SY, Shaban M (2025) Integrating machine learning for precision agriculture waste estimation and sustainability enhancement. Comput Electron Agric 230: 109933

Lal R (2004) Soil carbon sequestration impacts on global climate change and food security. Science 304(5677):1623–1627

Lazcano C, Zhu-Barker X, Decock C (2021) Effects of organic fertilizers on the soil microorganisms responsible for N_2O emissions: a review. Microorganisms 9(5):983

Leharwan M, Kumar Y, Kumar R, Kumar Saraswat P, Kumar R, Kumar Thaliyil Veetil A, Bhattacharjee S, Kumar A, Kumar S (2023) Assessing the effects of conservation tillage and in-situ crop residue management on crop yield and soil properties in rice–wheat cropping system. Sustainability 15(17):12736

Li D, Li H, Chen D, Xue L, He H, Feng Y, Ji Y, Yang L, Chu Q (2021) Clay-hydrochar composites mitigated CH_4 and N_2O emissions from paddy soil: a whole rice growth period investigation. Sci Total Environ 1(780):146532. https://doi.org/10.1016/j.scitotenv.2021.146532. Epub 2021 Mar 18

Li L, Awada T, Shi Y, Jin VL, Kaiser M (2025) Global greenhouse gas emissions from agriculture: pathways to sustainable reductions. Glob Chang Biol 31(1):e70015

Marku D, Minga A, Sosoli I (2024) Circular economy perspective and implications for livestock farming in Albania. Open Agric J 18(1). https://doi.org/10.2174/0118743315312132240611074625

Meena RS, Pradhan G (2023) Industrial garbage-derived biocompost enhances soil organic carbon fractions, CO_2 biosequestration, potential carbon credits, and sustainability index in a rice-wheat ecosystem. Environ Res:116525. https://doi.org/10.1016/j.envres.2023.116525

Meena RS, Pradhan G, Kumar S, Lal R (2023) Using industrial wastes for rice-wheat cropping and food-energy-carbon-water-economic nexus to the sustainable food system. Renew Sust Energ Rev 187:113756. https://doi.org/10.1016/j.rser.2023.113756

Meena RS, Pradhan G, Singh K, Kumar S, Singh AK, Shashidhar KS, Mina KK, Rao CS (2024a) Agriculture models for restoring degraded land to enhance CO_2 biosequestration and carbon credits in the Vindhyan region of India. Sci Total Environ 929:172661. https://doi.org/10.1016/j.scitotenv.2024.172661

Meena RS, Lal R, Kumar S, Pradhana G, Rao CS, Singh AK, Pathak H, Abhilash PC, Kumar A, Sharma SK, Jat ML, Singh S (2024b) Potential of Indian agriculture for capturing atmospheric CO_2 and monetizing carbon credits to the farmers: an overview and policy framework. Adv Agron 188:101–206. https://doi.org/10.1016/bs.agron.2024.06.001

Meena RS, Pradhan G, Nalani SK, Singh AK, Verma SK, Mina KK, Kumar S, Chaturvedi RK (2024c) Diversified cropping modules designed for soil restoration, CO_2 sequestration, and generating carbon credits. Land Degrad Dev 35. https://doi.org/10.1002/ldr.5224

Meena RS, Pradhan G, Kumar S (2025) Energy flow, eco-efficiency, and economic circulation with recycled industrial waste compost application in wheat and subsequent rice farming. Sci Total Environ 967:178779. https://doi.org/10.1016/j.scitotenv.2025.178779

Minh Khue NT, Tran QP, Anh NTQ, Cuong LK, Du NC, Cuong CV, Thuong VT, Anh DH, Vu NA (2024) Examining the factors influencing the level of circular economy adoption in agriculture: insights from Vietnam. Res World Agric Econ 5(1):48–58

Montagnini F (2024) Introduction. Challenges and achievements in agroforestry in the new millennium. In: Integrating landscapes: agroforestry for biodiversity conservation and food sovereignty. Springer, Cham, pp 3–19

Mostefaoui M, Ciais P, McGrath MJ, Peylin P, Patra PK, Ernst Y (2024) Greenhouse gas emissions and their trends over the last 3 decades across Africa. Earth Syst Sci Data 16(1):245–275

National Oceanic and Atmospheric Administration (NOAA) (2022) Carbon dioxide now more than 50% higher than pre-industrial levels, June 3. https://www.noaa.gov/news-release/carbon-dioxide-now-more-than-50-higher-than-pre-industrial-levels. Accessed on 4 Dec 2024

Nazir MJ, Li G, Nazir MM, Zulfiqar F, Siddique KH, Iqbal B, Du D (2024) Harnessing soil carbon sequestration to address climate change challenges in agriculture. Soil Tillage Res 237:105959

Nedelkov K, Angelova T, Krastanov J, Mihaylova M (2024) Feeding strategies to reduce methane emissions: a review. Bulgarian J Agr Sci 30(1):28–36

Nunes MR, Karlen DL, Moorman TB (2020) Tillage intensity effects on soil structure indicators—a US meta-analysis. Sustainability 12(5):2071

O'Mara FP (2011) The significance of livestock as a contributor to global greenhouse gas emissions today and in the near future. Anim Feed Sci Technol 166–167:7–15. https://doi.org/10.1016/j.anifeedsci.2011.04.074

Osumba JJ, Recha JW, Oroma GW (2021) Transforming agricultural extension service delivery through innovative bottom–up climate-resilient agribusiness farmer field schools. Sustainability 13(7):3938

Ouatahar L, Bannink A, Zentek J, Amon T, Deng J, Hempel S, Janke D, Beukes P, van der Weerden T, Krol D, Lanigan GJ, Amon B (2024) An integral assessment of the impact of diet and manure management on whole-farm greenhouse gas and nitrogen emissions in dairy cattle production systems using process-based models. Waste Manag 187:79–90

Pacheco-Torgal F (2024) Introduction to carbon dioxide sequestration through innovative cementitious construction materials. In: Carbon dioxide sequestration in cementitious construction materials. Woodhead Publishing, pp 1–9

Pal P, Singh AK, Srivastava RK, Rathore SS, Sahoo UK, Subudhi S, Sarangi PK, Prus P (2024) Circular bioeconomy in action: transforming food wastes into renewable food resources. Foods 13(18):3007

Paramesh V, Sreekanth GB, Chakurkar EB, Chethan Kumar HB, Gokuldas P, Manohara KK, Ramdas Mahajan G, Rajkumar RS, Ravisankar N, Panwar AS (2020) Ecosystem network analysis in a smallholder integrated crop–livestock system for coastal lowland situation in tropical humid conditions of India. Sustainability 12(12):5017

Pradhan G, Meena RS (2023a) Utilizing waste compost to improve the atmospheric CO_2 capturing in the rice-wheat cropping system and energy-cum-carbon credit audit for a circular economy. Sci Total Environ 164572. https://doi.org/10.1016/j.scitotenv.2023.164572

Pradhan G, Meena RS (2023b) Interaction impacts of biocompost on nutrient dynamics and relations with soil biota, carbon fractions index, societal value of CO_2 equivalent, and ecosystem services in the wheat-rice farming. Chemosphere:139695. https://doi.org/10.1016/j.chemosphere.2023.139695

Pradhan G, Meena RS, Kumar S, Lal R (2023) Utilizing industrial wastes as compost in wheat-rice production to improve the above and below-ground ecosystem services. Agric Ecosyst Environ 358:108704. https://doi.org/10.1016/j.agee.2023.108704

Priyadarshini P, Abhilash PC (2020) Fostering sustainable land restoration through circular economy-governed transitions. Restor Ecol 28(4):719–723

Punia H, Tokas J, Malik A, Satpal, Rani A, Gupta P, Kumari A, Mor VS, Bhuker A, Kumar S (2020) Solar radiation and nitrogen use efficiency for sustainable agriculture. In: Resources use efficiency in agriculture. Springer, Singapore, pp 177–212

Qin J, Duan W, Zou S, Chen Y, Huang W, Rosa L (2024) Global energy use and carbon emissions from irrigated agriculture. Nat Commun 15(1):3084

Raina N, Zavalloni M, Viaggi D (2024) Incentive mechanisms of carbon farming contracts: A systematic mapping study. J Environ Manag 352:120126

Rami Y, Allouhi A (2024) Design, economic, and environmental accounting assessment of a solar-powered cold room for fish storage in traditional markets. Sustainability 16(7):3080

Rauw WM, Gomez-Raya L, Star L, Øverland M, Delezie E, Grivins M, Hamann KT, Pietropaoli M, Klaassen MT, Klemetsdal G, Gil MG (2023) Sustainable development in circular agriculture: an illustrative bee⍰ legume⍰ poultry example. Sustain Dev 31(2):639–648. https://doi.org/10.1002/sd.2435

Ravindranath NH, Mauvie R, Fargione J, Canadell JG, Berndes G, Woods J, Watson H, Sathaye J (2009) Greenhouse gas implications of land use change and land conversion to biofuel crops. Cornell University Library's Initiatives in Publishing (CIP), Ithaca

Richardson K, Steffen W, Lucht W, Bendtsen J, Cornell SE, Donges JF, Drüke M, Fetzer I, Bala G, Von Bloh W, Feulner G (2023) Earth beyond six of nine planetary boundaries. Sci Adv 9(37): eadh2458

Rockström J, Gupta J, Qin D, Lade SJ, Abrams JF, Andersen LS, Armstrong McKay DI, Bai X, Bala G, Bunn SE, Ciobanu D (2023) Safe and just Earth system boundaries. Nature 619(7968): 102–111

Rodino S, Pop R, Sterie C, Giuca A, Dumitru E (2023) Developing an evaluation framework for circular agriculture: a pathway to sustainable farming. Agriculture 13(11):2047

Roy P, Mohanty AK, Misra M (2023) Prospects of carbon capture, utilization and storage for mitigating climate change. Environ Sci Adv 2(3):409–423

Santiago BDS, Scavarda LF, Gusmão Caiado RG, Santos RS, Mattos Nascimento DLD (2024) Corporate social responsibility and circular economy integration framework within sustainable supply chain management: building blocks for industry 5.0. Corp Soc Responsib Environ Manag 32(1):269–290

Sanyaolu M, Sadowski A (2024) The role of precision agriculture technologies in enhancing sustainable agriculture. Sustainability 16(15):6668

Sarkar S, Jaswal A, Singh A (2024) Sources of inorganic nonmetallic contaminants (synthetic fertilizers, pesticides) in agricultural soil and their impacts on the adjacent ecosystems. In: Bioremediation of emerging contaminants from soils. Elsevier, Amsterdam, pp 135–161

Seifu M, van Paassen A, Klerkx L, Leeuwis C (2022) A state-initiated multi-stakeholder platform as an instrument to build agricultural innovation system capacity: a case study from Ethiopia. Innov Dev:1–22. https://doi.org/10.1080/2157930X.2022.2064959

Selvakumar P (2024) Reasons for biodiversity loss. In: Impact of societal development and infrastructure on biodiversity decline. IGI Global, Hershey, pp 37–49

Selvan T, Panmei L, Murasing KK, Guleria V, Ramesh KR, Bhardwaj DR, Thakur CL, Kumar D, Sharma P, Digvijaysinh Umedsinh R, Kayalvizhi D (2023) Circular economy in agriculture: unleashing the potential of integrated organic farming for food security and sustainable development. Front Sustain Food Syst 7:1170380

Sharma M, Kaushal R, Kaushik P, Ramakrishna S (2021) Carbon farming: prospects and challenges. Sustainability 13(19):11122

Sikiru A, Michael AO, John MO, Egena SSA, Oleforuh-Okoleh VU, Ambali MI, Muhammad IR (2024) Methane emissions in cattle production: biology, measurement and mitigation strategies in smallholder farmer systems. Environ Dev Sustain:1–24. https://doi.org/10.1007/s10668-024-04939-1

Silvius J, Hoogstra AG, Candel JJ et al (2023) Determining the transformative potential of circular agriculture initiatives. Ambio 52:1968–1980

Singh S, Kiran BR, Mohan SV (2024) Carbon farming: a circular framework to augment CO 2 sinks and to combat climate change. Environ Sci Adva 3(4):522–542

Sobanaa M, Prathiviraj R, Selvin J, Prathaban M (2024) A comprehensive review on methane's dual role: effects in climate change and potential as a carbon–neutral energy source. Environ Sci Pollut Res 31(7):10379–10394

Soto RL, de Vente J, Padilla MC (2021) Learning from farmers' experiences with participatory monitoring and evaluation of regenerative agriculture based on visual soil assessment. J Rural Stud 88:192–204

Sreekumar NM, Sudheep NM, Radhakrishnan EK (2024) Framework for implementing circular economy in agriculture. In: The potential of microbes for a circular economy. Academic Press, London, pp 25–52

Tamba Y, Wafula J, Magaju C, Aynekulu E, Winowiecki L, St-Jacques B, Stiem-Bhatia L, Arias-Navarro C (2021) A review of the participation of smallholder farmers in land-based carbon payment schemes. TMG and ICRAF working paper. https://tmg-thinktank.com/220214

Tan L, Ge Z, Zhou X, Li S, Li X, Tang J (2020) Conversion of coastal wetlands, riparian wetlands, and peatlands increases greenhouse gas emissions: a global meta-analysis. Glob Chang Biol 26(3):1638–1653

Taron A, Gebrezgabher S (2024) Circular bioeconomy: a pathway to sustainable development in an age of global crisis. In: International trade, economic crisis and the sustainable development goals. Emerald, Leeds, pp 99–117

Tebbiche I, Mocellin J, Huong LT, Pasquier LC (2021) Circular economy and carbon capture, utilization, and storage. In: Biomass, biofuels, biochemicals. Elsevier, Amsterdam, pp 813–851

Thirunagari BK, Kancheti M, Kumar R, Kota SH (2024) Managing synthetic N-fertilizer emissions in India: insights from field surveys across 102 districts. J Environ Manag 366:121909

Tubiello FN, Salvatore M, Cóndor Golec RD, Ferrara A, Rossi S, Biancalani R, Federici S, Jacobs H, Flammini A (2014) Agriculture, forestry and other land use emissions by sources and removals by sinks. FAO, Rome

Tuomisto HL, Hodge ID, Riordan P, Macdonald DW (2012) Does organic farming reduce environmental impacts? A meta-analysis of European research. J Environ Manag 112:309–320

Valve H, Ekholm P, Luostarinen S (2020) The circular nutrient economy: needs and potentials of nutrient recycling. In: Handbook of the circular economy. Edward Elgar Publishing, Cheltenham, pp 358–368

Chung D, Duy L, Loan L (2024) Circular Agriculture: A General Review of Theories, Practices, and Policy Recommendations. Vietnam Journal of Agricultural Sciences, 7(2), 2173–2184. https://doi.org/10.31817/vjas.2024.7.2.07

Vaneeckhaute C, Meers E, Michels E, Ghekiere G, Accoe F, Tack FM (2013) Closing the nutrient cycle by using bio-digestion waste derivatives as synthetic fertilizer substitutes: a field experiment. Biomass Bioenergy 55:175–189

Velasco-Muñoz JF, Mendoza JMF, Aznar-Sánchez JA, Gallego-Schmid A (2021) Circular economy implementation in the agricultural sector: definition, strategies and indicators. Resour Conserv Recycl 170:105618

Vellingiri A, Kokila R, Nisha P, Kumar M, Chinnusamy S, Boopathi S (2025) Harnessing GPS, sensors, and drones to minimize environmental impact: precision agriculture. In: Designing sustainable internet of things solutions for smart industries. IGI Global, Hershey, pp 77–108

Verhoeven JT, Setter TL (2010) Agricultural use of wetlands: opportunities and limitations. Ann Bot 105(1):155–163

Vijay V, Pimm SL, Jenkins CN, Smith SJ (2016) The impacts of oil palm on recent deforestation and biodiversity loss. PLoS One 11(7):e0159668

Waghorn GC, Hegarty RS (2011) Lowering ruminant methane emissions through improved feed conversion efficiency. Anim Feed Sci Technol 166:291–301

Walling E, Vaneeckhaute C (2020) Greenhouse gas emissions from inorganic and organic fertilizer production and use: A review of emission factors and their variability. J Environ Manag 276: 111211

Wang Z, Huang W, Wang H, Gao J, Zhang R, Xu G, Wang Z (2024) Research on the improvement of carbon neutrality by utilizing agricultural waste: based on a life cycle assessment of biomass briquette fuel heating system. J Clean Prod 434:140365

World Meteorological Organization (WMO) (2021) WMO greenhouse gas bulletin. World Meteorological Organization, Geneva

Xing Y, Wang X (2024) Impact of agricultural activities on climate change: a review of greenhouse gas emission patterns in field crop systems. Plants 13(16):2285

Yan X, Ying Y, Li K, Zhang Q, Wang K (2024) A review of mitigation technologies and management strategies for greenhouse gas and air pollutant emissions in livestock production. J Environ Manag 352:120028

Yang C, Zhang Y, Xue Y, Xue Y (2022) Toward a socio-political approach to promote the development of circular agriculture: a critical review. Int J Environ Res Public Health 19: 13117. https://doi.org/10.3390/ijerph192013117

Yang M, Chen L, Wang J, Msigwa G, Osman AI, Fawzy S, Rooney DW, Yap P (2023) Circular economy strategies for combating climate change and other environmental issues. Environ Chem Lett 21:55–80. https://doi.org/10.1007/s10311-022-01499-6

Yin Q, Wang Q, Du M, Wang F, Sun W, Chen L, Tang H (2024) Promoting the resource utilization of agricultural wastes in China with public-private-partnership mode: an evolutionary game perspective. J Clean Prod 434:140206

Yrjälä K, Ramakrishnan M, Salo E (2022) Agricultural waste streams as resource in circular economy for biochar production towards carbon neutrality. Curr Opin Environ Sci Health 26: 100339

Zarb S, Taylor K (2023) Uneven local implementation of federal policy after disaster: policy conflict and goal ambiguity. Rev Policy Res 40(1):63–87

Zoboli O, Zessner M, Rechberger H (2016) Supporting phosphorus management in Austria: potential, priorities and limitations. Sci Total Environ 565:313–323. https://doi.org/10.1016/j.scitotenv.2016.04.171

Chapter 3
Policies and Technological Intervention for Regenerative Agriculture and Carbon Credits to Achieve Carbon-Negative Agriculture

K. Srikanth Reddy, G. Sreeja Reddy, C. M. Parihar, Panneerselvam Peramaiyan, K. Sreenivasa Reddy, Kiranmoy Patra, Ayan Sarkar, Alok Sinha, B. Mahesh Naik, and M. Karthika

Abstract Regenerative agriculture (RA) offers a transformative pathway to tackle climate change, restore biodiversity, and strengthen rural economies. Integrating RA practices with carbon credit mechanisms provides strong financial incentives to accelerate widespread adoption. Key strategies such as conservation agriculture, agroforestry, cover cropping, crop rotation, livestock integration, and precision resource management significantly improve soil carbon sequestration, reduce greenhouse gas emissions, and build climate-resilient farming systems. Strengthening Payment for Ecosystem Services (PES) frameworks, along with bundling carbon credits with yield insurance and productivity gains, can further offer a robust approach to scaling adoption, particularly among smallholder farmers. In the countries from the Global South, there is an urgent need to develop dedicated and tailored

K. S. Reddy (✉) · K. S. Reddy
ICAR-Indian Agricultural Research Institute, New Delhi, India

International Rice Research Institute (IRRI), South Asia Regional Centre, Varanasi, Uttar Pradesh, India
e-mail: srikant_12001@iari.res.in; sreenivasa_11722@iari.res.in

G. S. Reddy · C. M. Parihar · K. Patra · A. Sarkar · A. Sinha
ICAR-Indian Agricultural Research Institute, New Delhi, India
e-mail: gongalla_12592@iari.res.in; ayan_12403@iari.res.in; alok_12631@iari.res.in

P. Peramaiyan
International Rice Research Institute (IRRI), South Asia Regional Centre, Varanasi, Uttar Pradesh, India
e-mail: p.panneerselvam@irri.org

B. M. Naik
Malla Reddy University, Hyderabad, Telangana, India
e-mail: mahesh.b@mallareddyuniversity.ac.in

M. Karthika
Amrita Vishwa Vidyapeetham, Coimbatore, Tamil Nadu, India
e-mail: m_karthika@cb.amrita.edu

R. S. Meena et al. (eds.), *Carbon-Negative Agriculture*, Sustainability Sciences in Asia and Africa, https://doi.org/10.1007/978-981-95-1483-0_3

methodologies for measuring and verifying carbon outcomes, specifically from RA practices. Research communities must prioritize creating standardized protocols, validated through pilot projects implemented on farmers' fields, and demonstrating real-world feasibility and tangible benefits will build trust and drive adoption. Adopting technological innovations, including artificial intelligence, geo-spatial analytics, Internet of Things (IoT)-based soil sensing, mobile applications, and real-time carbon mapping, enables accurate measurement, monitoring, and verification of carbon dynamics. Agro-industries can support this transition through insulating mechanisms, providing farmers with more stable and predictable markets compared to the often-volatile voluntary carbon markets. Furthermore, linking carbon credits with associated ecosystem services such as water conservation, soil fertility improvement, and biodiversity enhancement can increase the quality and market value of credits, attracting premium prices nationally and internationally. Aligning financial incentives with ecological restoration firmly establishes RA as a pivotal strategy for achieving carbon neutrality, fostering sustainable food systems, and building resilient agroecological landscapes.

Keywords Agricultural sustainability · Carbon markets · Carbon sequestration · Ecosystem services · Greenhouse gas mitigation

Abbreviations

AI	Artificial intelligence
CA	Conservation agriculture
DSS	Decision support system
ETS	Emission trading system
FPO	Farmer producer organizations
GHGs	Greenhouse gases
GIS	Geographic information system
GPS	Global Positioning System
ML	Machine learning
MRV	Monitoring, reporting, and verification
N_2O	Nitrous oxide
NDC	Nationally Determined Contributions
NMSA	National Mission for Sustainable Agriculture
PES	Payment for ecosystem services
PGPR	Plant growth-promoting rhizobacteria
RA	Regenerative agriculture
SDGs	Sustainable Development Goals
SHG	Self-help groups
SOC	Soil organic carbon
SOM	Soil organic matter
VCMs	Voluntary carbon markets

VCS	Verified carbon standard
VRT	Variable rate technology
WOTR	Watershed Organization Trust
ZBNF	Zero Budget Natural Farming

3.1 Introduction

The degradation of natural ecosystems, particularly soils, has long been identified as a key driver behind the collapse of past civilizations (Lowdermilk 1950; Montgomery 2012). Today, the world confronts a comparable crisis. Since agriculture emerged nearly 12,000 years ago, an estimated 133 gigatons (Gt) of carbon have been lost from global soils, primarily due to poor land management and unsustainable practices (Sanderman et al. 2017). Industrial agriculture has exacerbated this loss through intensive tillage, monocultures, and heavy dependence on synthetic inputs, all of which weaken the soil's carbon storage potential. Soil erosion rates are now 10–100 times faster than natural soil formation, especially in systems dominated by conventional tillage (UNEP 2020). If these trends persist, by 2050, only about one-quarter of the arable land per person that was available in 1960 will remain productive (FAO 2011). In addition to degrading soils, industrial agricultural systems are major contributors to environmental pollution and climate change. The Agriculture, Forestry, and Other Land Use (AFOLU) sector was responsible for an average of 11.9 Gt of CO_2 equivalent emissions per year from 2010 to 2019, making up approximately 21% of global anthropogenic greenhouse gas (GHG) emissions (IPCC 2023). Poor management of synthetic fertilizers is a major contributor to nitrogen pollution, which contaminates freshwater systems and is the primary cause of impairment in rivers, streams, lakes, and wetlands (Lindwall 2021). Pesticide residues are now found in 88% of surface waters and have been detected in the bodies of more than 90% of populations in developed countries like the United States and China (Chiu et al. 2018). This agricultural model, focused on scale and uniformity, has also led to significant biodiversity loss, with global wildlife populations having dropped by nearly 60% since 1970 (WWF 2018). These interconnected challenges are highlighted in the Planetary Boundaries 2.0 framework by Rockström et al. (2023). This systems-based framework identifies nine essential Earth system processes that regulate the planet's stability, and it defines thresholds that should not be crossed to avoid catastrophic environmental change. Agriculture plays a central role in breaching several of these boundaries. It disturbs biogeochemical fluxes, especially the nitrogen and phosphorus cycles, by over-application of nutrients. Cropland expansion is still the primary cause of land-system change, and monocultures and chemical use compromise the integrity of the biosphere by hastening the loss of biodiversity. Furthermore, the world's freshwater system is under tremendous strain due to unsustainable water withdrawals for irrigation. These transgressions of planetary boundaries show how unsustainable the existing agricultural trend is, and the urgency of transformative change.

This ecological crisis highlights the urgent need to shift toward more resilient and restorative food production systems. Regenerative agriculture (RA) has emerged as a powerful alternative, focusing on healing rather than harming the land. It is not just a set of techniques but a holistic philosophy of land stewardship that views farms as complex ecological and social systems (Brown 2018). Rooted in indigenous knowledge and agroecological science, RA emphasizes principles like minimizing soil disturbance, maintaining permanent soil cover, fostering biodiversity, preserving continuous living roots, and integrating livestock into farming (Fuhrer 2016; Khangura et al. 2023; Meena et al. 2025). These principles work together to increase soil organic carbon (SOC), restore degraded ecosystems, and enhance resilience to climate extremes. By aligning food production with the Earth's ecological boundaries, RA presents a path to restore landscapes, mitigate climate change, and ensure long-term food and environmental security for future generations (Kumar et al. 2022b; Rehberger et al. 2023). Landscapes under regenerative practices, which are biologically diverse, have been shown to support greater biodiversity, improve soil health, and increase SOC (Kremen and Merenlender 2018; Karp 2022). These systems also enhance water infiltration, nutrient cycling, and resilience to extreme weather events like droughts and floods (Basche and DeLonge 2019; Meena et al. 2024c). Most importantly, RA supports the transition from carbon-emitting to carbon-sequestering systems, offering a natural climate solution with global implications. Despite its ecological promise, the transition to RA is complex and often challenging for farmers. The shift from input-intensive industrial systems to labor- and knowledge-intensive regenerative models typically requires time, resources, and adaptation to new knowledge systems. Many regenerative practices yield benefits only after several years of sustained implementation, creating financial and psychological stress for farmers during the transition period (Jacobs et al. 2022; Sheoran et al. 2022). Additionally, the shift toward RA often reshapes farmers' identities and their roles within both their communities and agricultural landscapes, as highlighted in research on farmer experiences and values (Burton and Wilson 2006; Cusworth 2020). From a climate standpoint, RA offers significant untapped potential. Soils are among the planet's largest carbon reservoirs, holding approximately 3170 Gt of carbon, three times the amount stored in the atmosphere (Lal 2020). Of this, 1550 Gt are stored as SOC and 950 Gt as inorganic soil carbon. However, the conversion of forests and grasslands into agricultural land, along with the continual disturbance of soils, has released 50–100 Gt of carbon into the atmosphere since the Industrial Revolution. Carbon farming, an integral part of RA, provides a systemic approach to carbon sequestration. Practices such as agroforestry, cover cropping, reduced tillage, and rotational grazing not only capture atmospheric CO_2 but also offer additional benefits like improved soil fertility, reduced erosion, increased biodiversity, and enhanced farm resilience (Jat et al. 2022; Meena et al. 2024a). It is estimated that widespread adoption of these practices could sequester up to 20 Gt of carbon over the next 25 years, which is equivalent to more than 10% of global anthropogenic CO_2 emissions. Connecting these regenerative practices to carbon credit systems not only provides economic incentives for farmers but also supports the broader vision of green agroecology, where food systems are climate-resilient, ecologically sustainable, and socially inclusive (Cariappa and Krishna 2024). By acknowledging and rewarding ecosystem services beyond carbon credits, the transition to RA can be expanded more effectively and equitably. Policymakers and private sector stakeholders are increasingly recognizing this synergy. Programs that reward carbon sequestration through carbon credits and payment for ecosystem services (PES) can provide the necessary incentives for farmers to adopt regenerative practices. This chapter examines how RA combines ecological principles with farming methods, creating intersections with carbon credit systems and ecosystem services. Technological advancements, adaptive policies, and performance-based incentives are essential for driving regenerative transitions that enhance ecosystem integrity, improve farmers' livelihoods, and strengthen global climate resilience.

3.2 Regenerative Agriculture

Regenerative agriculture has its roots in traditional indigenous farming systems, where communities practiced diversified cropping, natural soil enrichment, and ecosystem-based management long before industrial agriculture emerged (Sher et al. 2024; Meena et al. 2024b). Historical RA principles are evident in practices like the "Three Sisters" cropping systems of Native American tribes and integrated agroforestry methods in tropical areas, which simultaneously optimized biodiversity, soil health, and productivity (Hill 2020). However, it was in the late 1970s and early 1980s that RA was formally defined as a distinct agricultural philosophy, primarily by pioneers at the Rodale Institute in the United States (Rodale 1983). In contrast to sustainability models, which often focus on maintaining the status quo, RA actively restores degraded ecosystems, enhances ecological integrity, and builds resilience against climate and market disruptions. It uses outcome-driven methods, focusing on measurable ecological improvements such as increased SOC, improved microbial biodiversity, enhanced water cycles, and greater ecosystem service provision (LaCanne and Lundgren 2018). These ecological improvements not only benefit farm productivity and profitability but also contribute to broader global goals like food security, climate stabilization, and the revitalization of rural economies. A regenerative farming system is defined by several key principles that work together to create a more resilient and sustainable agricultural model (Sher et al. 2024; Meena et al. 2023). These principles, which include pluralism, protection, purity, permanence, peace, potential, and progress, serve as a guiding framework for farmers aiming to restore the ecological health of their land while ensuring long-term productivity. Pluralism emphasizes the value of plant species diversity, which enhances ecosystem resilience and pest management. Protection involves using cover crops to reduce soil erosion, boost microbial populations, and enhance soil fertility. Purity focuses on eliminating synthetic fertilizers and pesticides, instead relying on natural processes to maintain soil and plant health. Permaculture/Permanence advocates for using perennials and deep-rooted plants, which improve soil structure, prevent erosion, and support long-term sustainability. Peace represents working in harmony with nature, fostering a balanced relationship between farming practices and the environment. Potential refers to the natural rise of nutrients to the soil surface, supporting plant growth. Progress involves the ongoing improvement of soil quality, enhancing water retention, and maintaining optimal soil composition for sustained agricultural productivity (Sher et al. 2024). These principles are central to RA, providing farmers with increased resilience to climatic challenges and market fluctuations, enabling them to sustain both productivity and profitability in difficult conditions.

3.2.1 Definition and Principles

Regenerative agriculture describes holistic farming systems that, among other benefits, improve water and air quality, enhance ecosystem biodiversity, produce nutrient-dense food, and store carbon to help mitigate the effects of climate change (FAO 2022). According to the Royal Agricultural Society of England (RASE 2023), the five core principles of RA (Fig. 3.1) include:

3.2.1.1 Minimize Soil Disturbance

Fostering a resilient soil microbiome requires maintaining the biological integrity and structure of the soil. Soil carbon reserves are preserved, fungal networks like mycorrhizae are protected, and soil aggregates are not disturbed by reduced or zero tillage techniques. Regenerative systems improve the efficiency of nutrient cycling

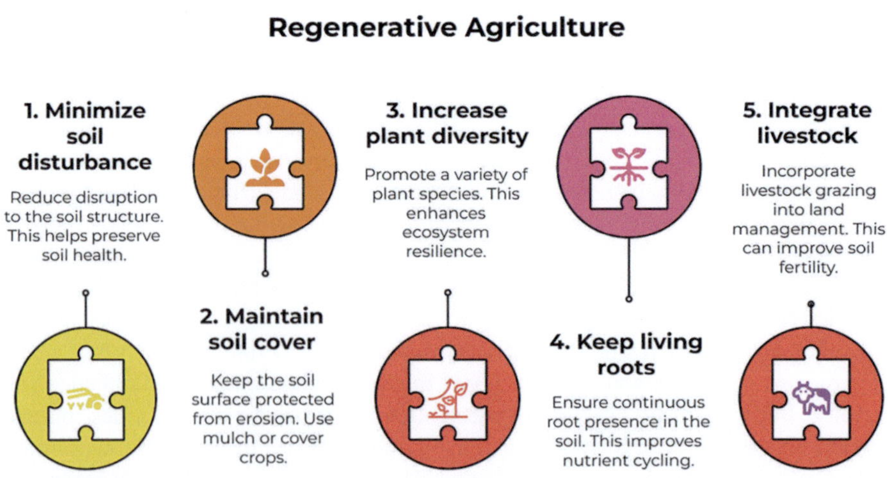

Fig. 3.1 Five core principles of regenerative agriculture

and lessen reliance on artificial fertilizers and pesticides by reducing mechanical, chemical, and biological disturbances (Punia et al. 2020).

3.2.1.2 Maintain Soil Cover

Using cover crops, agricultural residues, and perennial plants to provide permanent soil cover protects the soil from erosion, reduces weed pressure, controls soil temperature, and enhances water infiltration (Leharwan et al. 2023). Additionally, organic leftovers on the soil surface provide microbial communities with a substrate, enhancing the fertility and organic matter of the soil.

3.2.1.3 Increase Plant Diversity

Abiotic and biotic stressors are less likely to affect biodiverse agricultural systems. Techniques like agroforestry, varied crop rotations, and intercropping improve the usage of ecological niches, disturb the cycles of pests and diseases, and increase below-ground microbial diversity. Landscape-scale ecosystem services, water use efficiency, and nutrient dynamics are all enhanced by more plant diversity.

3.2.1.4 Keep Living Roots in the Soil

Sustaining healthy roots encourages the rhizosphere's biological activity, drives carbon into the soil, and helps soil particles aggregate and stabilize. Long-term soil health depends on practices like relay cropping, cover crops during fallow seasons, and using perennial plants, which provide year-round photosynthetic activity and soil biological vitality.

3.2.1.5 Integrate Animals into the System

Holistic livestock integration and adaptive multi-paddock grazing replicate natural grazing ecosystems. Managed cattle encourage grassland renewal, encourage variability in vegetation structures, and aid in nutrient recycling through depositing dung. In addition to improving water retention and soil carbon sequestration, well-managed grazing systems can help farmers diversify their revenue sources.

3.3 Prospects of Regenerative Agriculture Toward Carbon Credit and Green Agroecology

3.3.1 GHG Mitigation and Carbon Sequestration Strategies for Carbon Credit Generation

Regenerative agriculture has been gaining increasing attention from policymakers, farmers, and investors worldwide due to its vast potential to tackle critical global issues like climate change and food security. Beyond addressing these challenges, RA also focuses on revitalizing degraded soils, boosting biodiversity, and reducing reliance on synthetic inputs. These strategies together help transform agriculture from a major GHG emitter into a key carbon sink.

Conservation agriculture (CA), a central strategy of RA, prioritizes minimal soil disturbance, maintaining soil cover, and implementing diversified crop rotations. These practices reduce CO_2 emissions by preserving soil structure and organic matter. On a global scale, CA could sequester an estimated 25 Gt of carbon over the next 50 years (Lal 2015). Zero tillage, a key practice, helps prevent carbon loss and enhances the aggregation of micro and macro soil particles, which improves soil carbon storage (Rana et al. 2023; Meena and Pradhan 2023). Another important practice involves the use of organic inputs like farmyard manure, compost, and green manures. These inputs enhance soil microbial activity, increase SOC, and improve nutrient efficiency (Kumar et al. 2020). When combined with mineral fertilizers in an integrated nutrient management (INM) approach, they can reduce nitrous oxide (N_2O) emissions by up to 50% while maintaining or even improving crop yields (Bhattacharya et al. 2023; Kumar et al. 2022a, 2023). Cover cropping and green manuring also play an essential role in GHG mitigation. These practices increase biomass inputs, improve nitrogen fixation, and suppress weeds, besides adding organic matter to the soil. Studies show that cover crops can significantly increase SOC and improve nitrogen use efficiency, which lowers N_2O emissions (Chahal and Van Eerd 2019). When combined with no-till systems, cover crops can enhance SOC by up to 60 million tons annually worldwide, depending on regional adoption rates and soil types (Averill et al. 2014; Bai et al. 2019). Agroforestry systems, which integrate trees with crops and livestock, are among the most effective RA practices for long-term carbon sequestration (Jhariya et al. 2022). These systems not only capture carbon in both above- and below-ground biomass but also provide additional benefits such as regulating microclimates, controlling erosion, and enhancing biodiversity. On a global scale, agroforestry has the potential to sequester between 1.1 and 2.2 Gt of carbon over the next 50 years (Ghimire et al. 2020). Another promising strategy is the use of biochar, a stable form of carbon produced by pyrolyzing biomass. When added to soil, biochar improves water retention, nutrient cycling, and microbial activity, while also significantly reducing methane (CH_4) and nitrous oxide (N_2O) emissions from both upland and flooded systems (Bai et al. 2019). Its resistance to decomposition makes it an ideal solution for long-term carbon storage. Figure 3.2 illustrates GHG mitigation potential of various RA practices across

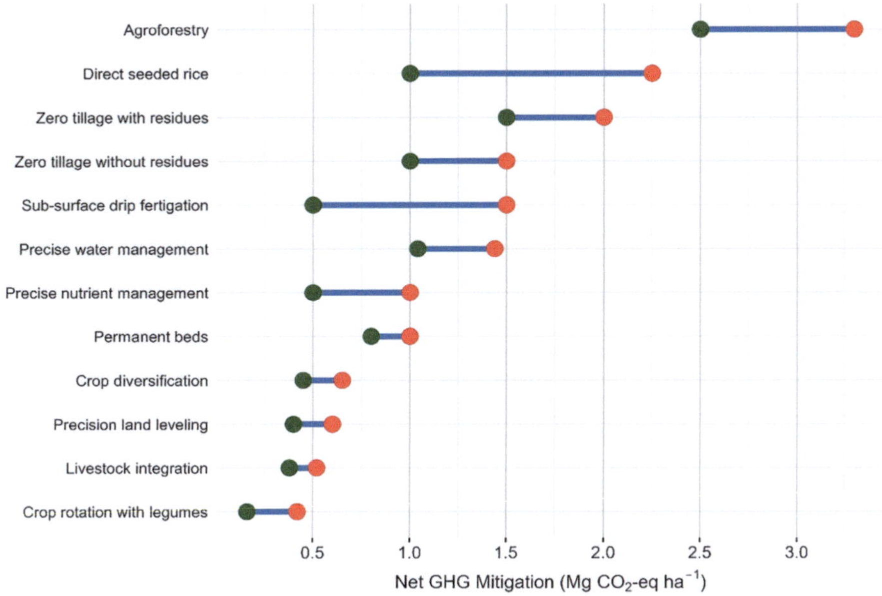

Fig. 3.2 Potential of different regenerative agriculture practices on mitigating greenhouse gas emissions. The green and red dots indicate the minimum and maximum mitigation potential, respectively. (*Source*: Data pooled from multiple publications)

different regions. In flooded rice systems, which are significant sources of methane, practices such as Direct Seeded Rice and the System of Rice Intensification have demonstrated the potential to reduce CH_4 emissions by up to 35%, while also improving water productivity (Fig. 3.2, Reddy et al. 2025a). The use of Alternate Wetting and Drying irrigation further aids in methane reduction. Technological advancements in precision nutrient and water management play a key role in GHG mitigation by applying inputs with high spatial and temporal accuracy, thus minimizing resource waste and associated emissions. Other RA strategies, such as subsurface drip irrigation, nitrogen inhibitors, and controlled-release fertilizers, can reduce emissions by up to 1.5 Mg CO_2-eq per hectare annually (Reddy et al. 2025b). Additionally, the development and adoption of improved crop varieties with deep rooting systems, higher biomass, and greater resilience to stress can enhance belowground carbon storage and reduce the need for frequent tillage or resowing (Bhattacharya et al. 2023; Pradhan and Meena 2023b). Together, these regenerative practices offer a comprehensive solution to both climate change and agricultural sustainability. Moreover, these practices create economic opportunities through carbon credit systems, where farmers are compensated for sequestering CO_2 by increasing SOC and implementing emission-mitigation strategies.

3.3.2 Carbon Credit Generation from Regenerative Agriculture

There is growing international agreement that addressing ecological crises necessitates widespread involvement across sectors, particularly agriculture, thanks to initiatives like the Paris Agreement. In order to financially encourage farmers to embrace sustainable methods, carbon markets were developed. According to Cariappa and Krishna (2024), carbon credits are tradable certificates that signify the removal or reduction of one metric ton of CO_2 or its equivalent. Carbon credits provide farmers using RA a game-changing possibility. Farmers raise SOC levels and reduce emissions by putting the RA techniques mentioned above into practice (Nozaki 2023). Farmers can obtain carbon credits and access new revenue sources while preserving and even enhancing the long-term productivity and resilience of their land when these climate-positive effects are precisely quantified and independently confirmed.

Farmers collaborate with carbon project developers to start the process of creating carbon credits (Fig. 3.3). Based on the farmer's intended actions, these specialists assist in estimating the farm's potential for SOC sequestration and evaluating the baseline carbon levels. Following the adoption of regenerative techniques, improvements in soil carbon are tracked over time through regular data collection and monitoring. After that, the results are audited by third-party verifiers to make sure the carbon gains are real, extra, and long-lasting. Farmers receive carbon credits after verification, which they can then sell on voluntary carbon markets (VCMs) (Fig. 3.3). Customers buy these credits to make up for their own environmental footprints, ranging from businesses looking to offset emissions to individuals backing climate-positive projects. Through this system, regenerative farmers not only contribute meaningfully to global climate goals but also achieve greater economic resilience (Salma et al. 2024). The growing interest in carbon credits represents a

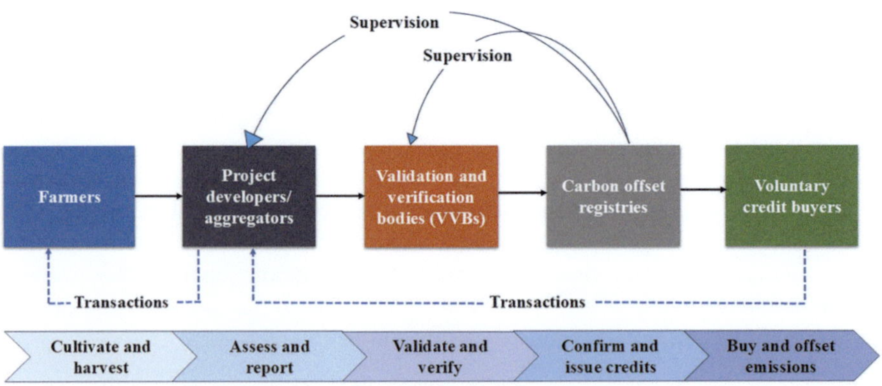

Fig. 3.3 The operational framework of the carbon market and the key stakeholders within the carbon credit ecosystem

major shift in how agriculture is perceived within climate strategies. RA has the potential to become a solution rather than just an issue that needs to be solved. Carbon credits from RA foster a positive feedback loop that benefits farmers and the environment by rehabilitating the soil and bolstering rural economies.

3.3.3 *Ecosystem Services from Regenerative Agriculture*

Beyond carbon sequestration, RA provides a multitude of ecosystem services that support green agroecology, a model that focuses on sustainability, biodiversity, and ecosystem health. According to Farmland (2017), RA provides a collection of ecosystem services as discussed below (Fig. 3.4):

:

3.3.3.1 Biodiversity Enhancement

Three levels of biodiversity are greatly increased by regenerative farming: soil biota, crop variety, and wider ecological habitat. High water infiltration, strong aggregate stability, and low compaction are all results of healthy soils supporting diverse microbial populations, such as fungi, protozoa, and earthworms. In addition to preventing pest accumulation, diverse crop rotations support beneficial creatures

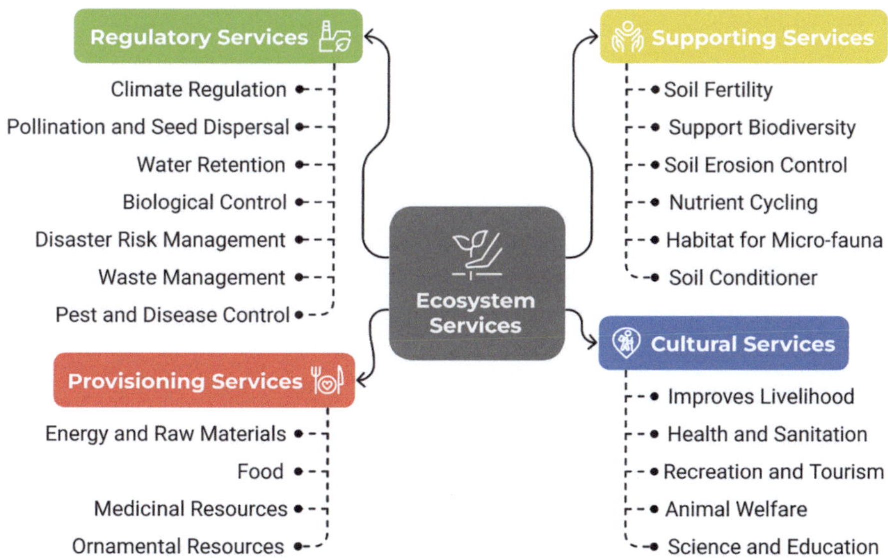

Fig. 3.4 The figure describes the four categories, each with distinct ecosystem services, through the adoption of regenerative agriculture practices

such as pollinators and pest predators. In addition to the agricultural fields, nearby forests, marshes, and riparian zones provide wildlife with essential habitat, facilitating ecosystem services like pollination and pest control. Planting hedgerows and restoring habitat are two examples of strategic biodiversity planning that supports these organic relationships.

3.3.3.2 Soil Health and Nutrient Cycling

Regenerative systems increase soil organic matter (SOM) and decrease erosion by simulating natural grasslands and introducing multi-year grazing rotations. Mycorrhizal fungi are fostered by techniques like cover crops and limited tillage, which enhance soil structure and nutrient uptake. Glomalin, a substance secreted by this fungus, is essential for the stability of soil carbon and water retention. This kind of management improves soil quality and economic resilience by lowering dependency on chemical fertilizers.

3.3.3.3 Water Regulation and Purity

Superior water infiltration and storage are made possible by high-organic-matter soils in regenerative systems, which lowers the demand for irrigation and shields water bodies from pollution from runoff. Erosion and nutrient leaching are avoided by cover crops and no-till techniques. By filtering impurities, these methods directly improve water quality and aid in water capture and conveyance, which is particularly important in areas that are intensively watered and susceptible to drought.

3.3.3.4 Carbon Sequestration and Climate Regulation

By storing atmospheric CO_2 in SOM, RA slows down climate change. Perennial pastures, organic inputs, and less tillage increase sequestration while reducing carbon emissions. To improve net carbon stocks, models such as COMET-Farm support methods including keeping agricultural residues, avoiding synthetic nitrogen, and incorporating legumes (USDA COMET-Farm, 2013). Significant co-benefits in lowering the overall carbon footprint of the system are also provided by minimizing N_2O emissions through biological nutrient management (Jangir et al. 2024).

3.3.3.5 Social and Cultural Services

Regenerative farming practices improve the general well-being of the community. Healthy living conditions are promoted by better air quality, flood prevention, and aesthetic improvements. In addition to adding scientific and cultural value, public

Table 3.1 Quantifiable metrics and indicators for ecosystem service assessment

Impact area	Ecosystem service metric	Underlying biophysical indicator
Social	Air quality	Particulates, NOx
	Aesthetic information	Housing prices, scenic quality
	Disaster risk reduction	Damage costs from flooding or storms
	Food provision	Yield (human and animal food crops)
Biodiversity	Biological control	Predation pressure, pest regulation
	Habitat	Acreage of native/non-cultivated areas
	Pollination and seed dispersal	Floral resources, bee counts, and seed set data
Climate and energy	Climate stability	Soil carbon stock (C), avoided emissions (CO_2, N_2O)
	Soil formation	Organic matter buildup, microbial biomass
Water systems	Water capture and conveyance	Infiltration rates, retention basin data
	Water quality	Nitrate/phosphate runoff, turbidity, sedimentation levels
Soils	Soil retention	Erosion rates (tons/acre/year)
	Soil quality	pH, CEC, organic carbon%, nutrient levels

Source: Farmland (2017)

involvement and educational research on restored lands turn farms into living labs and historical landmarks. Farmers can get reimbursement for these essential services through ecosystem service payments, which advances the larger objectives of green agroecology and sustainable agriculture. RA provides a solution that benefits both farmers and the environment by balancing financial incentives with ecological advantages. Ecosystem services have long been undervalued in conventional economic planning, despite their vital role in sustaining life and livelihoods. Strong techniques for evaluating these services using trustworthy and context-specific metrics are desperately needed in order to properly appreciate their worth and direct wise investment. Ecosystem services run the risk of being underestimated or disregarded in the absence of such measures, which could result in their deterioration and expensive replacement. Natural services can be meaningfully evaluated, compared, and integrated into frameworks for economic, agricultural, and environmental decision-making by establishing standard indicators (Table 3.1).

3.3.4 Synergies Between Carbon Credit Mechanisms and Ecosystem Services

The integration of carbon credit systems with ecosystem service payments creates a powerful synergy that benefits both farmers and the environment. Carbon credits provide financial incentives for regenerative practices by compensating farmers for the carbon they sequester in soils. Through certification standards like the Verified

Carbon Standard (VCS) and the Gold Standard, farmers can monetize these credits, encouraging sustainable land management. Simultaneously, ecosystem service payments reward farmers for delivering broader environmental benefits such as enhanced biodiversity, improved soil fertility, and better environmental quality (Lévesque et al. 2021). Together, these two mechanisms offer farmers multiple income streams from the same suite of regenerative practices. By linking carbon markets with ecosystem service payments, farmers are motivated to adopt practices that not only restore ecosystems and mitigate climate change but also advance the larger goals of green agroecology (Farmland 2017; Pradhan and Meena 2023a). This integrated approach ensures that RA is both environmentally sound and economically viable.

3.4 Plans for Enhancing Carbon Credit and Green Agroecology Through Regenerative Agriculture

For RA to effectively contribute to carbon credit generation and sustainable agroecology, practices must be adapted to the specific agro-climatic and socio-economic contexts of each region. A uniform approach is unlikely to succeed, given the wide variability in farming systems and environmental conditions across landscapes. Designing region-specific RA models that integrate robust scientific research with indigenous and traditional knowledge enhances their relevance, acceptance, and impact. For example, in mountainous regions, agroforestry systems are especially effective at reducing soil erosion while boosting carbon sequestration (Gosnell et al. 2019). Conversely, in arid and semi-arid zones, the use of drought-tolerant cover crops, mulching, and CA practices helps conserve soil moisture and increase SOC stocks (Kenne and Kloot 2019). Tools such as geographic information system (GIS), remote sensing, and soil health cards can aid district-level planning by helping policymakers and practitioners select appropriate interventions and track progress. At the farm level, creating tailored carbon sequestration plans is critical. These detailed blueprints outline specific practices designed to enhance carbon storage and ecosystem services (Jacobs et al. 2022). Moreover, implementing dynamic cropping systems, such as adaptive rotations, legume–cereal intercropping, and the inclusion of perennial species, can further strengthen soil carbon capture by improving nitrogen fixation, expanding root biomass, and providing a continuous organic carbon supply (Liu et al. 2022; Parihar et al. 2018). However, there is an urgent need to create regional carbon credit approaches that can precisely capture the effects of various RA practices in addition to creating region-specific regeneration models. Existing frameworks for verifying carbon credits are frequently generic and might not take into consideration regional variations in soil types, agricultural methods, and ecological roles. As a result, it will be crucial to create regionally specialized carbon accounting procedures that are based on baselines, monitoring metrics, and socio-economic circumstances. These approaches must guarantee that community-based businesses and smallholder farmers may contribute to global climate mitigation goals, earn just compensation, and engage effectively in carbon markets. Essentially, localized practice and localized measurement must come together to form a customized, reliable, and inclusive framework for the creation of carbon credits if RA is to succeed in the future.

3.4.1 Establish Baseline Carbon Stock Assessments

A carbon baseline must be established before beginning any carbon sequestration project. It acts as a benchmark for monitoring development and proving influence over time. Using field sampling, remote sensing, and modeling techniques like Roth C, DNDC, and ORYZA, a well-documented baseline assessment should measure SOC, biomass carbon, and above-ground carbon (Jordon et al. 2022). Using certified carbon verifiers or third-party auditors increases market trust and guarantees credibility. In order to create the groundwork for carbon credits and ecosystem service payments, these evaluations should also incorporate co-benefit indicators such as biodiversity, water retention, and nutrient cycling.

3.4.2 Farmer Training and Carbon Literacy Programs

Adoption requires awareness and comprehension. Many farmers do not know how carbon credits operate or how RA activities affect the carbon cycle (Fig. 3.3). Through the use of digital storytelling, community radio, and visual aids, carbon literacy projects should simplify these ideas, close the knowledge and technological gaps, and translate them into local languages. Community ownership can be promoted through participatory training techniques such as farmer field schools, demonstration plots, and experiential learning (Arunachalam 2024). Composting, green manuring, biofertilizers, and inexpensive soil health testing are examples of regenerative approaches that should be covered in training courses. Self-help groups (SHGs) and farmer producer organizations (FPOs) can act as platforms for group decision-making and peer learning.

3.4.3 Integrate Monitoring, Reporting, and Verification Systems

Obtaining carbon money requires credible monitoring, reporting, and verification (MRV). Real-time insights into carbon increases can be obtained by digital MRV techniques such as drone-enabled crop surveillance, Internet of Things (IoT)-based soil sensors, and satellite imaging (Doshi et al. 2019). The sections that follow provide further technical details about these tools. Blockchain technology can lower fraud while guaranteeing carbon credit traceability and transparency. However, mobile-based dashboards, automatic alerts, and inexpensive measurement procedures are needed to make these systems easier for smallholders to utilize (Rajak et al. 2023). Using young people in the area as "carbon scouts" can help gather data and create green jobs in rural areas.

3.4.4 Facilitate Access to Carbon Markets

Smallholder farmers frequently lack the resources or expertise necessary to participate in voluntary carbon markets. Pooled participation and negotiating power are made possible by combining them into cooperatives or FPOs. Collaborations with digital carbon platforms such as Regen Network, Nori, and Indigo Ag help expedite project paperwork, verification, and market connections. These platforms frequently provide packaged services, such as trading facilitation, MRV tools, and baseline assessments. Introducing carbon wallet applications can facilitate micro-payouts for carbon-positive behavior, track credits, and connect them to prizes (Spilker and Nugent 2022).

3.4.5 Linkage of Regenerative Agriculture with Nationally Determined Contributions and Sustainable Development Goals

As a component of the Paris Agreement, the Nationally Determined Contributions (NDCs) establish challenging climate goals for nations to cut emissions and increase climate resilience. RA is essential to accomplishing these objectives because it reduces GHG emissions, improves ecosystem resilience, and stores carbon in soils, all of which directly mitigate climate change (Crumpler et al. 2019). Countries are recognizing this and using regenerative methods more frequently in their national climate action programs. A comprehensive approach to sustainable agricultural development is promoted by RA, which not only lowers emissions but also makes a substantial contribution to a wider range of global goals by directly supporting eight of the Sustainable Development Goals (SDGs) (Fig. 3.5).

Linking RA with both NDCs and SDGs provides a clear, structured framework for accelerating the adoption of regenerative practices, besides offering farmers an innovative revenue stream from carbon credits, encouraging the widespread adoption of sustainable practices (Crumpler et al. 2019). As these techniques become more popular, they help nations fulfil their international climate pledges while tackling urgent issues with biodiversity, food security, and economic sustainability.

Fig. 3.5 SDGs addressed through the adoption of regenerative agriculture practices and carbon credit generation

3.5 Technological Intervention in Regenerative Agriculture for Carbon Credit

Artificial intelligence (AI) and machine learning (ML) have become pivotal in accurately modeling and predicting carbon sequestration in RA. These technologies analyze vast datasets including weather patterns, soil health indicators, crop rotation practices, and biomass data to generate predictive models for carbon fluxes (Deng et al. 2018). AI-based tools such as decision tree algorithms and neural networks help identify key variables influencing GHG emissions and SOC dynamics, allowing for tailored interventions that enhance carbon capture and mitigate emissions (Table 3.2). Furthermore, ML models may learn and adapt from fresh datasets continuously, increasing their predicted accuracy over time (Were et al. 2015). Table 3.2 displays the use of several AI models in various geographical locations to forecast the SOC. Furthermore, by identifying discrepancies in reports and offering clear, data-driven validation processes, AI can aid in the verification of carbon credits (Ross et al. 2020). For agribusinesses and farmers involved in carbon markets, such technologies are essential. MRV procedures are streamlined, and carbon claims are made more credible.

Table 3.2 Examples of the application of artificial intelligence (AI) methods for predicting gridded soil organic carbon, with studies selected to represent diverse geographic soil scopes, sample sizes, regional extents, and AI techniques

Location	Units	Depth (cm)	Min.	Mean	Max.	Soil samples	R^2	RMSE	AI method	References
Kenya	Mg ha^{-1}	0–30	41.99	103.15	193.42	220	0.64	14.88	Artificial Neural Network (ANN), Random Forest (RF), Support Vector Machine (SVM)	Were et al. (2015)
Denmark	%	0–30	0.7	3.7	31.6	328	0.66	1.7	Cubist Model (Cub) (upland model C)	Peng et al. (2015)
Eastern Australia	Mg ha^{-1}	0–30	5.08	24.8	88.23	564	0.48	7.5	Boosted Regression Trees (All Predictors) (BRT-All), Boosted Regulation Trees (Generic Algorithm - Selected Predictors) (BRT-GA), RF	Wang et al. (2018)
Santa Fe River Watershed, USA	kg m^{-2}	0–100	1.84	11.79	268.91	554	0.65	3.69	Regression Kriging / Regression Trees (RK/RT)	Vasques et al. (2010)
Zhejiang Province, China	t C ha^{-1}	0–20	1.18	49.74	213.55	29,927	0.76	10.63	Boosted Random Trees (BoRT), RF	Deng et al. (2018)
Florida, USA	%	0–20	0.13	2.68	38.57	850	0.78	–	Partial Least Squares Regression (PLSR), RF, Special Boosted Interpolation Framework Model (SBIFmod)	Adi and Grunwald (2020)
Florida, USA	kg m^{-2}	0–20	0.45	4.98	34.15	1080	0.63	2.64	Bayesian Regression Trees (BaRT), BoRT, Cub, RF	Xiong et al. (2014)
Florida, USA	kg m^{-2}	0–20	0.45	4.74	34.15	1014	0.72	2.39	Classification and Regression Trees (CaRT), RF, Regression Kriging with Regression Trees (RK-RT), SVM	Keskin et al. (2019)

South-Eastern USA	kg m^{-2}	0–20	1.1	3.7	12.6	2564	0.79	1.29	RF	Ross et al. (2020)
Australia	%	0–5	0.001	1.73	36.32	70,803	–	–	Cub, Cubist with Regression Kriging (cub-RK)	Viscarra Rossel et al. (2015)
USA	Mg ha^{-1}	0–20	0.26	56.87	524.83	3303	0.35	28.15	RF, Quantile Regression Forest (QRF)	Cao et al. (2019)
Europe	%	–	0	1.68	43.84	7142	0.24	1.52	BoRT	Safanelli et al. (2020)
Latin America	kg m^{-2}	–	0	6.85	573.76	11,268	–	–	PLSR, KK, QRF, SVM	Guevara et al. (2018)

3.5.1 Carbon Calculators and Mobile Applications

Carbon calculators and mobile apps have become practical, user-friendly tools that help farmers and stakeholders estimate, monitor, and manage carbon emissions and assess their carbon footprint (Andersson 2020). Within RA, these digital tools enable users to input detailed farm data, such as crop types, tillage methods, fertilizer application, and cover crop use, to calculate overall GHG emissions or potential carbon offsets (Tsesmelis et al. 2025). Well-known examples include the Cool Farm Tool, a globally recognized platform for estimating emissions and sequestration across various farm systems, and COMET-Farm, developed by the United States Department of Agriculture (USDA), which models GHG impacts based on specific management practices (USDA COMET-Farm, 2013). Additionally, mobile apps like Carbon Mapper and Soil Mentor combine Global Positioning System (GPS) and soil health data for real-time tracking and clear visual summaries of progress. These technologies not only help farmers make informed management choices but also provide essential documentation for entering carbon credit markets. Increasingly, calculators are being localized for regional conditions and languages, improving accessibility for smallholder farmers. Cloud integration further enhances these tools by enabling seamless data updates, long-term storage, and greater transparency in carbon MRV processes.

3.5.2 Drones and IoT-Based Soil Sensors

By offering high-resolution, real-time data that is crucial for improving soil health and carbon sequestration, drones and Internet of Things (IoT)-based soil sensors are transforming RA (Doshi et al. 2019). These technologies are essential for tracking and confirming changes in carbon stocks, which is a crucial component of carbon credit schemes. Aerial imagery is captured by drones fitted with thermal and multispectral cameras to evaluate crop vigor, biomass, canopy cover, and SOC indicators (Paul et al. 2024). Finding regions that are appropriate for regenerative methods, charting variability across fields, and spotting indications of soil deterioration are all made easier with the use of this data. IoT-based soil sensors monitor soil parameters such as moisture, temperature, pH, electrical conductivity, and CO_2 fluxes. These devices provide continuous, site-specific data that inform decisions on irrigation, nutrient application, and carbon input optimization (Rajak et al. 2023). Integration with wireless networks and cloud platforms allows farmers and researchers to analyze trends, predict carbon fluxes, and validate soil improvement efforts. By offering precision insights and automating data collection, these technologies contribute to more robust MRV systems, thereby strengthening the credibility and scalability of carbon credit mechanisms.

3.5.3 Variable Rate Technologies

Variable rate technologies (VRTs) play a pivotal role in cutting carbon emissions and boosting soil carbon sequestration, two key requirements for generating carbon credits. VRT enables the precise application of seeds, fertilizers, pesticides, and irrigation based on the spatial variability of soil and crop conditions (Bramley 2009). By applying inputs only where and when they are needed, VRT reduces waste, cuts emissions, and prevents soil degradation. Reduced reliance on chemical fertilizers further lowers GHG emissions, improving the overall carbon footprint of the farm (Schimmelpfennig and Ebel 2011). Within RA, where preserving soil health and organic matter is fundamental, VRT helps optimize the use of organic inputs like compost and biochar maximizing carbon additions to the soil while minimizing nutrient loss through leaching or runoff, thereby supporting long-term sequestration. Moreover, VRT can be combined with GIS, GPS, and remote sensing technologies to enable automated, data-driven management (Zhang 2016). These capabilities position VRT as a critical tool not only for improving agricultural productivity but also for aligning with the technical standards required under carbon credit frameworks.

3.5.4 Nutrient and Water Management

Data-driven nutrient and water management systems use real-time information from soil sensors, weather stations, satellite imagery, and crop models to provide precise recommendations tailored to crop needs (Uchimiya 2024). By avoiding the excessive application of nitrogen fertilizers, these systems can greatly cut N_2O emissions, a major GHG linked to conventional farming practices. Tools like Decision Support Systems (DSS) and AI-based models process diverse datasets to design site-specific fertilization and irrigation strategies (Huang et al. 2022). This approach not only enhances yields and resource use efficiency but also supports the foundational goals of RA by safeguarding soil health and lowering emissions. Additionally, data-driven irrigation solutions, such as smart drip systems, help conserve both water and energy, indirectly reducing carbon emissions (Benes et al. 2020; Reddy et al. 2025b; Pradhan et al. 2023; Kumar et al. 2025). These technologies are especially vital in dryland and semi-arid zones, where water scarcity and the need for sustainable farming intersect. In the context of carbon credit generation, precision management ensures measurable and verifiable GHG reductions, making data-driven nutrient and water management a key strategy for climate-smart agriculture.

3.5.5 Real-Time Carbon Footprint Mapping

Real-time carbon footprint mapping is designed to both reduce carbon emissions and enhance carbon sequestration by leveraging geospatial technologies like remote sensing, GIS, and real-time data analytics (Andersson 2020). This approach allows continuous tracking and assessment of emissions tied to various agricultural processes, including soil management, tillage, irrigation, fertilization, and transportation. A key advantage of real-time mapping is its ability to deliver immediate feedback on carbon emissions from different farm operations. Using networks of sensors that monitor soil carbon levels, water use, and energy consumption, farmers can receive instant alerts highlighting areas with elevated emissions, allowing them to act swiftly. These systems also apply carbon intensity indices to pinpoint high-emission activities, helping guide farmers' decisions (Kremmydas et al. 2025). For instance, low-emission practices like zero tillage and cover cropping can be seamlessly incorporated into farm management plans, lowering overall carbon outputs. Critically, for carbon credit generation, real-time mapping enables precise measurement and validation of carbon sequestration, ensuring more accurate and transparent carbon credit reporting. This strengthens the connection between RA practices and verified carbon credits, giving farmers strong financial incentives to adopt climate-friendly methods.

3.5.6 Microbial Bio-stimulants

Microbial bio-stimulants, which include beneficial microorganisms like bacteria, fungi, and actinomycetes, are vital for improving soil health by promoting plant growth, enhancing nutrient availability, and supporting the formation of stable SOC (Kaushal et al. 2023). By nurturing a robust soil microbiome, these bio-stimulants help regenerate degraded soil ecosystems and significantly contribute to carbon sequestration. Microbial inoculants such as mycorrhizal fungi, rhizobacteria, and plant growth-promoting rhizobacteria (PGPR) enhance soil structure, improve root penetration, and stabilize organic matter, thereby boosting carbon storage (Srivastava et al. 2016). These microbes accelerate the decomposition of organic residues, enriching SOM and transforming it into more persistent carbon forms. For example, mycorrhizal fungi establish symbiotic relationships with plant roots, improving nutrient and water uptake, particularly in low-fertility soils (Macik et al. 2023). This symbiosis not only stimulates root development but also increases carbon transfer from plants to soil, where fungi help build stable aggregates that lock in organic carbon. Furthermore, microbial bio-stimulants can lower the dependence on synthetic fertilizers and pesticides, promoting more sustainable farming practices and reducing GHG emissions (Macik et al. 2023). Given their ability to improve soil health and secure long-term carbon storage, microbial bio-stimulants represent a promising strategy for advancing carbon credit generation within RA systems.

3.5.7 *Biochar Production and Application*

Biochar, a carbon-rich material produced through the pyrolysis of organic biomass under low-oxygen conditions, has attracted considerable interest as a soil amendment in RA for improving soil health and advancing carbon sequestration (Yadav et al. 2023). Its incorporation into soils can enhance fertility, physical structure, and microbial dynamics, while locking carbon away in stable forms for centuries to millennia. Biochar is typically made by thermally decomposing materials like crop residues, forestry by-products, or agricultural waste, resulting in a highly stable product with long-lasting carbon retention potential (Gwenzi et al. 2015). This inherent stability makes biochar particularly valuable in carbon credit systems. Once added to soil, biochar improves water-holding capacity, aeration, and structural integrity, while boosting the soil's cation exchange capacity, which enhances nutrient retention. Additionally, biochar provides microhabitats for beneficial soil microbes and acts as a slow-release nutrient reservoir, reducing fertilizer leaching and fostering microbial diversity (Galinato et al. 2011). By stimulating both plant root development and microbial activity, biochar indirectly supports further carbon storage through increased organic matter accumulation and the stabilization of organic compounds. Importantly, biochar use aligns with RA's ethos by recycling agricultural waste back into the soil, effectively closing the carbon loop and reducing atmospheric carbon losses. This dual advantage, improving soil function while sequestering carbon, positions biochar as a potent strategy for climate change mitigation and an attractive avenue for generating carbon credits (Salma et al. 2024).

3.6 Rewarding Regenerative Practices with Carbon Credits and Ecosystem Services

3.6.1 *Carbon Credit Trading Platforms and Farmer Access*

3.6.1.1 Global and Regional Carbon Markets

Carbon credit trading markets provide a structured system for exchanging certified emission reductions, creating an economic incentive to advance climate change mitigation. These markets are generally divided into two main types: compliance markets and voluntary markets, both of which are increasingly incorporating credits from RA (FAO 2021). Compliance markets, such as the European Union Emissions Trading System (EU ETS), are government-regulated, setting legally binding emission caps for sectors like energy and manufacturing. Although traditionally focused

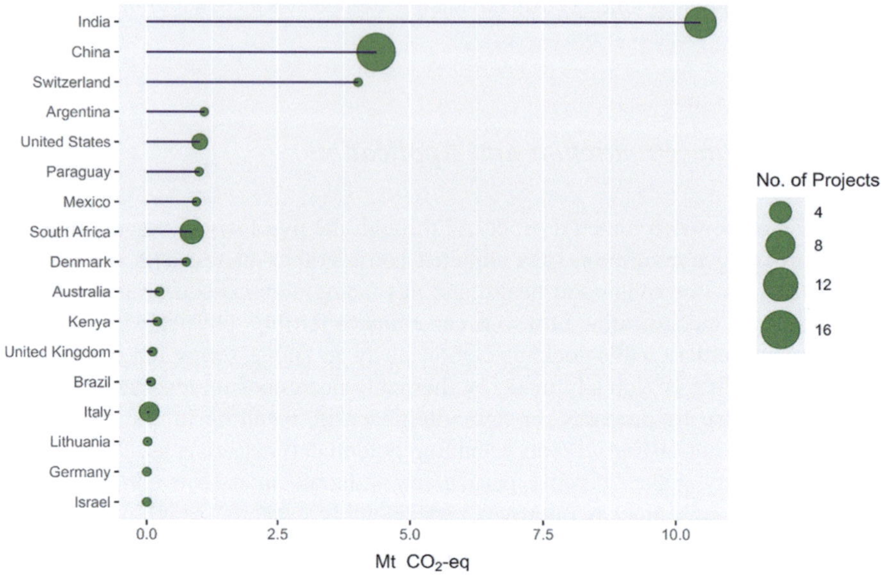

Fig. 3.6 The projected carbon sequestration and credits generation from registered projects in various countries. (*Source*: Nozaki 2023)

on industrial emissions, some programs, like California's Cap-and-Trade scheme, have started accepting agricultural carbon offsets under rigorous verification standards (European Commission 2021). In contrast, Voluntary Carbon Markets (VCMs) are more accessible to agricultural stakeholders. In VCMs, companies or individuals purchase carbon credits to meet self-imposed sustainability or corporate social responsibility targets rather than regulatory requirements.

VCS and the Gold Standard oversee the registration, verification, and issuance of soil carbon projects from farms worldwide. The emergence of specialized soil carbon markets, such as those developed by Indigo Ag and Nori, reflects the surging interest in nature-based solutions. These platforms help farmers generate, certify, and sell credits derived from carbon sequestration practices linked to sustainable land management (Fig. 3.6). Nonetheless, despite the promising opportunities, challenges like high transaction costs, technical complexity, and limited farmer awareness hinder broader participation. Efforts to streamline protocols and establish regional support hubs are underway to overcome these barriers and help unlock the full carbon credit potential of RA (FAO 2011).

3.6.1.2 Access to Voluntary Markets for Smallholders

VCMs offer significant opportunities for smallholder farmers engaged in RA, providing a pathway to monetize ecosystem services such as soil carbon sequestration. However, despite this potential, smallholder access remains constrained by systemic obstacles like high verification expenses, complex technical requirements, and inadequate digital infrastructure. These farmers typically manage small, fragmented plots and lack the scale advantages needed to engage effectively in standard carbon markets (Barbato and Strong 2023). Additionally, many smallholders are unfamiliar with the workings of carbon markets or the demanding MRV protocols necessary to produce tradable credits. To overcome these barriers, initiatives like Plan Vivo and the Cool Farm Alliance are developing simplified, smallholder-focused methodologies. Plan Vivo's participatory framework centers on community-led project design and long-term sustainability, empowering farmer groups to tap into carbon finance while retaining control over their land and practices (Plan Vivo Foundation n.d.). Similarly, initiatives such as Soil Capital and REDD+ pilot projects across Africa and Asia are combining digital tools, participatory training, and cooperative models to close the access gap, helping smallholders engage in and benefit equitably from the global carbon market (Soil Capita n.d.; UNFCCC n.d.-b; UN-REDD n.d.).

3.6.1.3 Case Studies

Several innovative platforms are transforming the landscape of RA by connecting farmers to carbon markets through cutting-edge technologies and business models. For instance, Indigo Ag, a US-based ag-tech firm, introduced the Indigo Carbon program, which compensates farmers for measurable increases in SOC through RA practices. Indigo leverages satellite imagery, soil sampling, and machine learning to ensure precise carbon quantification and verification (Indigo Ag n.d.). Similarly, Nori, a blockchain-based carbon removal marketplace, enables direct transactions between carbon farmers and buyers, emphasizing transparency by tracking carbon removal certificates on a public ledger. Nori simplifies the process for US row crop farmers to generate and sell carbon credits based on soil improvement practices (Nori n.d.). Regen Network, another key player, blends ecology with Web3 technologies, offering a decentralized marketplace for ecological credits. It allows land stewards, particularly in the Global South, to tokenize and sell verified ecosystem services, with a focus on inclusivity and long-term ecological monitoring (Regen Network n.d.). These pilot programs provide valuable insights into the importance of robust MRV systems, farmer education, user-friendly platforms, and equitable profit-sharing. Their success highlights RA as a promising pathway for climate mitigation and rural income generation. Impact reports show that farmers involved in these programs often experience improved soil health, higher yields, and diversified incomes. For example, Indigo Ag's early reports indicated an average of 0.3–0.5 tons of CO_2 sequestered per acre annually, leading to additional payments for farmers. Similarly, the Kenya agricultural carbon project led by Vi Agroforestry sequestered over 60,000 tons of CO_2 equivalents, improving food security for over 30,000 smallholders (Vi Agroforestry n.d.). However, these studies also underscore challenges like inconsistent carbon pricing, high MRV costs, and delays in credit issuance, which can discourage farmer participation. Some programs initially underestimated the need for capacity-building, particularly in regions with low digital or technical literacy. Others found that trust, transparency, and clear benefit-sharing mechanisms are crucial for long-term success. The success of these pilots suggests that scalable, low-cost verification methods, supportive policies, and collaborative governance are key to the sustainable growth of carbon-rewarding RA systems, with social equity and environmental integrity at the core of their design.

3.6.2 Designing Incentive Mechanisms

3.6.2.1 Carbon Credit Bundling with Yield Insurance

Bundling carbon credits with yield insurance presents an innovative incentive model that enhances farmers' economic resilience while promoting RA. This approach offers dual benefits: farmers earn carbon credits for sequestering carbon and receive protection against yield losses caused by climate variability. By combining these incentives, the model addresses two key concerns for farmers' income uncertainty and long-term sustainability (Lévesque et al. 2021).

Companies like Indigo Ag in the United States are exploring the integration of carbon credits with agricultural insurance to encourage the adoption of RA practices (Indigo Ag n.d.). These combined products make RA more appealing to risk-averse farmers and open up access to climate finance. The success of these initiatives relies on precise MRV of both yield and carbon outcomes. Technology-driven platforms that track carbon sequestration and automate insurance claims are essential for ensuring scalability and transparency in these bundled incentive programs.

3.6.2.2 Insetting Versus Offsetting Programs

Insetting and offsetting are two complementary approaches to utilizing carbon markets for RA, each with distinct characteristics in terms of scope, impact, and stakeholder involvement. Offsetting involves purchasing carbon credits from external projects to compensate for emissions produced elsewhere. For example, a company in Europe may buy credits from a reforestation project in Kenya to offset its own carbon footprint. Agricultural offset projects, such as those supported by Verra or Gold Standard, reward farmers for RA practices that help remove or avoid CO_2 emissions (Barbato and Strong 2023). Insetting, in contrast, focuses on reducing emissions within a company's own supply chain. For instance, a tea manufacturer may support regenerative farming practices among its suppliers to sequester carbon and lower its indirect emissions. Companies like Nestlé and Unilever are increasingly adopting insetting to enhance sustainability, build stronger supplier relationships, and improve product transparency. Insetting is often viewed as more sustainable and ethically aligned with corporate social responsibility because it promotes long-term transformation within production systems. While offsetting can be effective, it may risk being a more transactional approach if not integrated with broader emission reduction strategies (Barbato and Strong 2023). In RA, blending insetting with carbon market access offers a powerful route to finance and scale ecosystem-positive farming.

3.6.3 Payment for Ecosystem Services

3.6.3.1 Models and Case Studies in Agriculture

Payment for Ecosystem Services (PES) schemes provide financial incentives to farmers for adopting RA practices that enhance ecosystem functions. Land stewards are compensated by these models for water retention, biodiversity preservation, and carbon sequestration (Lévesque et al. 2021). The national PES program in Costa Rica is a prime example; it incentivizes farmers to engage in sustainable agroforestry, reforestation, and forest protection (UNFCCC n.d.-a). By connecting upstream watershed management with downstream water users, the Watershed Organization Trust (WOTR n.d.) in India adopted PES in agriculture, guaranteeing both ecological restoration and livelihood assistance. Another successful case is the Kenya Agricultural Carbon Project (KACP), which rewarded smallholder farmers for improved soil management that increased carbon sequestration and crop yields (Vi Agroforestry n.d.). These projects are effective when based on clear land tenure rights, simple verification mechanisms, and collaborative governance.

3.6.3.2 Institutional Frameworks to Support PES

Effective implementation of PES in agriculture depends heavily on robust institutional frameworks that provide legitimacy, oversight, and support for ecosystem service transactions. These frameworks include policy guidelines, legal backing, financial mechanisms, and institutional actors such as government bodies, NGOs, and local cooperatives. By allocating gasoline taxes to fund forest-based carbon sequestration, governments play a crucial role in creating the enabling conditions that support their national PES plan. For PES participants, institutional intermediaries like Rainforest Alliance, World Agroforestry (ICRAF), or the Watershed Organization Trust in India provide technical support, monitoring, and capacity-building. These organizations guarantee openness, fair benefit distribution, and community involvement. Multilateral finance and structure for PES initiatives in developing nations are also provided by international frameworks as REDD+ (Reducing Emissions from Deforestation and Forest Degradation).

3.7 Policies for Promoting Regenerative Agriculture

3.7.1 Need for Developing a Sound Regenerative Agricultural Policy

There is an urgent need for strong national and international policies to promote the adoption of RA practices, especially in areas experiencing ongoing soil degradation and reduced productivity, often due to limited access to modern technologies and resources (Montanarella et al. 2016). These policies should be based on solid agroecological principles, utilizing key indicators such as land-use patterns, cropping systems, soil fertility management, pest and disease control, and evaluations of soil health and biological quality. Additionally, economic sustainability must be a central focus, including aspects like land productivity, yield consistency, profitability, and the associated co-benefits for soil and ecosystem services. Equally important is the integration of social sustainability into policy design. The widespread adoption and effectiveness of RA depend heavily on social acceptance, which is underpinned by principles of self-reliance, equity, food security, and the capacity to manage risks and uncertainties in agricultural production (Rasul and Thapa 2004).

3.7.2 Policies and Institutional Frameworks for Regenerative Agriculture

It is now more important than ever to promote RA practices as awareness of climate change, soil degradation, and unsustainable farming grows worldwide. This section provides a chronological and structural summary of the policies, such as institutional roles, carbon trading schemes, and policy reforms, that have either directly or indirectly aided RA.

3.7.2.1 National and International Policies

India's agricultural policy framework has gradually evolved over the past few decades, but explicit support for RA remains limited. A significant step was the introduction of the National Mission for Sustainable Agriculture (NMSA) in 2010, part of the broader National Action Plan on Climate Change (Anand et al. 2022). This mission included initiatives like Soil Health Management and Rainfed Area Development, which aimed to increase resilience and reduce the carbon footprint of Indian agriculture. However, these programs often did not fully incorporate regenerative principles such as enhancing biodiversity, multi-tier cropping systems, or integrated livestock management (Chaudhari and Mishra 2016). The National Agroforestry Policy of 2014 was a pioneering effort globally, promoting tree-based farming practices. By encouraging tree intercropping and ecosystem diversification, it naturally aligned with regenerative practices, though its adoption has been hindered by issues like land tenure and regulatory obstacles related to timber species management. The introduction of the Soil Health Card Scheme in 2015 emphasized soil quality and the balanced application of fertilizers based on scientific analysis. While not explicitly rooted in regenerative practices, it contributed to reducing chemical dependency and improving fertilizer management (Chenyang et al. 2021). The policy landscape began to align more clearly with RA principles with the emergence of Zero Budget Natural Farming (ZBNF n.d.), which was initially piloted in Andhra Pradesh and Himachal Pradesh in 2016 and later supported by NABARD in 2019 (Veluguri et al. 2021). ZBNF's use of bio-inputs like Jeevamrutha and Bijamrita, alongside practices such as mulching and zero synthetic inputs, directly supports RA. This momentum was further institutionalized in 2020 with the introduction of the Bhartiya Prakritik Krishi Paddhati under the Paramparagat Krishi Vikas Yojana, which focuses on cow-based natural farming and on-farm biomass recycling (Ramdas and Pimbert 2024). On the global stage, the UNFCCC (2016) created a push for nations to include agriculture in their NDCs. While India's 2022 NDC update emphasizes enhancing forest carbon sinks, it does not fully integrate RA within its core commitments. Meanwhile, the "4 per 1000" initiative, launched at COP21, encourages a 0.4% annual increase in SOC stocks as a climate mitigation strategy, aligning closely with RA practices. Similarly, the European Union's Green Deal and its Farm to Fork Strategy offer global models for integrating biodiversity, food safety, and climate mitigation through agroecological and regenerative transitions (Dröge 2022; Falkner 2016).

3.7.2.2 Institutional Support Mechanisms

The institutional framework in India has played a crucial role in fostering regenerative practices. National Bank for Agriculture and Rural Development (NABARD) has been a key player, particularly through its Watershed Development Fund and its support for climate-resilient projects under the Adaptation Fund and Green Climate Fund (Arunachalam 2024). These initiatives have focused on soil and water conservation, which are essential elements of RA. Recently, NABARD has also started exploring carbon finance pilot projects aimed at aggregating smallholder farmers for participation in VCMs (Pandey et al. 2024). The Indian Council of Agricultural Research (ICAR) has made significant contributions through its specialized institutions, such as Central Research Institute for Dryland Agriculture (CRIDA), Central Soil Salinity Research Institute (CSSRI), and Indian Institute of Farming Systems Research (IIFSR), which have developed CA protocols suited to various agroecological zones. Since 2011, ICAR's National Innovations in Climate Resilient Agriculture project has supported climate-smart villages, many of which have adopted regenerative practices like cover cropping, organic amendments, and water harvesting. Since 2020, ICAR has also advanced soil carbon monitoring protocols across over 100 agro-climatic zones, setting the stage for potential participation in carbon credit markets (Chauhan et al. 2022).

3.7.2.3 Carbon Market Dynamics and Regulation

Regenerative farmers now have more financial options thanks to the growth of carbon trading platforms, especially in the VCMs. Carbon crediting for agroecological interventions like composting, agroforestry, and low-emission cropping systems has been made possible by crucial approaches like Verra's VM0042 and Plan Vivo's community-centered frameworks (Spilker and Nugent 2022). Early examples of RA-linked loan projects in India are the organic basmati value chain in Uttarakhand (2023) and the agroforestry carbon project in Madhya Pradesh (2022) (Datta and Behera 2024).

3.7.2.4 Incentivize Carbon Farming through Public Schemes and Startups

Public policy can play a crucial role in enabling the adoption of RA and carbon farming by integrating these practices into existing rural support programs such as Pradhan Mantri Kisan Samman Nidhi (PM-KISAN), Mahatma Gandhi National Rural Employment Guarantee Act (MGNREGA), and the National Mission for Sustainable Agriculture (NMSA). For example, farmers practicing reduced tillage or using green manures could receive additional benefits like seed kits or payments through these schemes. Governments could also introduce performance-based incentives linked to verified carbon sequestration, similar to payment-for-ecosystem-service models. Startups are driving innovation in carbon monitoring, advisory services, and financing, and collaboration with these companies could help scale up efforts and improve data management. Startups such as SatSure, CropIn, and KrishiTantra are offering satellite analytics, crop mapping, and soil diagnostics that could be integrated into FPO operations or village-level dashboards, empowering local stakeholders to track their carbon status. These startups are also developing carbon calculators to quantify farm-level carbon footprints and suggest pathways for improvement, making carbon farming more data-driven and accessible.

3.8 Conclusion

Regenerative agriculture represents a transformative approach to address the interconnected challenges of soil degradation, climate change, food security, and biodiversity loss. When integrated with carbon credit systems, RA not only provides sustainable land management but also offers viable economic opportunity, especially for smallholder farmers, by providing financial rewards for environmentally responsible practices. Strengthening frameworks like Payment for Ecosystem Services (PES) can provide long-term financial stability for farmers, especially when carbon credits are linked to ecosystem services such as improved water quality and enhanced biodiversity, boosting both the premium of produce and carbon credits. While the benefits of RA are evident, its widespread adoption faces significant challenges, particularly in developing regions. Issues such as insecure land tenure, inadequate financing, and poor infrastructure limit implementation. Additionally, technological and knowledge gaps hinder effective practice. Addressing these challenges requires strong policy support, improved access to funding, and streamlined carbon credit verification systems. Emerging technologies, including AI, IoT, and real-time data analytics, can enhance transparency in monitoring carbon sequestration, enabling farmers to earn verified carbon credits. To unlock the full potential of RA, ecological restoration must be aligned with economic incentives. Integrating RA into existing rural development programs, supported by multilevel strategies and context-specific regional models, can drive large-scale adoption. With the right combination of policies, market mechanisms, and community engagement, RA can transform agriculture into a powerful tool for both environmental regeneration and economic resilience, shaping a more sustainable future.

References

Adi SH, Grunwald S (2020) Integrative environmental modeling of soil carbon fractions based on a new latent variable model approach. Sci Total Environ 711:134566. https://doi.org/10.1016/j.scitotenv.2019.134566

Anand M, Pandey R, Kedia S (2022) Mission-oriented innovation systems in the climate change context: the case of agriculture and renewable energy sector in India. In: Innovation systems, economic development and public policy. Routledge, New Delhi, pp 352–370

Andersson D (2020) A novel approach to calculate individuals' carbon footprints using financial transaction data: app development and design. J Clean Prod 256:120396. https://doi.org/10.1016/j.jclepro.2020.120396

Arunachalam RS (2024) Climate adaptation and mitigation finance for low-income populations in India: a comprehensive. Inclusive Finance India report 2024, p 115

Averill C, Turner BL, Finzi AC (2014) Mycorrhiza-mediated competition between plants and decomposers drives soil carbon storage. Nature 505. https://doi.org/10.1038/nature12901

Bai X, Huang Y, Ren W, Coyne M, Jacinthe PA, Tao B, Hui D, Yang J, Matocha C (2019) Responses of soil carbon sequestration to climate-smart agriculture practices: a meta-analysis. Glob Change Biol 25(8):2591–2606. https://doi.org/10.1111/gcb.14658

Barbato CT, Strong AL (2023) Farmer perspectives on carbon markets incentivizing agricultural soil carbon sequestration. npj Clim Action 2:26. https://doi.org/10.1038/s44168-023-00055-4

Basche AD, DeLonge MS (2019) Comparing infiltration rates in soils managed with conventional and alternative farming methods: a meta-analysis. PLoS One 14(9):e0215702. https://doi.org/10.1371/journal.pone.0215702

Benes B, Guan K, Lang M et al (2020) Multiscale computational models can guide experimentation and targeted measurements for crop improvement. Plant J 103(1):21–31. https://doi.org/10.1111/tpj.14722

Bhattacharya P, Bandyopadhyay KK, Krishnan P, Maity PP, Purakayastha TJ, Bhatia A, Chakraborty B, Kumar SN, Adak S, Tomer R (2023) Impact of tillage and residue management on greenhouse gases emissions and global warming potential of winter wheat in a semi-arid climate. J Agrometeorol 25:503–509

Bramley RGV (2009) Lessons from nearly 20 years of precision agriculture research, development, and adoption as a guide to its appropriate application. Crop Pasture Sci 60(3):197–217. https://doi.org/10.1071/CP08304

Brown G (2018) Dirt to soil: one family's journey into regenerative agriculture, 1st edn. Chelsea Green Publishing, White River Junction

Burton RJF, Wilson GA (2006) Injecting social psychology theory into conceptualisations of agricultural agency: towards a post-productivity farmer self-identity? J Rural Stud 22(1): 95–115. https://doi.org/10.1016/j.jrurstud.2005.07.004

Cao B, Domke GM, Russell MB, Walters BF (2019) Spatial modeling of litter and soil carbon stocks on forest land in the conterminous United States. Sci Total Environ 654:94–106. https://doi.org/10.1016/j.scitotenv.2018.10.359

Cariappa AAG, Krishna VV (2024) Carbon farming in India: are the existing projects inclusive, additional, and permanent? Clim Policy. https://doi.org/10.1080/14693062.2024.2416497

Chahal I, Van Eerd LL (2019) Quantifying soil quality in a horticultural-cover cropping system. Geoderma 352. https://doi.org/10.1016/j.geoderma.2019.05.039

Chaudhari VR, Mishra A (2016) Multilevel policy responses to mainstream climate adaptation through watershed development in rainfed farming systems of India. Clim Dev 8(4):324–335

Chauhan JS, Pal S, Choudhury PR, Singh BB (2022) All India coordinated research projects and value for cultivation and use in field crops in India: genesis, outputs and outcomes. Indian J Agric Res 50(6):501–510

Chenyang L, Currie A, Darrin H, Rosenberg N (2021) Farming with trees: reforming US farm policy to expand agroforestry and mitigate climate change. Ecol Law Quart 48:1

Chiu Y-H, Williams PL, Mínguez-Alarcón L, Gillman M, Sun Q, Ospina M, Calafat AM, Hauser R, Chavarro JE (2018) Comparison of questionnaire-based estimation of pesticide residue intake from fruits and vegetables with urinary concentrations of pesticide biomarkers. J Expo Sci Environ Epidemiol 28(1):31–39. https://doi.org/10.1038/jes.2017.22

Crumpler K, Bloise M, Meybeck A, Salvatore M, Bernoux M (2019) Linking nationally determined contributions and the sustainable development goals through agriculture: a methodological framework. Env. Nat. Resour. Manag. Working Paper No. 75. FAO, Rome, 40pp

Cusworth G (2020) Falling short of being the 'good farmer': losses of social and cultural capital incurred through environmental mismanagement, and the long-term impacts agri-environment scheme participation. J Rural Stud 75:164–173. https://doi.org/10.1016/j.jrurstud.2020.01.021

Datta P, Behera B (2024) India's approach to agroforestry as an effective strategy in the context of climate change: an evaluation of 28 state climate change action plans. Agric Syst 214:103840. https://doi.org/10.1016/j.agsy.2023.103840

Deng X, Chen X, Ma W et al (2018) Baseline map of organic carbon stock in farmland topsoil in East China. Agric Ecosyst Environ 254:213–223. https://doi.org/10.1016/j.agee.2017.11.022

Doshi J, Patel T, Bharti SK (2019) Smart farming using IoT: a solution for optimally monitoring farming conditions. Procedia Comput Sci 160:746–751. https://doi.org/10.1016/j.procs.2019.11.016

Dröge S (2022) The Paris Agreement 2015: turning point for the international climate regime (No. RP 4/2016). SWP Research Paper

European Commission (2021) EU soil strategy for 2030: reaping the benefits of healthy soils for people, food, nature and climate (Issue November)

Falkner R (2016) The Paris Agreement and the new logic of international climate politics. Int Aff 92(5):1107–1125. https://doi.org/10.1111/1468-2346.12708

Farmland LP (2017) Valuing the ecosystem service benefits from regenerative agriculture practices: Farmland LP 2017 impact report [PDF]. Delta Institute & Earth Economics. Retrieved from https://delta-institute.org/wp-content/uploads/2018/09/Valuing-the-Ecosystem-Service-Benefits-from-Regenerative-Agriculture-Practices-_-Farmland-LP-Impact-Report.pdf (Accessed 15 February 2025)

Food and Agriculture Organization (FAO) (2011) The state of the world's land and water resources for food and agriculture, p 308. https://doi.org/10.4324/9780203142837

Food and Agriculture Organization (FAO) (2022) What is regenerative agriculture? https://www.cbf.org/issues/agriculture/regenerative-agriculture.html
Food and Agriculture Organization (FAO) (2021) Measuring and monitoring soil organic carbon stocks and changes. https://doi.org/10.4060/cb2934en
Fuhrer J (2016) Soil health…our shared journey [PowerPoint slides]. National Resources Conservation Services. https://www.nacdnet.org/wp-content/uploads/2016/06/JayFuhrer.pdf
Galinato SP, Yoder JK, Granatstein D (2011) The economic value of biochar in crop production and carbon sequestration. Energy Policy 39(10):6344–6350. https://doi.org/10.1016/j.enpol.2011.07.035
Ghimire R, Lamichhane S, Acharya BS, Bista P, Sainju UM (2020) Soil organic carbon dynamics under conservation agriculture in Nepal. Agric Syst 182:102855. https://doi.org/10.1016/j.agsy.2020.102855
Gosnell H, Gill N, Voyer M (2019) Transformational adaptation on the farm: processes of change and persistence in transitions to 'climate-smart' regenerative agriculture. Glob Environ Change 59:101965. https://doi.org/10.1016/j.gloenvcha.2019.101965
Guevara M, Olmedo GF, Stell E, Yigini Y, Montanarella L, Vargas R (2018) No silver bullet for digital soil mapping: country-specific soil organic carbon estimates across Latin America. Soil 4:173–193. https://doi.org/10.5194/soil-4-173-2018
Gwenzi W, Chaukura N, Mukome FND, Machado S, Nyamasoka B (2015) Biochar production and applications in sub-Saharan Africa: opportunities, constraints, risks and uncertainties. J Environ Manag 150:250–261. https://doi.org/10.1016/j.jenvman.2014.11.027
Hill CG (2020) Returning the "three sisters"—corn, beans, and squash—to Native American farms nourishes people, land, and cultures. Conversation. Retrived from: https://theconversation.com/global (Accessed on 12 February 2025)
Huang Y, Wang X, Xiang W et al (2022) Forward-looking roadmaps for long-term continuous water quality monitoring: bottlenecks, innovations, and prospects in a critical review. Environ Sci Technol 56(9):5334–5354. https://doi.org/10.1021/acs.est.1c07857
Indigo Ag (n.d.) Indigo Carbon: help remove carbon and get paid for it. Retrieved April 20, 2025, from https://www.indigoag.com/carbon
IPCC (2023) Summary for policymakers. In: Core Writing Team, Lee H, Romero J (eds) Climate Change 2023: synthesis report. Contribution of Working Groups I, II and III to the sixth assessment report of the Intergovernmental Panel on Climate Change. IPCC, Geneva, pp 1–34
Jacobs AA, Evans RS, Allison JK, Garner ER, Kingery WL, McCulley RL (2022) Cover crops and no-tillage reduce crop production costs and soil loss, compensating for lack of short-term soil quality improvement in a maize and soybean production system. Soil Tillage Res 218:105310. https://doi.org/10.1016/j.still.2021.105310
Jangir CK, Sangwan PS, Panghaal D, Kumar S, Meena RS, Bharti, Jat RD, Singh N (2024) Spatial variability and statistical analysis of soil properties in the rice wheat-based systems of north-west India. Commun Soil Sci Plant Anal 55(8):1205–1223
Jat ML, Chakraborty D, Ladha JK, Parihar CM, Datta A, Mandal B, Nayak HS, Maity P, Rana DS, Chaudhari SK, Gerard B (2022) Carbon sequestration potential, challenges, and strategies towards climate action in smallholder agricultural systems of South Asia. Crop Environ 1:86–101
Jhariya MK, Raj A, Banerjee A, Meena RS, Bargali SS, Kumar S, Nema S, Poonam, Oraon PR (2022) Plan and policies for soil organic carbon management under agroforestry system. In: Plans and policies for soil organic carbon management in agriculture. Singapore, Springer, pp 191–219
Jordon MW, Smith P, Long PR, Bürkner PC, Petrokofsky G, Willis KJ (2022) Can regenerative agriculture increase national soil carbon stocks? Simulated country-scale adoption of reduced tillage, cover cropping, and ley-arable integration using RothC. Sci Total Environ 825:153955. https://doi.org/10.1016/j.scitotenv.2022.153955
Karp D (2022) Biodiversity in working landscapes. Karp Lab. https://karp.ucdavisedu/research/biodiversity-in-working-landscape/

Kaushal P, Ali N, Saini S, Pati PK, Pati AM (2023) Physiological and molecular insight of microbial biostimulants for sustainable agriculture. Front Plant Sci 14:1041413. https://doi.org/10.3389/fpls.2023.1041413

Kenne GJ, Kloot RW (2019) The carbon sequestration potential of regenerative farming practices in South Carolina, USA. Am J Clim Change 8(2):157–172. https://www.scirp.org/journal/paperinformation?paperid=91637

Keskin H, Grunwald S, Harris WG (2019) Digital mapping of soil carbon fractions with machine learning. Geoderma 339:40–58. https://doi.org/10.1016/j.geoderma.2018.12.037

Khangura R, Ferris D, Wagg C, Bowyer J (2023) Regenerative agriculture—a literature review on the practices and mechanisms used to improve soil health. Sustainability 15:2338. https://doi.org/10.3390/su15032338

Kremen C, Merenlender AM (2018) Landscapes that work for biodiversity and people. Science 362. https://doi.org/10.1126/science.aau6020

Kremmydas D, Beber CL, Baldoni E, Ciaian P, Fellmann T, Gocht A, Hristov J, Pignotti D, Rey Vicario D, Stepanyan D, Tillie P (2025) The EU target for organic farming: Potential economic and environmental impacts of two alternative pathways. Applied Economic Perspectives and Policy 47(2):602–623. https://doi.org/10.1002/aepp.13470

Kumar R, Punia H, Kumar S (2020) Nanomaterials for agriculture input use efficiency. In: Resources use efficiency in agriculture. Springer, Singapore. https://doi.org/10.1007/978-981-15-6953-1_5

Kumar S, Sharma SK, Thakral SK, Bhardwaj KK, Jhariya MK, Meena RS, Jangir CK, Bedwal S, Jat RD, Gaber A, Atta AA (2022a) Integrated nutrient management improves the productivity and nutrient use efficiency of *Lens culinaris* Medik. Sustainability 14(3):1284

Kumar S, Bamboriya SD, Rani K, Meena RS, Sheoran S, Loyal A, Kumawat A, Jhariya MK (2022b) Grain legumes: a diversified diet for sustainable livelihood, food, and nutritional security. In: Advances in legumes for sustainable intensification. Academic Press, London, pp 157–178

Kumar S, Sharma SK, Dhaka AK, Bedwal S, Sheoran S, Meena RS, Jangir CK, Kumar D, Kumar R, Jat RD, Meena AK (2023) Efficient nutrient management for enhancing crop productivity, quality and nutrient dynamics in lentil (*Lens culinaris* Medik.) in the semi-arid region of northern India. PLoS One 18(2):e0280636

Kumar S, Meena RS, Kumar S, Pradhan G, Jangir CK, Ghosh S, Punia H, Sheoran P, Meena R, Ahmad MA, Goyal SK (2025) Designing a diversified Indian mustard production system for energy-carbon-cum-heat use efficiency and sowing dates assessment. GCB Bioenergy 17(6):e70044

LaCanne CE, Lundgren JG (2018) Regenerative agriculture: merging farming and natural resource conservation profitably. PeerJ 6:e4428. https://doi.org/10.7717/peerj.4428

Lal R (2015) Sequestering carbon and increasing productivity by conservation agriculture. J Soil Water Conserv 70(3):55A–62A

Lal R (2020) Regenerative agriculture for food and climate. J Soil Water Conserv 75(5):123A–124A. https://doi.org/10.2489/jswc.2020.0620A

Leharwan M, Kumar Y, Kumar R, Kumar Saraswat P, Kumar R, Kumar Thaliyil Veetil A, Bhattacharjee S, Kumar A, Kumar S (2023) Assessing the effects of conservation tillage and in-situ crop residue management on crop yield and soil properties in rice–wheat cropping system. Sustainability 15(17):12736

Lévesque A, Kermagoret C, Poder TG, L'Ecuyer-Sauvageau C, He J, Sauvé S, Dupras J (2021) Financing on-farm ecosystem services in southern Quebec, Canada: A public call for pesticides reduction. Ecol Econ 184:106997. https://doi.org/10.1016/j.ecolecon.2021.106997

Lindwall C (2021) Industrial agricultural pollution 101. NRDC. https://www.nrdc.org/stories/industrial-agricultural-pollution-101

Liu X, Tan S, Song X, Wu X, Zhao G, Li S, Liang G (2022) Response of soil organic carbon content to crop rotation and its controls: a global synthesis. Agric Ecosyst Environ 335:108017. https://doi.org/10.1016/j.agee.2022.108017

Lowdermilk WC (1950) Conquest of the land through 7,000 years. U.S. Department of Agriculture Soil Conservation Service, pp 32, 38. https://www.ebay.com/itm/394216646102

Macik M, Gryta A, Sas-Paszt L, Frac M (2023) New insight into the soil bacterial and fungal microbiome after phosphorus biofertilizer application. Appl Soil Ecol 189:104941. https://doi.org/10.1016/j.apsoil.2023.104941

Meena RS, Pradhan G (2023) Industrial garbage-derived biocompost enhances soil organic carbon fractions, CO_2 biosequestration, potential carbon credits, and sustainability index in a rice-wheat ecosystem. Environ Res 116525. https://doi.org/10.1016/j.envres.2023.116525

Meena RS, Pradhan G, Kumar S, Lal R (2023) Using industrial wastes for rice-wheat cropping and food-energy-carbon-water-economic nexus to the sustainable food system. Renew Sust Energ Rev 187:113756. https://doi.org/10.1016/j.rser.2023.113756

Meena RS, Pradhan G, Singh K, Kumar S, Singh AK, Shashidhar KS, Mina KK, Rao CS (2024a) Agriculture models for restoring degraded land to enhance CO_2 biosequestration and carbon credits in the Vindhyan region of India. Sci Total Environ 929:172661. https://doi.org/10.1016/j.scitotenv.2024.172661

Meena RS, Lal R, Kumar S, Pradhana G, Rao CS, Singh AK, Pathak H, Abhilash PC, Kumar A, Sharma SK, Jat ML, Singh S (2024b) Potential of Indian agriculture for capturing atmospheric CO_2 and monetizing carbon credits to the farmers: an overview and policy framework. Adv Agron 188:101–206. https://doi.org/10.1016/bs.agron.2024.06.001

Meena RS, Pradhan G, Nalani SK, Singh AK, Verma SK, Mina KK, Kumar S, Chaturvedi RK (2024c) Diversified cropping modules designed for soil restoration, CO_2 sequestration, and generating carbon credits. Land Degrad Dev 35. https://doi.org/10.1002/ldr.5224

Meena RS, Pradhan G, Kumar S (2025) Energy flow, eco-efficiency, and economic circulation with recycled industrial waste compost application in wheat and subsequent rice farming. Sci Total Environ 967:178779. https://doi.org/10.1016/j.scitotenv.2025.178779

Montanarella L, Pennock DJ, McKenzie N, Badraoui M, Chude V, Baptista I et al (2016) World soils under threat. Soil 2(1):79–82. https://doi.org/10.5194/soil-2-79-2016

Montgomery DR (2012) Dirt: the erosion of civilizations, 2nd edn. University of California Press, Berkeley

Nori (n.d.) Nori: a carbon removal marketplace. Retrieved April 19, 2025, from https://nori.com/

Nozaki Y (2023) Rise of carbon farming in India: world's largest agrarian country expected to become the leading market for carbon farming credits. Mitsui & Co. Glob. Strateg. Stud. Inst. Mon. Rep., February 2023

Pandey VK, Anshu, Sharma KK (2024) Climate change adaptation policies and governance: a critical appraisal. In: Climate change and human adaptation in India: sustainability and public policy. Cham, Springer, pp 199–206

Parihar CM, Parihar MD, Sapkota TB, Nanwal RK, Singh AK, Jat SL, Nayak HS, Mahala DM, Singh LK, Kakraliya SK, Stirling CM (2018) Long-term impact of conservation agriculture and diversified maize rotations on carbon pools and stocks, mineral nitrogen fractions, and nitrous oxide fluxes in Inceptisol of India. Sci Total Environ 640:1382–1392. https://doi.org/10.1016/j.scitotenv.2018.05.405

Paul RAI, Palanisamy MA, Peramaiyan P et al (2024) Spray volume optimization with UAV-based herbicide application for effective droplet deposition and weed control in direct-seeded rice. Front Agron 6:1491842. https://doi.org/10.3389/fagro.2024.1491842

Peng Y, Xiong X, Adhikari K, Knadel M, Grunwald S, Greve MH (2015) Modeling soil organic carbon at regional scale by combining multi-spectral images with laboratory spectra. PLoS One 10(11):e0142295. https://doi.org/10.1371/journal.pone.0142295

Plan Vivo Foundation (n.d.) The Plan Vivo Standard. Retrieved April 4, 2025, from https://www.planvivo.org/the-plan-vivo-standard

Pradhan G, Meena RS (2023a) Utilizing waste compost to improve the atmospheric CO_2 capturing in the rice-wheat cropping system and energy-cum-carbon credit audit for a circular economy. Sci Total Environ 164572. https://doi.org/10.1016/j.scitotenv.2023.164572

Pradhan G, Meena RS (2023b) Interaction impacts of biocompost on nutrient dynamics and relations with soil biota, carbon fractions index, societal value of CO_2 equivalent, and ecosystem services in the wheat-rice farming. Chemosphere 139695. https://doi.org/10.1016/j.chemosphere.2023.139695

Pradhan G, Meena RS, Kumar S, Lal R (2023) Utilizing industrial wastes as compost in wheat-rice production to improve the above and below-ground ecosystem services. Agric Ecosyst Environ 358:108704. https://doi.org/10.1016/j.agee.2023.108704

Punia H, Tokas J, Malik A, Satpal, Rani A, Gupta P, Kumari A, Mor VS, Bhuker A, Kumar S (2020) Solar radiation and nitrogen use efficiency for sustainable agriculture. In: Resources use efficiency in agriculture. Springer, Singapore, pp 177–212

Rajak P, Ganguly A, Adhikary S, Bhattacharya S (2023) Internet of things and smart sensors in agriculture: scopes and challenges. J Agric Food Res 14:100776. https://doi.org/10.1016/j.jafr.2023.100776

Ramdas SR, Pimbert MP (2024) A cog in the capitalist wheel: co-opting agroecology in South India. J Peasant Stud 51(5):1251–1273. https://doi.org/10.1080/03066150.2024.2345337

Rana B, Parihar CM, Jat ML, Patra K, Nayak HS, Reddy KS, Sarkar A, Anand A, Naguib W, Gupta N (2023) Combining sub-surface fertigation with conservation agriculture in intensively irrigated rice under rice-wheat system can be an option for sustainably improving water and nitrogen use-efficiency. Field Crop Res 302:109074

Rasul G, Thapa GB (2004) Sustainability of ecological and conventional agricultural systems in Bangladesh: an assessment based on environmental, economic and social perspectives. Agric Syst 79(3):327–351. https://doi.org/10.1016/S0308-521X(03)00090-8

Reddy KS, Parihar CM, Panneerselvam P, Sarkar A, Patra K, Bharadwaj S, Sena DR, Reddy GS, Sinha A, Dhakar R, Kumar V, Nayak HS (2025a) Meta-analysis of yield-emission trade-off in direct seeded vs. puddled transplanted rice: towards a cleaner and sustainable production. Clean Environ Syst 16:100238. https://doi.org/10.1016/j.cesys.2024.100238

Reddy KS, Parihar CM, Panneerselvam P, Sarkar A, Nayak HS, Patra K, Sena DR, Reddy GS, Sinha A, Bharadwaj S, Kumar S, Kumar V (2025b) Eco-optimizing rice-wheat system of Eastern Indo-Gangetic plains of India through resource conservation technologies: insights from field experiments and modeling. Front Sustain Food Syst 9:1499425. https://doi.org/10.3389/fsufs.2025.1499425

Regen Network (n.d.) Regen Network: regenerating the planet through ecological data and Web3. Retrieved April 14, 2025, from https://www.regen.network/

Rehberger E, West PC, Spillane C, McKeown PC (2023) What climate and environmental benefits of regenerative agriculture practices? An evidence review. Environ Res Commun 5(5):1–14. https://doi.org/10.1088/2515-7620/acd6dc

Rockström J, Gupta J, Qin D et al (2023) Safe and just Earth system boundaries. Nature 619(7968):102–111. https://doi.org/10.1038/s41586-023-06083-8

Rodale R (1983) Breaking new ground: the search for a sustainable agriculture. Futurist 1:15–20

Ross CW, Grunwald S, Vogel JG et al (2020) Accounting for two-billion tons of stabilized soil carbon. Sci Total Environ 703:134615. https://doi.org/10.1016/j.scitotenv.2019.134615

Royal Agricultural Society of England (RASE) (2023) Farm of the future: journey to net zero. https://www.rase.org.uk/content/large/documents/reports/fotf_farmers_guide_update_30_aug_23_digital.pdf

Safanelli JL, Chabrillat S, Ben-Dor E, Demattê JAM (2020) Multispectral models from bare soil composites for mapping topsoil properties over Europe. Remote Sens 12(9):1369. https://doi.org/10.3390/rs12091369

Salma A, Fryda L, Djelal H (2024) Biochar: a key player in carbon credits and climate mitigation. Resources 13(2):31. https://doi.org/10.3390/resources13020031

Sanderman J, Hengl T, Fiske GJ (2017) Soil carbon debt of 12,000 years of human land use. Proc Natl Acad Sci USA 114(36):9575–9580. https://doi.org/10.1073/pnas.1706103114

Schimmelpfennig D, Ebel R (2011) On the doorstep of the information age: recent adoption of precision agriculture. USDA Econ. Res. Serv. https://www.ers.usda.gov/publications/pub-details?pubid=44576

Sheoran S, Gupta M, Kumari S, Kumar S, Rakshit S (2022) Meta-QTL analysis and candidate genes identification for various abiotic stresses in maize (*Zea mays* L.) and their implications in breeding programs. Mol Breed 42(5):26

Sher A, Li H, Ullah A et al (2024) Importance of regenerative agriculture: climate, soil health, biodiversity, and its socioecological impact. Discov Sustain 5:462. https://doi.org/10.1007/s43621-024-00662-z

Soil Capital (n.d.) Helping farmers transition to regenerative agriculture. Retrieved March 19, 2025, from https://www.soilcapital.com/

Spilker G, Nugent N (2022) Voluntary carbon market derivatives: growth, innovation & usage. Borsa Istanbul Rev 22:S109–S118. https://doi.org/10.1016/j.bir.2022.07.005

Srivastava S, Bist V, Srivastava S et al (2016) Unraveling aspects of *Bacillus amyloliquefaciens*-mediated enhanced production of rice under biotic stress of *Rhizoctonia solani*. Front Plant Sci 7. https://doi.org/10.3389/fpls.2016.00587

Tsesmelis DE, Gkotsis I, Saltogiannis C, Reppas S, Panagakis S, Zervas E (2025) An integrated approach of carbon footprint calculation for agricultural sector through smart-farming. J Clean Prod 486:144556. https://doi.org/10.1016/j.jclepro.2024.144556

Uchimiya M (2024) Big data-driven water research towards metaverse. Water Sci Eng 17(2): 101–107. https://doi.org/10.1016/j.wse.2024.02.001

UN Environment Programme (2020) Climate change and land: summary for policymakers. Intergovernmental Panel on Climate Change, p 41. https://www.ipcc.ch/site/assets/uploads/sites/4/2020/02/SPM_Updated-Jan20.pdf

United States Department of Agriculture (USDA) (2013) COMET-Farm: conservation calculation. https://www.usda.gov/about-usda/news/blog/comet-farmtm-conservation-calculation (Accessed 23 January 2025)

United Nations Framework Convention on Climate Change (UNFCCC) (n.d.-a) Payments for Environmental Services (PES) Program. https://unfccc.int/climate-action/momentum-for-change/financing-for-climate-friendly-investment/payments-for-environmental-services-program

United Nations Framework Convention on Climate Change (UNFCCC) (n.d.-b) What is REDD+? https://unfccc.int/topics/land-use/workstreams/redd/what-is-redd

United Nations Framework Convention on Climate Change (UNFCCC). (2016). The Paris Agreement [PDF]. Retrieved from https://unfccc.int/sites/default/files/resource/parisagreement_publication.pdf (Accessed 31 January 2025)

United Nations REDD Programme (UN-REDD) (n.d.) REDD+ projects and results. Retrieved April 14, 2025, from https://www.un-redd.org/

Vasques GM, Grunwald S, Comerford NB, Sickman JO (2010) Regional modelling of soil carbon at multiple depths within a subtropical watershed. Geoderma 156(3–4):326–336. https://doi.org/10.1016/j.geoderma.2010.03.002

Vi Agroforestry (n.d.) Kenya Agricultural Carbon Project (KACP)

Viscarra Rossel RA, Chen C, Grundy MJ, Searle R, Clifford D, Campbell PH (2015) The Australian three-dimensional soil grid: Australia's contribution to the GlobalSoilMap project. Soil Res 53(8):845–864. https://doi.org/10.1071/SR14366

Wang B, Waters C, Orgill S, Cowie A, Clark A, Liu DL, Simpson M, McGowen I, Sides T (2018) Estimating soil organic carbon stocks using different modelling techniques in the semi-arid rangelands of eastern Australia. Ecol Indic 88:425–438. https://doi.org/10.1016/j.ecolind.2018.01.049

Watershed Organisation Trust (WOTR) (n.d.) Climate resilience for rural India. https://wotr.org/

Were K, Bui DT, Dick ØB, Singh BR (2015) A comparative assessment of support vector regression, artificial neural networks, and random forests for predicting and mapping soil

organic carbon stocks across an Afromontane landscape. Ecol Indic 52:394–403. https://doi.org/10.1016/j.ecolind.2014.12.028
World Wildlife Foundation (2018) Living planet report—2018: aiming higher, p 145. https://c402277.ssl.cf1.rackcdn.com/publications/1187/files/original/LPR2018_Full_Report_Spreads.pdf
Xiong X, Grunwald S, Myers DB, Kim J, Harris WG, Comerford NB (2014) Holistic environmental soil-landscape modeling of soil organic carbon. Environ Model Softw 57:202–215. https://doi.org/10.1016/j.envsoft.2014.03.002
Veluguri D, Bump JB, Venkateshmurthy NS, Mohan S, Pulugurtha KT, Jaacks L (2021) Political analysis of the adoption of the Zero-Budget Natural Farming program in Andhra Pradesh, India. Agroecology and Sustainable Food Systems 45(6):907–930. https://doi.org/10.1080/21683565.2021.1901832
Yadav SPS, Bhandari S, Bhatta D et al (2023) Biochar application: a sustainable approach to improve soil health. J Agric Food Res 11:100498. https://doi.org/10.1016/j.jafr.2023.100498
Zero Budget Natural Farming (ZBNF) (n.d.) First piloted in Andhra Pradesh and Himachal Pradesh in 2016 and subsequently gaining momentum with support from NABARD in 2019
Zhang Q (2016) Precision agriculture technology for crop farming. CRC Press, Boca Raton

Chapter 4
Agronomical and Technological Advances for Carbon-Negative Agriculture and Carbon Credit

Tarun Sharma, Arkaprava Roy, Priyanka Saha, Alekhya Gunturi, Sunil Kumar Prajapati, Nilutpal Saikia, Lalit Kumar Mainway, and Alok Sinha

Abstract Global greenhouse gas (GHG) emissions have reached 53.0 Gt CO_2eq in 2023, marking a 62% increase since 1990. Agriculture contributes a significant one-third of total anthropogenic GHG emissions. The land-based activities, enteric fermentation and supply chain processes primarily constitute the agri-food system's emissions, which totalled 16 billion tonnes in 2020. Methane (CH_4) emissions from enteric fermentation, rice cultivation and nitrous oxide (N_2O) emissions owing to the use of nitrogenous fertilizers and soil reactions are of serious concern because of their higher global warming potential (GWP) than carbon dioxide (CO_2) and a significant proportion of total GHG emissions from the agriculture sector. The sector's reliance on fossil fuels, less efficient irrigation systems, synthetic fertilizers and monoculture practices exacerbates these emissions. Climate change issues are escalated through increased GHG emissions that pose challenges, affecting agricultural production through rising temperatures and shifting weather patterns. To minimize the impacts of climate change intensified from global warming and to meet the Paris Agreement's goals of limiting rise in atmospheric temperature below 1.5 °C, agriculture must adopt best management practices (BMPs) that improve resource use efficiency (RUE), enhance carbon sequestration and lower carbon footprints (CF) in the lifecycle of agricultural produce. Precision agriculture (PA), regenerative agriculture (RA), conservation agriculture (CA), organic farming (OF), balanced fertilisation, limited irrigation, waste recycling, crop and livestock improvement through genome editing and the integration of renewable energy systems are essential for reducing the carbon footprints (CF) and achieving carbon credits (CCs) goals through enhanced carbon sequestration. The use of climate-smart technologies having the potential to turn GHG emission sources into sinks by

T. Sharma · A. Roy (✉) · A. Gunturi · S. K. Prajapati · N. Saikia · L. K. Mainway · A. Sinha
Division of Agronomy, ICAR-Indian Agricultural Research Institute, New Delhi, India

P. Saha
Division of Agronomy, ICAR-Indian Agricultural Research Institute, New Delhi, India

ICAR-Krishi Vigyan Kendra, West Garo Hills, Tura, Meghalaya, India

R. S. Meena et al. (eds.), *Carbon-Negative Agriculture*, Sustainability Sciences in Asia and Africa, https://doi.org/10.1007/978-981-95-1483-0_4

enhanced soil organic carbon (SOC) sequestration deserves incentivisation for its promotion and widespread adoption., There is a need for robust standards and regulatory principles for the development of effective and transparent systems for measuring and verifying carbon credits (CCs). Awareness of citizens and suitable policy support for the adoption of sustainable farming practices and Research and Development (R&D) for advanced green technologies are essential for balancing the interplay between carbon emissions, climate actions and global food security.

Keywords Best management practices · Conservation agriculture · Global warming potential · Monitoring · Regenerative agriculture · Verification and reporting

Abbreviations

AI	Artificial intelligence
BMPs	Best management practices
BNI	Biological nitrification inhibition
C	Carbon
CA	Conservation agriculture
CCs	Carbon credits
CF	Carbon footprints
CH_4	Methane
CO_2	Carbon dioxide
DI	Drip irrigation
ETs	Electric tractors
GHGs	Greenhouse gas
GPS	Global positioning system
Gt	Gigatonnes
GWP	Global warming potential
HTs	Hybrid tractors
K	Potassium
LCA	Lifecycle assessment
Mg	Megagram
MRV	Monitoring, reporting, and verification
Mt.	Million tonnes
N	Nitrogen
N_2O	Nitrous oxide
NDCs	Nationally determined contributions
NDVI	Normalized difference vegetation index
NIR	Near infrared
NUE	Nitrogen use efficiency
P	Phosphorus
PA	Precision agriculture
Pg	Petagrammes

Pg	Petagrams
PM	Particulate matter
PV	Photovoltaic
RA	Regenerative agriculture
RUE	Resource use efficiency
SDGs	Sustainable Development Goals
SDI	Sub-surface drip irrigation
SOC	Soil organic carbon
SOM	Soil organic matter
SWIR	Short wave infrared
Tg	Teragram
VRT	Variable rate technology
WSN	Wireless sensor networks
WUE	Water-use efficiency

4.1 Introduction

Greenhouse gas (GHGs) emissions are a major driver of climate change. The post-industrial era has faced a considerable increase in global GHG emissions, which led to an increase in mean tropospheric temperature. From 1990 to 2023, global GHG emissions have increased by 62%, which totalled 53.0 Gt CO_2eq (Crippa et al. 2021). The major sectors contributing to GHG emissions are energy, transportation, agriculture and industry (Filonchyk et al. 2024; Meena et al. 2025). Among the major forms of GHG emissions, fossil carbon dioxide (CO_2) constitutes the maximum (73.7%), followed by methane (CH_4) (18.9%), nitrous oxide (N_2O) (4.7%) and fluorinated gases (F-gases) (2.7%) (Crippa et al. 2021). The agriculture sector is both a major contributor and sufferer of climate change. Agrifood systems comprising on-farm and off-farm processes, from production to consumption, including crop production, processing, transport and packaging, account for one-third of total anthropogenic GHG emissions. Among these processes, the land-based sector, that is, agriculture and associated land use and land-use change activities, contributes 71% of GHG emissions, while the rest comes from post-harvest and supply chain activities (Crippa et al. 2021). In 2020, global agri-food systems generated 16 billion tonnes of GHG emissions, out of which 7.4 billion tonnes were produced within the farm gate, 3.1 billion tonnes resulted from land use changes and 5.6 billion tonnes came from supply-chain activities (FAO 2022a). The GHG emissions from the agriculture sector consist of CO_2 (52%), CH_4 (35%), N_2O (10%) and F-gases (2%) (Crippa et al. 2021; Das et al. 2021). The farming practices involved in crop and livestock production influence GHG emissions, and major sources have been identified by FAO (2022a) as enteric fermentation and manure management, agricultural soils, crop residue burning, deforestation and rice cultivation (Leharwan et al. 2023). To meet the United Nations agenda of Sustainable Development Goals (SDGs) by

2030, and net zero emission by 2070, it is essential to incorporate cleaner and sustainable production practices, shifting to renewable energy, and enhancing resource use efficiency (RUE; water, energy and nutrients) (Filonchyk et al. 2024; Meena et al. 2024).

The major bottleneck in reducing agricultural carbon footprints (CF) is the heavy reliance on farm machinery that requires fossil fuel energy and, use of industrial inputs like nitrogenous fertilizers. Farm machinery using diesel and oil emits GHGs during operations such as ploughing, sowing, irrigation, crop protection, harvesting and transportation of agricultural products (Sharma et al. 2023; Meena et al. 2024a). The nitrogenous fertilizers are a source of N_2O emission, which is a potent GHGs with a global warming potential (GWP) 273 times that of CO_2 over a 100-year time scale (IPCC 2024). Moreover, inefficient irrigation systems lead to energy wastage, and the high-water footprint in many agricultural regions exacerbates the problem. In regions dependent on monocultures, such as rice growing regions, practices like continuous flooding maintain anaerobic conditions that emit CH_4 as GHG having 24 times the GWP than $CO_{2,}$ and further intensive tillage degrades soil health by emitting the stores soil organic carbon (SOC) back to the atmosphere, reducing the carbon sequestration rates (Pathak 2015). Therefore, intensive tillage, over-application of fertilizers, pesticides and water contribute to GHG emissions and result in environmental degradation and resource depletion. Additionally, climate change will affect agricultural production with rising average temperatures, shifts in seasonal patterns, combined heat and drought stress, extreme rainfall events, water shortages, changes in pest and disease occurrences, sea level rise and ocean acidification (FAO 2022b). The targets for reducing GHG emissions are embodied in nationally determined contributions (NDCs) to meet Paris Agreement mandates of limiting atmospheric temperature increase below 2 °C, and ideally to 1.5 °C, above pre-industrial levels (IPCC 2019). To meet the national and international goals, agriculture must evolve towards sustainability by adopting cleaner technologies that not only improve resource use efficiency but also contribute to carbon sequestration. The climate smart practices like development and conservation of multifunctional landscapes, conservation agriculture (CA), regenerative agriculture (RA), organic farming (OF), precision nutrient management, drip irrigation (DI), direct seeded rice, agroforestry, modified dietary behaviours, improved livestock management and green supply chains are imperative to mitigate GHGs emission, improve carbon capture and SOC sequestration, and achieve carbon credits (CCs) (Dijkstra et al. 2011; Bozal-Leorri et al. 2022; Rosa and Gabrielli 2023; Brummitt et al. 2024; Cariappa et al. 2024).

A carbon credit is defined as the reduction of one metric tonne of CO_2 equivalent GHGs from the atmosphere. It can be achieved by practicing cleaner production practices with lower carbon footprints (by reducing and avoiding emissions) and enhancing the sequestration of the atmospheric CO_2 into plant biomass and soil (carbon capture) (Hartman and Broom 2020). Using climate adaptation strategies has the potential to turn agricultural lands back into carbon sinks by enhancing SOC sequestration that deserves incentivisation for its promotion and widespread adoption. For an authentic carbon credit generation, the GHGs reductions or removals should be real (conservatively and adequately measured), measurable (standard methodology for quantification of reductions/removals), additional (improvement over business-as-usual for reductions/removals), permanent (GHGs removals are not released back), independently verified (third-party auditing for validation and verification), conservatively estimated (not overstated), uniquely numbered, and transparently listed (Liu 2022; Verra 2025; Meena et al. 2024b). However, there are certain challenges regarding the implementation of robust systems with standard methodology, flexible eligibilities for measuring, reporting, and verifying (MRV) carbon credits in the agriculture sector and a lack of awareness among farmers (Cariappa et al. 2024). The impacts of climate-smart practices on carbon removals and sequestration are spatially and temporally variable that pose challenges for scalable quantification of emissions reduction and deployment of incentives (Brummitt et al. 2024). For stakeholders and verification bodies, the additional challenges are heterogeneous farm fields, diverse cropping patterns, and soil variability that ultimately affect the credibility of carbon credit programmes (Verra 2025). Technological innovations in agriculture are vital to reduce carbon footprint and enhance input use efficiency for achieving carbon credits, but the lack of awareness among farmers, expensive, and sound technicalities makes it equally difficult for farmers to participate in carbon credit markets (Liu 2022). Hence, improving the stakeholders' awareness, suitable policy for a permanent adoption of cleaner production practices, and development and implementation of robust MRV protocols are crucial for a country to meet NDCs.

4.2 Carbon Footprints from the Agriculture Sector

Agroecosystems can both emit and fix carbon. Carbon footprint (CF) measurement from agriculture includes both direct and indirect GHGs emissions of a product or service during its lifecycle. The CF based on lifecycle assessment (LCA) is considered an appropriate methodology that determines emissions from multiple levels, namely raw material extraction, production and processing, storage and transportation and waste disposal (Miao et al. 2023; Meena et al. 2023). The LCA assessment of agricultural products helps in measuring the environmental impact of an individual production system. GHGs emissions occur in the form of CO_2, CH_4 and N_2O in crop and livestock management. The measurement of CF and overall carbon efficiency for LCA in agroecosystems involves calculations of inputs/outputs of the production system with corresponding coefficients of CO_2equivalent (CO_2eq) (Litskas et al. 2020). The non-CO_2 emissions (CH_4 and N_2O) are converted into the CO_2equivalent emissions by using equivalency factors, that is, GWP values 27 and 273 for CH_4, and N_2O for a 100-year time horizon respectively. (ISO 2018; IPCC 2024). Carbon footprints vary across farming systems, crops, management practices and are mainly determined by the type of planting methods, N application levels, irrigation levels and farm waste recycling (Miao et al. 2023). Of the total GHGs emissions, agriculture has been projected to be accountable for about 80% of anthropogenic N_2O emissions, 70% of NH_3 emissions (N_2O and NH_3 emissions owing to N-fertilizer use and soil reactions) and 40% of CH_4 emissions (owing to enteric fermentation by methanogens and organic matter decomposition under anaerobic conditions). According to the CGIAR (2019) projections for 2030, 41% of non-CO_2 emissions (5756 $MtCO_2e$) from the agricultural sector will come from enteric fermentation (CH_4), whereas 38.1% will be N_2O emissions from agricultural soils and 8.7 and 7.2% are projected from rice cultivation and manure management respectively. The projections clearly indicate the importance of crop cultivation practices, nutrients and livestock management practices in managing the carbon footprints of agricultural soils. The estimation of CF from different production practices across the regions could identify the emission hotspots, and input management in such regions could be optimized to lower CF. Fertilizer use is the major reason for CF hotspots under intensive cultivation systems because of its energy-intensive production, transportation processes and N_2O emissions after application. Among field and horticultural crops, median CF was observed to be highest for rice cultivation (2.55 kg CO_2eq. kg^{-1} product) while meat production (4.12–5.85 kg CO_2eq. kg^{-1} product) had the highest GWP (Clune et al. 2017). Globally, emissions from meat production (54.7%), milk production (17.9%) and rice cultivation (17.9%) constitute around 88% of total GHG emissions from the agriculture sector which is even higher in developing counties (91%), and 79.6% in developed countries. The market demand for products as driven by dietary behaviours also influences the CF. Precision agriculture (PA) technologies, including satellite imaging, remote sensing and drones, enable farmers to apply inputs like water, fertilizers and pesticides more efficiently. The advances in fertilizer application technology, such as slow-release fertilizers (SRFs) and nitrification inhibitors and recycling of farm waste along with precision irrigation systems, such as drip irrigation or micro-sprinklers, reduce energy consumption and lower carbon emissions from agricultural systems. Furthermore, agroecosystems also act as a sink for GHGs by offering plans and SOC sequestration through various ways. The practices with higher carbon sequestration potential include agroforestry systems (planting perennial trees with field crops), biochar application (carbon-rich material from pyrolysis), cover cropping, OF, CA and RA (Jhariya et al. 2022). The integration of sustainable agriculture practices like zero tillage or reduced tillage with no overuse of fertilizers, no residue burning and limited irrigation can help in earning up to US$ 18 ha^{-1} and US$ 30 ha^{-1} CCs from wheat cultivation in Bihar and Punjab respectively (Cariappa et al. 2024). These practices need to be scaled up at the global level to achieve CCs across diverse farming systems (Rosa and Gabrieli 2023; Brummitt et al. 2024; Meena and Pradhan 2023). Moreover, scaling up the right resource-efficient technology having lower CF at the right place/location, particularly in the emission hotspots, requires policy support. Subsidies on agricultural inputs reduce the cost of production, which leads to increased agricultural output without addressing or reducing domestic consumption of outputs. This results in an oversupply of goods, which can alter global prices by influencing demand-supply equations and encourage higher consumption of goods in other regions. Eventually, this increased production and consumption cycle contribute to rising global emissions by promoting input-intensive agriculture, and ultimately, the associated environmental challenges. Therefore, mitigating the significant proportion of GHGs emissions from agriculture by implementing resource-efficient technologies or best management practices (BMPs) by optimizing input management and promoting sustainable practices driven by suitable policies are essential for reducing carbon footprints and achieving climate action goals.

4.3 Best Management Practices for Reducing Carbon Footprint

Agriculture accounts for 18% of GHGs emissions, with CO_2, CH_4 and N_2O accounting for the majority of these emissions (Maier et al. 2022; Pradhan and Meena 2023). One way to reduce GHGs emissions and subsequently climate change is to evaluate the agricultural CF and control agricultural management accordingly. Recent studies have looked at the agricultural CF under various management practices, including cropping patterns, tillage and nitrogen fertilizer rates (Ozlu and Kumar 2018; Lal 2022). Above all, though, improving input use efficiency is a great approach to lessen the carbon footprint (Fig. 4.1). The operations involved in agricultural practices require a substantial amount of energy. Consequently, reducing crop production-related CF and increasing energy use efficiency are crucial.

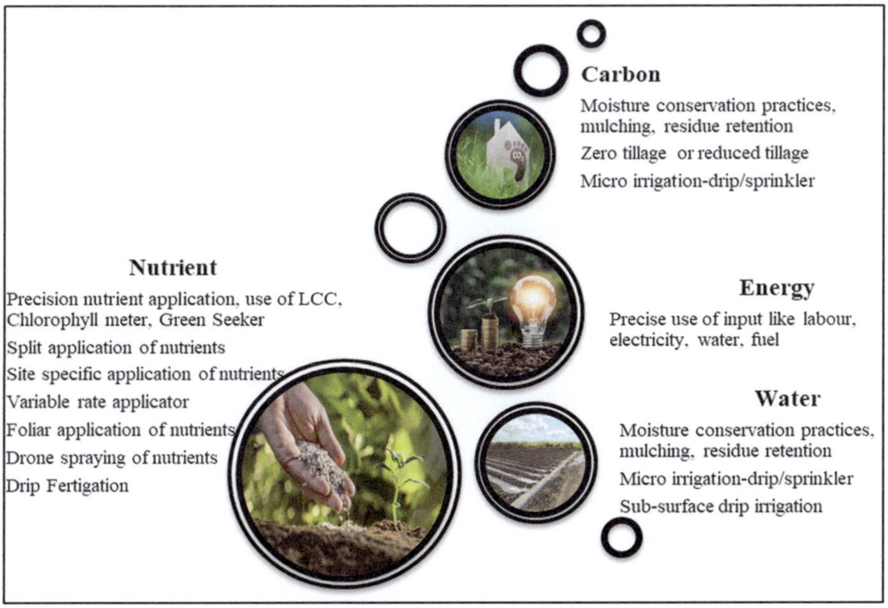

Fig. 4.1 Major inputs in crop production and ways to increase their use efficiency

Because soils are the source of GHGs emissions and biological activity in soils is sensitive to soil parameters (Baldock et al. 2012; Pradhan and Meena 2023a), there is an increased focus on both reducing carbon footprints and raising production efficiency. Global farming practices and management approaches have made the soil a significant source and sink of atmospheric carbon.

Soils contain a large amount of organic carbon and nitrogen, which vary throughout the soil profile and contribute to GHGs emissions through decomposition and microbial activity. If all other C cycle variables stayed the same, a 1% change in soil C content would result in an 8-ppm change in atmospheric CO_2. However, this 8 ppm CO_2 response could be reduced by taking into account potential mediating responses from photosynthesis and oceanic exchange (Baldock et al. 2012). As a result, soil management and worldwide soils can either increase or decrease atmospheric GHGs concentrations and climate change. Therefore, it is crucial to consider how agricultural management practices impact the connection between photosynthesis and GHGs emissions, with a focus on soil and related processes. The BMPs aim to minimize adverse environmental consequences while increasing crop productivity. These techniques focus on improving soil health, increasing productivity and making efficient and sustainable use of resources. The finest management practices, such as CA, balanced fertilization, adding biosolids or organic amendments, manuring, crop rotation and optimal residue management, can increase the amount of SOC sequestered in agricultural areas (Fig. 4.2). By taking into consideration the mechanistic processes that result in the natural sequestration of soil C, the crop density, crop type and consequently photosynthesis become significant aspects in CF evaluations (Table 4.1). More research is required, but naturally occurring processes that trap carbon in soil may be more stable than traditional methods.

4.3.1 Conservation Agriculture

Conservation agriculture (CA) is a broad phrase that has gained significant global traction. It is defined by the Food and Agriculture Organization (FAO) as a technique to conserve natural resources through minimal soil disturbance, permanent organic soil cover and crop diversification. Adoption of CA methods promotes SOC sequestration and improves the physical qualities of soil, such as soil aggregation and bulk density reduction. Conservation tillage techniques alone have the potential to store carbon at rates of between 0.4 and 0.8 petagrammes (Pg) annually worldwide. The switch from conventional to no-till farming may result in an additional 0.1–1.0 mg of carbon per hectare of SOC every year (Lal 2022).

Fig. 4.2 Best management practices for reducing carbon footprint

Table 4.1 Potential of best management practices for reducing carbon footprint

Farming practices	Carbon sequestration potential (kg C ha^{-1} yr^{-1})	Reference
Organic farming practices (average)	287 ± 102	Blanco-Canqui et al. (2017)
Green manure (alfalfa)	180	Blanco-Canqui et al. (2017)
Agroforestry system: *Alnus nepalensis* + wheat, mustard, lentil in winter season; *Alnus nepalensis* + pigeon pea, cow pea in rainy season	256	Maikhuri et al. (2000)
Crop rotation and diversification	50–250	Lal (2004), Shah and Das (2012)
Cover cropping and mulching	150–250	Ngosong et al. (2019)
Zero till or reduced tillage	70–330	Robertson et al. (2000)
Manure management	50–150	
Biochar application[a]	Up to 2	Blanco-Canqui et al. (2020): Yadav et al. (2023)

[a]*Note*: Biochar can potentially sequester 0.7–1.8 Gt of CO_2 (C eq.) yr^{-1}

4.3.2 Organic Cover

Raindrops on bare soil in conventional farming cause runoff and erosion because they break up aggregates, obstruct pores and decrease water infiltration (Yadav et al. 2023, 2024). By retaining nutrients and lowering soil erosion, cover crops like sweet potatoes and cowpeas serve to minimize these problems (Yadav et al. 2024). Grasses reduce nitrogen loss and leaching, while legume covers crops fix N through biological nitrogen fixation and supply N for subsequent crops (Singh et al. 2021). By supplying consistent organic inputs and lowering mineralization rates, organic mulches dramatically raise SOC (Saroa and Lal 2003; Zahed et al. 2022). The adoption of regenerative farming practices can contribute to reducing atmospheric CO_2 levels, raising soil productivity and quality, and lowering CO_2 emissions.

4.3.3 Organic Farming

Organic farming has a major influence on SOC stocks, especially when considering the effects of climate change. Rising temperatures and regular precipitation patterns are often associated with increased soil carbon turnover and soil organic matter mineralization. OF methods have a strong emphasis on providing organic matter through crop residues, cover crops and regulated grazing to maintain or raise SOC levels. Organic inputs integrated with conservation tillage also minimize soil cultivation to minimize SOC mineralization (Chenu et al. 2019: Pradhan et al. 2023). Organic cultivation uses on-farm preparations, FYM (farmyard manure), vermicompost, and other decomposed products for nutrient supplementation, which are environmentally friendly techniques to maintain nutrient availability while boosting SOC sequestration. These practices lower carbon footprints in the lifecycle assessment of agricultural production by excluding the use of chemical fertilizer inputs, which are energy and carbon-intensive, as well as relying on non-renewable energy in their production and transport.

4.3.4 Precision Agriculture

Water and nutrient management are of paramount importance to PA in crop production. Reducing N_2O emissions from soil N management and improving agricultural efficiency to prevent future land conversion should be the main goals of agricultural GHG emission mitigation (Eory and Moran 2012; Kumar et al. 2022a). In addition to lowering GHGs emissions, PA might raise or sustain agricultural productivity and profitability. Use of sensor-based technologies and global positioning system (GPS) maximizes input applicability. Reduced waste and increased efficiency by precise application of water, insecticides, herbicides and fertilizers using variable rate technology (VRT), following the needs of the field, lower the CF of agricultural production.

4.3.5 Crop Rotation and Diversification

Agricultural systems can reduce the carbon footprints of their produce by improving soil health and reducing the need for energy-intensive inputs like synthetic pesticides and fertilizers through crop diversification and rotation (Kebede 2021). Growing a variety of crops in a sequence on the same plot of land, or crop rotation, breaks pest and disease cycles, reduces weed infestation and increases soil fertility, particularly when legumes are included. Legumes reduce synthetic nitrogen fertilizers and fix atmospheric N_2, which greatly reduces and mitigates GHGs emissions (Lal 2004). The benefits of diversification through intercropping or intensification of sustainable cropping systems include increased resilience to climate change and reduced reliance on chemical interventions. Continuous ground cover reduces soil erosion and increases soil organic matter, which ultimately improves SOC sequestration and reduces soil CO_2 flux (Smith et al. 2008). Crop diversification can also increase energy efficiency by spreading the labour and input requirements over different crop types and growing seasons. Because they lower overall GHGs emissions associated with production, input use and energy consumption, crop rotation and diversification are essential for climate-smart and sustainable agriculture (Gan et al. 2011).

4.3.6 Balanced Fertilizer

Fertilizers, particularly N, constitute a significant proportion of the CF of agricultural production. It is essential to lower the CF of agricultural systems through balanced fertilization, which delivers nutrients in the proper amounts to meet crop needs. Using fertilizers unevenly or excessively, especially N, leads to increased emissions of N_2O (IPCC 2007). Optimizing nutrient uptake through synchronizing crop nutrient needs with soil supply and reducing excess nutrient runoff or leaching through fertilizer usage based on crop requirements and soil tests reduce N_2O emissions. Crop growth and yield efficiency are improved by balanced fertilization and the retention of plant debris in the soil, which also increase soil organic matter. This reduces the GHGs emissions associated with the manufacture, distribution and use of synthetic fertilizers. Additionally, balanced fertilization can improve energy efficiency by reducing the need for repeated fertilizer applications as well as fuel consumption and emissions from agricultural machinery (Smith et al. 2008). Therefore, in addition to guaranteeing optimum crop output, the use of balanced fertilization techniques encourages more sustainable and ecologically friendly agriculture.

4.4 Technologies for Reduction in Carbon Footprint, Enhancing Input Use Efficiency and Carbon Credit in Agriculture

4.4.1 Adopting Input-Efficient Crops and Varieties

As the global population continues to rise and climate change poses increasing threats to food security, the development and adoption of sustainable agricultural practices have become essential. Among these practices, the development of input-efficient crops and varieties is of paramount importance for reducing the CF, enhancing input use efficiency and generating CCs. Input-efficient crops are developed with the ability to thrive with lower quantities of inputs (water, fertilizers and pesticides) while maintaining or even improving yield and to withstand abiotic stresses like drought (Kumar et al. 2020a). These crops can be genetically modified or conventionally bred to exhibit traits governing drought tolerance, nitrogen-use-efficiency (NUE), pest resistance and improved photosynthetic capacity. Breeding objectives have been set to develop climate resilient traits in different crops, namely stay green in pearl millet and sorghum; and nutrient enrichment in pearl millet, groundnut and bean. Stay-green genotypes have narrow root angles and deep rooting that allows more water extraction, and increase drought tolerance, improve water and NUE by extending the photosynthetic period (Pixley et al. 2023). Root architecture also influences the nutrient uptake in low-nutrient soils. Increasing the nutrient uptake allows effective use of resource inputs and carbon capture. Drought-tolerant crops (Table 4.2) exhibit a range of adaptations, such as deeper root systems, improved stomatal control and enhanced osmotic adjustment, making them capable of using water more efficiently and maintaining yields under water-stressed conditions. By using less of resource inputs, these crops reduce the cost of cultivation and the environmental impact of agricultural production.

Table 4.2 Drought-tolerant varieties of maize and wheat

Traits	Varieties	Impact	References
Drought-tolerant maize	DT maize	Up to a 35% yield increase under drought conditions compared to traditional varieties.	Lunduka et al. (2019)
Drought-tolerant wheat	Australia-'Drysdale' India-'PBW 550'	These varieties have outperformed conventional wheat varieties in water-limited environments by 20–30%.	Kumar et al. (2020)
Drought-tolerant rice	Pusa rice DST1	Lower stomatal density, tolerance to drought and stress	Priyadarshini (2025)
	DRR Dhan 100	A 19% higher yield, 15–20 days earlier maturity. Perform better under drought and lower fertilizer conditions	Priyadarshini (2025)

Nitrogen-use-efficient crops are designed to maximize nitrogen uptake and utilization, thereby enhancing the NUE by reducing losses. The development of nitrogen-efficient rice varieties has been such an example of reducing nitrogen fertilizer dependence and lowering associated emissions. A gene NRT1.1B, encoding nitrate transporters, was isolated from an indica (IR24) × japonica (Nipponbare). The introgression of NRT1.1 B-indica into japonica rice has led to an increase in grain yield and NUE (~by 30%) under low-nitrogen conditions. This enhancement is due to improved nitrate uptake and transport, along with elevated expression of nitrate reductase genes, which facilitate efficient nitrogen assimilation (Hu et al. 2015). Biological nitrification inhibition (BNI) is the phenomenon associated with plants capable of lowering nitrification rates (conversion of NH_4^+ to NO_3^-) in soil by releasing root exudates that allow retention of soil N in NH_4^+-N form for an extended period. BNI isogenic lines of wheat, that is, ROELFS-BNI and MUNAL-BNI, reduce ammonia-oxidizing bacteria abundance by 45% and lower nitrification rate (~by 60%) and N-fertilizer use, and enhance NUE (Bozal-Leorri et al. 2022). BNI wheat cultivation could reduce the use of N fertilizers (~by 15–20%) without sacrificing yield or grain quality. It has been estimated that BNI has the potential to reduce GHGs emissions up to 0.016–0.19 $GtCO_2eq\ yr^{-1}$, which corresponds to 0.4–6% of total global N_2O emissions annually, besides mitigating nitrate pollution in groundwater. In rice, root exudates containing 1,9-decanediol inhibits nitrification by 37.0–58.1% and lowers N_2O emission by 90% (Wang et al. 2021). Such lines can be upscaled for higher N-use efficiency and lower emissions (Punia et al. 2020).

The use of CRISPR-Cas9 technology has been widely recognized to enhance resource use efficiency. Using CRISPR-Cas9 technology, the genes responsible for nitrogen uptake in wheat have been edited, resulting in the development of varieties that can assimilate more nitrogen under low-N application conditions. Such wheat genotypes can reduce N fertilizer application by 40% without compromising yield (Zhang et al. 2021). The varieties developed through genome editing by ICAR institutes in India, by editing the cytokinin oxidase gene (OsCKX2), that is, IARI, New Delhi and IIRR, Hyderabad, are Pusa Rice DST 1 (drought and salinity tolerant) and DRR Dhan 100 (drought tolerant and early maturity). These varieties need less water, which reduces the number of irrigations and withstand drought, low fertilizer and salinity situations, thereby increasing input-use efficiency and lowering GHGs emissions (Kumar et al. 2025).

Pesticides, while essential for controlling pests and diseases, contribute to environmental pollution and carbon emissions during production and application. By adopting pest- and disease-resistant crop varieties, farmers can significantly reduce pesticide use, thereby lowering GHGs emissions and enhancing input efficiency. In India, Bt cotton adoption has reduced pesticide use by 50–90%, leading to significant reductions in the carbon footprint associated with pesticide production and application (Peshin et al. 2021). Climate change has increased the frequency and severity of abiotic stresses such as heat waves and soil salinity. Heat-tolerant wheat varieties like HD3385 exhibit high temperature stress tolerance and escape terminal heat tolerance owing to a mild vernalization requirement that shortens their maturity duration (Damodaran 2023, The Indian Express). Therefore, developing heat- and salt-tolerant crop varieties is critical for maintaining productivity in affected regions while reducing resource inputs.

4.4.2 Conservation Agriculture

Conservation agriculture has emerged as a key strategy for sustainable agricultural development, aimed at reducing the CF of farming practices while enhancing input use efficiency and fostering soil health (Jat et al. 2020). The approach is founded on three interrelated principles: minimal soil disturbance, permanent soil cover and crop diversification. By adopting these principles, CA has demonstrated significant potential to reduce GHG emissions, sequester carbon in soils and improve resource use efficiency, especially in terms of water and fertilizers (Das et al. 2020). Minimum mechanical soil disturbance practice greatly reduces the mechanical energy required for intensive ploughing, thereby lowering fossil-fuel consumption and associated CO_2 emissions (Table 4.3). No-till/zero-till practices allow for greater accumulation of SOC, as plant residues are left on the surface to decompose, and SOC is protected from oxidation, thus sequestering atmospheric CO_2 (Dey et al. 2016; Dutta et al. 2023). Maintaining surface cover by retaining crop residues or cover crops protects the soil from erosion, regulates hydrothermal regimes and promotes soil biological activity (earthworms and microorganisms). These factors contributed to SOC sequestration and protection by reducing GHGs emission flux. Crop diversification, through rotations and intercropping, improves soil health and fertility by enhancing nutrient cycling and reducing pest and disease pressures by disrupting their lifecycle. A global meta-analysis of 97 research papers related to CA revealed that using these three principles, an increase in SOC and total nitrogen (TN) stock at 0–15 and 0–30 cm has been found over conventional practices. The highest increase in SOC (21.39%) and TN (54.34%) was observed at 0–15 cm, the highest sequestration rate was at 0–30 cm depth, whereas there was no significant difference at 0–60 cm depth for these parameters (Bohoussou et al. 2022).

Table 4.3 Impact of no-till wheat on fuel use and carbon sequestration (Jat et al. 2019)

Practice	Fuel use (lit. ha^{-1})	CO_2 emission reduction (%)	Carbon sequestration (t ha^{-1} yr^{-1})
Conventional tillage	60	0	0.1
No-till wheat	20	0.4	0.4

4.4.3 *Efficient Farm Machinery*

Farm machinery is an important component of the cropping system. A variety of implements are used in farming, ranging from tractors used for ploughing, harrowing, cultivators, sprayers, to harvesters that consume significant amounts of fossil fuel. Inefficient use of this machinery can lead to higher energy consumption, increased CF and excessive operational costs. Efficient management of farm machinery involves optimizing its use through regular maintenance, minimizing idle time and employing technologies that reduce fuel consumption. By ensuring that machinery operates at optimal efficiency, farmers can significantly reduce their CF while also lowering operational costs (Bacenetti et al. 2018; Galli et al. 2022). For instance, adopting newer, fuel-efficient tractors and engines, precision farming tools and proper machinery calibration can greatly enhance input use efficiency.

4.4.3.1 Fuel-Efficient Tractors

Modern tractors are equipped with fuel-efficient engines that enhance engine performance and reduce fuel consumption and emissions. These tractors utilize advanced engine technologies such as turbocharging, common-rail fuel injection and electronic control systems, which optimize fuel delivery and improve combustion efficiency. Tractors equipped with engines that meet Tier 4 emission standards in Europe and North America contain setup for Exhaust Gas Recirculation and Selective Catalytic Reduction and diesel particulate filters that significantly reduce nitrogen oxide (NOx) and particulate matter (PM) emissions (Bacenetti et al. 2018; Galli et al. 2022). These technologies reduce fuel consumption by 10–20% compared to older models.

4.4.3.2 Hybrid and Electric Tractors

The development of hybrid tractors (HTs) and fully electric tractors (ETs) has emerged as viable alternatives to traditional diesel-powered machinery. The HTs are operated on both diesel and electric power, while ETs operate on electric motors powered by batteries. The ETs produce zero emissions during operation and significantly reduce CF in agriculture. The electric energy used in these engines can be generated from solar or wind resources, providing an emission-free alternative to traditional fossil fuel-based diesel-powered equipment. Integrating photovoltaic solar energy (PV) technology with electric farm tractors and agricultural robots can offer more sustainable energy use options in farm work with a significant reduction in CF (Gorjian et al. 2021). Using electric tractors reduces CO_2 emissions by nearly 100% during operation, contributing significantly to carbon credit generation (Table 4.4).

Table 4.4 Comparison of diesel and electric tractors on emission and operating costs (Liu et al. 2024)

Tractor type	Fuel cost (US$ ha^{-1})	CO_2 emissions (t ha^{-1})	Maintenance cost (US$ ha^{-1})
Diesel tractors	50	0.2	15
Electric tractors	30	0.0	5

4.4.3.3 Precision Agriculture and Smart Machinery Utilization

Precision agriculture, combined with smart machinery, is a transformative farming approach that optimizes input use and enhances operational efficiency. PA employs GPS-guided tractors, remote sensing, drones, and VRT to optimize the application of inputs such as seeds, fertilizers and pesticides, based on spatial and temporal variability. This reduces wastage and ensures that resources are used only when and where needed, leading to lower CF and higher RUE. GPS-guided tractors use satellite navigation systems to control tractor movements with extreme precision (up to sub-inch accuracy), reducing overlap in field operations, such as ploughing, planting and spraying. Auto-steering systems enable tractors to follow predefined paths, ensuring accurate field coverage with minimal operator intervention. The GPS-guided tractors reduce fuel consumption by 10–15% and cut fertilizer and pesticide applications by 10–20%, which helps in lowering CF (Sharma et al. 2024). VRT enables precise application of inputs based on soil variability (nutrient and moisture status) and crop requirements (normalized difference vegetation index [NDVI] values, pest infestation). VRT-equipped machines work on the right time, right place, right quantity and right way approach based on sensor or map data (Sharma et al. 2024). This ensures that inputs are applied only where needed, reducing overuse and environmental impact. The use of VRT in corn fields has been shown to reduce nitrogen fertilizer use by 20–30%, leading to lower N_2O emissions (Thakur et al. 2024).

4.4.3.4 Renewable Energy-Powered Farm Machinery

The use of renewable energy sources, such as solar PV technology for electric energy, wind energy for electric energy and bioenergy (bioethanol and biofuels), with farm machinery is a key development in reducing CF from machinery used in agricultural operations (Gorjian et al. 2021). Solar-powered irrigation pumps, solar dryers, wind-powered grain dryers and biofuel-driven engines are examples of renewable energy applications that could serve as alternative, cleaner production tools in the lifecycle of agricultural produce. Solar-powered irrigation reduces dependence on diesel, cutting CO_2 emissions by 80–90% (Aliyu et al. 2018; Shirsath et al. 2020) (Table 4.5). Furthermore, farmers also have an option to sell excess electricity back to the grid, creating additional income. Biodiesel made from soybean or corn oil is increasingly used as an alternative fuel for machinery. Using biodiesels

Table 4.5 Comparison of diesel and solar-powered irrigation systems

Irrigation system	Fuel cost (US\$ ha^{-1})	CO_2 emissions (t ha^{-1})	Maintenance cost (US\$ ha^{-1})
Diesel-powered	40	0.25	10
Solar-powered	10	0.00	05

Xie et al. (2021)

creates a circular economy that benefits farmers financially by reducing input costs on fossil fuels (Rajendran 2021) and lowering CO_2 emissions by 50–70% compared to conventional diesel (Tomić et al. 2013).

4.4.4 Sensor-Driven Input Management

India's agricultural sector still engages approximately 42% of the workforce and remains a critical element of the nation's socioeconomic development. The sustainability of agricultural practices is deeply linked to their CF, which involves the balance between carbon emissions and sequestration (Shabir et al. 2023). Research by Crippa et al. (2021) reported that 27% of global GHGs emissions come from developed countries, while 73% come from developing nations, with over 71% of these emissions directly linked to agricultural land use, whereas Renner et al. (2020) argue that agriculture contributes nearly half of the world's total carbon emissions. Agricultural land use, farm energy consumption, product transportation and food waste management play an immense role in global GHGs emissions (Yu et al. 2024). Amidst high CF, productivity per hectare in developing countries remains below the global average, highlighting the urgent need to improve agricultural output while optimizing resource use. The eco-friendly technologies and machinery, encouraging low-resource use, could be a sustainable and cost-effective solution. The overuse of nitrogenous fertilizers has caused significant environmental damage, including water pollution, biodiversity loss and global warming effects, along with escalating the cost of cultivation (Paul et al. 2022). By continuously monitoring soil nutrient levels of N, P and K through sensor-based technologies, as well as tracking climatic factors like humidity, temperature and sunlight, farmers can maintain soil fertility, increase productivity and reduce emissions (Gavade and Bhoi 2017). These practices are essential for minimizing the CF of agriculture and ensuring long-term food security (Kumar et al. 2022).

A sensor is an electronic device designed to detect specific environmental variables and convert them into electrical signals and useful outputs. In simpler terms, a sensor responds to inputs like light, motion, pressure, temperature or humidity, transforming them into signals that can be analysed by humans. Sensors are equipped to handle various functions, such as data transmission, processing, storage, display, recording and control (Teja 2021). Sensors can collect information about climatic parameters, plant growth and health, converting this data into electrical

Table 4.6 Different types of sensors used in agriculture and their potential applications

Type of sensor	Potential applications	References
Location and position sensors	Used in PA Used to locate the latitude, longitude, and altitude of a field from signals of GPS satellites Used in monitoring livestock Used in vehicle guidance systems for harvesting and tilling operations	Schriber (2021)
Optical/photometric sensors	Uses properties of light through measurements of light at different wavelengths to characterize soil properties Used in remote sensing to know soil moisture content, organic matter, etc.	Law et al. (2020)
Electrochemical sensors	Used in precision agriculture to estimate soil pH and nutrient levels	Lin et al. (2008) and Ali et al. (2020)
Mechanical sensors	Used in tensiometers to detect soil water tension/potential Also measures the mechanical resistance of soil with the help of a probe	Schriber (2021) and Lakhiar et al. (2018)
Dielectric/ electromagnetic sensors	Used in soil moisture sensors Measures dielectric constant using electromagnetic field properties	Stacheder et al. (2009) and Manoj and Udupa (2015)
Frequency domain (FD) sensors	Uses dielectric conductance property to measure changes in soil water content	Chen et al. (2012) and Zhu et al. (2019)
Airflow sensors	Measures soil air permeability Used in determining the structure of soil and its compaction	Ayaz et al. (2019)
Photosynthetically active radiation (PAR) sensors	Used in quantum sensor instruments to measure the light intensity required for plant growth	Sharma et al. (2019)
Robotic sensors	Constructed with colour-based and greyscale vision systems to use in mechatronics for spraying herbicides and other pesticides without damaging crop	Song et al. (2015), Chang and Lin (2018), and Azeta et al. (2019)

signals for analysis. The implementation of advanced sensor technologies in farming practices can collect high-resolution data on soil and crop characteristics, down to the centimetre level, across an entire field. The information gathered gives inputs for smart farming practices and helps farmers in making well-informed decisions throughout the growing season, leading to better resource use, increased yields and quality (Adamides et al. 2020; Kumar et al. 2023). The use of Internet of Things (IoT)-based sensor networks in agriculture has seen rapid growth in recent years, with sensors now being used for data collection, storage and transmission (Surai et al. 2018; Miranda et al. 2019). Key sensors with notable agricultural applications are listed in Table 4.6.

The precision, accuracy and efficiency of equipment used in smart agriculture can be enhanced by the sensors of artificial intelligence (AI) and Internet of Things (IoT).

The sensors used in these tools are more versatile and are used in various sectors, namely automation, robotics, land navigation, underwater imaging, remote sensing and others (Ullo and Sinha 2021). Sensors are essential for maximizing crop quality and yield, as they monitor and interpret a range of crop growth-related factors. These are also capable of monitoring soil parameters like pH, N, P and K. Precision agriculture also has a key part in sensor-based input management with the use of real-time nitrogen management tools such as GreenSeeker, SPAD (Soil Plant Analysis Development)/Chlorophyll metre, the use of digital crop advisor and automated irrigation systems. A variety of sensors now facilitate real-time data collection, such as microfluidic sensors that measure soil nitrate levels and plant nutrient uptake (Sharma et al. 2019), and optical sensing tools that assess nutrient stress by analysing leaf-level reflectance spectra, enabling precise nutrient management (Asundi et al. 2006). The integration of smart practices with advanced digital information technologies, such as GPS/GNSS-based guided soil sampling, control systems, and in vertical farming, hydroponics, autonomous vehicles and robotics, can improve farm management to achieve higher yields, profitability while reducing CF (Saiz-Rubio and Rovira-Más 2020). IoT-based sensor networks integrated with 5G technology connect farms to comprehensive smart farming solutions that offer services like data analytics, security and application facilitation (Tang et al. 2021). The generated information empowers farmers to make well-informed decisions, improving crop management throughout the growing season (from planting to harvest) (Surai et al. 2018; Miranda et al. 2019; Adamides et al. 2020). The real-time data could help farmers in fine-tuning essential practices like reduced tillage, optimizing fertilizer application dose and time, water and pesticide use, which lowers CF.

GreenSeeker is a handheld device and cost-effective solution that is used to optimize N fertilizer usage by providing mid-season recommendations based on normalized difference vegetation index (NDVI) value of plants. It consists of an optical sensor alongside a nitrogen-rich reference strip and crop-specific algorithms that guide location-specific fertilizer recommendations. This approach allows farmers to achieve or even surpass crop yields over business as usual while minimizing the overall fertilizer application (Bruulsema et al. 2016). This device emits both red and infrared light onto plants, measuring the amount of light reflected. The data collected is then processed into an NDVI value, which is proportional to the health and vitality of the plants (Trimble 2012). NDVI readings are used to determine the mid-season N dose for the crop in a specific area, enabling higher NUE, which usually ranges from 35 to 65%. GreenSeeker can significantly reduce fertilizer costs, increase or sustain crop yields, farm profitability and reduce CF (mitigating N_2O emission) and fertilizer runoff (Matson 2012; Langholtz et al. 2021). GreenSeeker is a valuable tool for promoting sustainable farming practices that support both economic growth and environmental stewardship in the long run.

The SPAD metres work on a non-destructive sampling approach used to assess nitrogen (N) levels in plants based on chlorophyll content and offer an alternative to conventional tissue analysis. The Minolta SPAD-502, a portable model, is the most commonly used SPAD metre, created by Minolta in Osaka, Japan. It works by clamping unharvested leaves between two LEDs that emit red and infrared light

(650 and 940 nm) The sensor measures the light absorbed and transmitted by the leaf, providing a chlorophyll content reading in arbitrary units (0–99.9) that must be calibrated against actual chlorophyll or nitrogen levels (Huang et al. 2008).

Another notable chlorophyll metre is the Field Scout CM 1000, which uses point-and-shoot technology to assess relative chlorophyll content over a 1.27 cm diameter area. It uses powerful lasers and an ambient light sensor to make precise measurements (Maiti et al. 2004). Two common strategies for managing nitrogen fertilizer with SPAD metres include the Fixed Threshold Value Approach, where fertilizer is applied when SPAD readings fall below a set threshold limit, and using critical SPAD values tailored to crop needs. Critical values were developed by researchers across various rice and wheat cultivars, for example, critical SPAD values of 35 were used for rice by Huan et al. (2002) and were found to considerably reduce nitrogen fertilizer use, thereby decreasing the CF. The other associated benefits are reduced nitrogen runoff and increased NUE. Optical sensors used for guiding N applications can significantly enhance crop yields and NUE (Singh et al. 2013).

Yara N-Sensor is a passive multispectral scanner developed by Yara International ASA to assess crop chlorophyll content and biomass. It uses two diode array spectrometers: one spectrometer directed towards the crop and the other capturing skyward irradiance to measure crop reflectance over a 50 m^2 area. It can be mounted over a tractor and employs fibre optics with a 12° field of view on each side. It operates across five selectable wavelengths (450–900 nm) and adjusts fertilizer application rates in real-time based on the crop N levels without being influenced by soil properties (Portz et al. 2012; Raper and Varco 2015; Jangir et al. 2024).

4.4.5 Improved Irrigation Management System

Water application plays a major part in CF. It is important to increase water-use efficiency (WUE) and lower the CF of water application. Several improved systems, like sprinkler, surface drip irrigation (DI) and sub-surface drip irrigation (SDI), are used in agriculture to increase WUE. SDI lowered total CH_4 emissions (~by 80%) and ecosystem respiration fluxes (~by 41%) over DI (Guardia et al. 2023). In apple cultivation, SDI reduced the cumulative CO_2 and N_2O emission flux by 5.3–6.5% and 14.41–18.51% respectively and increased the cumulative CH_4 absorption flux and GHGs intensity by 14.82–26.51% and 29.41–31.58% respectively over DI (Liu et al. 2025). In comparison to traditional irrigation methods, sensor-based systems are effective in reducing the overapplication and increasing WUE. Wireless sensor networks (WSN) are employed in agriculture for improving the WUE of irrigation practices. These networks have real-time monitoring and control systems that manage water application more efficiently (Prathibha et al. 2017; Sushanth and Sujatha 2018). GSM (Global System for Mobile Communications)-based irrigation systems are equipped with rain guns and automatic microcontrollers and can be efficiently used for crops with high water requirements. Furthermore, the integration of mobile apps with GPRS (General Packet Radio Service) functionality in irrigation systems allows farmers to receive alerts regarding potential over- or under-irrigation (Gutierrez et al. 2014; Roselin and Jawahar 2017). WSN enables precise, large-scale monitoring, generating real-time data, overcoming the limitations of traditional wired setups. These technologies create decision-support systems that use field-specific data to optimize water and chemical inputs. Abbasi et al. (2014) developed a smart sensor system for lawns and gardens by using soil moisture sensors to assess field moisture and soil characteristics. This system used ZigBee technology for communication between sensor nodes and wireless stations, which consists of four primary components: control unit, water controller, security system (password protection) and mobile messaging (Liang 2007; Mulla 2013). The water controller regulates water levels in the field and connects with microcontrollers to automatically activate pumps. Furthermore, integrating solar energy-powered irrigation systems with temperature and moisture sensors can improve water management while minimizing environmental impacts (Lamb et al. 2017). Remote sensing can also be utilized in irrigation management, aided by satellite and aerial imagery, to assess soil moisture, crop health and soil moisture distribution. The sensors that operate in bands like near infrared (NIR) and short-wave infrared (SWIR) give real-time insights into soil moisture and ET, which are inputs for scheduling irrigation (Li et al. 2019). Thermal sensors assess crop canopy temperature, which detects crop stress from moisture deficiency (Shah et al. 2012). Therefore, advanced irrigation systems integrated with sensor technologies can enhance WUE while reducing CF and environmental impacts in agriculture.

4.4.6 Artificial Intelligence-Based Input Management

There is no denying that aberrant weather phenomena such as increasing global temperatures, extreme weather, biodiversity loss and risks to human health and well-being are the consequences of climate change (Bellard et al. 2012). It is a need of the modern agricultural system to find a way to overcome the negative effects of climate change to become more productive and sustainable. Governments, organizations and individuals are increasingly realizing that prompt and effective action is necessary to address these concerns. In recent years, artificial intelligence (AI) has emerged as a powerful tool that could significantly contribute to mitigating climate change and achieving the ambitious goal of net-zero carbon emissions (Kumari and Pandey 2023).

The introduction of AI in agriculture would be a prominent option to combat the loss of natural resources due to climate change. By examining soil data, weather patterns, crop characteristics and past yield data, AI systems can optimize irrigation, fertilizer use, and pest control. Furthermore, AI can assist farmers to use precision farming by providing precise recommendations, which lowers pollution and waste of resources while improving crop yields (Bhagat et al. 2022). Based on past data, machine learning (ML) and AI can lower CF in agriculture by giving precise estimates of their carbon emissions and increasing the effectiveness of farming methods.

4.4.7 *Improved Livestock Management Operations*

According to the Food and Agriculture Organization of the United Nations (FAO 2021), animal husbandry is responsible for 9% of human-caused carbon emissions, 37% of CH_4 emissions and 65% of N_2O emissions. To mitigate these environmental impacts, several management strategies can be implemented as follows.

4.4.7.1 Boost Feeding Techniques

Enhancing animal productivity through food modifications is one of the most promising approaches to reducing emissions in livestock management. This might involve adding various dietary supplements (certain oils) to the feed or making low-quality fodder more digestible, which speeds up the enteric fermentation process because ruminant animals have trouble breaking it down, for instance, changing a cow's diet from grass silage to maize silage (Mills et al. 2001; Sheoran et al. 2022). According to Dijkstra et al. (2011), intestinal CH_4 levels are reduced by around 8% when 50% of the grass silage is substituted with maize silage in a diet that typically contains 30% concentrates and 70% grass silage. This approach, which focuses on animals, has the potential to lower GHGs emissions.

4.4.7.2 Use a Strategy for Managing Anaerobic Manure

About 7% of worldwide agriculture emissions (361 $MtCO_2e$) and 10% of global livestock GHGs emissions (5% N_2O and 4% CH_4) are attributable to manure management (FAO 2021). Through the process of anaerobic manure management, microorganisms break down the manure in the absence of oxygen, producing digestate and biogas (mostly CH_4 and CO_2). Biogas collectors can be used to capture CH_4 in situations when production is done on a large, intense scale and manure is kept in anaerobic conditions. The CH_4 that has been caught can either be flared or utilized to power electric generators, lights or heat. Thus, application of anaerobic digestion probably could avoid most of the estimated GHGs emissions (CH_4 and N_2O) from manure management on the studied farms if the manure and digestate are stored in covered tanks and AD takes place in a closed tank (Kaparaju and Rintala 2011).

4.4.7.3 Separate Solid and Liquid Manure Content

Manure processing technologies can use gravity or mechanical equipment like centrifuges or filter presses to separate solid and liquid manure. This method aerates the manure storage conditions, limiting the amount of CH_4 released. Anaerobic digestion and solid-liquid separation may both reduce GHGs emissions, and combining the two can cut GHGs emissions by 41% (Zhang et al. 2023).

4.4.7.4 Composting of Manures and Storage Time

Composting is the aerobic breakdown of manure or other organic waste by microorganisms in a controlled setting. Air, moisture and organic material with a high nitrogen and carbon content are required for composting. The process might take many weeks or months, depending on the extent of turning or aeration management. Composting manure emits less CH_4 than exposed anaerobic lagoons or liquid/slurry manure treatment techniques. Composting all cow dung stored as slurry and stockpile results in a decrease of 0.70 Tg CO_2eq yr^{-1} (Pattey et al. 2005). The amount of time manure is held in anaerobic conditions can be shortened by processing it, transferring it out of a storage facility using methods such as slatted floor pit storage, or spreading it on land regularly during excellent weather and soil conditions. Daily spreading of manure produces the biggest decreases in CH_4 generation, although lowering storage duration from months to weeks can also have a major impact. Daily spread management involves removing manure from a barn and applying it to crops or pastures daily.

4.4.7.5 Enhance Pasture Management

Grazing pressure may be changed to improve pasture quality and raise soil carbon. The following changes can be made to improve pasture management: introducing species (like legumes), inoculating plants, utilizing new technologies (like solar-powered electrical fences) to balance the temporal and spatial presence of livestock, enhancing fertilization and nutrient management, increasing animal mobility in pastoral and agropastoral systems and integrating trees and pastures. GHGs emissions per kilogram of beef decreased from 2 to 57% as a result of intensifying beef production through better pasture and herd management (Mazzetto et al. 2015).

4.4.7.6 Implement Rotational Grazing

With rotational grazing, a big pasture is divided into smaller fields, and over time, cattle are moved between the smaller areas. This approach keeps carbon in the ground rather than releasing it into the atmosphere, enables cattle to receive the nutrients they require and sustains the long-term health of the grass and soil. Rotational grazing also produces better-quality feed that is simpler for animals to digest, potentially lowering GHG emissions. The continuous grazing treatment consistently produced greater CO_2 emissions than the rotational grazing treatment (Baronti et al. 2022).

4.4.7.7 Boost the Health and Supervision of Animals

Individual animals' lifetime performance can be increased, and GHG emission intensities can be decreased by increasing reproductive efficiency and prolonging the reproductive life of animals. Higher production and efficiency, fewer losses and fewer inefficient animals that release GHGS are the outcomes of lowering the prevalence and effects of illnesses, parasites and insect loads. Productivity increases and emission intensity reductions are the results of genetic advancements. Improved health has the potential to decrease culling because of illness and injury, which would increase the herd's average productive lifespan. It has been proposed that raising the average lifespan of dairy cattle in the herd can boost animal production and lower GHG emissions per kilogram of output (Llonch et al. 2017).

4.4.7.8 Prevent Rangelands and Grasslands from Being Turned Into Agriculture or Other Types of Land Use

Soils and grasses retain a lot of organic carbon, which would mostly be released as CO_2 emissions if they were exposed to the environment (e.g. by tillage). The sustainability of livestock production systems that significantly rely on rangeland ecosystems is under growing strain due to the worldwide issue of accelerated climate change. There is little chance that rangeland management techniques will absorb GHGs emissions. Steer clear of choosing grasslands for afforestation without carefully considering the ecological services, biodiversity, carbon sequestration potential, and the cultural, economic, and other benefits that grasslands provide.

4.4.8 Improved and Renewable-Energy-Based Value-Added Chain and Food Processing

Energy is essential at every stage of agricultural production, from manufacturing fertilizers and operating machinery to planting, harvesting, processing and adding value to food products. Fluctuating energy prices and unstable markets significantly impact the cost of agricultural inputs. Moreover, heavy reliance on fossil fuel-based energy results in higher CF of agricultural produce. However, by adopting energy-efficient practices and harnessing on-farm renewable resources such as solar, wind, biomass and geothermal energy, producers can reduce operational expenses and improve resilience (Majeed et al. 2023). Striking a balance between energy use and economic feasibility is key to advancing sustainable agriculture. Yet, in many developing countries, energy constraints hinder the modernization of food systems, limit the production of higher-value goods and contribute to postharvest losses. These regions must decide whether to continue depending on fossil fuels, worsening climate impacts or move towards cleaner, renewable alternatives.

Table 4.7 Comparison of renewable energy solutions in agriculture

Energy source	Applications	Advantages	Challenges
Solar energy	For irrigation, food drying (Fig. 4.3), cold storage	Renewable energy sources, abundant, cost-effective, and require minimal maintenance	Operational challenges, as it cannot generate power at night, produces lower output during cloudy weather and becomes ineffective under sandstorm conditions
Wind energy	For electrical water pumps, greenhouse operation, grinder mills (Majeed et al. 2023)	Environment friendly, as no GHG emissions, renewable	Intermittent availability (Ren et al. 2017)
Hydropower	For irrigation, milling, and lighting (Majeed et al. 2023)	Reliable power supply	Limited by geography
Biomass	For heating, electricity, fertilizer production, cooking and drying (Majeed et al. 2023)	Renewable and abundant resource	Food vs fuel conflict, logistics and storage issues (Saleem 2022)

Agri-food value chains include a wide array of interconnected activities, spanning input supply, production, processing, distribution, consumption and waste management. These chains not only facilitate the movement of food but also create income opportunities for diverse stakeholders (Michel et al. 2024). Value addition can take place at multiple points along the chain, depending on product characteristics and market needs. Industrial sectors need to shift away from fossil fuel dependency and embrace low-carbon energy alternatives by integrating solar, wind and biomass energy, along with adopting circular economy practices in their operations (Table 4.7).

4.4.9 Food Processing Technologies Powered by Renewable Energy

4.4.9.1 Solar Drying Units

Drying of food is one of the most ancient and the most common preservation techniques, necessary for shelf-life extension, minimizing post-harvest losses and retaining nutritional quality (Bradford et al. 2020). Solar drying provides a clean and energy-saving alternative to traditional drying technologies, including electrical and fuel-fired dryers, through the use of renewable solar energy.

A solar dryer harnesses sunlight, turning it into heat energy, which is then used to heat food products and evaporate moisture within them. The main principle consists

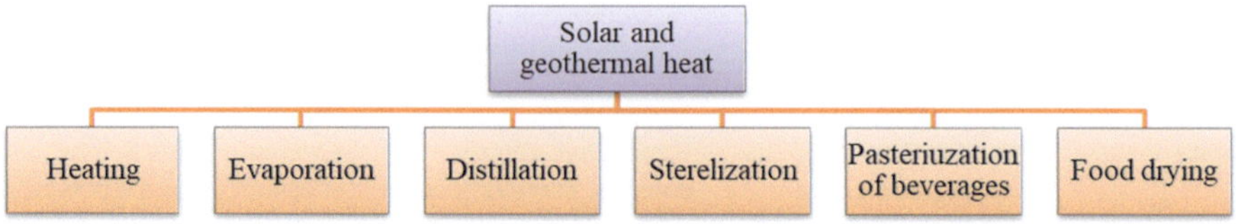

Fig. 4.3 Use of solar and geothermal heat in food processing

of drying to a point where microbial growth and enzymatic processes that lead to spoilage are restrained (Fernandes and Tavares 2024). Several factors control the drying process, such as air temperature, humidity, air velocity and the period of exposure. Solar dryers can operate in direct, indirect or hybrid modes based on the heat capture and transfer mechanisms (Nnamchi et al. 2025). Rakshamuthu et al. (2021) created a greenhouse solar dryer that includes zinc nitrate hexahydrate as phase change material (PCM) with the intent of increasing food drying efficiency. With the use of this PCM, the system was able to maintain higher temperatures and more efficiently remove moisture, enabling gooseberries to be dried at a much faster rate. This achievement brought about an increase in the reduction of the total drying period. The compact solar dryer, enhanced with PCM, not only provided better results in terms of drying but also offered a cleaner and more controlled atmosphere for food storage and preservation. Rehman et al. (2023) investigated a new design of solar dryer with thermal storage consisting of paraffin wax as PCM in another study. Adding phase change material (PCM) to the dryer made it noticeably more efficient. During the hottest part of the day, the wax absorbed excess heat and melted, storing energy in the process. Later, when sunlight was weak or absent, the stored heat was slowly released, helping the drying process continue without interruption. The dryer was efficient with a variety of agricultural products, producing 69.6% moisture loss for onions, 65% for apricots and 75% for peas of their initial weights. Solar drying is also used to conserve fruits during periods with low market prices or to enable access to the export market. These systems can improve the post-harvest process (Udomkun et al. 2020). Solar dryers offer several economic advantages, such as lower costs for fossil fuel and combustion equipment. An active dryer powered by PV can be an attractive option, especially in rural areas where access to commercial energy sources is limited (Udomkun et al. 2020). The solar dryer can decrease conventional energy consumption by 27–80% (Arata et al. 1993). Moreover, by providing a controlled drying environment, solar dryers help retain the nutritional content, natural colour, aroma and texture of food more effectively than traditional open-air drying methods.

4.4.9.2 Biomass-Powered Boilers

Biomass-powered boilers offer a sustainable energy alternative by using organic materials such as farm waste, wood chips and pellets for fuel. This approach helps cut down reliance on fossil fuels, reduces GHGs emissions and supports the principles of a circular economy (Ali et al. 2024). Widely used across sectors like agriculture, food processing and district heating, they generate steam or hot water essential for various operations (Majeed et al. 2023). Agricultural and food industry by-products can be repurposed as energy sources for activities like processing, storage and cooking (Xia et al. 2016). Biomass, such as wood waste, paper, paddy husk, straw, and agricultural or forestry residues, is one of the most commonly used renewable low-grade fuels. Biomass is converted into biofuel typically by biological or thermochemical processes. Though the thermal efficiency of power generation using biomass combustion (20–40%) is inferior to that of gas turbines, biomass has environmental benefits (Nami et al. 2021). Importantly, CO_2 released during the burning of biomass is less than what was absorbed during biomass development, generating a more level CF (Nami et al. 2021). Awan et al. (2019) pointed out that rice husk has a combustion ability equal to other expensive and ecologically detrimental energy sources, and so biomass is an alternative that can be used for power generation. Rice husks are expected to produce more than 100 MW of electricity per year in Pakistan by 2030 (Awan et al. 2019). Its burning also produces comparatively low levels of carbon monoxide (CO), NO_x and sulphur dioxide (SO_2). While rice husk is successful for cogeneration in fixed-bed boilers, the research suggests that fluidized bed boilers should be used exclusively for biomass fuels. conducted studies on combining solar PV and biomass systems. A 1.3 Mw fixed-axis solar PV system for a poultry processing plant can save 1876 MWh of electricity per year, cutting down on CO_2 emissions by 966 tonnes and diverting 18.4% of the plant's electricity demand away from the national grid. Payback time is 3.4 years. Moreover, a biomass boiler utilizing 1462 t of wood per year saves 15.5% in fuel costs and emissions by 2784 t of CO_2, with a payback period of 1.9 years. A mixture of 20% wood and coal further saves emissions by 557 t of CO_2 and fuel costs by 3.1%, with a payback period of 0.93 years. Switching completely from coal to wood fuel in boilers can decrease emissions by 91% and cost savings of US$ 25,724. Using both solar PV and biomass systems can lower CO_2 emissions by 3751 t yr^{-1}, proving the possibility of a sustainable energy solution.

4.4.9.3 Solar-Powered Cold Storage

Solar-powered cold storage facilities use solar power to store perishable commodities (Munir et al. 2021), providing a creative solution to tackle the issues of food preservation in areas where there is no access to reliable electricity. Solar-powered cold storage facilities are important for minimizing post-harvest losses, improving food security and promoting sustainable agriculture by offering off-grid or grid-independent refrigeration (Jarman et al. 2023). Takeshima et al. (2023) reported that the use of cold storage facilities significantly improved the sales volumes and revenues of horticultural market agents in northeast Nigeria. Preliminary estimates indicate that the rise in net revenues for these agents could be high enough to repay the investments and operating costs involved in cold storage within a reasonable timeframe. Moreover, the application of cold storage reduced the percentage of food loss by increasing the shelf life of products, resulting in increased prices received by farmers and market agents. This improvement in product quality was linked to the expansion of value-adding activities by market agents and a greater reliance on advance payments. These increases in quantity sold by market agents are not simply due to reduced loss but also supply responses to higher prices that market agents were able to command when using cold storage and were then able to pass through to farmers selling products to them (Takeshima et al. 2023). These benefits were realized without negative spillovers to non-user market agents in the same intervention markets. Higher prices were also likely to have provided economic incentives to improve the quality of horticulture products farmers sell to market agents and to engage in other value-adding practices by market agents (Takeshima et al. 2023). These improvements characterize essential elements of agrifood system modernization. Solar-powered cold storage can be one of the potentially viable technological triggers for such a process, particularly in developing regions like northeast Nigeria (Yamauchi et al. 2023). In India, nearly 40% of total horticultural produce is lost annually, with fruits and vegetables accounting for around 70% of this wastage due to their highly perishable nature (Mohan et al. 2023; Biswas and Tortajada 2014; Chauhan 2013). Much of this wastage takes place post-harvest, primarily due to a lack of proper infrastructure for storage and transportation. Also, since most of these products move over long distances through different climatic zones, they are exposed to varying environmental conditions that can deteriorate their quality, as evidenced in tomato by Mohan et al. (2023). To curb post-harvest losses, there is a need to put up on-farm cold storage facilities. The conventional energy to drive cold storage comes from grid systems or diesel generators, both of which are not energy efficient. It is therefore important to embrace cold storage technologies that are energy efficient and have low CF, especially in rural agricultural areas showed that the use of a phase change material (PCM)-based solar cold storage system led to a notable reduction in energy usage by 17.9% compared to the traditional cold storage systems. The experimental results suggest that the PCM-based solar cold storage system is a reliable alternative to the conventional systems for rural agricultural locations with improved energy efficiency and zero carbon emissions.

4.4.9.4 Solar and Wind Energy in Pumping and Milling

Solar and wind energy currently meet only 9% of the global energy demand (Majeed et al. 2023). Conjoining renewable resources of energy with water pumping systems is the way forward to reduce dependence on conventional energy and mitigate environmental impacts, particularly in irrigation. Considering the richness of solar and wind resources in India, creating a hybrid solar PV–wind power system for water pumping is a viable option (Gopal et al. 2013). Similarly, solar energy is also efficiently applied in milling created a solar-powered, small-scale maize mill specifically for remote areas with poor access to electricity and low, household-scale demand for flour. The system consists of a PV panel, mounting frame, energy storage unit, inverter, charge controller and a lightweight domestic mill able to process 8–10 kg of maize per hour. Field testing at Udaipur, Rajasthan, was encouraging: the system earned a net present value of ₹1,63,486, a benefit-cost ratio of 1.92, and recouped its investment in about four years. It also saved ₹19,430.4 yr^{-1} in comparison to conventional electricity-based milling.

4.4.9.5 Integrated Food-Energy Systems and Energy Sovereignty

Integrated food-energy systems promote the co-existence of food production and renewable energy generation. An example is agriphotovoltaics (APV) systems, which allow farmers to grow crops under PV panels, reducing water demand while generating electricity. APV holds significant potential for improving land-use efficiency, promoting renewable energy and fostering climate resilience. Reher et al. (2024) experimented Belgium's temperate maritime climate with Agriphotovoltaic systems and reported that beets grown between vertical and tracked APV systems showed 11–19% yield reduction and 11% land loss, generating 376–560 MWh ha^{-1} of electricity over two seasons. Land Equivalent RatioS (LERS) for the vertical system were 1.00 and 1.18, and for the tracked system, 1.17 and 1.22 in 2021 and 2022 respectively. Wheat grown under a horizontal APV system in 2022 and 2023 saw yield reductions of 33% and 46%, with an 8% land loss, though grain quality and protein content were unaffected. The horizontal system achieved LERs of 1.09 in 2022 and 0.97 in 2023. These results highlight that, despite yield challenges from shading and construction, open-field AV systems can deliver positive LERs under supporting renewable energy integration with agriculture when carefully designed. Jo et al. (2022) found that the APV system had no significant effect on rye and corn forage yields or rye forage quality, indicating its suitability for livestock feed production. In potato crops, most growth and yield parameters were similar between APV and control (full sun exposure) plots, except for increased plant height under APV (Lee et al. 2022). The future of sustainable food production lies in renewable value chains that not only reduce environmental harm but also regenerate social and economic systems. By adopting renewable energy technologies, the food industry can balance food security, environmental health and economic growth.

4.5 R&D Towards Enhancing Input Use Efficiency and Reducing Carbon Footprints

Innovations and research in agriculture and the environment are widely acknowledged as effective strategies for improving the efficiency of resource use while minimizing CF and addressing environmental challenges. Facing the dual challenges of domestic environmental degradation and international climate negotiations, the Indian government has made a series of ambitious commitments to reducing carbon emissions. Currently, several well-known measures are in place to reduce carbon emissions. One of the most effective strategies is to enhance energy efficiency, which has been proven to significantly lower CF. Farm mechanization is essential for large-scale agricultural production systems, playing a crucial role in ensuring timely operations, improving the quality of work and boosting overall productivity.

However, agricultural productivity, such as grain and biomass yield per unit area, is increasingly subject to erratic fluctuations in weather phenomena and unpredictable negative impacts of global climate change (Mall et al. 2017). The current challenges in agricultural production and future growth will depend on the adoption of mechanization. In India, the overall farm mechanization level is at 40%, with an average farm power availability of 2.24 kw ha^{-1} (Modi et al. 2020). This figure varies across the states and is notably higher in Punjab (4.4 kw ha^{-1}), while lesser in the hilly regions of Uttarakhand (1.05 kw ha^{-1}) (Modi et al. 2020). Moreover, climate-smart technology-based farm mechanization such as precision farming (GIS, GPS, VRT and Drones); smart irrigation (IoT-based drip irrigation and solar-powered pumps), conservation agriculture (no-till seeders and strip-till machines), renewable energy machines (electric tractors and biofuel harvesters), AI & robotics (weeding robots and AI harvesters) and IoT & data analytics (soil sensors and automated grain storage) offers a strategic approach to the sustainable agricultural development and resilient food value chains (Modi et al. 2020; Balyan et al. 2024; Dhanaraju et al. 2022; Petrovic and Csambalik 2025; Kumar et al. 2023a). It increases incomes and ensures food security through higher agricultural productivity and profitability (Vishnoi and Goel 2024). By adopting practices such as CA, PA or CSA, there is huge potential for enhancing SOC sequestration and RUE, and thereby long-term environmental sustainability (Prajapati et al. 2023a; Kumar et al. 2023). While farm mechanization is not a new concept, its adaptation to CSA, CA and PA represents an innovative pathway towards more sustainable and productive farming practices (Mehta et al. 2014). This approach would enhance the capability and efficiency of farm machinery to adapt and mitigate the effects of global climate change by reducing non-renewable energy use, input use and CF (Modi et al. 2020). CSA involves actions that improve or sustain food production, albeit climate variability, through the judicious integration of agricultural machinery and inputs such as nutrient, weed and water management (Kumar et al. 2023a; Prajapati et al. 2023b). Such integrations build a more climate-resilient agricultural system, while promoting long-term sustainable farming practices. This variation in mechanization levels across the country remains inadequate to fully address climate variability and improve RUE in agriculture. Climate-smart mechanization incorporates an extensive range of farm machinery in various farm operations starting from seedbed preparation, sowing/transplanting, intercultural operations, irrigation, nutrient management, plant protection, harvesting and threshing, or even in post-harvesting of crops, that is, straw management which enhances the input use efficiency and reduces CF (Modi et al. 2020). Green technology is generally heralded as a crucial tool in achieving energy conservation, emission reductions and economic growth, and such technology is recognized as a significant factor in efforts to improve carbon emissions efficiency. Renewable energy technological innovation is an important way to reduce carbon emissions and achieve sustainable development.

Promoting agricultural practices aimed at restoring SOC, augmenting nutrient mineralization through crop rotation and enhanced microbial activity, and lowering the use of nitrogenous fertilizers could serve the functions of emission reduction. The four key farming practices that can be integrated into alternative rice-wheat or cereal based production systems to reduce C emissions (Gan et al. 2014): (i) integration of grain legumes or pulses in crop rotations to fix atmospheric nitrogen (N_2), providing plant-available N and reducing the need for inorganic nitrogen fertilizers in rice–wheat production system; (ii) perform annual soil testing to assess the residual soil fertility and productivity, the potential for mineralized nitrogen management from organic matter, helping determine nutrient availability for succeeding crops; (iii) appropriate inorganic fertilizer application to meet their nutrient needs without under and over fertilization, reducing unnecessary GHGs emissions; (iv) diversified crop rotation systems to reduce the frequency of summer fallowing by adopting more frequent crop rotations, increase CO_2 sequestration, improving the systems overall carbon balance. Integrating improved agricultural practices like fertilization based on soil tests, reducing summer fallow frequencies and rotating cereals with grain legumes lower wheat CF effectively, averaging 256 kg CO_2eq. ha^{-1} yr^{-1}.

Scientific regulation of carbon flows under conservation tillage is of great significance for mitigating carbon emissions and for increasing carbon sequestration potential in soils. Chen et al. (2015) investigated conventional tillage without residue retention (CT), conventional tillage with residue retention (CTS), rotary tillage with residue retention (RTS) and no-till with residue retention (NTS) in China. Carbon budgeting of these practices by calculating carbon input of machinery, irrigation, herbicides, pesticides, fertilizer, seeds and other farm inputs was taken into account and converted into equivalent carbon emissions. NTS treatment had the lowest net carbon flux with 1246 kg Ceq. ha^{-1} yr^{-1}. Taking CT as the baseline, the relative net C flux under RTS and NTS was 918 and 1976 kg Ceq. ha^{-1} yr^{-1} respectively. This means that the widespread adoption of conservation tillage would be beneficial in reducing carbon emissions from these types of agricultural production. Visser et al. (2015) showed that CA could generate an emissions credit of 48 kg CO_2eq. per bale at the farm level, which will amount to a 27% reduction in the farm emissions/ footprint and a 15% reduction in the whole farm-to-ship CF. Riano and García-Gonzalez (2015) estimated the carbon emissions reduction of a swine manure treatment plant in Spain. Compared with conventional storage in anaerobic tanks, implementing the manure treatment plant could lead to a total annual carbon emission reduction of 62%, including a CO_2 emission reduction of 72% and a CH_4 emission reduction of 69%. In India, the total amount and rate of fertilizer use surged dramatically between 1950 and 2020. Consequently, emissions from the agricultural sector have risen since the 1950s, partly due to the inefficient use of fertilizers, such as nitrogen and imbalanced fertilizer application. Sah and Devakumar (2018) reported a 161% increase in agricultural emissions, rising from 14.81 Tg Ceq. yr^{-1} (0.12 Mg Ceq. ha^{-1} yr^{-1}) in 1960 to 38.71 Tg Ceq. (0.28 Mg

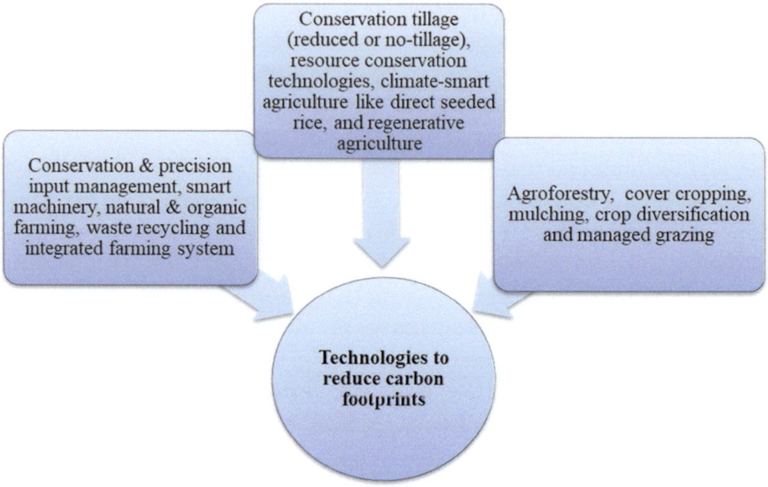

Fig. 4.4 Farming practices enhancing input use efficiency and reducing carbon footprint

Ceq. ha^{-1} yr^{-1}) in 2010 due to changes in cropping patterns and a shift from coarse-cereals towards rice cultivation. In addition to reducing emissions from crop production, GHG emissions from agroecosystems can be offset by C sequestration in soil as SOC and SIC. Therefore, with policies oriented towards enhancing SOC sequestration, lowering N_2O emissions from cultivation practices, and lowering CH_4 emissions from livestock production, overall CF from agriculture can be lowered (Fig. 4.4).

PA, a blend of sustainable and CSA practices, holds immense potential to transform Indian agriculture towards food security, productivity, carbon sequestration and aid farmers to adapt and mitigate the global climate change effects. Data-driven crop management practices can mitigate the impacts of climate change by enhancing RUE (water and nutrient), and reduce the vulnerability of agriculture to extreme weather events such as drought, flood, heat wave and hailstorm. To reduce the CF in agriculture, a holistic approach is needed, combining policy measures that promote energy-efficient technologies (Organic farming), use of renewable energy and eco-friendly farming practices (Ahuja and Mehra 2024). Slow-release fertilizer (SRF) has been found to reduce N leaching, ammonia volatilization and N_2O emissions compared to urea with the same nitrogen application rate and reduce CF. Han et al. (2023) reported that the use of slow-release fertilizer combined with plastic film mulching (PM-S) increased average grain yield by 12.01–37.89%, WUE by 9.19–23.33%, nitrogen accumulation by 27.07–66.19% and net return by 6.21–29.57%. N fertilizer was the primary contributor to total GHG emissions (≥71.97%), and PM-S reduced the CF by 12.87–44.31%, and CF per net return increased by 14.25–41.16%. Over the year, PM-S also increased SOC storage in the 0–40 cm soil depth by 2.46%. Nano-fertilizers, as a non-conventional and economically sound chitosan-based technology, can play a pivotal role in advancing future agriculture. They have the potential to sustain or improve crop yields and minimizing the overuse of conventional chemical fertilizers that ultimately reduces CF (Jakhar et al. 2022).

Cropping systems, based on cover cropping and the use of legume-based rotations, can also reduce CF. Gan and colleagues also found that durum wheat grown in rotation with an oilseed crop (*Brassica* spp.) had a CF of 0.33 kg CO_2eq. per kilogram of grain, which was 7% lower than the CF in a cereal-cereal durum system. The CFP of durum wheat was reduced by 17% when grown in rotation with a nitrogen-fixing pulse crop and by 34% in a pulse-pulse-durum system, bringing the CF down to 0.27 kg CO_2eq. per kilogram of grain. These findings raise questions regarding the accounting system boundaries used to measure the carbon footprint in different cropping systems. The importance of crop rotations and residue management in reducing the CF of cropping systems has been highlighted in several studies. These practices are considered part of 'climate-smart agriculture'. reported that ignoring crop rotation could result in an underestimation of annual GHGs savings from EU-28 rapeseed biodiesel by 1.67 TgCO_2eq. They argued that crop rotations and straw harvest should be accounted for when calculating the product CFP of bread, milk and both first- and second-generation biofuels. These recommendations are applicable across regions worldwide.

Livestock is an important component of agriculture, especially in the case of small landholders and resource-poor farmers. Globally, the livestock sector is estimated to account for 15% of GHGs emissions, and 80% of these emissions originate from ruminant animal systems involving enteric fermentation and manure management (Persson et al. 2015). Thus, emissions from the livestock sector have raised numerous questions regarding the possibility of reducing animal-based diets as a strategy to reduce emissions of GHGs. In 2012, India's livestock sector produced 304, 31 and 4 Tg in CO_2eq of enteric CH_4, manure CH_4 and N_2O respectively (Patra 2017). The relative contribution (% of total) of different species to GHGs was 55, 37, 4, 2.1 and 1 for cattle, buffalo, sheep, goat, pig and other animals respectively. Enteric fermentation was the major contributor and accounted for 89.7% of the total emissions from India's livestock sector, followed by manure CH_4 (9.2%) and N_2O (1.2%). The CF (kg CO_2eq. kg^{-1} of the produce) was estimated at 1.21 for cross-bred cattle to 2.96 for indigenous cattle.

The food system, being a major contributor to anthropogenic emissions (Röös et al. 2014), must be critically assessed concerning the source of GHGs. The primary GHGs involved in FSs are CO_2 from deforestation, land use conversion, ploughing and erosional processes. Erosion-induced transport of SOC can be a major source of all three gases. N_2O is primarily contributed by soil through the use of inorganic and organic fertilizers. CH_4 is contributed from livestock, rice paddies and wetlands. The U.N. FSS of 2021 identified some specific strategies to advance SDGs or the Agenda 2030 of the United Nations (Lal et al. 2021). The environmental footprint of the diet of 7.8 billion people, projected to be 9.7 B by 2050 and 11 B by 2100 (UN 2019), is an important factor in the sustainable management of soil and water resources but also in addressing the issues of global warming along with quality and renewability of water resources. Leach et al. (2016) reported that the EFP (Environmental

Footprints) of food production spans multiple dimensions, including CF based on GHGs emissions, nitrogen footprint and water use (both green and blue), and land use. It is, thus, important to identify diets that can minimize each of these FP values, especially those diets that have synergies rather than trade-offs among different low FP diets.

Conventional versus organic farming is another important factor affecting the CF of bread. Treu et al. (2017) computed the CF of conventional versus organically grown food products. Treu and colleagues observed that the CF of the conventional versus organic diets is essentially similar at ~1250 kg CO_2eq per capita per year. However, the land area used to provide food is 40% greater for organic versus conventional diet at 1900 and 2750 m^2 of land per capita per year respectively. Furthermore, the average conventional diet contains 45% more meat, and the organic diet contains 40% more vegetables, fruits and legumes (combined). Treu and colleagues also observed that animal-based food products dominate the CFP in both diets and land use and account for 70–75% of CF in both diets. Treu et al. (2017) concluded that diet-related CFP can be reduced by shifting towards diets with less animal-based food products, and there is not much difference in CF among organic versus conventional systems of food production.

4.6 Investment and Financing for Carbon and Input-Efficient Technologies for Carbon Credit Generation

Carbon sequestration in agricultural soils through the formation of humus or recalcitrant SOC is a vital aspect of long-term sustainable agriculture and climate change adaptation and mitigation. Soil is the largest terrestrial carbon pool and has a carbon storage capacity three times greater than the atmospheric carbon pool. The annual carbon storage potential of agricultural soils is estimated to be 1.5–15% of global anthropogenic carbon emissions (Lal 2004). Programmes that incentivize farmers to increase humus content in their soils are relatively new, and as a result, there is still limited understanding of what drives farmers to participate in such initiatives. One option is the formation of humus in agricultural soils, which is expressed in particular in the concept of sustainable soil management, for example, in organic farming and specific humus-building measures (Sharma et al. 2021). Farmers can increase the humus content of their soils by using humus-building measures, also known as carbon farming, such as intercropping, no-tillage and agroforestry systems (Sharma et al. 2021). In Germany, humus programmes are managed by non-governmental organizations (CarboCert 2022) and Positerra (2022). Besides humus programmes, more general non-governmental carbon farming initiatives exist in several countries, using different approaches (Climate-KIC 2022). One of the leading platforms for SOC storage is 'Agreena-Carbon', which utilizes remote sensing technology to monitor land management practices such as tillage, cover cropping and crop rotation and diversification, and issue CO_2 certificates to farmers in 17 European markets (Agreena 2023). Verra's Verified Carbon Standard (VCS) programme enables farmers to earn CCs by adopting sustainable agricultural practices that reduce GHGs and enhance SOC storage. VCS programme uses Improved Agricultural Land Management (IALM) (VM0042) methodology (developed in collaboration with Indigo Ag and TerraCarbon LLC) to incentivize practices such as reduced fertilizer application, improved residue management and regulated grazing techniques, VM0044 (for Biochar utilization) and VM0047 (for afforestation/reforestation), to engage farmers in voluntary carbon markets. VCS programme gives CCs after third party verification for realness, measurability, additionality, permanence and independent registries (Verra 2025). There are also some government-led carbon farming projects, such as in Australia (Australian Government Clean Energy Regulator, 2023) and Europe (Interreg North Sea Region Carbon Farming initiative). To upscale the adoption of such programmes, it is essential to design them with flexibility to match the farmers' needs, which requires a deep understanding of farmers' decision-making processes. The transparent and timely generation of financial incentives is a key driver of farmers' participation in carbon farming programmes (Bartkowski and Bartke 2018), whereas behavioural factors, including cognitive, emotional, personal and social aspects, also significantly influence their adoption (Schaub et al. 2023). Although studies such as Hou and Hou (2019) have examined Chinese farmers' decision-making regarding low-carbon production, there is no existing research specifically exploring the adoption of carbon sequestration programmes and the impact of behavioural factors. Since humus programmes are still relatively new and not widely recognized by farmers, it is especially important at this stage to understand what motivates their participation. This understanding will help address farmers' needs and increase adoption rates (Fig. 4.5).

4.7 Policy Perspective for Enhancing Input Use Efficiency Towards the Management of Carbon Footprint

The essentiality of mitigating global climate change has led to an interest in developing and adopting cleaner production innovations. Rapid economic growth in several major developing countries has highlighted the importance of green technologies in boosting carbon productivity, which is crucial for improving living standards without exacerbating global warming. The attention of policymakers towards agricultural emissions is necessary to upscale the innovations. According to Honle et al. (2018), 38% of agricultural emissions are currently addressed by nationally determined contributions under the Paris Agreement. In terms of policy implications, the results highlight the need to differentiate between the development and adoption of cleaner technologies. For less developed countries, adopting cleaner technologies is more critical for enhancing total-factor carbon productivity and

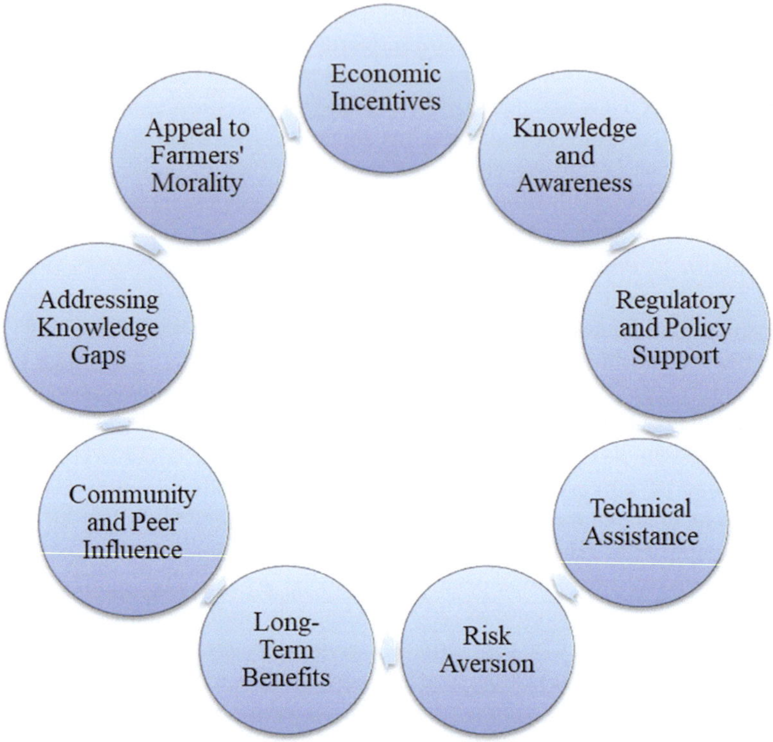

Fig. 4.5 Key factors that influence farmers' participation in carbon sequestration programmes

reducing CF. These nations invest heavily to improve living standards, and such investments can be aligned towards cleaner innovations with lower CF. Otherwise, adopting conventional technologies depending on fossil fuels could result in long-term reliance on outdated, polluting technologies and such short-term gains could lock them into environmentally harmful practices for future decades. GHG emissions create a negative effect on the environment through global warming, and one approach to combat climate change could be the pricing of environmental damage caused by GHG emissions. It would encourage individuals and businesses to factor the pollution costs into their production processes and would influence the selection of inputs/processes to adopt cleaner technologies. Cleaner innovations remain limited in less developed countries as compared to developed economies. Simply adopting cleaner technologies created locally is insufficient to significantly boost total-factor carbon productivity. Since climate change is a global issue, there is a need for policies that promote international cooperation. This includes speeding up the transfer of technologies to developing countries through imports, foreign direct investment and licensing agreements. The MRV standards should align with the needs of farmers and on-ground situations to ensure optimum inclusion of farm practices under verification standards. Regulations for transparency, timely verification and transactions are also important for better adoption of carbon credit programmes. A policy framework focusing on the adoption of climate-resilient agriculture can improve farm profitability, lower emissions, increase carbon sequestration and supplement farm income through CCs.

4.8 Future Input Use Efficiency in an Agroecosystem for Footprint Management

The high-quality development in the area of AI, with the advancements in big data, cloud computing and deep learning, offers environmentally suitable options for agricultural transformation. These technologies could optimize resource use, sustain crop productivity and lower GHGs emissions through suitable landscape planning, resource allocation and monitoring crop performance. Precision agriculture is crucial for increasing input efficiency by using sensor-based technologies and GPS to optimize the application of water, fertilizers, pesticides and herbicides. VRT applies inputs according to field-specific needs, reducing waste, lowering GHGs emissions and enhancing efficiency. Precise input application enhances agricultural productivity and profitability by minimizing unnecessary input use of synthetic fertilizers and pesticides, thus reducing the carbon footprint of agricultural production. The advancements in soil-centred carbon farming practices (CA, RA and OF) coupled with digital platforms (AI/ML) can facilitate farmers' access to carbon markets. Blockchain technology is used to improve transparency and traceability in carbon credit trading (Boumaiza and Maher et al. 2024). This technology ensures that carbon sequestration and emission avoidance/reductions are accurately recorded, and incentivizing farmers to adopt sustainable practices. Additionally, tools like remote sensing and ML are improving the accuracy of MRV systems. Novel innovations in plant science are also contributing to the reduction in the agricultural carbon footprint. The development of crop varieties that require lower fertilizer inputs, that is, varieties having traits for biological nitrification inhibition that moderates nitrification and reduces N losses, or are drought-tolerant (Pusa Rice DST 1 and DRR Dhan 100 from IARI New Delhi, India) reduces the number for irrigations and withstand drought situations, thereby lowering GHG emissions.

In animal science, genome editing in ruminants to regulate enteric fermentation will be of paramount importance to combat CH_4 emissions. Additionally, the development of vaccines and innovative feed additives can reduce CH_4 production in ruminants by inhibiting the process of enteric fermentation, lowering livestock-related emissions.

Carbon farming practices, especially regenerative agriculture and conservation agriculture, which focus on soil health, soil cover, living roots year-round, reduced tillage and crop diversification, have gained attention for their potential to offset carbon and enhance soil carbon sequestration. Integration of these technologies to minimize the CF without compromising yields is imperative to achieve CCs

(Cariappa et al. 2024). Enhancing consumers' awareness is important owing to the demand-driven production of crops/meat, and consequently, promoting the use of CF or product environmental footprint labels on food products is crucial for informing consumers about the environmental impact of their diets and encouraging sustainable choices. The use of renewable energy in food production by making the shift away from fossil fuels is feasible and essential for reducing emissions. Additionally, the development of policies oriented towards decarbonization and energy efficiency within the food sector and in input/product supply chains through fuel-efficient smart machinery is essential from a mitigation perspective. Integration of advanced technologies, precision agriculture and on and off-farm innovative practices in both crop and livestock production is a futuristic area of research for reducing GHGs emissions and transforming agriculture into a more resource-efficient.

4.9 Recommendations

1. Adoption of resource conservation technologies like conservation agriculture, organic farming and regenerative agriculture practices by shifting towards minimal soil disturbance techniques, recycling of farm waste and crop residues is effective to optimize resource inputs, sequester and protect SOC. Implementing strategies like rotational grazing, regulated feeds and anaerobic digestion of manure is efficient for CO_2 capture and reduces CH_4 emissions. AI-based tools to optimize input application, scaling-up precision agriculture using GPS-guided tractors and variable-rate technology can reduce farm wastage and enhance resource use efficiency. Implementing balanced fertilization strategies (4Rs) from efficient fertilizer sources based on soil tests to synchronize crop needs reduces excess nitrogen application and mitigates N_2O emissions. Use of precision agriculture tools like sensor-based technologies enhances nutrient uptake efficiency.
2. Lifecycle assessment of agricultural produce considers a sequence of events from raw material extraction to ultimate disposal of produce. Besides on-farm cleaner production processes, the use of smart machinery that operates on renewable energy (wind or electrical energy) reduces reliance on fossil fuels and mitigates GHGs significantly. These machineries can be employed for farm operations, food processing and supply chain management.
3. Development of standard methodologies is important to quantify and verify the cleaner production practices, focusing on lower GHGs emissions and higher carbon sequestration. However, MRV protocols should effectively monitor the principles of realness, completeness, consistency, transparency, accuracy, conservativeness, leakage, additionality, permanence and ownership of CCs over a defined time scale.
4. Reorientation of farm subsidies given for fertilizer, water and other agri-inputs with GHGs mitigation and ensuring timely incentives for such practices to assure long-term commitment of farmers.

5. Genome editing in crops and ruminants for improvements in resource use efficiency and lower **emissions.** .
6. General lifestyle awareness and capacity building of citizens for BMPs and climate-smart agriculture.

4.10 Conclusion

Agriculture is both a source and sink for atmospheric CO_2. The global GHGs emissions have reached 53 Gt CO_2eq in 2023, and agriculture contributes a significant portion, particularly through CO_2, CH_4 and N_2O. The proportion of agri-food systems is one-third of global emissions, with 71% coming from land-based activities. To address the growing challenge of climate change, it is essential to reduce the carbon footprint in the lifecycle of agricultural produce through the adoption of sustainable farming practices. Adopting resource conservation technologies like precision agriculture, conservation agriculture, regenerative agriculture, novel genome editing tools focusing on resource conservation in crop improvement, and the use of renewable energy technologies can reduce resource use, GHGs emissions and sequester carbon. Optimum input use by integrating advanced technologies can lower the emissions generated within the supply chain and reduce carbon footprints. Implementing these measures, along with scaling CCs through improved transparent and principled verification systems, can help meet climate targets and promote sustainable agriculture.

References

Abbasi AZ, Islam N, Shaikh ZA (2014) A review of wireless sensors and networks' applications in agriculture. Comput Stand Interfaces 36(2):263–270

Adamides G, Kalatzis N, Stylianou A, Marianos N, Chatzipapadopoulos F, Giannakopoulou M, Papadavid G, Vassiliou V, Neocleous D (2020) Smart farming techniques for climate change adaptation in Cyprus. Atmos 11(6):557. https://doi.org/10.3390/atmos11060557

Agreena (2023) AgreenaCarbon. https://agreena.com/de/

Ahuja S, Mehra P (2024) Sustainable artificial intelligence solutions for agricultural efficiency and carbon footprint reduction in India. IntechOpen. https://doi.org/10.5772/intechopen.112996

Ali MA, Dong L, Dhau J, Khosla A, Kaushik A (2020) Perspective-electrochemical sensors for soil quality assessment. J Electrochem Soc 167(3):037550. https://doi.org/10.1149/1945-7111/ab69fe

Ali F, Dawood A, Hussain A, Alnasir MH, Khan MA, Butt TM, Janjua NK, Hamid A (2024) Fueling the future: biomass applications for green and sustainable energy. Discov Sustain 5(1): 156

Aliyu M, Hassan G, Said SA, Siddiqui MU, Alawami AT, Elamin IM (2018) A review of solar-powered water pumping systems. Renew Sust Energ Rev 87:61–76

Arata A, Sharma VK, Spagna G (1993) Performance evaluation of solar assisted dryers for low temperature drying application—II. Experimental results. Energy Con Manag 34(5):417–426

Asundi A, Liew O-W, Li B, Chong J (2006) Optical sensors enhance plant nutrient monitoring. https://spie.org/news/0096-optical-sensors-enhance-plant-nutrient-monitoring?SSO=1

Awan MB, Iqbal T, Yaseen S, Nawaz S, Ali CH (2019) Techno-economic sustainability analysis of biomass fired industrial boiler: biomass evolution as heat and power generation source. IET Renew Power Gener 13(4):650–658

Ayaz M, Ammad-Uddin M, Sharif Z, Mansour A, Aggoune EH (2019) Internet-of-things (IoT)-based smart agriculture: toward making the fields talk. IEEE Access 7:129551–129583

Azeta J, Bolu CA, Alele F, Daranijo EO, Onyeubani P, Abioye AA (2019) Application of mechatronics in agriculture: a review. J Phys Conf Ser 1378(3):032006. https://doi.org/10.1088/1742-6596/1378/3/032006

Bacenetti J, Lovarelli D, Facchinetti D, Pessina D (2018) An environmental comparison of techniques to reduce pollutants emissions related to agricultural tractors. Biosyst Eng 171:30–40

Baldock J, Wheeler I, McKenzie N, McBrateny A (2012) Soils and climate change: potential impacts on carbon stocks and greenhouse gas emissions, and future research for Australian agriculture. Crop Pasture Sci 63:269–283

Balyan S, Jangir H, Tripathi SN, Tripathi A, Jhang T, Pandey P (2024) Seeding a sustainable future: navigating the digital horizon of smart agriculture. Sustainability 16(2):475

Baronti S, Ungaro F, Maienza A, Ugolini F, Lagomarsino A, Agnelli AE, Vaccari FP (2022) Rotational pasture management to increase the sustainability of mountain livestock farms in the alpine region. Reg Environ Chang 22(2):50

Bartkowski B, Bartke S (2018) Leverage points for governing agricultural soils: a review of empirical studies of European farmers' decision-making. Sustainability 10:3179. https://doi.org/10.3390/su10093179

Bellard C, Bertelsmeier C, Leadley P, Thuiller W, Courchamp F (2012) Impacts of climate change on the future of biodiversity. Ecol Lett 15(4):365–377

Bhagat PR, Naz F, Magda R (2022) Artificial intelligence solutions enabling sustainable agriculture: a bibliometric analysis. PLoS One 17(6):e0268989

Biswas AK, Tortajada C (2014) India must tackle food waste. Available online: https://www.weforum.org/agenda/2014/08/india-perishable-food-waste-population-growth/. Accessed 30 Apr 2025

Blanco-Canqui H, Sindelar M, Wortmann CS, Kreikemeier G (2017) Aerial interseeded cover crop and corn residue harvest: soil and crop impacts. Agron J 109(4):1344–1351

Blanco-Canqui H, Laird DA, Heaton EA, Rathke S, Acharya BS (2020) Soil carbon increased by twice the amount of biochar carbon applied after 6 years: Field evidence of negative priming. Glob Change Biol Bioenergy 12(4):240–251

Bohoussou YN, Kou YH, Yu WB, Lin BJ, Virk AL, Zhao X, Dang YP, Zhang HL (2022) Impacts of the components of conservation agriculture on soil organic carbon and total nitrogen storage: a global meta-analysis. Sci Total Environ 842:156822. https://doi.org/10.1016/j.scitotenv.2022.156822

Boumaiza A, Maher K (2024) Leveraging blockchain technology to enhance transparency and efficiency in carbon trading markets. Int J Electr Power Energy Syst 162:110225

Bozal-Leorri A, Subbarao GV, Kishii M, Urmeneta L, Kommerell V, Karwat H, Braun H, Aparicio-Tejo PM, Ortiz-Monasterio I, González-Murua C, González-Moro MB (2022) Biological nitrification inhibitor-trait enhances nitrogen uptake by suppressing nitrifier activity and improves ammonium assimilation in two elite wheat varieties. Frontiers Plant Scien, vol 13

Bradford KJ, Dahal P, Van Asbrouck J, Kunusoth K, Bello P, Thompson J, Wu F (2020) The dry chain: reducing postharvest losses and improving food safety in humid climates. In: Food industry wastes. Academic Press, pp 375–389

Brummitt CD, Mathers CA, Keating RA, O'Leary K (2024) Solutions and insights for agricultural monitoring, reporting, and verification (MRV) from three consecutive issuances of soil carbon credits. Journal of Environmental Management 369(1):122284. https://doi.org/10.1016/j.jenvman.2024.122284

Bruulsema TW, Fixen PE, Sulewski GD (2016) 4R plant nutrition: a manual for improving the management of plant nutrition. International Plant Nutrition Institute

CarboCert (2022) Humusaufbauprogramm. https://www.carbocert.de/fuer-unsere-landwirte-in-de-und-ch/carbocert-humusaufbauprogramm

Cariappa AAG, Konath NC, Sapkota TB et al (2024) Evaluating the potential and eligibility of conservation agriculture practices for carbon credits. Sci Rep 14:9193. https://doi.org/10.1038/s41598-024-59262-6

CGIAR (2019) Direct agricultural emissions. Available at: https://ccafs.cgiar.org/bigfacts/#theme¼food-emissions&subtheme¼direct-agriculture worldwide; global carbon project; expert(s) (Robbie Andrew; Friedlingstein et al. (2023)); Statista; 1940 to 2023; *Projection

Chang CL, Lin KM (2018) Smart agricultural machine with a computer vision based weeding and variable-rate irrigation scheme. Robotics 7(3):38. https://doi.org/10.3390/robotics7030038

Chauhan C (2013) India wastes more farm food than China: UN. Hindustan Times 12

Chen R, Chen Y, Chen W, Chen Y (2012) Time domain reflectometry for water content measurement of municipal solid waste. Environ Eng Sci 29(6):486–493. https://doi.org/10.1089/ees.2010.0489

Chen Z, Dikgwatlhe SB, Xue JF, Zhang HL, Chen F, Xiao X (2015) Tillage impacts on net carbon flux in paddy soil of the southern China. J Clean Prod 103:70e76. https://doi.org/10.1016/j.jclepro.2014.05.014

Chenu C, Angers DA, Barré P, Derrien D, Arrouays D, Balesdent J (2019) Increasing organic stocks in agricultural soils: knowledge gaps and potential innovations. Soil Tillage Res 188:41–52

Climate-KIC (2022) Carbon farming project of EIT climate-KIC. https://www.climate-kic.org/innovation-spotlight/carbon-farming-is-innovating-land-use/

Clune S, Crossin E, Verghese K (2017) Systematic review of greenhouse gas emissions for different fresh food categories. J Clean Prod 140:766e783. https://doi.org/10.1016/j.jclepro.2016.04.082

Crippa M, Solazzo E, Guizzardi D, Monforti-Ferrario F, Tubiello FN, Leip AJNF (2021) Food systems are responsible for a third of global anthropogenic GHG emissions. Nature Food 2(3): 198–209

Damodaran H (2023) ICAR develops wheat that can beat the heat. The Indian Express. https://indianexpress.com/article/india/icar-scientists-wheat-that-can-beat-the-heat-8459154/

Das TK, Nath CP, Das S, Biswas S, Bhattacharyya R, Sudhishri S, Raj R, Singh B, Kakralia SK, Rathi N, Sharma AR, Dwivedi BS, Biswas AK, Chaudhari SK (2020) Conservation agriculture in rice-mustard cropping system for five years: impacts on crop productivity, profitability, water-use efficiency, and soil properties. Field Crop Res 250:107781

Das TK, Ghosh S, Das A, Sen S, Datta D, Ghosh S, Raj R, Behera B, Roy A, Vyas AK, Rana DS (2021) Conservation agriculture impacts on productivity, resource-use efficiency and environmental sustainability: a holistic review. Indian J Agron 66:S111–S127

Dey A, Dwivedi BS, Bhattacharyya R, Datta SP, Meena MC, Das TK, Singh VK (2016) Conservation agriculture in a rice-wheat cropping system on an alluvial soil of north-western indo-Gangetic plains: effect on soil carbon and nitrogen pools. J Indian Soc Soil Sci 64(3):246–254

Dhanaraju M, Chenniappan P, Ramalingam K, Pazhanivelan S, Kaliaperumal R (2022) Smart farming: internet of things (IoT)-based sustainable agriculture. Agriculture 12(10):1745

Dijkstra J, Oenema O, Bannink A (2011) Dietary strategies to reducing N excretion from cattle: implications for methane emissions. Curr Opin Environ Sustain 3(5):414–422

Dutta SK, Laing A, Kumar S, Shambhavi S, Kumar S, Kumar B, Verma DK, Kumar A, Singh RG, Gathala M (2023) Sustainability, productivity, profitability and nutritional diversity of six cropping systems under conservation agriculture: a long term study in eastern India. Agric Syst 207:103641

Eory V, Moran D (2012) Review of potential measures for RPP2-agriculture. Available online: https://www.climatexchange.org.uk/files/3413/7338/8148/Review_of_Potential_Measures_for_RPP2_-_Agriculture.pdf. Accessed 15 Sep 2024

FAO (2021) Emissions due to agriculture. Global, Regional and Country Trends 2000–2018. FAOSTAT Analytical Brief Series No 18 (Rome) available at: www.fao.org/3/cb3808en/cb3808en.pdf

FAO (2022a) Greenhouse gas emissions from agrifood systems. Global, regional and country trends, 2000–2020. FAOSTAT Analytical Brief Series No. 50. Rome, FAO

FAO (2022b) FAO strategy on climate change 2022–2031. Rome

Fernandes L, Tavares PB (2024) A review on solar drying devices: heat transfer, air movement and type of chambers. In: Solar, vol 4, No. 1. MDPI, pp 15–42

Filonchyk M, Hurynovich V, Yan H (2024) Greenhouse gas emissions and reduction strategies for the world's largest greenhouse gas emitters. Science of The Total Environment 944:173895. https://doi.org/10.1016/j.scitotenv.2023.173895

Galli LE, Facchinetti D, Gibin M, Pessina D (2022) Environmental impact of real gaseous pollutants emission of agricultural tractors. In: Conference of the Italian Society of Agricultural Engineering. Springer, Cham, pp 555–563

Gan Y, Liang C, May W (2011) Carbon footprint of canola and field pea crops in semiarid regions. Agron J 103(4):1193–1199

Gan Y, Liang C, Chai Q, Lemke RL, Campbell CA, Zentner RP (2014) Improving farming practices reduces the carbon footprint of spring wheat production. Nat Commun 5(1):1–13. https://doi.org/10.1038/ncomms6012

Gavade LC, Bhoi AD (2017) N, P, K detection and control for agriculture applications using PIC controller: a review. Int J Eng Res Technol 6(4):638–641

Gopal C, Mohanraj M, Chandramohan P, Chandrasekar P (2013) Renewable energy source water pumping systems—a literature review. Renew Sustain Energy Rev 25:351–370

Gorjian S, Ebadi H, Trommsdorff M, Sharon H, Demant M, Schindele S (2021) The advent of modern solar-powered electric agricultural machinery: a solution for sustainable farm operations. J Clean Prod 292:126030

Guardia G, Monistrol-Arcas A, Montoya M, Garcia-Gutierrez S, Abalos D, Vallejo A (2023) Subsurface drip irrigation reduces CH 4 emissions and ecosystem respiration compared to surface drip irrigation. Agric Water Manag 285. https://doi.org/10.1016/j.agwat.2023.108380

Gutierrez J, Villa-Medina JF, Nieto-Garibay A, Porta-Gandara MA (2014) Automated irrigation system using a wireless sensor network and GPRS module. IEEE Trans Instrum Meas 63(1): 166–176

Han X, Dang P, Liao L, Song F, Zhang M, Zhang M, Siddique KH (2023) Combining slow-release fertilizer and plastic film mulching reduced the carbon footprint and enhanced maize yield on the loess plateau. J Environ Sci 147:359–369. https://doi.org/10.1016/j.jes.2023.12.001

Hartman T, Broom D (2020) What are carbon credits and how can they help fight climate change? In: Nature and Biodiversity. World Economic Forum. https://www.weforum.org/stories/2020/11/carbon-credits-what-how-fight-climate-change/

Hönle SE, Heidecke C, Osterburg B (2018) Climate change mitigation strategies for agriculture: an analysis of nationally determined contributions, biennial reports and biennial update reports. Clim Pol 1–15. https://doi.org/10.1080/14693062.2018.1559793

Hou J, Hou B (2019) Farmers' adoption of low-carbon agriculture in China: an extended theory of the planned behavior model. Sustainability 11:1399. https://doi.org/10.3390/su11051399

Hu B, Wang W, Ou S et al (2015) Variation in NRT1.1B contributes to nitrate-use divergence between rice subspecies. Nat Genet 47:834–838. https://doi.org/10.1038/ng.3337

Huan TTN, Tan PS, Hiraoka H (2002) Optimum application rates of nitrogen fertiliser for high yielding rice cultivars based on growth diagnosis in the Mekong Delta. Japan Int Res Center Agri Sci 26:3–9

Huang J, He F, Cui K, Bures RJ, Xu B, Gong W, Peng S (2008) Determination of optimal nitrogen rate for rice varieties using a chlorophyll meter. Field Crop Res 105:70–80

Intergovernmental Panel on Climate Change (IPCC) (2007) Climate change 2007: the scientific basis. In: Solomon S et al (eds) Contribution of working group I to the fourth assessment report of the intergovernmental panel on climate change. Cambridge Univ. Press, New York

IPCC (2019) Climate change and land. An IPCC special report on climate change, desertification, land degradation, sustainable land management, food security, and greenhouse gas fluxes in terrestrial ecosystems

IPCC (2024) Greenhouse gas protocol. Accessed from https://ghgprotocol.org/sites/default/files/2024-08/Global-Warming-Potential-Values%20(August%202024).pdf

ISO (2018) Greenhouse gases – carbon footprint of products – requirements and guidelines for quantification and communication. ISO, Geneva. https://www.iso.org/standard/71206.html

Jakhar AM, Aziz I, Kaleri AR, Hasnain M, Haider G, Ma J, Abideen Z (2022) Nano-fertilizers: a sustainable technology for improving crop nutrition and food security. NanoImpact 27:100411. https://doi.org/10.1016/j.impact.2022.100411

Jangir CK, Sangwan PS, Panghaal D, Kumar S, Meena RS, Bharti, Jat RD, Singh N (2024) Spatial variability and statistical analysis of soil properties in the rice wheat-based systems of North-West India. Commun Soil Sci Plant Anal 55(8):1205–1223

Jarman A, Thompson J, McGuire E, Reid M, Rubsam S, Becker K, Mitcham E (2023) Postharvest technologies for small-scale farmers in low-and middle-income countries: a call to action. Postharvest Biol Technol 206:112491

Jat SL, Parihar CM, Singh AK, Kumar B, Choudhary M, Nayak HS, Meena BR (2019) Energy auditing and carbon footprint under long-term conservation agriculture-based intensive maize systems with diverse inorganic nitrogen management options. Sci Total Environ 664:659–668

Jat ML, Chakraborty D, Ladha JK, Rana DS, Gathala MK, McDonald A, Gerard B (2020) Conservation agriculture for sustainable intensification in South Asia. Nature Sustain 3(4): 336–343

Jhariya MK, Raj A, Banerjee A, Meena RS, Bargali SS, Kumar S, Nema S, Poonam, Oraon PR (2022) Plan and policies for soil organic carbon management under agroforestry system. In: Plans and policies for soil organic carbon management in agriculture. Springer, Singapore, pp 191–219

Jo H, Asekova S, Bayat MA, Ali L, Song JT, Ha YS, Hong DH, Lee JD (2022) Comparison of yield and yield components of several crops grown under agro-photovoltaic system in Korea. Agriculture 12(5):619

Kaparaju P, Rintala J (2011) Mitigation of greenhouse gas emissions by adopting anaerobic digestion technology on dairy, sow and pig farms in Finland. Renew Energy 36(1):31–41

Kebede E (2021) Contribution, utilization, and improvement of legumes-driven biological nitrogen fixation in agricultural systems. Front Sustain Food Syst 5:767998

Kumar A, Bharti B, Kumar J, Singh GP, Jaiswal JP, Prasad R (2020) Evaluation of drought tolerance indices for identification of drought tolerant and susceptible genotypes in wheat (Triticum aestivum L.). Electronic journal of. Plant Breed 11(3):727–734

Kumar R, Punia H, Kumar S (2020a) Nanomaterials for agriculture input use efficiency. In: Resources use efficiency in agriculture. Springer, Singapore. https://doi.org/10.1007/978-981-15-6953-1_5

Kumar S, Bamboriya SD, Rani K, Meena RS, Sheoran S, Loyal A, Kumawat A, Jhariya MK (2022) Grain legumes: a diversified diet for sustainable livelihood, food, and nutritional security. In: Advances in legumes for sustainable intensification. Academic Press, pp 157–178

Kumar S, Sharma SK, Thakral SK, Bhardwaj KK, Jhariya MK, Meena RS, Jangir CK, Bedwal S, Jat RD, Gaber A, Atta AA (2022a) Integrated nutrient management improves the productivity and nutrient use efficiency of *Lens culinaris* Medik. Sustainability 14(3):1284

Kumar S, Sharma SK, Dhaka AK, Bedwal S, Sheoran S, Meena RS, Jangir CK, Kumar D, Kumar R, Jat RD, Meena AK (2023) Efficient nutrient management for enhancing crop productivity, quality and nutrient dynamics in lentil (*Lens culinaris* Medik.) in the semi-arid region of northern India. PLoS One 18(2):e0280636

Kumar S, Yadav A, Kumar A, Hasanain M, Shankar K, Karan S, Rawat S, Sinha A, Kumar V, Gairola A, Prajapati SK, Dayal P (2023a) Climate smart irrigation practices for improving Water productivity in India: a comprehensive review. Int J Environ Climate Change 13(12):333–348. https://doi.org/10.9734/ijecc/2023/v13i123689

Kumar S, Meena RS, Kumar S, Pradhan G, Jangir CK, Ghosh S, Punia H, Sheoran P, Meena R, Ahmad MA, Goyal SK (2025) Designing a diversified Indian mustard production system for energy-carbon-cum-heat use efficiency and sowing dates assessment. GCB Bioenergy 17(6): e70044

Kumari N, Pandey S (2023) Application of artificial intelligence in environmental sustainability and climate change. In: Visualization techniques for climate change with machine learning and artificial intelligence, pp 293–316

Lakhiar IA, Jianmin G, Syed TN, Chandio FA, Buttar NA, Qureshi WA (2018) Monitoring and control systems in agriculture using intelligent sensor techniques: a review of the aeroponics system. Hindawi J Sensors. https://doi.org/10.1155/2018/8672769

Lal R (2004) Soil carbon sequestration impacts on global climate change and food security. Science 304(5677):1623–1627

Lal R (2022) Reducing carbon footprint of agriculture and food systems. Carbon Footprints 1:3. https://doi.org/10.20517/cf.2021.05

Lal R, Bouma J, Brevik E, Dawson L, Field DJ, Glaser B, Zhang J (2021) Soils and sustainable development goals of the United Nations: an International Union of Soil Sciences perspective. Geoderma Reg 25:e00398. https://doi.org/10.1016/j.geodrs.2021.e00398

Lamb JB, van de Water JAJM, Bourne DG et al (2017) Seagrass ecosystems reduce exposure to bacterial pathogens of humans, fishes, and invertebrates. Science 355(6326):731–733

Langholtz M, Davison BH, Jager HI, Eaton L, Baskaran LM, Davis M, Brandt CC (2021) Increased nitrogen use efficiency in crop production can provide economic and environmental benefits. Sci Total Environ 758:143602. https://doi.org/10.1016/j.scitotenv.2020.143602

Law CS, Marsal LF, Santos A (2020) 9 – Electrochemically engineered nanoporous photonic crystal structures for optical sensing and biosensing. In: Handbook of nanomaterials in analytical chemistry, modern trends in analysis. Elsevier. ISBN 9780128166994, pp 201–226. https://doi.org/10.1016/B978-0-12-816699-4.00009-8

Leach AM, Emery KA, Gephart J, Davis KF, Erisman JW, Leip A, Galloway JN (2016) Environmental impact food labels combining carbon, nitrogen, and water footprints. Food Policy 61: 213–223. https://doi.org/10.1016/j.foodpol.2016.03.006

Lee HJ, Park HH, Kim YO, Kuk YI (2022) Crop cultivation underneath agro-photovoltaic systems and its effects on crop growth, yield, and photosynthetic efficiency. Agronomy 12(8):1842

Leharwan M, Kumar Y, Kumar R, Kumar Saraswat P, Kumar R, Kumar Thaliyil Veetil A, Bhattacharjee S, Kumar A, Kumar S (2023) Assessing the effects of conservation tillage and in-situ crop residue management on crop yield and soil properties in rice–wheat cropping system. *Sustainability 15*(17):12736

Li H, Issaka Z, Jiang Y, Tang P, Chen C (2019) Overview of emerging technologies in sprinkler irrigation to optimize crop production. International Journal of Agricultural and Biological Engineering 12(3):1–9

Liang Q (2007) A cross-layer transmission scheduling scheme for wireless sensor networks. Computer Communications 30(14–15):2987–2994

Lin J, Wang M, Zhang M, Zhang Y, Chen L (2008) Electrochemical sensors for soil nutrient detection: opportunity and challenge. In: Li D (ed) Computer and computing Technologies in Agriculture, volume II, pp 1349–1353

Litskas VD, Platis DP, Anagnostopoulos CD, Tsaboula AC, Menexes GC, Kalburtji KL, Stavrinides MC, Mamolos AP (2020) Climate change and agriculture: carbon footprint estimation for agricultural products and labeling for emissions mitigation. Sustain Food Sys:33–49. https://doi.org/10.1016/B978-0-12-818293-2.00003-3

Liu S (2022) Restoring carbon sinks in agriculture through carbon credits. Stanford Sustainable Finance Initiative. https://sfi.stanford.edu/sites/sfi/files/media/file/restoring_carbon_sinks_in_agriculture_through_carbon_credits_whitepaper_0_0_0.pdf

Liu W, Yang R, Li L, Zhao C, Li G (2024) Energy and environmental evaluation and comparison of a diesel-electric hybrid tractor, a conventional tractor, and a hillside mini-tiller using the life cycle assessment method. J Clean Prod 469:143232

Liu J, Gaob X, Songa J, Wena M, Wanga J, Caib Y, Zhao X (2025) Subsurface drip irrigation mitigated greenhouse gas emission and improved root growth and yield in apple in semi-arid region. Agric Water Manag 308:109290

Llonch P, Haskell MJ, Dewhurst RJ, Turner SP (2017) Current available strategies to mitigate greenhouse gas emissions in livestock systems: an animal welfare perspective. Animal 11(2): 274–284

Lunduka RW, Mateva KI, Magorokosho C, Manjeru P (2019) Impact of adoption of drought-tolerant maize varieties on total maize production in south eastern Zimbabwe. Clim Dev 11(1): 35–46

Maier R, Hörtnagl L, Buchmann N (2022) Greenhouse gas fluxes (CO_2, N_2O and CH_4) of pea and maize during two cropping seasons: drivers, budgets, and emission factors for nitrous oxide. Sci Total Environ 849:157541

Maikhuri RK, Semwal RL, Rao KS, Singh K, Saxena KG (2000) Growth and ecological impacts of traditional agroforestry tree species in central Himalaya. India Agroforestry Syst 48:257–271

Maiti D, Das DK, Karak T, Banerjee M (2004) Management of nitrogen through the use of leaf color chart (LCC) and soil plant analysis development (SPAD) or chlorophyll meter in rice under irrigated ecosystem. The Science World Journal 4:838–846

Majeed Y, Khan MU, Waseem M, Zahid U, Mahmood F, Majeed F, Sultan M, Raza A (2023) Renewable energy as an alternative source for energy management in agriculture. Energy Rep 10:344–359

Mall RK, Gupta A, Sonkar G (2017) Effect of climate change on agricultural crops. In: Current developments in biotechnology and bioengineering. Elsevier, pp 23–46

Manoj HG, Udupa NGS (2015) Application of soil moisture sensor in mixed farming. Int Res J Eng Technol 2(4):311–314

Matson PA (2012) Seeds of sustainability: lessons from the birthplace of the green revolution in agriculture. Island Press, Washington, DC

Mazzetto AM, Feigl BJ, Schils RL, Cerri CEP, Cerri CC (2015) Improved pasture and herd management to reduce greenhouse gas emissions from a Brazilian beef production system. Livest Sci 175:101–112

Meena RS, Pradhan G (2023) Industrial garbage-derived biocompost enhances soil organic carbon fractions, CO_2 biosequestration, potential carbon credits, and sustainability index in a rice-wheat ecosystem. Environ Res:116525. https://doi.org/10.1016/j.envres.2023.116525

Meena RS, Pradhan G, Kumar S, Lal R (2023) Using industrial wastes for rice-wheat cropping and food-energy-carbon-water-economic nexus to the sustainable food system. Renew Sustain Energy Rev *187*:113756. https://doi.org/10.1016/j.rser.2023.113756

Meena RS, Pradhan G, Nalani Singh K, Singh AK, Verma SK, Mina KK, Kumar S, Chaturvedi RK (2024) Diversified cropping modules designed for soil restoration, CO_2 sequestration, and generating carbon credits. Land Degrad Dev 35. https://doi.org/10.1002/ldr.5224

Meena RS, Pradhan G, Singh K, Kumar S, Singh AK, Shashidhar KS, Mina KK, Rao CS (2024a) Agriculture models for restoring degraded land to enhance CO_2 biosequestration and carbon credits in the Vindhyan region of India. Sci Total Environ 929:172661. https://doi.org/10.1016/j.scitotenv.2024.172661

Meena RS, Lal R, Kumar S, Pradhan G, Rao CS, Singh AK, Pathak H, Abhilash PC, Kumar A, Sharma SK, Jat ML, Singh S (2024b) Potential of Indian agriculture for capturing atmospheric CO_2 and monetizing carbon credits to the farmers: an overview and policy framework. Adv Agron 188:101–206. https://doi.org/10.1016/bs.agron.2024.06.001

Meena RS, Pradhan G, Kumar S (2025) Energy flow, eco-efficiency, and economic circulation with recycled industrial waste compost application in wheat and subsequent rice farming. Sci Total Environ 967:178779. https://doi.org/10.1016/j.scitotenv.2025.178779

Mehta CR, Chandel NS, Senthilkumar T (2014) Status, challenges and strategies for farm mechanization in India. Agricultural Mechanization Asia Africa Latin America 45(4):43–50

Miao Z, Zhao Z, Long T, Chen X (2023) Carbon footprint in agriculture sector: A literature review. Carbon Footprints 2:13. https://doi.org/10.20517/cf.2023.29

Michel M, Eldridge AL, Hartmann C, Klassen P, Ingram J, Meijer GW (2024) Benefits and challenges of food processing in the context of food systems, value chains and sustainable development goals. Trends Food Sci Technol:104703

Mills JAN, Dijkstra J, Bannink A, Cammell SB, Kebreab E, France J (2001) A mechanistic model of whole-tract digestion and methanogenesis in the lactating dairy cow: model development, evaluation, and application. J Anim Sci 79:1584–1597

Miranda J, Ponce P, Molina A, Wright P (2019) Sensing, smart and sustainable technologies for Agri-food 4.0. J Computers Ind 108:21–36. https://doi.org/10.1016/j.compind.2019.02.002

Modi RU, Manjunatha K, Gauam PV, Nageshkumar T, Sanodiya R, Chaudhary V, Rao CS (2020) Climate-smart technology based farm mechanization for enhanced input use efficiency. In: Climate change and indian agriculture: challenges and adaption strategies, pp 325–358

Mohan A, Krishnan R, Arshinder K, Vandore J, Ramanathan U (2023) Management of postharvest losses and wastages in the Indian tomato supply chain—a temperature-controlled storage perspective. Sustainability *15*(2):1331

Mulla DJ (2013) Twenty five years of remote sensing in precision agriculture: key advances and remaining knowledge gaps. Biosyst Eng 114(4):358–371

Munir A, Ashraf T, Amjad W, Ghafoor A, Rehman S, Malik AU, Hensel O, Sultan M, Morosuk T (2021) Solar-hybrid cold energy storage system coupled with cooling pads backup: a step towards decentralized storage of perishables. *Energies 14*(22):7633

Nami H, Anvari-Moghaddam A, Nemati A (2021) Modeling and analysis of a solar boosted biomass-driven combined cooling, heating and power plant for domestic applications. Sustain Energy Technol Assessments 47:101326

Ngosong C, Okolle JN, Tening AS (2019) Mulching: a sustainable option to improve soil health. Soil Fertility Manag Sustain Develop:231–249

Nnamchi OA, Tom CN, Godwin A, Umunna MF, Offiong UD, Mathew I, Adindu LC, Akuwueke L, Nnamchi SN, Ben AE, Ndukwu MC (2025) Solar dryers: a review of mechanism, methods and critical analysis of transport models applicable in solar drying of product. Green Energy Res:100118

Ozlu E, Kumar S (2018) Response of surface GHG fluxes to long-term manure and inorganic fertilizer application in corn and soybean rotation. Sci Total Environ 626:817–825

Pathak H (2015) Greenhouse gas emission from Indian agriculture: trends, drivers and mitigation strategies. Proc Indian Natn Sci Acad 81:1133–1149

Patra AK (2017) Accounting methane and nitrous oxide emissions, and carbon footprints of livestock food products in different states of India. J Cleaner Prod 162:678–686. https://doi.org/10.1016/j.jclepro.2017.06.096

Pattey E, Trzcinski MK, Desjardins RL (2005) Quantifying the reduction of greenhouse gas emissions as a result of composting dairy and beef cattle manure. Nutr Cycl Agroecosyst 72: 173–187

Paul K, Chatterjee SS, Pai P, Varshney A, Juikar S, Prasad V, Bhadra B, Dasgupta S (2022) Viable smart sensors and their application in data driven agriculture. Comput Electron Agric 198: 107096

Persson UM, Johansson DJ, Cederberg C, Hedenus F, Bryngelsson D (2015) Climate metrics and the carbon footprint of livestock products: where's the beef? Environ Res Lett 10(3):034005. https://doi.org/10.1088/1748-9326/10/3/034005

Peshin R, Hansra BS, Singh K, Nanda R, Sharma R, Yangsdon S, Kumar R (2021) Long-term impact of Bt cotton: an empirical evidence from North India. J Clean Prod 312:127575

Petrovic B, Csambalik L (2025) Enhancing precision agriculture for climate change mitigation in Visegrad countries: factors shaping adaptation. Land 14(2):399

Pixley KV, Cairns JE, Lopez-Ridaura S, Ojiewo CO, Dawud MA, Drabo I, Mindaye T, Nebie B, Asea G, Das B et al (2023) Redesigning crop varieties to win the race between climate change and food security. Mol Plant:16

Portz G, Molin JP, Jasper J (2012) Active crop sensor to detect variability of nitrogen supply and biomass on sugarcane fields. Precis Agric 13:33–44

Positerra (2022) Humusaufbau. https://positerra.org/humus-aufbau

Pradhan G, Meena RS (2023) Interaction impacts of biocompost on nutrient dynamics and relations with soil biota, carbon fractions index, societal value of CO_2 equivalent, and ecosystem services in the wheat-rice farming. Chemosphere:139695. https://doi.org/10.1016/j.chemosphere.2023.139695

Pradhan G, Meena RS (2023a) Utilizing waste compost to improve the atmospheric CO_2 capturing in the rice-wheat cropping system and energy-cum-carbon credit audit for a circular economy. Sci Total Environ:164572. https://doi.org/10.1016/j.scitotenv.2023.164572

Pradhan G, Meena RS, Kumar S, Lal R (2023) Utilizing industrial wastes as compost in wheat-rice production to improve the above and below-ground ecosystem services. Agric Ecosyst Environ 358:108704. https://doi.org/10.1016/j.agee.2023.108704

Prajapati SK, Choudhary S, Kumar V, Dayal P, Srivastava R, Gairola A, Borate RB (2023a) Carbon sequestration: a key strategy for climate change mitigation towards a sustainable future. Emerging Trends Climate Change 2(2):1–14. https://doi.org/10.18782/2583-4770.128

Prajapati SK, Verma S, Yadav PK (2023b) Agricultural transformation through precision farming: using agrochemicals efficiently to enhance productivity. Food Sci Rep 4(5):36–50. E-ISSN 2582-5437

Prathibha S, Hongal A, Jyothi M (2017) IoT based monitoring system in smart agriculture. In: Proceedings of the 2017 International Conference on Recent Advances in Electronics and Communication Technology (ICRAECT). IEEE, Bangalore

Priyadarshini S (2025) India approves first genome-edited rice varieties. Nature India. https://doi.org/10.1038/d44151-025-00078-2

Punia H, Tokas J, Malik A, Satpal R, Rani A, Gupta P, Kumari A, Mor VS, Bhuker A, Kumar S (2020) Solar radiation and nitrogen use efficiency for sustainable agriculture. In: Resources use efficiency in agriculture, pp 177–212

Rajendran S (2021) A comparative study of performance and emission characteristics of neat biodiesel operated diesel engine: a review. J Therm Anal Calorim 146(3):1015–1025

Rakshamuthu S, Jegan S, Benyameen JJ, Selvakumar V, Anandeeswaran K, Iyahraja S (2021) Experimental analysis of small size solar dryer with phase change materials for food preservation. J Energy Storage 33:102095

Raper TB, Varco JJ (2015) Canopy-scale wavelength and vegetation index sensitivities to cotton growth parameters and nitrogen status. Precis Agric 16:62–76

Reher T, Lavaert C, Willockx B, Huyghe Y, Bisschop J, Martens JA, Diels J, Cappelle J, Van de Poel B (2024) Potential of sugar beet (*Beta vulgaris*) and wheat (*Triticum aestivum*) production in vertical bifacial, tracked, or elevated agrivoltaic systems in Belgium. Appl Energy 359:122679

Rehman HU, Naseer F, Ali HM (2023) An experimental case study of solar food dryer with thermal storage using phase change material. Case Stud Therm Eng 51:103611

Ren G, Liu J, Wan J, Guo Y, Yu D (2017) Overview of wind power intermittency: impacts, measurements, and mitigation solutions. Appl Energy 204:47–65

Renner A, Cadillo-Benalcazar JJ, Benini L, Giampietro M (2020) Environmental pressure of the European agricultural system: anticipating the biophysical consequences of internalization. Ecosyst Serv 46:101195

Riano B, García-Gonzalez MC (2015) Greenhouse gas emissions of an on-farm swine manure treatment plant-comparison with conventional storage in anaerobic tanks. J Clean Prod 103:542e548. https://doi.org/10.1016/j.jclepro.2014.07.007

Robertson GP, Paul EA, Harwood RR (2000) Greenhouse gases in intensive agriculture: contributions of individual gases to the radiative forcing of the atmosphere. Science 289:1922–1925

Röös E, Sundberg C, Hansson PA (2014) Carbon footprint of food products. Assess Carbon Footprint Diff Indus Sect 1:85–112. https://doi.org/10.1007/978-981-4560-41-2_4

Rosa L, Gabrielli P (2023) Achieving net-zero emissions in agriculture: a review. Environ Res Lett 18:063002. https://doi.org/10.1088/1748-9326/acd5e8

Roselin AR, Jawahar A (2017) Smart agro system using wireless sensor networks in proceedings of the 2017 international conference on intelligent computing and control systems (ICICCS). IEEE, Madurai

Sah D, Devakumar AS (2018) The carbon footprint of agricultural crop cultivation in India. Carbon Manag 9(3):213–225. https://doi.org/10.1080/17583004.2018.1457908

Saiz-Rubio V, Rovira-Más F (2020) From smart farming towards agriculture 5.0: a review on crop data management. Agronomy 10(2):207. https://doi.org/10.3390/agronomy10020207

Saleem M (2022) Possibility of utilizing agriculture biomass as a renewable and sustainable future energy source. *Heliyon 8*(2)

Saroa GS, Lal R (2003) Soil restorative effects of mulching on aggregation and carbon sequestration in a Miamian soil in central Ohio. Land Degrad Dev 14(5):481–493

Schaub S, Ghazoul J, Huber R, Zhang W, Sander A, Rees C, Banerjee S, Finger R (2023) The role of behavioural factors and opportunity costs in farmers' participation in voluntary agri-environmental schemes: a systematic review. J Agric Econ 74:617–660. https://doi.org/10.1111/1477-9552.12538

Schriber S (2021) Smart agriculture sensors: helping small farmers and positively impacting global issues, too. Mouser electronics https://www.mouser.co.uk/applications/smart-agriculture-sensors. Accessed 24 Mar 2022

Shabir I, Dash KK, Dar AH, Pandey VK, Fayaz U, Srivastava S, Nisha R (2023) CF evaluation for sustainable food processing system development: a comprehensive review. Future Foods 7: 100215

Shah N, Das I (2012) Precision irrigation sensor network based irrigation. Problems, perspectives and challenges of agricultural water management. InTech Rijeka Croatia 13:1034219. https://doi.org/10.3389/fpls.2022.1034219

Sharma N, Pant BD, Mathur J (2019) MEMS devices used in agriculture – a review. J Biosens Bioelectron 10(1):267. https://doi.org/10.4172/2155-6210.1000267

Sharma M, Kaushal R, Kaushik P, Ramakrishna S (2021) Carbon farming: prospects and challenges. Sustainability 13:11122. https://doi.org/10.3390/su131911122

Sharma T, Das TK, Maity PP, Biswas S, Sudhishri S, Govindasamy P, Raj R, Sen S, Singh T, Paul AK et al (2023) Long-term conservation agriculture influences weed diversity, Water productivity, grain yield, and energy budgeting of wheat in North-Western indo-Gangetic Plains. Sustainability 15(9):7290. https://doi.org/10.3390/su15097290

Sharma A, Prakash A, Bhambota S, Kumar S (2024) Investigations of precision agriculture technologies with application to developing countries. Environ Dev Sustain 27:1–37

Sheoran, S., Gupta, M., Kumari, S., Kumar, S. and Rakshit, S., 2022. Meta-QTL analysis and candidate genes identification for various abiotic stresses in maize (*Zea mays* L.) and their implications in breeding programs. Molecular Breeding, *42*(5), p.26

Shirsath PB, Saini S, Durg N, Senoner D, Ghose N, Verma S, Sikka A (2020) Compendium on solar powered irrigation systems in India. In: CGIAR research program on climate change. Agriculture and Food Security (CCAFS)

Singh B, Singh V, Singh Y, Thind HS, Kumar A, Singh S, Choudhary OP, Gupta RK, Vashistha M (2013) Supplementing fertiliser nitrogen application to irrigated wheat at maximum tillering stage using chlorophyll meter and optical sensor. Agric Res 2:81–89

Singh R, Babu S, Avasthe RK, Yadav GS, Das A, Mohapatra KP, Kumar A, Singh VK, Chandra P (2021) Crop productivity, soil health, and energy dynamics of Indian Himalayan intensified organic maize-based systems. Int Soil Water Conserv Res 9(2):260–270. https://doi.org/10.1016/j.iswcr.2020.11.003

Smith P, Martino D, Cai Z, Gwary D, Janzen H, Kumar P, McCarl B, Ogle S, O'Mara F, Rice C, Scholes B (2008) Greenhouse gas mitigation in agriculture. Philos Trans R Soc Lond B Biol Sci 363(1492):789–813

Song Y, Sun H, Li M, Zhang Q (2015) Technology application of smart spray in agriculture: a review. Intell Automation Soft Comput 21(3):319–333. https://doi.org/10.1080/10798587.2015.1015781

Stacheder M, Koniger F, Schuhmann R (2009) New dielectric sensors and sensing techniques for soil and snow moisture measurements. Sensors (Basel) 9(4):2951–2967. https://doi.org/10.3390/s90402951

Surai S, Kundu R, Ghosh R, Bid G (2018) An IoT based smart agriculture system with soil moisture sensor. J Innovation Res 1(1):39–42

Sushanth G, Sujatha S (2018) IOT based smart agriculture system. In: Proceedings of the 2018 international conference on wireless communications, signal processing and net working (WiSPNET). IEEE, Chennai

Takeshima H, Yamauchi F, Edeh HO, Hernandez MA (2023) Solar-powered cold-storage and agrifood market modernization in Nigeria. Agric Econ 54(2):234–255

Tang Y, Dananjayan S, Hou C, Guo Q, Luo S, He Y (2021) A survey on the 5G network and its impact on agriculture: challenges and opportunities. Comput Electron Agric 180:105895. https://doi.org/10.1016/j.compag.2020.105895

Teja R (2021) What is a sensor? Different Types of Sensors and their Applications ElectronicsHuborg Published April 2, 2021. Accessed from: https://www.electronicshub.org/different-types-sensors/

Thakur AK, Mandal KG, Mohanty RK, Sarangi A (2024) Next-gen rice farming: ways to achieve food, nutritional and economic security under changing climatic conditions. Curr Sci 126(4): 426

Tomić MD, Savin LĐ, Mićić RD, Simikić MĐ, Furman TF (2013) Effects of fossil diesel and biodiesel blends on the performances and emissions of agricultural tractor engines. Therm Sci 17(1):263–278

Treu H, Nordborg M, Cederberg C, Heuer T, Claupein E, Hoffmann H, Berndes G (2017) Carbon footprints and land use of conventional and organic diets in Germany. J Clean Prod 161:127–142. https://doi.org/10.1016/j.jclepro.2017.05.041

Trimble (2012) GreenSeeker handheld crop sensor: Quick reference card. http://www.farmworks.com/files/pdf/GreenSeeker%20HCS/GreenSeekerQRC_91500-00-ENG_Screen.pdf

Udomkun P, Romuli S, Schock S, Mahayothee B, Sartas M, Wossen T, Njukwe E, Vanlauwe B, Müller J (2020) Review of solar dryers for agricultural products in Asia and Africa: an innovation landscape approach. J Environ Manag 268:110730

Ullo SL, Sinha GR (2021) Advances in IoT and smart sensors for remote sensing and agriculture applications. Remote Sens 13(13):2585. https://doi.org/10.3390/rs13132585

UN (2019) World Population Prospects (2019): Highlights (ST/ESA/SER. A/423). Available from: https://population.un.org/wpp/publications/files/wpp2019_highlights.pdf

Verra (2025) Verified Carbon Standard – Verra. The world's leading greenhouse gas crediting program. https://verra.org/programs/verified-carbon-standard/

Vishnoi S, Goel RK (2024) Climate smart agriculture for sustainable productivity and healthy landscapes. Environ Sci Pol 151:103600

Visser F, Dargusch P, Smith C, Grace PR (2015) Application of the crop carbon progress calculator (CCAP) in a 'farm to ship' cotton production case study in Australia. J Clean Prod 103:675e684. https://doi.org/10.1016/j.jclepro.2014.09.093

Wang X, Bai J, Xie T, Wang W, Zhang G, Yin S, Wang D (2021) Effects of biological nitrification inhibitors on nitrogen use efficiency and greenhouse gas emissions in agricultural soils: a review. Ecotoxicol Environ Saf 220:112338. https://doi.org/10.1016/j.ecoenv.2021.112338

Xia G, Zhou L, Freedman JM, Roy SB, Harris RA, Cervarich MC (2016) A case study of effects of atmospheric boundary layer turbulence, wind speed, and stability on wind farm induced temperature changes using observations from a field campaign. Clim Dyn 46:2179–2196

Xie H, Ringler C, Mondal MAH (2021) Solar or diesel: a comparison of costs for groundwater-fed irrigation in sub-Saharan Africa under two energy solutions. Earth's Future 9(4):2020EF001611

Yadav D, Singh D, Babu S, Madegowda M, Singh DV, Mandal D, Rathore AC, Sharma VK, Singhal V, Kumawat A, Yadav DK, Yadav RK, Kumar S (2023) Intensified cropping reduces soil erosion and improves rainfall partitioning and soil properties in the marginal land of the Indian Himalayas. Int Soil Water Conserv Res 12. https://doi.org/10.1016/j.iswcr.2023.10.002

Yadav D, Singh D, Madhu M, Singh DV, Islam S, Kumar G, Gupta AK, Mandal D, Babu S, Barh A, Kumawat A, Dubey R, Pati K, Bhargavi B, Yadav DK (2024) Interplanted crops and rainfall variability regulate soil erosion in rainfed maize of the Indian Himalayas. Land Degrad Dev 35(14):4271–4287

Yamauchi F, Takeshima H, Estudillo JP, Kijima Y, Sonobe T (2023) Solar power to transform food systems and rural livelihoods: evidence from a solar-powered cold storage intervention in Nigeria. In: Agricultural development in Asia and Africa: emerging-economy state and international policy studies. Springer, Singapore, pp 329–341

Yu B, Bi X, Liu X, Sun H, Buysse J (2024) Exploring the application and decision optimization of climate-smart agriculture within land-energy-food-waste nexus. Sustain Prod Consum 50:536–555

Zahed Z, Mufti S, Kumar SS, Wani OA, Mushtaq F, Rasool R, Babu S, Abidi I, Gaber A, Hossain A (2022) Organic and inorganic mulches combination improve the productivity, quality and profitability of rainfed potato in the temperate Himalayan region. Gesunde Pflanz 74:1109–1122. https://doi.org/10.1007/s10343-022-00650-1

Zhang J, Zhang H, Li S, Li J, Yan L, Xia L (2021) Increasing yield potential through manipulating of an ARE1 ortholog related to nitrogen use efficiency in wheat by CRISPR/Cas9. J Integr Plant Biol 63(9):1649–1663

Zhang Y, Bo Q, Ma X, Du Y, Du X, Xu L, Yang Y (2023) Solid–liquid separation and its environmental impact on manure treatment in scaled Pig farms—evidence based on life cycle assessment. Agriculture 13(12):2284

Zhu W, Zhang L, Peng Y, Zhang Q, Gao M, Fu Y (2019) Principle analysis of soil electrical conductivity test based on pulsed eddy current. IOP Conf Ser Earth Environ Sci 242(4):042018. https://doi.org/10.1088/1755-1315/242/4/042018

Chapter 5
Crop Diversification for Enhancing Carbon Sequestration in Cropping Systems

Anusmita Saha, Shambhunath Ghosh, Sandeep Kumar, and Ram Swaroop Meena

Abstract To address the increasing food demand, intensified crop production has become a prevalent agricultural practice, primarily relying on cereal-based cropping systems. While these systems are effective in alleviating hunger, they often lead to soil exhaustion and increased reliance on external inputs. Numerous issues arise from this monotonous cropping approach, including soil health degradation, soil deterioration, water pollution, and heightened disease and pest infestations. However, the most pressing concern is climate change, which is linked to the production of greenhouse gases (GHG) at both the field and input manufacturing levels. Agricultural emissions have reached 9.3 billion tons of carbon dioxide equivalent (CO_2eq), with 57% originating from on-farm activities and the remainder from land use changes. A significant portion of these emissions, approximately 52%, is attributed to rice cultivation, which exceeds 60% when wheat is included in the rice-wheat cropping system. While these practices contribute to carbon loss from the soil, agriculture can also function as a carbon sink. Implementing several modified practices can enhance carbon storage in the soil and reduce GHG emissions from the agricultural sector. A diversified cropping system is a vital step towards enhancing the carbon sink in the field. This can be achieved through the incorporation of legumes, cover crops, intercropping with diverse crop families, and agroforestry practices. The inclusion of pigeon pea in cereal-based cropping systems significantly boosts aboveground carbon stock, exceeding 2 Mg C ha^{-1}. The agroforestry system plays a crucial role in soil organic carbon (SOC) sequestration, accounting for 12% of global terrestrial carbon. Additionally, the widespread use of cover crops shows promise in reducing agricultural GHG emissions by 10%. Beyond these benefits,

A. Saha (✉) · S. Ghosh
ICAR-Indian Institute of Wheat and Barley Research, IARI Karnal Hub, Karnal, Haryana, India

S. Kumar
ICAR-Indian Agricultural Research Institute, Regional Station, Karnal, Haryana, India

R. S. Meena
Department of Agronomy, Institute of Agricultural Sciences, Banaras Hindu University, Varanasi, Uttar Pradesh, India
e-mail: meenars@bhu.ac.in

R. S. Meena et al. (eds.), *Carbon-Negative Agriculture*, Sustainability Sciences in Asia and Africa, https://doi.org/10.1007/978-981-95-1483-0_5

crop diversification enhances soil health, disrupts prevalent pest cycles, mitigates water pollution, and supports biodiversity within agroecosystems. However, the successful adoption of new cropping systems hinges on various factors, including climate conditions, land characteristics, resource availability, market demand, and the socioeconomic status of farmers. This chapter delves into the potential methods, mechanisms, benefits, and challenges associated with diversified cropping systems aimed at carbon sequestration and the sustainability of agricultural practices.

Keywords Climate change · Crop diversification · Carbon sequestration · Organic carbon

Abbreviations

AF	Agroforestry
CF	Carbon footprint
CH_4	Methane
CO_2	Carbon dioxide
CO_2eq	Carbon dioxide equivalent
GHGs	Greenhouse gases
GT	Gigatons
LCA	Life cycle assessment
MBC	Microbial biomass carbon
N_2O	Nitrous oxide
NPP	Net primary productivity
SLM	Sustainable land management
SOC	Soil organic carbon
SOM	Soil organic matter

5.1 Introduction

Intensified crop production to meet the burgeoning food demand is a common practice of agriculture, which mainly relies on the cereal-based cropping system. While effective in combating hunger, such practices exhaust the soil, increasing its dependency on external inputs, including synthetic fertilizers, crop protection, and chemicals, leading to several problems associated with this monotonous cropping practice, including degradation of soil health, soil deterioration, water pollution, more disease, and pest infestation. But climate change is the most crucial problem associated with this due production of greenhouse gases (GHG), [carbon dioxide (CO_2), methane (CH_4), and nitrous oxide (N_2O)] either at the field level or at the input manufacturing level that ultimately influences the global warming, which causes extreme weather events such as droughts, floods, typhoons, and intense precipitation (Hussain et al. 2015; Choudhary et al. 2017). Agriculture is a major source of carbon emissions, accounting for roughly 14% of total anthropogenic emissions (Meena et al. 2025). Emissions from agriculture reached 9.3 billion tons of carbon dioxide equivalent (CO_2eq), with 57% from on-farm activities and the remaining from land use and change. The majority of these emissions come from paddy fields (52%), rising to over 60% when wheat is included in the rice–wheat cropping system. Global warming has triggered a series of ecological and environmental crises, posing a major challenge, that humanity must urgently address (Fu et al. 2015; Meena et al. 2024). If current trends are followed, global average surface temperatures are projected to increase by 2.2–3.5 °C by the middle of this century (IPCC 2023). Climate change is already exerting serious impacts on agricultural production and sustainability. Without any action to reduce GHG emissions, global food security will face critical threats (IPCC 2021). Rising temperatures have led to a higher frequency of crop failures and significant yield losses ranging from 5 to 47% (IPCC 2023). As a result, emission reduction and capturing and storing more atmospheric carbon have received growing attention as essential tools for mitigating climate change (Kumar et al. 2025b).

Managing the CO_2 concentration at this level is a tough task. The production of CO_2 cannot be stopped, but it can be controlled. Another option is to capture the atmospheric carbon in soil by carbon sequestration, that is, plants absorb CO_2 from the atmosphere through photosynthesis. Sequestering soil organic carbon (SOC) is vital for crop ecosystem sustainability. It enhances organic matter and nitrogen content in soil (Khormali et al. 2009; Meena et al. 2024a; Jangir et al. 2024), improves water and nutrient retention capacity by improving soil structural stability (Tisdall and Oades 1982; Ayoubi et al. 2012; Zeraatpisheh et al. 2021; Kumar et al. 2022a), improves crop productivity (Plaza-Bonilla et al. 2016), reduces soil erosion risks (Rahimi et al. 2013; Li et al. 2016), and enhances surface and groundwater quality (Laird 2008). While agriculture acts as an emitting agent side by side, it also serves as a major carbon store within terrestrial landscapes, holding more than 1550 gigatons (GT) of carbon globally (Wackernagel and Rees 1998; Batjes 1996). Several approaches have already been developed to combat this problem. One of the most recent approaches is crop diversification.

Crop diversification, meaning the cultivation of multiple crop species across space or over time, has traditionally served to improve yields, stabilize production, and reduce farming risks. Recent research shows that diversification can substantially enhance SOC accumulation and decrease net GHG emissions. This improvement is mainly attributed to greater biomass contributions, extensive root growth, better soil coverage, and stimulated microbial processes, all of which foster a positive carbon balance in farming systems (Yogi et al. 2023; Kumar et al. 2024; Meena et al. 2024b). In conventional monoculture systems, the heavy dependence on a limited number of crop species often results in soil health deterioration, a decline in organic matter content, and increased sensitivity to climate-related stresses. Alternatively, a diversified cropping system can be a crucial step towards enhancing the carbon sink in the field. This can be done through the inclusion legume, cover crops, intercropping with crops of different families, agroforestry practices, etc. (Ghosh et al. 2024). Introducing pigeon pea in cereal-based cropping systems contributes the highest levels of aboveground carbon stock, exceeding 2 Mg C ha^{-1}. The agroforestry system significantly contributes to SOC sequestration, accounting for 12% of terrestrial carbon globally. On the other hand, the extensive utilization of cover crops holds promise for reducing agricultural GHG output by 10%. Li et al. (2020) reported that the reduction in N_2O emissions is greater in legume crop rotations than in cereal crops. Rigon et al. (2018) found that crop rotation can reduce CO_2 emissions. Research on GHG fluxes under corn and soybean rotation demonstrated that soil CH_4 fluxes did not vary within corn–soybean rotation (Wegner et al. 2018).

Apart from these, crop diversification also helps in improving soil health, breaks the most prevalent pest cycle, reduces water pollution, and maintains biodiversity in the agroecosystem. But the adoption of a new cropping system depends on several factors, including climate and land situation, resource availability, market demand, and socioeconomic condition of farmers. In this context, this chapter explores the possible ways, mechanisms, benefits, and challenges of diversified cropping systems towards carbon sequestration and sustainability of the system.

5.2 Crop Diversification Strategies to Enhance Carbon Sequestration

Monocropping of cereal-cereal system is mostly practiced in the majority of the cultivated area. This not only has depleted the resources but brought people to face critical issues like extreme climatic adversities, particularly greenhouse gases. So, it is urgently needed to change the repeating cereal cropping system to a diversified system by incorporating crops according to the agroecological regions, crop morphology, and various other physiological parameters like foliage density, rooting depth, nitrogen fixing capacity, and carbon sequestration ability (Punia et al. 2020). The break in the monotonous cropping cycle can be brought into practice by alternating cereals with pulses, legumes, oilseeds, cover crops, agroforestry, etc. (Jhariya et al. 2022). This section examines a range of diversification initiatives capable of substantially increasing carbon sinks across cropping systems situated in tropical, subtropical, and temperate regions.

5.2.1 Cereal-Legume-Based Diversification

One effective strategy for boosting carbon sinks in cereal-dominated systems is incorporating leguminous crops into rotations. Including legumes is a key approach to sustainable agriculture, as it offers numerous ecosystem services that are essential for food production, soil fertility, greenhouse gas reduction, and pest and weed control (Leoni et al. 2022; Kumar et al. 2022; Notz et al. 2023; Meena et al. 2023). Legumes contribute to the accumulation of SOC through nitrogen fixation, residue addition, and enhanced microbial activity. Including legumes in rotations has been shown to increase the resistance and resilience of wheat yields. After fertilizer removal, wheat yield following maize dropped by 69%, whereas it decreased by only 37–52% when preceded by legumes (Liu et al. 2023a, b). This is also reflected in the higher resistance index of wheat yield after legumes. SOC is crucial for providing a range of ecosystem functions, influencing soil-based ecosystem services, and is commonly used as a measure of soil quality (Krause et al. 2022).

Including legumes in the cropping system increases the nitrogen-fixing capability of the soil (Jensen et al. 2020), along with increasing the soil fertility. Not only this, legumes having higher root biomass enhance the microbial activity and return an optimum amount of resources into the soil and increase the SOC quantity (Lal 2017). More often, the released exudates from the soil roots are found to have provided a favorable environment for various other nutrients, which quickens the mineralization of organic matter (Brooker et al. 2015; Ding et al. 2021; Kumar et al. 2023). Additionally, legumes consume a very small amount of inputs for their growth, thus restricting wastage of resources, proving beneficial as a cereal-legume cropping system (Jensen et al. 2020).

In cereal-legume intercropping systems, an average of 33.6% of weight, 28.1% of carbon, and 22.8% of nitrogen were decomposed, compared to only 24.5, 20.1, and 17.8% in cereal monocultures. As a result, the losses of weight, carbon, and nitrogen were 37.1, 40.2, and 28.1% higher in cereal-legume intercropping than in cereal monocultures, respectively (Wang et al. 2025).

5.2.2 *Integration of Agroforestry Systems*

India's forest cover is accounted to be 70.17 M ha, which is 21.34% of the geographical area. and 9.26 M ha (2.82% of the geographic area) is occupied by other tree cover, which makes the 79.42 M ha, occupying 24.16% of its geographical area. In India, the agroforestry systems include trees on farms, community forestry, and a variety of local forest management practices (Pandey 2007). In India, the agroforestry systems include trees on farms, community forestry, and a variety of local forest management practices (Pandey 2007). These additions have a positive effect on soil biodiversity, soil functionality, productivity, and the soil's capacity to sequester carbon (Oelbermann et al. 2004). Consequently, tree-based land use systems, including agroforestry, were recognized as carbon offset activities under the Kyoto Protocol. Chhattisgarh is a state dominated by tribes. The main crop in this area is rice. Traditional farming systems are incredibly inefficient. Farmers permit Acacia nilotica to grow naturally on rice field bunds either alone or in conjunction with Terminalia arjuna and Butea monosperma under traditional agroforestry (Puri 2007).

Table 5.1 Agroforestry-based carbon sequestration potential in various agroclimatic zones of India (Ajit et al. 2017)

Agroclimatic zone	States/UTs covered	Agroforestry system	Area ('000 ha)	Carbon stock (tons ha^{-1} yr^{-1})	Carbon sequestration potential ('000 tons CO_2eq yr^{-1})
Southern plateau and hills	Andhra Pradesh, Karnataka	Agri-silviculture	579.9	1.18	2511.4
Indo-Gangetic plains	West Bengal, Bihar, UP, Haryana, Punjab	Agri-silviculture	671.5	0.76	1872.9
Eastern plateau and hills	Delhi, Bihar	Agri-silviculture	185.8	2.21	1506.7
West coast plains and hills	Goa, Karnataka	Agri-silviculture	66.9	1.34	329.2
East coast plains and hills	Andhra Pradesh, Tamil Nadu, Puducherry	Agri-silviculture	61.1	1.10	246.7
Indo-Gangetic plains	Uttar Pradesh	Agri-horticulture	373.0	0.15	205.3
Western Himalayas	Himachal Pradesh	Silvi-pastoral	228.5	0.25	208.2
Western Himalayas	Uttarakhand	Agri-horticulture	265.1	0.17	162.5
Central plateau and hills	Rajasthan, Uttar Pradesh	Agri-horticulture	133.4	0.22	107.7
West coast plains and hills	Kerala, Maharashtra	Agri-silviculture	44.6	0.25	41.0
Western dry region	Rajasthan	Agri-silviculture	193.8	0.09	64.0
Eastern Himalayas	NE States and West Bengal (hill zones)	Agri-silviculture	68.7	0.10	25.7
Total (all zones)			2872.4		7281.3

Under agroforestry, CO_2 from the atmosphere can be sequestered in above ground biomass, that is, stems, branches, foliage and straw, and in below ground biomass, that is, roots, and in the soil (Meena and Pradhan 2023) increasing the carbon sequestration potential of integrated cropping system (Tables 5.1 and 5.2). In tropical India, accumulated C was reported to be 0.29–15.21 Mg ha^{-1} yr^{-1} (Dhyani et al. 2020). In temperate regions, silvo-pastoral systems in France had the highest aboveground tree C stock, up to 37 Mg C ha^{-1}, compared to silvo-arable sites (Cardinael et al. 2017).

Table 5.2 Carbon sequestration potential of agroforestry systems in India

Agroclimatic zone	Agroforestry system	Species/tree mix	Tree density (ha^{-1})	Age (yrs)	CSP (kg C m^{-2} yr^{-1})	Reference
All zones (Pan India)	mixed agroforestry	Diverse native and exotic species	2–51	30	0.03–5.13	
Humid/sub-humid tropics	Block plantation	*Acacia mangium*	625	6.5	0.0082	
Indo-Gangetic plains	Agri-silviculture	Unspecified	740	7.0	0.0162	Chauhan et al. (2010)
Humid/sub-humid tropics	Block plantation	*Gmelina arborea*	–	–	0.016–0.028	Swamy et al. (2003)
Indo-Gangetic plains	Agri-silviculture	*Populus deltoides*	38	–	0.00513	
Humid/sub-humid tropics	Block plantation	*Acacia mangium*	2500	6.5	0.0153	
Humid/sub-humid tropics	Block plantation	*Ceiba pentandra*	–	–	0.013–0.034	Swamy et al. (2003)
Indo-Gangetic plains	Agri-silviculture	Unspecified	357	6.0	0.0195	
Indo-Gangetic plains	Agri-silviculture	Unspecified	357	6.0	0.0263	
Humid/sub-humid tropics	Block plantation	*Acacia mangium*	1250	6.5	0.0036	
Humid/sub-humid tropics	Block plantation	*Acacia mangium*	5000	6.5	0.0109	

5.2.3 Cover Cropping and Green Manuring

According to Shackelford et al. (2019), cover crops can be produced as ground cover to prevent losses, nutrient capture crops, grazed or exported feed, and green manure when utilized as a source of nutrients. Mixtures of cover crops have higher SOC than individual species (Sainju et al. 2006). Increased occlusion of SOC in aggregates or chemically linked to mineral particles is a possible mechanism for the higher SOC with cover crops as opposed to no cover crop (Zhang et al. 2022). Another is the selective preservation of plant or microbial necrosis. The most significant predictor of SOC responses to cover cropping to enhance SOC sequestration is the length of time that cover crops grow (McClelland et al. 2021; Pradhan and Meena 2023). Interactions between the cover crop growth period and other elements such as residue quality (Clark et al. 1994; Alonso-Ayuso et al. 2014; Bana et al. 2023a, b), plant biomass (McClelland et al. 2021), and plant establishment are probably the cause of this significance. By boosting root biomass and hyphae, which entangle and cement soil particles together to create aggregates, cover crops can improve soil aggregation (Tisdall and Oades 1982; Verhulst et al. 2011). The percentage of big macroaggregates, particularly at 0–10 cm, was probably increased by the annual, continuous addition of above and belowground cover crop residue for five years. According to Trivedi et al. (2017), adding agricultural residue to the soil also boosts microbial activity, which, during the breakdown of organic matter, releases polysaccharides and organic acid, encouraging aggregation. Coarse, deep-rooted species can raise the bulk density of the soil next to their roots, whereas smaller-diameter roots can lower bulk density by improving soil porosity (Gyssels et al. 2005). Compared to broad taproot systems, species with denser, finer root systems—like grasses—bind soil more efficiently, increasing topsoil macro-aggregation and accumulating labile SOC in both macro. and microaggregates (Angers and Caron 1998).

5.2.4 Intercropping Systems

The secret to growing food on less land while preserving the natural ecology is sustainable crop intensification (Ausubel et al. 2013). The retention of organic matter is not improved by preferred crop production techniques like the solitary cropping system combined with traditional tillage (Bana et al. 2020; Ruis et al. 2022; Leharwan et al. 2023). Diversified planting, a strategy that enhances the soil carbon pool, has been successfully implemented in both agriculture and forestry through various systems, such as cover crop rotation (Garland et al. 2021), cereal-legume intercropping systems (Wang et al. 2016), intercropping with forage and other family crops (Zhou et al. 2021), and the "jujube-cotton" agriculture-forestry complex cropping system (Zhang et al. 2016). Intercropping, a typical diversified cropping system, offers substantial productivity benefits and carbon sequestration potential (Hombegowda et al. 2020; Saad et al. 2018; Scalise et al. 2017) due to the efficient use of agroecosystem resources such as light, heat, water, and nutrients (Stomph et al. 2020).

5.2.5 Crop–Livestock Integration

Although integration with livestock (e.g., the use of fodder legumes and grasses) is not always considered in traditional crop diversification, it plays a crucial role in recycling organic waste (manure), thereby enhancing soil carbon content. In integrated crop–livestock systems, agricultural and livestock operations are carried out together in the same location but at separate periods, either in succession, rotation, or consortia (Nunes et al. 2021). Re-integrating livestock into cropland holds significant potential to improve the functioning of semi-arid agroecosystems by shifting biogeochemical processes and facilitating various ecosystem services related to carbon and nutrient cycling, as well as resource-use efficiency. Annual C sequestration in farmland soil of farms integrating livestock with cereal-legume and forage crops was significantly higher (158.07 ± 1.79 kg C ha^{-1} yr^{-1}) followed by farms integrating livestock with cereal-legume crops (99.51 ± 0.95 kg C ha^{-1} yr^{-1}), which in turn had a higher value than farms practicing pastoral mobility (78.46 ± 0.70 kg C ha^{-1} yr^{-1}) (Idrissou et al. 2024). Soil carbon stocks rise in crop–livestock systems as a result of increased photosynthesis and crop residue decomposition, which raises soil metabolism overall (Carvalho et al. 2024; Silva et al. 2024).

Grazing on cash crop residues, cover crops, and understory biomass offers an opportunity to boost net primary productivity (NPP) by extending vegetation cover and optimizing forage quality and composition, particularly through using forage legumes. Grazing has been shown to affect NPP (Bardgett and Wardle 2003) in various ways: (i) by redirecting photosynthate allocation to roots; (ii) through defoliation, removal of dead plant material, and increased light for actively photosynthesizing vegetation; (iii) by altering the CtoN ratio of litter and speeding up residue transformation rates; and (iv) by changing plant and soil microbial community structure and biomass (Bardgett and Wardle 2003). Arable-only cropping, including potatoes, averaged 2.54% SOC, while mixed farms with cows and sheep had 3.47% SOC (Down to Earth 2024).

5.2.6 Perennial Cropping and Pasture Integration

Switching from annual to perennial crops (such as sugarcane, lucerne, or bioenergy crops) ensures continuous carbon input through their long-lasting root systems. This, in turn, makes soil organic matter and carbon crucial indicators of soil health. Perennial cereal grain crops can thrive for several years without the need for annual reseeding, significantly reducing soil disturbance that usually results from tillage (Chapman et al. 2022). Climate-smart agricultural systems may benefit from the use of perennial crops (Asbjornsen et al. 2013). Compared to annual crops, perennial crops have been shown to have the capacity to boost soil carbon stores (Ferchaud et al. 2016; Ledo et al. 2020; Pradhan and Meena 2023a). According to a three-year study conducted in the United States, the active soil C percentage in the 0–40 cm layer decreased when perennial grassland was converted to annual crops (DuPont et al. 2010). The C stocks in topsoil rose by an average of 1.4 Mg C ha^{-1} over nine years in three perennial cropping systems, while in two annual cropping systems, they fell by an average of 3.4 Mg C ha^{-1} (Shang et al. 2024).

Sustainable crop diversification goes beyond being just a strategy for enhancing yield resilience; it is a climate-smart approach that can turn agriculture into a net carbon sink. By choosing the right diversification models tailored to regional needs and supported by appropriate policy frameworks, it provides a multifunctional solution-boosting carbon storage, improving soil health, and ensuring long-term sustainability (Fig. 5.1).

Fig. 5.1 Soil organic carbon and nutrient cycling in the cropping system are regulated by crop species, diversity, and soil management practices. (Image concept: Rajan Ghimire)

5.3 Mechanisms of Carbon Sequestration in Diversified Cropping Systems

5.3.1 Enhanced Root Biomass and Soil Organic Carbon Accumulation

Sustainable crop production systems should focus on fostering diverse and robust soil microbial communities to improve biological soil processes, rather than relying solely on chemical treatments like pesticides. Crop rotation, as opposed to monoculture, generally increases below-ground biomass and promotes greater soil microbial diversity (Lupwayi et al. 2024; Pradhan et al. 2023). Liu et al. (2023a, b) observed that crop rotation led to a 13.43% increase in soil microbial biomass carbon and a 7.68% rise in bacterial Shannon's diversity index compared to monoculture. Different crops have distinct root systems, depths, and organic matter decomposition rates. For example, deep-rooted crops like alfalfa and lucerne bring organic matter from deeper layers to the surface to be decomposed or stored. When crops are rotated or intercropped, the organic matter leads to the formation of stable organic carbon. This leads to a positive loop that enhances carbon storage (Kumar et al. 2025a).

5.3.2 Improved Soil Structure and Reduced Erosion

Crop diversification helps develop a varied root network that stabilizes the soil, which minimizes erosion and runoff. This is needed to preserve the stored organic carbon in the soil, which binds the soil and protects it from erosion and runoff (Kumar et al. 2025). The ability of the soil to promote biological production, maintain environmental quality, and improve the health of plants and animals is known as soil health (Barbosa et al. 2023; Serafim et al. 2023; Carvalho et al., Frontiers in Plant Science 2024). As a result, crops cultivated in succession, like soybeans, have seen notable gains in production as soil health has improved. When compared to growing after maize in monoculture, Brito et al. (2023) found that growing soybeans in an integrated system after intercropping them with fodder increased their grain output by 21%. However, Silva et al. (2024) discovered that integrated systems enhance soil fertility, resulting in a 15% increase in soybean grain output over time. Additionally, including cover crops adds crop biomass into the soil, which enhances organic matter, resulting in the aggregation of the soil particles, thus preventing erosion and soil loss.

5.3.3 Biological Nitrogen Fixation and Enhanced Soil Fertility

Since legumes contain higher concentrations of N-rich compounds than other species (Fustec et al. 2010), their root exudates are particularly interesting. These compounds can be rapidly assimilated into microbial biomass, increasing the microbial CUE and MAOM content (de Neergaard and Gorissen 2004; Cotrufo et al. 2013; Kopittke et al. 2020). Greater SOC contents are typically the result of legumes' capacity to supply enough N needs for optimizing C storage, and SOC accumulation is commonly seen in macroaggregates and across contexts soon after legume development (Topps et al. 2021). C input from biomass and cover crop productivity is frequently reported to benefit from legumes in cover crop combinations with grasses (Fornara and Tilman 2008; Freund et al. 2021). Rhizobia in legumes can also promote aggregate formation and stability through the binding action of rhizobia exocellular polysaccharides (Haynes and Beare 1997; Alami et al. 2000).

5.3.4 Increased Photosynthetic Activity and Biomass Production

Research has shown that crops grown in conservation agriculture fields with residues may experience higher moisture levels during the growth cycle and adverse situations like drought (Silva et al. 2024), and protect the soil surface from the extensive solar radiation. Thus, stomatal conductance, transpiration flux, nutrient absorption, and photosynthesis can all be maintained at high levels by plants growing under these conditions. Furthermore, an increased amount of nutrients available in the soil as a result of plant waste decomposition could lead to increased nutrient absorption, which would likely boost biomass accumulation, photosynthesis, and chlorophyll biosynthesis. The carbon absorbed through photosynthesis eventually contributes to the soil carbon pool, promoting soil carbon sequestration (Ontl and Schulte 2012). When compared to conventional farming, no-till systems can also improve soil microbial conditions, especially in the short term (Mirzavand et al. 2022).

5.4 Sustainability Benefits of Crop Diversification

5.4.1 *Biodiversity Conservation*

To address food and nutritional security, crop diversification boosts agricultural biodiversity (genetic, species, and environment), enhances crop yields, and creates high-quality products. Furthermore, by disrupting insect and disease cycles, lowering weeds and soil erosion, and preserving soil moisture, this method can provide several agronomic advantages in pest control. Inclusion of flower crops in the crop diversification program may attract an ample number of pollinator species and various other biological control agents.

5.4.2 *Soil Health Improvement*

Crop diversification has turned out to be an excellent and sustainable way to enhance soil health in the present scenario. Various practices like legume incorporation, cover cropping, and green manuring have been found to add an ample amount of crop biomass to the soil. Mineralization of this crop biomass, in turn, releases various compounds which bind the soil particles more strongly, preventing runoff and soil loss. The released root exudates enhance the microbial diversity and hence decomposition in the long run, increasing the carbon sink of the combined system.

5.4.3 *Water Conservation*

Continuous cereal-cereal cropping guzzles a huge amount of water throughout its life cycle. This has led to water shortage and depletion of the water table. This can be solved by growing various crops, including pulses, oilseed, sorghum, and millets, over water-intensive crops like rice and sugarcane. These drought-tolerant crops thrive in limited water resources, easing the pressure on already depleted water

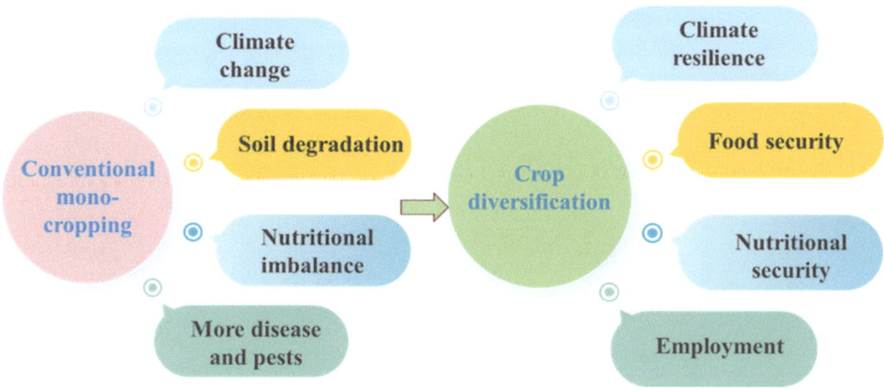

Fig. 5.2 Benefits of transition from conventional monocropping to diversified cropping system

resources. Additionally, millets are also adapted to dry conditions and offer resilience to farmers in low groundwater table areas. Another aspect that can be tackled is by changing the method of irrigation by which they irrigate the crops. Instead of flood irrigation, farmers should opt for precise application of water using drip or sprinkler irrigation, where both the amount of water and the timing of application are optimized. This will lead to achieving the target "per drop more crop."

5.4.4 Enhanced Climate Resilience

With the advancement of extreme climatic adversities like floods, droughts, landslides, crop diversification must be taken into account (Fig. 5.2). Diversified cropping systems have become more important and climatic hazards have increased drastically. Crop yields are significantly affected by the changing temperature and variable rainfall. Alternating the traditionally grown cereal crops with deep-rooted legumes, low water requirement crops can help the system fight against the adversities without hampering the yield and productivity. It should always be kept in mind that the population is never decreasing, even though limitations are increasing. Crop diversification will serve as a fruitful way to build back the resilience, serving the food of the population.

5.4.5 Economic Diversification and Stability

Crop diversification allows farmers to engage the field for year-round production and generate a continuous flow of income. Apart from crops, integrating the crops with allied sector animal husbandry, mushroom, fishery, poultry, and dairy will raise the scope for marketing. This, in turn, unlocks access to emerging markets for other

Fig. 5.3 Challenge and limitation of crop diversification

markets like organic, sustainable, and other markets (Ghosh and Nath 2024). Crop diversification has proved to be a crucial tool to stabilize income and ensure financial stability. By this, small farms are saved from the huge losses of production and the risks associated with agricultural production, food insecurity, and nutritional insecurity.

5.5 Constraints in Implementing Crop Diversification

Although crop diversification has several advantages, it depends on several factors like prevailing climatic conditions of a particular region, land situation, availability of resources, facilities of all technologies, demand, marketing conditions, socio-economic status of farmers, and opportunities from new diversified cropping systems over the existing one. The main drivers that hinder the adoption of a new crop in an area are discussed here (Fig. 5.3).

5.5.1 *Limited Adoption and Knowledge Barriers*

Adoption of any new technology at the farmers' level is quite low. Only a very few farmers (<10%) are willing to take the risks of the existing system's change. But as per the data, nearly 40% of farmers are refusing to take risks. However, the other follows the early adopters' groups only after seeing clear benefits from the new technology.

5.5.2 *Climatic Variability and Risks*

Crop diversification is needed to mitigate climate change, although climate plays a crucial role in adopting crop diversification. Every crop needs a different climatic situation. The current changing climatic situation is a challenge for farmers in adopting a new crop in the cropping system. As nearly 60% of crop area in India depends on rainfall, untimely and erratic distribution of rainfall due to climate change may hamper crop production.

5.5.3 *Eduphic Constraints*

Diversification of crops is also situation-specific. A more carbon-sequestering crop cannot be included in a cropping system if it is not fit for the existing land situation, like in lowland areas or coastal areas, where the inundation of land is a common condition, the inclusion of a carbon-sequestering crop like pulse is not possible to grow under normal conditions. Rice-based cropping system is the only option in this situation. But improving land modification (like a raised bed, bund cultivation, vertical cultivation, etc.) can help in growing other crops in this area.

5.5.4 *Financial Barriers*

As most of the farmers belong to small and marginal categories, they have poor agricultural resources, and always want to get more profit from their small land. Even if a crop has high carbon sequestering capacity, but it has no market demand in that particular area, farmers will not adopt the new crop. Not only that, if the new crop cannot surpass the profit of the pre-existing one, it will also be difficult for the farmer to take this crop in the next season. Even sometimes government policy also fails to meet the financial profitability of existing cropping systems. For example, under the Mera Pani-Meri Virasat scheme, started in Kharif 2020, the government of Haryana is providing financial assistance of Rs 7000 per acre to the farmers for diversifying the paddy crop with alternative crops like maize, cotton, millet, pulses, vegetables, and fruits (agriharyana.gov.in; Sheoran et al. 2022). But the profit from Basmati Rice is far more than Rs 7000 per acre, which is why the adoption rate of a new crop in place of Basmati Rice is very low.

5.5.5 *Lack of Awareness*

Most of the farmers are not aware of the beneficial role of crop diversification. Even there are lack of knowledge that crop diversification may reduce the cost of cultivation and improve soil health for future cultivation. Even carbon sequestration from crop diversification and their trading is still a myth to most farmers. More and more extension activities are required to create awareness among farmers for the future sustainability of the system through crop diversification. Besides this lack of technology and crop-based processing units in different regions is a big challenge behind the low rate of crop diversification.

5.6 Conclusion

This chapter has demonstrated that sustainable crop diversification plays a crucial role in enhancing carbon sequestration within agricultural systems. It significantly contributes to climate change mitigation while improving the resilience of farming practices. This approach offers numerous opportunities to promote system sustainability through improving soil health, breaking prevalent pest cycles, reducing water pollution, supporting beneficial microbial activity, and maintaining biodiversity in agroecosystems. Perennial agroforestry systems, with their extensive root structures, continuous ground cover, and prolonged growth seasons, provide notable advantages over annual crops. Research indicates that these systems can improve soil organic carbon (SOC) levels, showing a gradual increase of 20–50% in SOC over time compared to annual systems. This enhancement is driven by photosynthesis, the release of root exudates, and the addition of organic materials. Employing methods such as crop rotation, intercropping, cover cropping, and companion planting can boost plant material yield by 15–30% and improve soil quality, potentially leading to an annual increase in soil carbon storage of 2–5 tons of carbon per hectare. Introducing a diverse mix of crops, particularly perennials, legumes, and deep-rooted species, facilitates carbon storage in both above-ground plant biomass and soil organic matter. Despite the numerous advantages of a diversified cropping system, the practical implementation rate remains low. Raising awareness among farmers about the beneficial role of crop diversification for agricultural profitability and sustainability is essential. Additionally, focusing on alternative income generation from stored carbon as carbon credits and implementing supportive government policies is crucial. Policy measures aimed solely at increasing soil carbon have proven to be ineffective. Instead of large-scale policies focused exclusively on soil carbon, targeted site-specific measures that address multiple goals, including soil productivity, biodiversity, and water body impacts alongside soil carbon sequestration, are likely to be more justifiable.

References

Ajit S, Dhyani AK, Handa RN et al (2017) Estimating carbon sequestration potential of existing agroforestry systems in India. Agrofor Syst 91:1101–1118. https://doi.org/10.1007/S10457-016-9986-Z

Alami Y, Achouak W, Marol C, Heulin T (2000) Rhizosphere soil aggregation and plant growth promotionof sp. sunflowers strain by isolated Microbiol. 66, an from exopolysaccharide-producing sunflower. Roots Appl Rhizobium Environ:3393–3398. https://doi.org/10.1128/AEM.66.8.3393-3398

Alonso-Ayuso M, Gabriel JL, Quemada M (2014) The kill date as a management tool for cover cropping success. PLoS One 9:e109587. https://doi.org/10.1371/journal.pone.0109587

Angers DA, Caron J (1998) Plant-induced changes in soil structure: processes and feedbacks. Biogeochemistry 45:55–72. https://doi.org/10.1023/A:1005944025343

Asbjornsen H, Hernandez-Santana V, Liebman M, Bayala J, Chen J, Helmers M, Ong CK, Schulte LA (2013) Targeting perennial vegetation in agricultural landscapes for enhancing ecosystem services. Renew Agric Food Syst 29(2):101–125. https://doi.org/10.1017/s1742170512000385

Ausubel JH, Wernick IK, Waggoner PE (2013) Peak farmland and the prospect for land sparing. Popul Dev Rev 38(1):221–242

Ayoubi S, Mokhtari Karchegani P, Mosaddeghi MR, Honarjoo N (2012) Soil aggregation and organic carbon as affected by topography and land use change in western Iran. Soil Tillage Res 121:18–26. https://doi.org/10.1016/j.still.2012.01.011

Bana RS, Singh D, Nain MS, Kumar H, Kumar V, Sepat S (2020) Weed control and rice yield stability studies across diverse tillage and crop establishment systems under on-farm environments. Soil Tillage Res 204(2020):104729. https://doi.org/10.1016/j.still.2020.104729

Bana RS, Bamboriya S, Kumar V, Godara S, Gaydon DS, Laing AM, Dhakar RK, Shivay YS, Meena VS, Singh D (2023a) Identifying optimum residue levels under long-term conservation agriculture for greater crop and water productivity, sustainability and carbon sequestration of rice-wheat cropping system using APSIM model. Soil Tillage Res 232:105745. https://doi.org/10.1016/j.still.2023.105745

Bana RS, Grover M, Singh D et al (2023b) Enhanced pearl millet yield stability, water use efficiency and soil microbial activity using superabsorbent polymers and crop residue recycling across diverse ecologies. Eur J Agron 148:126876. https://doi.org/10.1016/j.eja.2023.126876

Barbosa JZ, Poggere G, Correa RS, Hungria M, Mendes IC (2023) Soil enzymatic activity in Brazilian biomes under native vegetation and contrasting cropping and management. Appl Soil Ecol 190:105014. https://doi.org/10.1016/j.apsoil.2023.105014

Bardgett RD, Wardle DA (2003) Herbivore-mediated linkages between aboveground and belowground communities. Ecology 84. https://doi.org/10.1890/02-0274

Batjes NH (1996) Total carbon and nitrogen in the soils of the world. Eur J Soil Sci 47:151–163. https://doi.org/10.1111/j.1365-2389.1996.tb01386.x

Brito LCR, Souza HA, Araujo Neto RB, Azevedo DMP, Sagrilo E, Vogado RF et al (2023) Improved soil fertility, plant nutrition and grain yield of soybean and millet following maize intercropped with forage grasses and crotalaria in the Brazilian savanna. Crop Pasture Sci 74: 438–448. https://doi.org/10.1071/CP22251

Brooker RW, Bennett AE, Cong W, Daniell TJ, George TS, Hallett PD, Hawes C, Iannetta PPM, Jones HG, Karley AJ (2015) Improving intercropping: a synthesis of research in agronomy, plant physiology and ecology. New Phytol 206:107–117

Cardinael R, Chevallier T, Cambou A, Bèral C, Barthès BG, Dupraz C, Chenu C (2017) Increased soil organic carbon stocks under agroforestry: a survey of six different sites in France. Agric Ecosyst Environ 236:243–255

Carvalho AM, Ramos MLG, Santos DCR, Oliveira AD, Mendes IC, Silva SB et al (2024) Understanding the relations between soil biochemical properties and N2O emissions in a long-term integrated crop–livestock system. Plants 13:365. https://doi.org/10.3390/plants13030365

Chapman EA, Thomsen HC, Tulloch S, Correia PMP, Luo G, Najafi J, DeHaan LR, Crews TE, Olsson L, Lundquist PO, Westerbergh A, Pedas PR, Knudsen S, Palmgren M (2022) Perennials as future grain crops: opportunities and challenges. Front Plant Sci 13. https://doi.org/10.3389/fpls.2022.898769

Chauhan SK, Sharma SC, Chauhan R, Gupta N, Srivastava R (2010) Accounting poplar and wheat productivity for carbon sequestration in an agri-silviculture system. Indian Forester 136(9): 174–182

Choudhary M, Rana KS, Bana RS, Ghasal PC, Choudhary GL, Jakhar P, Verma R (2017) Energy budgeting and carbon footprint of pearl millet—mustard cropping system under conventional and conservation agriculture in rainfed semi-arid agro-ecosystem. Energy 141:1052–1058. https://doi.org/10.1016/j.energy.2017.09.136

Clark AJ, Decker AM, Meisinger JJ (1994) Seeding rate and kill date effects on hairy vetch-cereal rye cover crop mixtures for corn production. Agron J 86:1065–1070. https://doi.org/10.2134/agronj1994.00021962008600060025x

Cotrufo MF, Wallenstein MD, Boot CM, Denef K, Paul E (2013) The Microbial Efficiency-Matrix Stabilization (MEMS) framework integrates plant litter decomposition with soil organic matter stabilization: do labile plant inputs form stable soil organic matter? Glob Chang Biol 19:988–995. https://doi.org/10.1111/gcb.12113

de Neergaard A, Gorissen A (2004) Carbon allocation to roots, rhizodeposits and soil after pulse labelling: a comparison of white clover (Trifolium repens L.) and perennial ryegrass (Lolium perenne L.). Biol Fertil Soils 39:228–234. https://doi.org/10.1007/s00374-003-0699-x

Dhyani SK, Ram A, Newaj R, Handa AK, Dev I (2020) Agroforestry for carbon sequestration in tropical India. In: Carbon management in tropical and sub-tropical terrestrial systems, pp 313–331

Ding W, Cong WF, Lambers H (2021) Plant phosphorus-acquisition and use strategies affect soil carbon cycling. Trends Ecol Evolution 36:899–906

DuPont ST, Culman SW, Ferris H, Buckley DH, Glover JD (2010) No-tillage conversion of harvested perennial grassland to annual cropland reduces root biomass, decreases active carbon stocks, and impacts soil biota. Agric Ecosyst Environ 137(1–2):25–32. https://doi.org/10.1016/j.agee.2009.12.021

Ferchaud F, Vitte G, Mary B (2016) Changes in soil carbon stocks under perennial and annual bioenergy crops. Glob Change Biol Bioenergy 8(2):290–306. https://doi.org/10.1111/gcbb.12249

Fornara DA, Tilman D (2008) Plant functional composition influences rates of soil carbon and nitrogen accumulation. J Ecol 96:314–322. https://doi.org/10.1111/j.1365-2745.2007.01345.x

Freund L, Mariotte P, Santonja M, Buttler A, Jeangros B (2021) Species identity, rather than species mixtures, drives cover crop effects on nutrient partitioning in unfertilized agricultural soil. Plant Soil 460:149–162. https://doi.org/10.1007/s11104-020-04782-z

Fu CH, Large S, Knight B, Richardson AJ, Bundy A, Reygondeau G (2015) Relationships among fisheries exploitation, environmental conditions, and ecological indicators across a series of marine ecosystems. J Mar Syst 148:101–111. https://doi.org/10.1016/j.jmarsys.2015.01.004

Fustec J, Lesuffleur F, Mahieu S, Cliquet J-B (2010) Nitrogen rhizodeposition of legumes. A review Agron Sustainable Dev 30:57–66. https://doi.org/10.1051/agro/2009003

Garland G, Edlinger A, Banerjee S, Degrune F, García-Palacios P, Pescador DS, Herzog C, Romdhane S, Saghai A, Spor A, Wagg C, Hallin S, Maestre FT, Philippot L, Rillig MC (2021) Crop cover is more important than rotational diversity for soil multifunctionality and cereal yields in European cropping systems. Nat Food 2(1). https://doi.org/10.1038/s43016-020-00210-8

Ghosh S, Nath R (2024) Assessment of Indian mustard varieties based on their phenological and yield attributing characteristics in Gangetic Plains of West Bengal. J Crop and Weed 20(1): 138–143

Ghosh S, Saha A, Kumar S, Pathania S (2024) Regenerative agriculture for climate change mitigation and food security. In: Regenerative agriculture for sustainable food systems. Singapore, Springer Nature Singapore, pp 83–133

Gyssels G, Poesen J, Bochet E, Li Y (2005) Impact of plant roots on the resistance of soils to erosion by water: a review. Prog Phys Geog Earth Environ 29:189–217. https://doi.org/10.1191/0309133305pp443ra

Haynes RJ, Beare MH (1997) Influence of six crop species on aggregate stability and some labile organic matter fractions. Soil Biol Biochem 29:1647–1653. https://doi.org/10.1016/S0038-0717(97)00078-3

Hombegowda HC, Adhikary PP, Jakhar P, Madhu M, Barman D (2020) Hedge row intercropping impact on run-off, soil erosion, carbon sequestration and millet yield. Nutr Cycl Agroecosys 116(1):103–116. https://doi.org/10.1007/s10705-019-10031-2. https://agriharyana.gov.in/data/SchemeDoc/mera_pani_meri_virasat.pdf

Hussain S, Peng SB, Fahad S, Khaliq A, Huang JL, Cui KH (2015) Rice management interventions to mitigate greenhouse gas emissions: a review. Environ Sci Pollut Res 22:3342–3360. https://doi.org/10.1007/s11356-014-3760-4

Idrissou Y, Vall E, Blanfort V et al (2024) Integrated crop-livestock effects on soil carbon sequestration in Benin. West Africa Heliyon 10. https://doi.org/10.1016/j.heliyon.2024.e28748

IPCC (2021) "Climate change 2021: the physical science basis" in Contribution of working group I to the sixth assessment report of the intergovernmental panel on climate change. Cambridge University Press, Cambridge

IPCC (2023) Climate change 2023: Synthesis report. Global warming of 1.5°C, climate change and land, the ocean and cryosphere in a changing climate. Cambridge University Press, Cambridge

Jangir CK, Sangwan PS, Panghaal D, Kumar S, Meena RS, Bharti J, R.D. and Singh, N. (2024) Spatial variability and statistical analysis of soil properties in the rice wheat-based systems of North-West India. Commun Soil Sci Plant Anal 55(8):1205–1223

Jensen ES, Carlsson G, Hauggaard-Nielsen H (2020) Intercropping of grain legumes and cereals improves the use of soil N resources and reduces the requirement for synthetic fertilizer N: a global-scale analysis. Agron Sustain Dev 40(5)

Jhariya MK, Raj A, Banerjee A, Meena RS, Bargali SS, Kumar S, Nema S, Poonam, Oraon PR (2022) Plan and policies for soil organic carbon management under agroforestry system. In: Plans and policies for soil organic carbon management in agriculture. Springer, Singapore, pp 191–219

Khormali F, Ajami M, Ayoubi S, Srinivasarao C, Wani SP (2009) Role of deforestation and hillslope position on soil quality attributes of loess-derived soils in Golestan province. Iran Agri Ecosys Environ 134:178–189. https://doi.org/10.1016/j.agee.2009.06.017

Kopittke PM, Dalal RC, Hoeschen C, Li C, Menzies NW, Mueller CW (2020) Soil organic matter is stabilized by organo-mineral associations through two key processes: the role of the carbon to nitrogen ratio. Geoderma 357:113974. https://doi.org/10.1016/j.geoderma.2019.113974

Krause HM, Stehle B, Mayer J (2022) Biological soil quality and soil organic carbon change in biodynamic, organic, and conventional farming systems after 42 years. Agron Sustain Develop 42:117. https://doi.org/10.1007/s13593-022-00843-y

Kumar S, Sharma SK, Thakral SK, Bhardwaj KK, Jhariya MK, Meena RS, Jangir CK, Bedwal S, Jat RD, Gaber A, Atta AA (2022a) Integrated nutrient management improves the productivity and nutrient use efficiency of *Lens culinaris* Medik. Sustainability 14(3):1284

Kumar S, Bamboriya SD, Rani K, Meena RS, Sheoran S, Loyal A, Kumawat A, Jhariya MK (2022b) Grain legumes: a diversified diet for sustainable livelihood, food, and nutritional security. In: Advances in legumes for sustainable intensification. Academic Press, pp 157–178

Kumar S, Sharma SK, Dhaka AK, Bedwal S, Sheoran S, Meena RS, Jangir CK, Kumar D, Kumar R, Jat RD, Meena AK (2023) Efficient nutrient management for enhancing crop productivity, quality and nutrient dynamics in lentil (*Lens culinaris* Medik.) in the semi-arid region of northern India. PLoS One 18(2):e0280636

Kumar S, Meena RS, Sheoran P, Jharia MK, Ghosh S (2024) Regenerative agriculture for sustainable food system: an overview. In: Regenerative agriculture for sustainable food systems. Springer, Singapore, pp 1–13

Kumar S, Meena RS, Ghosh S (2025) Agriculture toward net zero emissions: An overview. In: Agriculture Toward Net Zero Emissions. Academic Press Publications, pp 1–9

Kumar S, Meena RS, Kumar S, Pradhan G, Jangir CK, Ghosh S, Punia H, Sheoran P, Meena R, Ahmad MA, Goyal SK (2025a) Designing a diversified Indian mustard production system for energy-carbon-cum-heat use efficiency and sowing dates assessment. GCB Bioenergy 17(6): e70044

Kumar Y, Kumar S, Kaushik N, Prakash V, Garg K, Nalia M, Ghosh S (2025b) Recent advancements in climate change projections and socioeconomic scenarios used to evaluate climate impacts and adaptation measures. In: Agriculture Toward Net Zero Emissions. Academic Press publications, pp 425–440

Laird DA (2008) The charcoal vision: a win–win–win scenario for simultaneously producing bioenergy, permanently sequestering carbon, while improving soil and water quality. Agron J 100:178–181. https://doi.org/10.2134/agrojnl2007.0161

Lal R (2017) Improving soil health and human protein nutrition by pulses-based cropping systems. Adv Agron 145:167–204. https://doi.org/10.1016/bs.agron.2017.05.003

Ledo A, Smith P, Zerihun A, Whitaker J, Vicente-Vicente JL, Qin ZC, McNamara NP, Zinn YL, Llorente M, Liebig M, Kuhnert M, Dondini M, Don A, Diaz-Pines E, Datta A, Bakka H, Aguilera E, Hillier J (2020) Changes in soil organic carbon under perennial crops. Glob Change Biol 26(7):4158–4168. https://doi.org/10.1111/gcb.15120

Leharwan M, Kumar Y, Kumar R, Kumar Saraswat P, Kumar R, Veetil KT, A., Bhattacharjee, S., Kumar, A. and Kumar, S. (2023) Assessing the effects of conservation tillage and in-situ crop residue management on crop yield and soil properties in rice–wheat cropping system. Sustainability 15(17):12736

Leoni F, Lazzaro M, Ruggeri M (2022) Relay intercropping can efficiently support weed management in cereal-based cropping systems when appropriate legume species are chosen. Agron Sustain Dev 42:75. https://doi.org/10.1007/s13593-022-00787-3

Li Y, Jiao J, Wang Z, Cao B, Wei Y, Hu S (2016) Effects of revegetation on soil organic carbon storage and erosion-induced carbon loss under extreme rainstorms in the hill and gully region of the loess plateau. Int J Environ Res Public Health 13:456. https://doi.org/10.3390/ijerph13050456

Liu C, Feng X, Xu Y (2023a) Legume-based rotation enhances subsequent wheat yield and maintains soil carbon storage. Agron Sustain Dev 43

Liu Q, Zhao Y, Li T, Chen L, Chen Y, Sui P (2023b) Changes in soil microbial biomass, diversity, and activity with crop rotation in cropping systems: A global synthesis. Appl Soil Ecol 186: 104815. https://doi.org/10.1016/j.apsoil.2023.104815

Lupwayi NZ, Turkington TK, Tidemann B (2024) Crop rotation and fungicides impact microbial biomass, diversity and enzyme activities in the wheat rhizosphere. Front Aron 6. https://doi.org/10.3389/fagro.2024.1455448

McClelland SC, Paustian K, Schipanski ME (2021) Management of cover crops in temperate climates influences soil organic carbon stocks: a meta-analysis. Ecol Appl 31:e02278. https://doi.org/10.1002/eap.2278

Meena RS, Pradhan G (2023) Industrial garbage-derived biocompost enhances soil organic carbon fractions, CO_2 biosequestration, potential carbon credits, and sustainability index in a rice-wheat ecosystem. Environ Res 116525. https://doi.org/10.1016/j.envres.2023.116525

Meena RS, Pradhan G, Kumar S, Lal R (2023) Using industrial wastes for rice-wheat cropping and food-energy-carbon-water-economic nexus to the sustainable food system. Renew Sust Energ Rev 187:113756. https://doi.org/10.1016/j.rser.2023.113756

Meena RS, Pradhan G, Nalani Singh K, Singh AK, Verma SK, Mina KK, Kumar S, Chaturvedi RK (2024) Diversified cropping modules designed for soil restoration, CO_2 sequestration, and generating carbon credits. Land Degrad Dev 35. https://doi.org/10.1002/ldr.5224

Meena RS, Pradhan G, Singh K, Kumar S, Singh AK, Shashidhar KS, Mina KK, Rao CS (2024a) Agriculture models for restoring degraded land to enhance CO_2 biosequestration and carbon credits in the Vindhyan region of India. Sci Total Environ 929:172661. https://doi.org/10.1016/j.scitotenv.2024.172661

Meena RS, Lal R, Kumar S, Pradhan G, Rao CS, Singh AK, Pathak H, Abhilash PC, Kumar A, Sharma SK, Jat ML, Singh S (2024b) Potential of Indian agriculture for capturing atmospheric CO_2 and monetizing carbon credits to the farmers: an overview and policy framework. Adv Agron 188:101–206. https://doi.org/10.1016/bs.agron.2024.06.001

Meena RS, Pradhan G, Kumar S (2025) Energy flow, eco-efficiency, and economic circulation with recycled industrial waste compost application in wheat and subsequent rice farming. Sci Total Environ 967:178779. https://doi.org/10.1016/j.scitotenv.2025.178779

Mirzavand J, Asadi-Rahmani H, Moradi-Talebbeigi R (2022) Biological indicators of soil quality under conventional, reduced, and no-tillage systems. Arch Agron Soil Sci 68:311–324. https://doi.org/10.1080/03650340.2020.1832656

Notz I, Topp CF, Schuler J (2023) Transition to legume-supported farming in Europe through redesigning cropping systems. Agron Sustain Dev 43:12. https://doi.org/10.1007/s13593-022-00861-w

Nunes PAA, Laca EA, Carvalho PCF, Li M, Souza Filho W, Kunrath TR et al (2021) Livestock integration into soybean systems improves long-term system stability and profits without compromising crop yields. Sci Rep 11:1649. https://doi.org/10.1038/s41598-021-81270-z

Oelbermann M, Voroney RP, Gordon AM (2004) Carbon sequestration in tropical and temperate agroforestry systems: a review with examples from Costa Rica and Southern Canada. Agri Ecosys Environ 104:359–377. https://doi.org/10.1016/j.agee.2004.04.001

Ontl TA, Schulte LA (2012) Soil carbon storage. Nature Educ Know 3(10)

Pandey DN (2007) Multifunctional agroforestry systems in India. Curr Sci 92:455–463

Plaza-Bonilla D, Nolot JM, Passot S, Raffaillac D, Justes E (2016) Grain legume-based rotations managed under conventional tillage need cover crops to mitigate soil organic matter losses. Soil Tillage Res 156:33–43. https://doi.org/10.1016/j.still.2015.09.021

Pradhan G, Meena RS (2023) Interaction impacts of biocompost on nutrient dynamics and relations with soil biota, carbon fractions index, societal value of CO_2 equivalent, and ecosystem services in the wheat-rice farming. Chemosphere 139695. https://doi.org/10.1016/j.chemosphere.2023.139695

Pradhan G, Meena RS (2023a) Utilizing waste compost to improve the atmospheric CO_2 capturing in the rice-wheat cropping system and energy-cum-carbon credit audit for a circular economy. Sci Total Environ 164572. https://doi.org/10.1016/j.scitotenv.2023.164572

Pradhan G, Meena RS, Kumar S, Lal R (2023) Utilizing industrial wastes as compost in wheat-rice production to improve the above and below-ground ecosystem services. Agric Ecosyst Environ 358:108704. https://doi.org/10.1016/j.agee.2023.108704

Punia H, Tokas J, Malik A, Satpal, Rani A, Gupta P, Kumari A, Mor VS, Bhuker A, Kumar S (2020) Solar radiation and nitrogen use efficiency for sustainable agriculture. Resources use efficiency in agriculture:177–212

Puri S (2007) Agroforestry practices in Chhattisgarh—scope and potentials. In: Puri S, Panwar P (eds) Agroforestry systems and practices. New India Publishing Agency, New Delhi, pp 227–240

Rahimi MR, Ayoubi S, Abdi MR (2013) Magnetic susceptibility and Cs-137 inventory variability as influenced by land use change and slope positions in a hilly, semiarid region of west-central Iran. J Appl Geophys 89:68–75. https://doi.org/10.1016/j.jappgeo.2012.11.009

Rigon JPG, Calonego JC, Rosolem CA, La Scala N (2018) Cover crop rotations in no-till system: Short-term CO_2 emissions and soybean yield. Scientia Agricola, Brazil Scientia Agricola 75(1): 18–26

Ruis SJ, Blanco-Canqui H, Jasa PJ, Jin VL (2022) No-till farming and greenhouse gas fluxes: insights from literature and experimental data. Soil Tillage Res 220:105359

Saad RF, Kobaissi A, Goux X, Calusinska M, Echevarria G, Kidd P, Benizri E (2018) Soil microbial and Ni-agronomic responses to Alyssum murale interplanted with a legume. Appl Soil Ecol 132:60–73. https://doi.org/10.1016/j.apsoil.2018.08.019

Sainju UM, Singh BP, Whitehead WF, Wang S (2006) Carbon supply and storage in tilled and nontilled soils as influenced by cover crops and nitrogen fertilization. J Environ Qual 35:1507–1517. https://doi.org/10.2134/jeq2005.0189

Scalise A, Pappa VA, Gelsomino A, Rees RM (2017) Pea cultivar and wheat residues affect carbon/nitrogen dynamics in pea-triticale intercropping: a microcosms approach. Sci Total Environ 592:436–450. https://doi.org/10.1016/j.scitotenv.2017.03.012

Serafim ME, Mendes IC, Wu J, Ono FB, Zancanaro L, Valendorff JDP et al (2023) Soil physico-chemical and biological properties in soybean areas under notill systems in the Brazilian Cerrado. Sci Total Environ 862:160674. https://doi.org/10.1016/j.scitotenv.2022.160674

Shackelford GE, Kelsey R, Dicks LV (2019) Effects of cover crops on multiple ecosystem services: ten meta-analyses of data from arable farmland in California and the Mediterranean. Land Use Policy 88:104204. https://doi.org/10.1016/j.landusepol.2019.104204

Shang Y, Olesen JE, Lærke PE et al (2024) Perennial cropping systems increased topsoil carbon and nitrogen stocks over annual systems—a nine-year field study. 365. https://doi.org/10.1016/j.agee.2024.108925

Sheoran S, Gupta M, Kumari S, Kumar S, Rakshit S (2022) Meta-QTL analysis and candidate genes identification for various abiotic stresses in maize (*Zea mays* L.) and their implications in breeding programs. Mol Breed 42(5):26

Silva AA, Lacerda JJJ, Carvalho SP, Ferreira RS, Brito RR, Vogado RF et al (2024) Chemical and biological attributes of soil and soybean (Glycine max) yield in integrated systems in the Cerrado of north-east Brazil. Soil Res 62:SR23120. https://doi.org/10.1071/SR23120

Stomph T, Dordas C, Baranger A, de Rijk J, Dong B, Evers J, Gu C, Li L, Simon J, Jensen ES, Wang Q, Wang Y, Wang Z, Xu H, Zhang C, Zhang L, Zhang WP, Bedoussac L, van der Werf W (2020) Chapter one-designing intercrops for high yield, yield stability and efficient use of resources: are there principles. In: Sparks DL (ed) Advances in Agronomy, vol 160. Academic Press, pp 1–50. https://doi.org/10.1016/bs.agron.2019.10.002

Swamy SL, Mishra A, Puri S (2003) Biomass production and root distribution of Gmelia arborea under an agri-silviculture system in sub-humid tropics of central India. News Forest 26:167–186

Tisdall JM, Oades JM (1982) Organic matter and water-stable aggregates in soils. Eur J Soil Sci 33: 141–163. https://doi.org/10.1111/j.1365-2389.1982.tb01755.x

Topps D, Khabir MI, Abdelmagid H, Jackson T, Iqbal J, Robertson BK et al (2021) Impact of cover crop monocultures and mixtures on organic carbon contents of soil aggregates. Soil Syst 5:43. https://doi.org/10.3390/soilsystems5030043

Trivedi P, Delgado-Baquerizo M, Jeffries TC, Trivedi C, Anderson IC, Lai K, McNee M, Flower K, Pal Singh B, Minkey D, Singh BK (2017) Soil aggregation and associated microbial communities modify theimpact of agricultural management on carbon content. Environmental microbiology 19(8):pp.3070–3086

Verhulst N, Kienle F, Sayre KD, Deckers J, Raes D, Limon-Ortega A, Tijerina-Chavez L, Govaerts B (2011) Soil quality as affected by tillage-residue management in a wheat-maize irrigated bed planting system. Plant Soil 340:453–466. https://doi.org/10.1007/s11104-010-0618-5

Wackernagel M, Rees W (1998) Our ecological footprint: reducing human impact on the earth. New Society Publishers, United Kingdom

Wang JL, Liu KL, Zhao XQ, Zhang HQ, Li D, Li JJ (2016) Balanced fertilization over four decades has sustained soil microbial communities and improved soil fertility and rice productivity in red paddy soil. Sci Total Environ 793:148664. https://doi.org/10.1016/j.scitotenv.2021.148664

Wang W, Li M, Wen Q (2025) Cereal-legume intercropping stimulates straw decomposition and promotes soil organic carbon stability. Sci China-Life Sci 68. https://doi.org/10.1007/s11427-024-2683-y

Wegner BR, Chalise KS, Singh S, Lai L, Abagandura GO, Kumar S, Osborne SL, Lehman RM, Jagadamma S (2018) Response of soil surface greenhouse gas fluxes to crop residue removal and cover crops under a corn-soybean rotation. J Environ Qual 47(5):1146–1154

Yogi AK, Bana RS, Godara S, Sangwan S, Choudhary AK, Nirmal RC, Bamboriya SD, Shivay YS, Singh D, Singh T, Yadav A, Nagar S, Singh N (2023) Elucidating the interactive impact of tillage, residue retention and system intensification on pearl millet yield stability and biofortification under rainfed agro-ecosystems. Front Nutr 10:1205926. https://doi.org/10.3389/fnut.2023.1205926

Zeraatpisheh M, Ayoubi S, Mirbagheri Z, Mosaddeghi MR, Xu M (2021) Spatial prediction of soil aggregate stability and soil organic carbon in aggregate fractions using machine learning algorithms and environmental variables. Geoderma 27:e00440. https://doi.org/10.1016/j.geodrs.2021.e00440

Zhang X, Sallam A, Gao L, Kantarski T, Poland J, DeHaan LR, Wyse DL, Anderson JA (2016) Establishment and optimization of genomic selection to accelerate the domestication and improvement of intermediate wheatgrass. Plant Genome 9. https://doi.org/10.3835/plantgenome2015.07.0059

Zhang ZL, Kaye JP, Bradley BA, Amsili JP, Suseela V (2022) Cover crop functional types differentially alter the content and composition of soil organic carbon in particulate and mineral-associated fractions. Glob Chang Biol 28:5831–5848. https://doi.org/10.1111/gcb.16296

Zhou G, Chang D, Gao S, Liang T, Liu R, Cao W (2021) Co-incorporating leguminous green manure and rice straw drives the synergistic release of carbon and nitrogen, increases hydrolase activities, and changes the composition of main microbial groups. Biol Fertil Soils 57:547–561. mhb6m66 666666

Chapter 6
Measurement, Reporting, and Verification Methodologies for Carbon Credit Towards the Carbon-Negative Agriculture

Shambhunath Ghosh, Vikas Patidar, Anusmita Saha, Hemant Kumar Mina, and Sandeep Kumar

Abstract Accurate measurement and verification of carbon credits in agriculture are essential to uphold the credibility and effectiveness of agricultural carbon markets. These credits, derived from carbon sequestration practices like conservation tillage, cover cropping, and agroforestry, rely on precise assessment to quantify sequestration, build market confidence, and align with global climate objectives. Challenges remain due to the diverse nature of soil types, climatic conditions, and management practices, which make carbon accounting complex. However, emerging technologies such as remote sensing, machine learning, and blockchain are revolutionizing measurement and verification by increasing accuracy, lowering costs, and improving transparency. Innovations in soil sampling, modeling, and satellite-based monitoring provide more precise and scalable methods for assessing carbon stocks and fluxes. Blockchain technology improves traceability, safeguarding the integrity of credits throughout their lifecycle. Despite these advancements, establishing uniform standards remains challenging, particularly for smallholder farmers who often encounter barriers such as limited resources and technical complexities. To ensure their inclusion in carbon markets, it is essential to develop equitable frameworks, provide financial and technical assistance, and promote collaboration among stakeholders. Policy support plays a crucial role as governments and international organizations strive to build robust governance frameworks and align agricultural carbon credit standards with broader climate goals. Future initiatives should prioritize combining technological advancements with inclusive stakeholder engagement to promote equity and uphold environmental integrity. Reliable measurement and verification of agricultural carbon credits will

S. Ghosh (✉) · V. Patidar · A. Saha
ICAR-Indian Institute of Wheat and Barley Research, IARI Karnal Hub, Karnal, Haryana, India

H. K. Mina
Institute of Agricultural Sciences, Banaras Hindu University, Varanasi, Uttar Pradesh, India

S. Kumar
ICAR-Indian Agricultural Research Institute, Regional Station, Karnal, Haryana, India

R. S. Meena et al. (eds.), *Carbon-Negative Agriculture*, Sustainability Sciences in Asia and Africa, https://doi.org/10.1007/978-981-95-1483-0_6

not only strengthen market credibility but also advance sustainable development and global climate action.

Keywords Carbon credit · Carbon sequestration · Climate change · Verification

Abbreviations

C	Carbon
CAA	Clean Air Act
CDM	Clean Development Mechanism
CH_4	Methane
CO_2	Carbon dioxide
CO_2eq	Carbon dioxide equivalent
COP	Conference of the Parties
DOM	Dissolved organic matter
ETS	Emission trading system
EU	European Union
FAO	Food and Agriculture Organization
GHGs	Greenhouse gases
GWP	Global warming potential
IPCC	Intergovernmental Panel on Climate Change
LULUCF	Land use, land-use change and forestry
SOC	Soil organic carbon
SOM	Soil organic matter
VCM	Voluntary carbon market
VCS	Verified carbon standard
VERs	Verified emission reductions

6.1 Introduction

Climate change is widely acknowledged as a critical global challenge, posing significant hazards to sustainability across all sectors (Pan et al. 2022). It presents major risks to economic stability, disrupting agriculture, energy, tourism, and health sectors, while intensifying extreme weather events such as hurricanes, floods, and wildfires (Angwaomaodoko 2024; Kumar et al. 2025). The impacts of climate change are extensive, endangering human health and well-being, undermining assets, response capacities, and infrastructure, and significantly increasing the number of individuals and communities facing social, economic, and psychological vulnerabilities (Palinkas and Wong 2020). Climate change has widespread effects, jeopardizing human health and well-being, weakening assets, response systems, and infrastructure, and greatly amplifying the social, economic, and psychological vulnerabilities of individuals and communities (Chachar et al. 2023). Climate change intersects with population dynamics, migration, and urbanization, heightening risks for the most vulnerable populations (Sahoo and Rath 2023). Throughout history, climate change has played a pivotal role in shaping the trajectory of societies and states, contributing to instability, conflict, and collapse, while also fostering periods of expansion and reorganization (Hulme 2007). Climate change refers to long-term shifts or variations in typical weather patterns that last for years or even decades. While natural factors such as solar activity, volcanic eruptions, and changes in Earth's orbit can drive these changes, the current rapid and unprecedented shifts are largely the result of human activities.

Since the eighteenth century, industrial activities, especially burning fossil fuels like coal, oil, and gas, have accelerated climate change by releasing greenhouse gases (GHGs). Key GHGs, including carbon dioxide (CO_2), methane (CH_4), and nitrous oxide (N_2O), disrupt the climate system and drive global warming (Filonchyk et al. 2024; Meena et al. 2025). While all three gases trap heat, their impacts vary by global warming potential (GWP), a measure comparing their warming effects over 100 years. CO_2 has a GWP of 1, CH_4 is 25 times more potent, and N_2O is nearly 300 times stronger. The Intergovernmental Panel on Climate Change's (IPCC) Sixth Assessment Report highlights human activities, particularly GHG emissions, as the primary driver of global warming. This conclusion is supported by extensive scientific research, data, and climate modeling (IPCC 2023). Among these gases, CO_2 has the greatest impact due to its abundance and persistence in the atmosphere. CO_2 emissions, mainly from burning fossil fuels and deforestation, have steadily risen, increasing atmospheric concentrations from about 280 ppm pre-industrial levels to over 420 ppm in recent years—a significant rise in a short geological period. The relentless rise in GHG concentrations, particularly CO_2, acts as a key catalyst for climate change, driving a range of interconnected environmental impacts. These include rising global temperatures, sea-level rise, more frequent and intense extreme weather events, changes in precipitation patterns, ocean acidification, and the loss of biodiversity. The cascading effects of these changes pose significant risks to human health, food security, water resources, and economic stability, making climate change a multifaceted challenge that requires urgent and coordinated global action to mitigate its impacts and adapt to the changes already set in motion (Kumar et al. 2022). Managing CO_2 levels is challenging, but emissions can be reduced through carbon sequestration in plants or soil. Plants absorb CO_2 through photosynthesis and store it in their tissues, releasing oxygen back into the atmosphere, thus helping to mitigate rising CO_2 levels. Practices like agroforestry and conservation agriculture generate carbon credits, which play a key role in national and global efforts to reduce GHG emissions and fight global warming (Reichle 2020). A carbon credit allows a country or organization to emit a specific amount of CO_2 or equivalent GHG (CO_2eq), typically one metric ton per credit. Unused credits can be traded, creating financial incentives for emission reductions. These credits are exchanged through brokers, online markets, and trading platforms, supporting global efforts to reduce emissions by promoting resource efficiency and investment in low-carbon technologies. Once certified, carbon credits

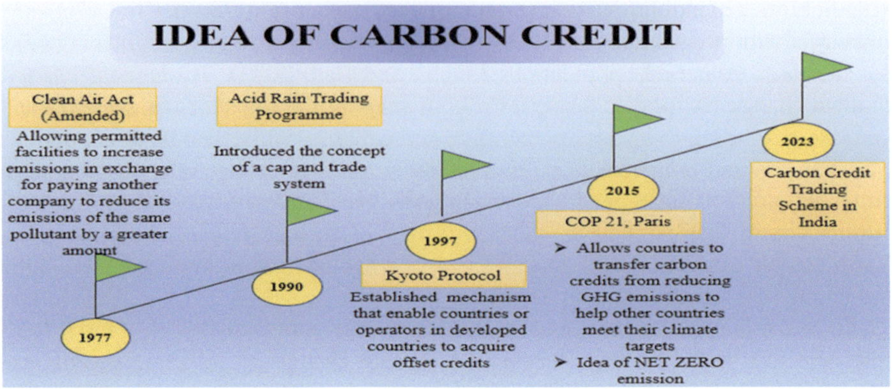

Fig. 6.1 The idea of carbon credit

can be bought and sold (Raihan et al. 2022; Punia et al. 2020). Each sector has a government-established cap on GHG emissions. Organizations that emit below this limit earn carbon credits, while those exceeding the cap must buy credits from those with a surplus. The concept of carbon credits emerged from global climate action efforts, starting with the Kyoto Protocol in 1997, which introduced the clean development mechanism (CDM) (Fig. 6.1).

The Clean Air Act (CAA) is the United States' primary federal air quality law, intended to reduce and control air pollution nationwide. Initially enacted in 1963 and amended many times, it is one of the United States' first and most influential modern environmental laws. This system allows industrialized countries to earn credits by funding cost-effective emission reduction projects in developing nations. In the European Union (EU's) emissions trading system, a single carbon credit is priced at US$ 61.3 (April 2024). For farmers, reducing GHG emissions could provide a sustainable income source through carbon credits, supporting both economic goals and the broader net-zero emission target (Weiskopf et al. 2020; Meena et al. 2024c).

6.2 The Methodology Used for Evaluating Carbon Credit

A carbon credit is a tradable unit that individuals and organizations can use to offset their GHG emissions. In the voluntary carbon market (VCM), each credit represents the reduction or prevention of one metric ton of GHGs from entering the atmosphere. Carbon credits are commonly used to meet voluntary emission reduction targets or corporate sustainability goals, but their value can fluctuate due to the unregulated nature of the market. Factors such as market trends, project costs, location, and the managing organization all impact the price, leading to variability in carbon credit values (Pathak et al. 2021). Therefore, the price of carbon credits can vary significantly depending on these factors, which should be considered when calculating and reporting them.

The UK Department for Environment, Food, and Rural Affairs (DEFRA) outlines a five-step process for calculating carbon credits to offset emissions:

1. **Identify emission activities**: Pinpoint activities responsible for GHG emissions, such as industrial processes or transportation.
2. **Quantify emissions**: Measure the emissions from each identified activity using data such as energy use or fuel consumption.
3. **Obtain emission factors**: Use emission factors to convert activity data into actual emissions (e.g., CO_2 per kilowatt-hour).
4. **Convert to CO_2 equivalents**: Standardize emissions by converting them to CO_2eq using each gas's GWP.
5. **Calculate total emissions and market rate**: Total emissions and multiply by the carbon credit market rate to determine the number of credits needed for offset.

This process ensures accurate carbon credit calculation and provides a standardized method for determining credit values in the market.

6.2.1 Carbon Pricing and Mechanisms

Carbon pricing is a policy tool designed to encourage organizations, companies, and individuals to reduce their GHG emissions, particularly CO_2. The goal is to assign a cost to emitting carbon, thus incentivizing emission reductions by making it more expensive to pollute. There are two primary mechanisms through which carbon pricing is implemented:

1. **Carbon tax**: A carbon tax imposes a fee on the production of each unit of GHG emissions. Under this system, every company, organization, or individual pays a specified tax based on the amount of CO_2 or other GHGs they emit. This tax serves as a financial deterrent, encouraging entities to lower their emissions to reduce their tax burden. Unlike other mechanisms, there is no specific emissions threshold or cap, meaning that every ton of GHG produced incurs the same tax rate.
2. **Emission trading system (ETS)**: The emission trading system (ETS) is a market-based approach where a cap is set on the total amount of CO_2 emissions that can be produced by participating entities. Regulatory bodies determine an overall emissions cap and allocate carbon credits to businesses or organizations within this limit. These credits can be traded, so companies that do not use all their allocated emissions allowances can sell the surplus credits to other companies that are likely to exceed their emissions limit. Over time, the cap is gradually lowered, which increases the price of carbon credits, creating a financial incentive for companies to decarbonize. This system promotes the reduction of overall emissions while providing flexibility in how businesses meet their emissions targets.

There are two main types of carbon markets where carbon trading occurs. A carbon credit market is a platform where carbon credits, representing emission reductions, are bought and sold. These markets allow companies, governments, or individuals to offset their emissions by investing in projects that reduce or capture carbon. The goal is to reduce global GHG emissions by exchanging carbon credits, providing financial incentives for adopting low-carbon alternatives, and raising awareness about the impacts of carbon pollution. The two key types of carbon markets are:

6.2.1.1 Compliance Carbon Markets (Allowance Markets)

Compliance carbon markets, also known as allowance markets, operate under a legally binding framework. In these markets, buyers—typically companies or industries—are legally required to reduce their GHG emissions to meet regulatory requirements. These markets operate through a Cap-and-Trade system, where a cap (limit) is set on the total emissions a company is allowed to produce. If a company exceeds this cap, it must purchase additional carbon credits to offset the excess emissions. Alternatively, companies can generate carbon credits by reducing their emissions below their allocated cap, and these credits can be sold in the market to other companies that have exceeded their limits. The cap is typically lowered over time to ensure continued progress in emission reductions.

6.2.1.2 Voluntary Carbon Market

The voluntary carbon market differs from the compliance market in that participants are not legally required to purchase carbon credits. Instead, buyers choose to buy credits voluntarily. These credits, known as verified emission reductions (VERs), are typically purchased by private companies seeking to reduce their carbon footprint or fulfill their corporate social responsibility goals. To ensure credibility, VERs must be independently verified by third parties. In the voluntary market, the primary motivation for buying credits is often to enhance the company's public image, demonstrate environmental stewardship, or contribute to global emission reduction efforts. This market plays an important role in financing carbon reduction projects that may not be covered under compliance mechanisms, thus supporting a broader range of environmental initiatives.

In both types of carbon markets, the goal is the same: to reduce overall GHG emissions by creating a financial incentive for emission reductions. However, while compliance markets are driven by legal obligations, voluntary markets are driven by corporate responsibility, environmental goals, and the desire to contribute to global climate mitigation efforts.

6.2.2 Determination of Carbon Credit

Accurately determining carbon (C) credits linked to soil carbon sequestration in carbon offset markets is a highly complex and evolving process. This intricate task requires a thorough evaluation of the methodologies employed, beginning with an in depth analysis of the foundational scientific principles underlying these approaches. It involves examining how these principles are applied in practice, assessing their limitations, and addressing the uncertainties inherent in each method (Das et al. 2018; Meena et al. 2024b). Such scrutiny aims to illuminate the diverse methodologies—from theoretical models to real-world applications—while identifying the barriers that may hinder precise C credit assessments. The primary methods for estimating soil C credits encompass empirical models, field-based direct measurements, and remote sensing paired with geospatial analytical tools. These techniques form the backbone of soil C credit calculation, providing the scientific basis for quantifying sequestration. However, they operate within the broader context of economic, policy, and technological frameworks that influence the valuation and verification of credits in carbon markets. Each method faces challenges, particularly concerning temporal and spatial variability in soil C stock assessments. Methodological uncertainties, such as variability in soil types, land management practices, and climatic influences, introduce complexities. To counteract the risks of overestimating carbon credits, conservative baselines and cautious assumptions are often adopted. This approach not only ensures greater reliability but also aligns with the overarching goal of fostering transparency and credibility in carbon offset markets. Generally, there are two ways for measuring the amount of carbon sequestered, which also helps to estimate the carbon credits, and those are measurement of soil carbon stock and measurement of GHG emission from soil (Fig. 6.2).

Direct methods for estimating C are more precise, involving steps to measure soil C stocks and convert them into C credits. This process starts with collecting soil

Fig. 6.2 Measurement of carbon sequestration in soil

samples from various depths and sites to account for spatial variability. The samples are analyzed to determine organic carbon (OC) content. While this method provides detailed, site-specific data, it is costly and labor-intensive, making it less feasible for large-scale projects. The carbon stock is then calculated using the laboratory results and the following equation:

$$\text{Total soil carbon stock}\ \left(\text{Mg ha}^{-1}\right) = \Delta SOC \times BD \times D \times A \tag{6.1}$$

where:

ΔSOC = change in soil organic carbon content, percentage or g kg^{-1}
A = area of land, hectares
BD = bulk density of the soil, g cm^{-3}
D = depth of soil layer considered, cm

This formula assumes a uniform distribution of soil organic carbon (SOC) changes across depth and area, but in reality, SOC levels and bulk density vary due to factors like soil type, land use, and management practices. These variations require careful consideration to ensure accurate soil carbon stock calculations. When estimating changes in soil carbon stocks (ΔSOC), including accumulation, transfers, and losses, data must be standardized to C units for consistency. Measurements in dry matter, like biomass growth or harvest volumes, must be converted into tons of C to comply with carbon accounting protocols.

To calculate changes in soil carbon stocks, two equally robust yet methodologically distinct approaches can be employed:

1. **Process-based method**: This approach focuses on quantifying the net balance of carbon inputs (e.g., organic amendments or plant residues) and outputs (e.g., decomposition and emissions) within a defined soil stock. It is particularly useful for understanding the dynamics of carbon cycling over time.
2. **Stock-based method**: This method evaluates changes by directly comparing SOC levels at two distinct points in time, relying on field measurements or laboratory analyses to assess the differences. It provides a straightforward snapshot of carbon gains or losses, making it a preferred choice for monitoring projects with clear start and end points.

Both methods have distinct strengths, chosen based on project scale, data availability, and objectives. Incorporating variability and contextual factors is crucial for accurate soil carbon stock calculations. The process-based approach, using the gain-loss method (Eq. 6.2), calculates yearly changes in C stocks by accounting for all C inputs (gains) and outputs (losses). Gains include biomass growth and C transfers between pools, while losses occur through activities like harvesting or decomposition. This method provides a comprehensive understanding of carbon fluxes, making it valuable for assessing land management, environmental changes, and disturbances at various scales.

$$\Delta C = \Delta C_G - \Delta C_L \tag{6.2}$$

where:

ΔC = annual C stock change in the pool, Mg C yr^{-1}
ΔCG = annual gain of carbon, Mg C yr^{-1}
ΔCL = annual loss of carbon, Mg C yr^{-1}

CO_2 removals refer to the absorption of CO_2 from the atmosphere into a C pool, while CO_2 emissions represent its release back into the atmosphere. Carbon transfers between pools are recorded as losses for the donor pool and gains for the recipient, without immediately affecting atmospheric CO_2. For example, when carbon moves from biomass to deadwood, it is documented as a decrease in biomass and an increase in deadwood, but not as an emission. The stock-difference method (Eq. 6.3) calculates carbon stock changes by measuring carbon in pools at two points in time, reflecting the net change in storage. Both methods are complementary: the process-based approach tracks fluxes, while the stock-difference method captures overall carbon storage changes.

$$\Delta C = \frac{(Ct2 - Ct1)}{(t2 - t1)} \tag{6.3}$$

where:

ΔC = annual C stock change in the pool, Mg C yr^{-1}
C_{t1} = C stock in the pool at time t_1, Mg C
C_{t2} = C stock in the pool at time t_2, Mg C

To calculate changes in C stocks per hectare, the value obtained is multiplied by the total area of each stratum to estimate the overall stock change. If aggregate data, such as national-level harvest statistics, is available, stock change is derived using conversion factors to express the results in carbon mass. For accurate results using the stock-difference method, it's crucial to maintain a consistent land area for each category at both time points (t_1 and t_2) to avoid inaccuracies in carbon stock change estimates.

The process method, which uses empirical coefficients, smooths annual fluctuations better than the stock change method, which tracks stock differences over time. Both methods are valid if they accurately capture disturbances and trends, and can be verified with real-world data. Soil C stocks must be converted into C credits by establishing baseline and project scenarios, assessing net sequestration, and converting the results into CO_2eq. The baseline scenario represents carbon stock before any intervention, while the project scenario measures change due to specific actions like reforestation or land management. By comparing these scenarios, the additional carbon sequestered is quantified to generate C credits.

The conversion of soil carbon to CO_2 equivalents uses a molecular weight factor (−44/12) to standardize reporting. Non-CO_2 emissions like CH_4 must be subtracted from CO_2 totals. Accurate baseline data and consistent monitoring are essential for effective carbon credit generation and climate change mitigation.

$$CO2 \text{ sequestered } (Mg\ ha-1) = C \text{ stock} \times \frac{\text{Molecular weight of CO2 } (44.01 \text{ g/mol})}{\text{Molecular weight of } C \ (12.01 \text{ g/mol})}$$

Or

$$CO_2 \text{ sequestered } \left(Mg\ ha^{-1}\right) = C \text{ stock } \left(Mg\ ha^{-1}\right) \times 3.67 \tag{6.4}$$

For example, 10 Mg ha^{-1} of carbon (*C*) sequestered in the soil equals 36.7 Mg ha^{-1} of CO_2 removed. This conversion is crucial for accurate carbon accounting in markets and climate assessments. In carbon trading, each carbon credit typically represents one metric ton of CO_2 equivalent sequestered or reduced. The conversion must follow established protocols and undergo verification to ensure accuracy and maintain the credibility of carbon credits in environmental mitigation efforts.

$$\text{Soil } C \text{ credit } \left(\text{US ha}^{-1}\right) = \text{Soil CO}_2 \text{ sequestration } \left(\text{Mg ha}^{-1}\right) \times \text{US } 50 \tag{6.5}$$

For example, a land restoration initiative sequestering 15 Mg of CO_2 per hectare would generate US$ 750 worth of carbon credits (15 Mg × US$ 50). However, to ensure the credibility of these credits, the project must undergo a rigorous verification process under recognized standards, such as the Verified Carbon Standard (VCS) or the Gold Standard. This process confirms that the sequestration is real, permanent, additional, and independently verified. Accurate measurement, documentation, and monitoring are essential to maintaining the integrity of the carbon credits and ensuring they can be reliably traded or used to offset emissions.

6.2.3 Carbon Credit Estimation in Croplands

Cropland includes areas like arable land, rice paddies, and agroforestry systems, which may not meet specific vegetation thresholds. It encompasses both annual crops (grains, oilseeds, vegetables, etc.) and perennial crops (trees, shrubs, orchards). Arable land, used for annual crops or mixed with fodder crops and grazing, is also classified as cropland. The carbon stored in cropland varies based on factors like crop type, management practices, and soil and climate conditions. Soil carbon stocks in cropland can be significant, influenced by farming techniques and environmental factors (Babu et al. 2020; Meena et al. 2023). Soil carbon levels are significantly influenced by factors such as crop rotation, tillage methods, drainage practices, residue management, and the addition of organic materials. Implementing effective soil management practices, like using cover crops or reducing tillage, can help to stabilize and increase soil carbon stocks, highlighting cropland's role in carbon sequestration (Jangir et al. 2024).

Estimating carbon changes in cropped areas involves evaluating annual variations in carbon stocks across all pools and sources, as well as calculating the annual emissions of non-CO_2 gases from these pools. Carbon stock changes within a specific stratum are determined by summing the changes from all collections, as described in Eq. 6.6.

$$\Delta C_{LUi} = \Delta C_{AB} + \Delta C_{BB} + \Delta C_{DW} + \Delta C_{LI} + \Delta C_{SO} + \Delta C_{HWP} \tag{6.6}$$

where ΔC_{LUi} = *C* stock changes for a stratum of a land-use category; *AB* = aboveground biomass; *BB* = belowground biomass; *DW* = deadwood; *LI* = litter; *SO* = soils; *HWP* = harvested wood products.

Cropland management plays a significant role in shaping soil carbon (C) stocks, with variations depending on how specific practices impact *C* inputs and outputs (Roy et al. 2022). Important factors influencing soil *C* levels include residue management, tillage methods, the application of mineral and organic fertilizers, crop choice, cropping intensity (such as continuous cropping or crop rotations with fallow periods), irrigation practices, and mixed systems that integrate cropping with pasture or hay (Leharwan et al. 2023). Additionally, practices like draining and cultivating organic soils can lead to a decline in soil C stocks. These relationships are supported by research from studies such as Bruce et al. (1999) and Ogle et al. (2005). The overall change in soil C stocks for cropland is calculated using Eq. 6.10, which incorporates changes in soil organic carbon stocks for both mineral and organic soils, as well as alterations in soil inorganic carbon pools.

$$\Delta C_{\text{Soils}} = \Delta C_{\text{Mineral}} - L_{\text{Organic}} + \Delta C_{\text{Inorganic}} \tag{6.7}$$

where:

ΔC_{Soils} = annual change in *C* stocks in soils, Mg *C* yr^{-1}
$\Delta C_{\text{Mineral}}$ = annual change in organic *C* stocks in mineral soils, Mg C yr^{-1}
L_{Organic} = annual loss of *C* from drained organic soils, Mg C yr^{-1}
$\Delta C_{\text{Inorganic}}$ = annual change in inorganic *C* stocks from soils, Mg C yr^{-1}

To estimate changes in soil organic carbon (SOC) stocks in mineral soils, subtract the carbon stock at the end of the period (SOC_o) from the starting stock (SOC_o-T), then divide by a time factor (D). This uses reference SOC values and default change factors for land use, management, and inputs. After calculating carbon sequestration in biomass, dissolved organic matter (DOM), and soil, the next step is converting the sequestered carbon into CO_2 equivalent (CO_2eq) and carbon credits. One carbon credit represents one ton of net CO_2 bio-sequestration, including plant parts like grain, straw, and roots (Meena et al. 2024a). This process is crucial in determining the carbon credit associated with a particular cropping system, as it helps quantify the climate benefits of agricultural practices.

To calculate the total carbon credited in a cropping system, sequestered carbon is linked to dry matter production using CO_2eq values. Aboveground biomass carbon (grain, straw, and stubble) is calculated by multiplying total dry matter by 0.44 (Choudhary et al. 2018; Meena and Pradhan 2023), reflecting the 44% carbon content of aboveground biomass. Belowground biomass carbon (roots) is determined by multiplying aboveground carbon by 0.15 (Mandal et al. 2007). The total CO_2 bio-sequestration is calculated by accounting for both above- and belowground biomass carbon. Finally, carbon credits are determined by multiplying CO_2 sequestration by the market price of one ton of CO_2 (Roy et al. 2022).

$$\text{Aboveground biomass } C = \text{Total aboveground dry matter } \left(Mg\ ha^{-1}\right) \times 0.44 \tag{6.8}$$

$$\text{Belowground biomass } C = \text{Aboveground biomass } \left(Mg\ ha^{-1}\right) \times 0.15 \tag{6.9}$$

$$\text{Total biomass } C \text{ stock} = \text{Aboveground biomass } C + \text{Belowground biomass } C \tag{6.10}$$

The total biomass carbon stock is converted into CO_2 bio-sequestration and then into biomass carbon credits using Eq. 6.4. Similarly, the soil carbon credit is determined by multiplying the SOC pool by the soil CO_2 sequestration. The SOC stock is calculated using the standard equation. The total carbon credit is the sum of both the biomass carbon credit and the soil carbon credit, as shown in Eq. 6.11.

$$\text{Total } C \text{ credit } \left(US\ ha^{-1}\right) = \text{Biomass } C \text{ credit} + \text{Soil } C \text{ credit} \tag{6.11}$$

6.2.4 Carbon Credit Estimation in Agroforestry Systems

The carbon credit of a tree in an agroforestry system is calculated by estimating CO_2 sequestration through biomass and its carbon content. Aboveground biomass carbon (leaves, branches, and trunks) is calculated by multiplying total dry matter by 0.50 (Devi and Singh 2021), reflecting that 50% of the biomass is carbon. Belowground biomass carbon (roots) is estimated by multiplying aboveground biomass by 0.26 (Roy et al. 2022), as roots contain 26% of the carbon in the aboveground parts. This comprehensive calculation provides the total carbon sequestration, which can be converted into carbon credits for trading or climate mitigation programs (Saimun et al. 2021; Pradhan and Meena 2023).

$$\text{Aboveground biomass } C \text{ in tree species} = \text{Total aboveground dry matter } \left(Mg\ ha^{-1}\right) \times 0.5 \tag{6.12}$$

$$\text{Belowground biomass } C \text{ in tree species} = \text{Aboveground biomass } \left(Mg\ ha^{-1}\right) \times 0.26 \tag{6.13}$$

The total biomass carbon stock in tree species is determined by combining the carbon in both aboveground and belowground biomass, as outlined in Eq. 6.13. Similarly, the CO_2 bio-sequestration of a tree species is estimated by multiplying the total biomass carbon stock by a factor of 3.67, as shown in Eq. 6.4. The biomass carbon credit for the tree species is then calculated by multiplying the CO_2 bio-sequestration by the tree species with a value of US$ 50, as described in Eq. 6.5. The soil carbon credit for the tree species can be estimated using the same method applied to field crops.

6.2.5 Carbon Credit Estimation from Grassland Ecosystems

CO_2 bio-sequestration is calculated using the methodology from the IPCC (Dondini et al. 2018). Carbon input from plant residues (CRes) is estimated by determining the total carbon content of both aboveground and belowground residues, including stems, leaves, seeds, and root biomass. The biomass carbon credit in a grassland ecosystem is then calculated by multiplying CO_2 bio-sequestration by the market price of one ton of CO_2 sequestration (Roy et al. 2022). This process quantifies the economic value of carbon sequestration in grasslands, offering financial incentives for sustainable land management and emphasizing the role of grasslands in climate change mitigation.

$$\begin{aligned}\text{Total biomass C stock } \left(Mg\ ha^{-1}\right) &= C \text{ in aboveground residues } \left(Mg\ ha^{-1}\right) \\ &+ C \text{ in belowground residues } \left(Mg\ ha^{-1}\right)\end{aligned} \tag{6.14}$$

Following the IPCC methodology, aboveground residues can be estimated based on the total yield and subsequently converted into their carbon content using the following approach (Intergovernmental Panel on Climate Change (IPCC) 2023):

$$C \text{ in aboveground residues } \left(Mg\ ha^{-1}\right) = (\text{Aboveground dry matter} \times 0.3) \times 0.475 \tag{6.15}$$

The same approach can be used to estimate the C content from the belowground biomass residues:

$$C \text{ in belowground residues } \left(Mg\ ha^{-1}\right) = \text{Belowground dry matter} \times 0.9 \quad (6.16)$$

To calculate CO_2 bio-sequestration from biomass carbon, the carbon stored is converted into CO_2 equivalents using the molecular weights of carbon and CO_2 (Eq. 6.4). This conversion estimates the CO_2 sequestered through biomass, aiding in understanding ecosystems' role in capturing atmospheric CO_2 and mitigating climate change. Estimating CO_2 sequestration also helps evaluate the effectiveness of land management practices. In grasslands, carbon is assessed across five pools: aboveground biomass, belowground biomass, dead wood, litter, and soil organic matter (SOM). Accurately measuring these pools highlights the carbon sequestration potential and guides management to enhance climate change mitigation.

6.3 The Importance of Carbon Credit Assessment in Different Cropping Systems

The importance of carbon credit assessment in different cropping systems lies in its potential to encourage sustainable agricultural practices, mitigate climate change, and support the global carbon market. By quantifying the carbon sequestration and emissions of various cropping systems, farmers, policymakers, and organizations can make informed decisions to reduce GHG emissions and receive financial incentives. Here's a closer look at the key benefits:

Agriculture is the second-largest source of anthropogenic GHG emissions. Identifying carbon-efficient cropping systems is key to reducing these emissions and improving environmental quality. Using life cycle assessment (LCA), Kantwa et al. (2024) evaluated four cropping systems from 2018 to 2022: fallow-chickpea (F–C), Sesbania-mustard (Ses–M), black gram-chickpea (B–C), and sorghum + cowpea-mustard (S + C–M). Energy use and carbon footprints were analyzed using three metrics: CFa (emissions per area), CFb (emissions per yield), and CFe (emissions per economic output). Non-renewable energy sources, particularly diesel and fertilizer, contributed over 80% of the total energy use. The highest total energy was required by S + C–M (16,972 MJ ha^{-1}), followed by Ses–M (14,365 MJ ha^{-1}), B–C (11,132 MJ ha^{-1}), and F–C (8679 MJ ha^{-1}). The S + C–M system also showed the highest energy efficiency (9.13), while F–C had the lowest carbon footprints but the highest carbon efficiency (10.7).

To mitigate climate change, adopting practices like crop rotation, intercropping, and cover cropping is essential for increasing carbon sequestration. Soil is a vital carbon reservoir, and sustainable agricultural practices improve soil quality, restore degraded land, enhance biodiversity, and reduce CO_2 emissions. Cover cropping, especially in semi-arid regions, has been recognized for improving SOC storage and enhancing soil health (Acharya et al. 2024).

Carbon sequestration in soils not only contributes to mitigating climate change but also enhances agricultural productivity and food security. A technical potential of 3 Peta-grams (Pg) C per year could reduce atmospheric CO_2 by 50 ppm by the end of the century, increasing food production in developing countries (Lal 2010). Incentivizing farmers with carbon credits for adopting sustainable practices also creates new income streams, supports carbon markets, and helps meet global climate goals.

However, the impact of management practices on soil carbon varies regionally. For example, no-till practices and cover cropping increase carbon content in the upper soil profile, but their effects at deeper soil depths are debated. Generally, conservation practices store more carbon in the top 10 cm, but deeper carbon storage may be influenced by soil type, climate, and management techniques. Despite uncertainties, conservation practices are crucial for increasing soil carbon and mitigating climate change.

6.4 Global Overview of Carbon Credit Potential in Agriculture

Agricultural carbon credits offer significant potential for reducing global GHG emissions, enhancing soil health, and providing economic opportunities for farmers. As the demand for carbon offsets increases, agriculture's role in carbon sequestration has gained attention from policymakers, investors, and the farming community. However, challenges related to measurement, verification, and market access must be addressed to fully realize this potential. Policy support and technological advances are crucial. Lal (2004) estimates that improved agricultural practices could sequester 0.4 to 1.2 billion metric tons of carbon annually. Zomer et al. (2017) highlight that increasing soil organic carbon could sequester 23–50 billion tons of CO_2 over 20 years, with the highest potential in regions like Africa, Asia, and Latin America, where degraded lands offer significant sequestration opportunities. Globally, agricultural soils could sequester 1.2 to 2 billion metric tons of CO_2 per year, with substantial regional variability.

6.4.1 Potential of Carbon Credits

Carbon credits offer a promising solution to the economic challenges faced by organic farmers (Yang et al. 2024). By monetizing the carbon sequestration benefits of organic farming, these credits provide an additional revenue stream, helping offset the higher costs associated with organic farming and making it more economically viable (Raina et al. 2024; Pradhan et al. 2023). This aligns with global climate change efforts by incentivizing practices that reduce GHG emissions (IPCC). As of September 26, 2023, 41 agricultural land management projects in the Verra registry are estimated to reduce emissions by 27 million tons of CO_2eq annually. This is equivalent to removing six million gasoline-powered passenger vehicles from the road for a year, generating 27 million carbon credits valued at approximately INR 2200 crores (@US$ 10 per credit). Assuming 60% of this amount reaches farmers, it totals around INR 1300 crores, roughly equivalent to the combined budget for initiatives like the creation of 10,000 Farmer Producer Organizations (FPOs) and the Agriculture Infrastructure Fund (AIF).

6.4.2 Regional Variability

Carbon credit potential varies across regions due to factors like climate, soil type, and agricultural practices. Farage et al. (2007) note that tropical regions, particularly in Latin America, Africa, and Southeast Asia, have high carbon sequestration potential due to degraded soils and low initial soil carbon. These areas can benefit from sustainable practices to enhance soil health. Paustian et al. (2016) highlight opportunities in sub-Saharan Africa and parts of Asia for soil carbon sequestration through climate-smart practices, such as agroforestry and rotational grazing. In North America and Europe, existing infrastructure, technology, and policies like reduced tillage and carbon markets, including the US cap-and-trade program and the EU Emissions Trading System (ETS), have enabled significant progress in carbon-friendly farming. Latin America, with its large agricultural land and ongoing deforestation, can sequester carbon through agroforestry and regenerative grazing, with emerging carbon markets like Colombia. Africa's agriculture holds high carbon sequestration potential due to degraded lands and low-input farming, but limited access to carbon markets and lack of infrastructure are challenges (Kumar et al. 2020). In Asia, intensive farming and emissions are a concern, yet opportunities for emissions reduction exist through practices like alternate wetting and drying in rice paddies, cover crops, and land restoration, particularly in India and China.

6.4.3 Economic and Environmental Potential

Agricultural carbon credits offer both income for farmers and climate benefits. Burney et al. (2010) showed that sustainable intensification can reduce emissions by improving yields on existing lands, reducing the need for land expansion and deforestation. Griscom et al. (2017) estimate that agricultural carbon sequestration could contribute over 20% of the emissions reductions needed to meet the Paris Agreement by 2030. Carbon credits create an income stream for farmers, with examples like Australia's Emissions Reduction Fund and US companies like Indigo Ag and Nori facilitating the sale of soil carbon credits. Practices that enhance soil health also improve water retention and resilience, leading to higher yields and long-term benefits. By absorbing CO_2, agriculture plays a key role in mitigating climate change and supporting global climate goals, including the Paris Agreement target of limiting warming to 1.5 °C and helping to achieve carbon credits (Table 6.1).

Table 6.1 Carbon credit potential of various agricultural activities

Technologies or methodologies	Crop/cropping system	GHG mitigation (t CO_2 eq. ha^{-1})	Earning from carbon credit (US$ ha^{-1})	References
No-tillage without residue retention	Wheat	1–1.5	61.3–91.95	Yadvinder-singh et al. (2014)
Zero tillage with residue retention	Wheat	1.5–2	91.95–122.6	Gathala et al. (2014) and Choudhary et al. (2018)
Laser land levelling	Rice-wheat; maize-wheat	0.4–0.6	24.52–36.78	Jat et al. (2015) and Bijarniya et al. (2020)
Direct-seeded rice (DSR)	Rice	1–2.25	61.3–137.92	Bijarniya et al. (2020) and Chakraborty et al. (2017)
Site-specific nutrient management (SSNM)	Rice-wheat; maize-wheat	0.5–1	30.65–61.3	Sapkota et al. (2014), Singh et al. (2020), Pradhan and Meena (2023a)
Sub-surface drip irrigation	Rice-wheat	0.5–1.5	30.65–91.95	Sidhu et al. (2019) and Jat et al. (2019)
Permanent bed	Maize-wheat	0.8–1	49.04–61.3	Das et al. (2018) and Jat et al. (2019)
Diversifying crops	Wheat and mustard-based cropping system	Varied significantly across various crops		Pathak et al. (2021)
Neem-coated urea	Rice	0.25	15.32	Pathak et al. (2021)
Integration of diverse climate-smart practices	Rice-wheat	2.8–3	171.64–183.9	Bijarniya et al. (2020)
Legume in the cropping system	Rice-wheat; maize-wheat	0.065–0.12	3.98–7.35	Gathala et al. (2014), Thind et al. (2019) and Sheoran et al. (2022)
Leaf color chart (LCC)	Rice	0.25	15.32	Pathak et al. (2021)
Integrated nutrient management (INM)	Rice-wheat	1.7	104.21	Pathak et al. (2021)

6.4.4 Barriers to Unlocking the Full Potential of Agricultural Carbon Credits

Reliable measurement and verification are crucial for carbon credits, but soil carbon sequestration is challenging to measure due to factors like spatial variability and time requirements. Smith et al. (2020) highlight the difficulties in measuring, reporting, and verifying soil carbon changes, noting that without accurate measurement tools, it is hard to ensure that carbon credits represent actual sequestration. They advocate for new technologies and standardization to enhance measurement accuracy. Zhou et al. (2023) emphasize the importance of baseline measurements and long-term monitoring to verify carbon sequestration, stressing the need for robust verification systems to meet market demands.

A major challenge in agricultural carbon credits is accurately tracking soil carbon changes over time, given the variability in soil types, climate, and farming practices. Developing cost-effective monitoring methods, such as remote sensing and soil sampling, is essential. Many farmers, particularly in developing regions, face challenges accessing carbon markets due to limited resources, infrastructure, and awareness of carbon credit programs. Policy support and investment are vital for integrating agriculture into carbon markets. Carbon credits must meet standards of additionality (proving reductions exceed what would have happened without intervention) and permanence (ensuring long-term sequestration). Changes in market or climate conditions can affect the long-term stability of soil carbon, raising concerns about permanence.

6.5 Case Studies on Carbon Credit Assessment in Field Crops

The use of carbon credit systems in field crop production is growing as interest in sustainable agriculture increases. Crops like corn, soybeans, wheat, and rice offer significant carbon sequestration potential, with various global case studies showing the effectiveness of different practices for generating carbon credits. Here are key examples:

1. **United States (corn and soybeans)**: No-till farming and cover cropping in the Midwest increased SOC by 0.3–0.5 tons per hectare per year. Farmers could earn US$10–40 per ton of CO_2, covering up to 30% of adoption costs (Baker et al. 2007).
2. **Brazil (soybeans)**: Agroforestry systems sequestered an additional 2.3 tons of CO_2 per hectare per year. Farmers could earn US$20–50 per ton of CO_2 while benefiting from improved biodiversity and weather resilience (Nair 2012).
3. **Australia (wheat)**: Reduced tillage in wheat fields sequestered 0.2 tons of CO_2 per hectare annually, providing farmers with credits worth AUD15–25 per ton of CO_2. This practice also improved soil quality and water retention (Button et al. 2022).
4. **China (rice)**: Alternate Wetting and Drying (AWD) in rice paddies reduced CH_4 emissions by 1.4 tons of CO_2eq per hectare per season. Carbon credits ranged from US$8 to 20 per ton of CO_2eq (Chen et al. 2024).

5. **India (wheat-rice)**: Using organic fertilizers and conservation tillage reduced CO_2eq emissions by 1.2 tons per hectare annually, with credits valued at US$ 10–30 per ton.

These examples highlight the potential for carbon credits to offset costs and encourage sustainable practices, although challenges with measurement and market access remain.

6.6 Research and Development Towards Carbon Credit in Field Crops

The market for land use, forestry, and agriculture carbon credits was estimated to be worth US$6.28 billion in 2023 and is expected to increase at a compound annual growth rate (CAGR) of 31.49% to reach US$97.1 billion by 2033 (Fig. 6.3). This growth is fueled by advancements in sustainable practices like precision agriculture, forest management, and eco-friendly land-use strategies. Increased collaboration among farmers, landowners, and forestry companies, along with regulatory changes, is driving the adoption of solutions that enhance carbon sequestration. The market is segmented by application (avoidance, removal, and combination projects) and product type (forestry, land use, and agriculture). Market analysis combines expert interviews and secondary research to assess technological innovations, regulatory impacts, and demand trends, offering strategic insights for industry players.

The carbon credit market for agriculture, forestry, and land use is growing rapidly, driven by the economic potential of carbon sequestration. Practices like reforestation, afforestation, and soil carbon sequestration are increasingly adopted to reduce emissions. Key players such as Indigo Ag, Carbon Credit Capital, Terra Global Capital, South Pole, and Cargill are expanding through strategic partnerships. Combination projects, which integrate carbon removal and emissions reduction,

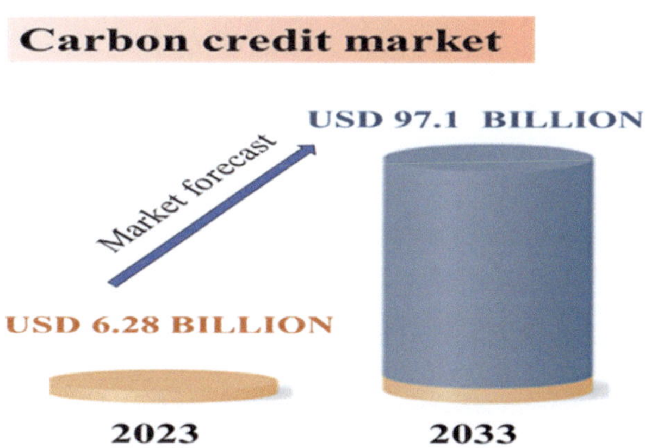

Fig. 6.3 Carbon credit market growth

Table 6.2 Advancements in the carbon credit market for agriculture, forestry, and land use

Organization	Year	Achievements
Indigo Ag, Inc.	2024 (February)	Completed its third carbon crop, generating 163,048 carbon credits, bringing the total CO_2 sequestered or avoided by participating farmers to nearly 300,000 metric tons
Nori, Inc.	2023 (December)	Launched the Nori Net Zero Tonne, a hybrid carbon removal credit that combines Nori Regenerative Tons with permanent carbon removal from Frontier to provide credible emissions offset solutions
Terra Global Capital	2023 (January)	Secured a US$640 million investment from Anew Climate to expand nature-based climate solutions with social and environmental benefits
Ecosystem Services Market Consortium and Benson Hill	2022 (March)	Completed the first year of a carbon credit pilot with Midwest soy and corn farmers, reducing emissions and boosting soil carbon
Tesla	2021 (May)	Generated US$518 million from carbon credit sales, contributing to its ongoing profitability

dominate the market due to their dual benefits. Nature-based solutions like afforestation and agroforestry, along with industrial carbon removal, are in high demand and command premium prices.

The forestry and land use segment, including REDD+, ARR, and IFM projects, plays a critical role in climate mitigation, with efforts to improve transparency through standards and regulations. In Asia-Pacific, countries like China, Japan, and India are driving market growth with government incentives and sustainable practices, helping to establish the region as a key player in the global carbon credit market. Recent developments in the carbon credit market for agriculture, forestry, and land use are shown in Table 6.2.

6.6.1 *Market Driver: Corporate Net-Zero Commitments Boost Carbon Credit Demand*

Corporate net-zero and sustainability commitments are increasing as businesses recognize their climate role. Over 140 countries, including China, the United States, and the EU, have set net-zero targets, covering 88% of global emissions. More than 9000 companies, 1000 cities, and 600 financial institutions are working towards halving emissions by 2030. Companies like PwC India and Ford have set ambitious goals, further driving demand for carbon offset projects, especially in agriculture, forestry, and land use, to reduce emissions and invest in carbon sequestration.

6.6.2 Market Challenge: Potential for Greenwashing

Carbon offset programs can undermine net-zero goals if not coupled with real emissions reductions. Critics like Greenpeace and WWF-UK warn against greenwashing, where companies focus on offsetting rather than cutting emissions, leading to temporary fixes and issues like double-counting. Stricter standards and verification processes are being introduced to ensure transparency, accountability, and genuine emissions reductions.

6.6.3 Market Opportunity: Growth of Agroforestry and Sustainable Farming Practices

The agricultural sector plays a key role in carbon trading, both emitting and sequestering carbon. Practices like tilling and livestock farming release GHG, while methods such as agroforestry and soil carbon sequestration remove carbon from the atmosphere (Jhariya et al. 2022). Carbon trading offers farmers financial incentives to adopt sustainable practices by selling carbon credits, reducing emissions, and promoting resource conservation. While individual farmers face challenges, FPOs and cooperatives simplify participation by aggregating credits and managing sales, making carbon trading more accessible and rewarding.

6.6.4 Research on Carbon Credits

Future research on carbon credit generation can be divided into primary and secondary research (Table 6.3). Primary research involved interviews with industry experts in agriculture, forestry, and land use, providing both qualitative and quantitative insights. Secondary research included sources like company websites, annual reports, and databases such as Hoovers, Bloomberg, and IRENA, offering a comprehensive market overview.

6.7 Policy Perspective for Enhancing Carbon Credit in Agriculture

The carbon credit consortium's policy recommendations, based on project findings, highlight ongoing challenges in mainstreaming carbon farming. Policymakers can boost carbon sequestration by closing knowledge gaps, promoting practical techniques, and offering financial incentives through both public programs and private markets, ideally combining these strategies for optimal results. A comprehensive

Table 6.3 Key data can be gathered from primary and secondary research

Primary research	Secondary research
Data validation and graphical representation verification	Market divisions and percentage allocations
Verification of market segmentation and key qualitative insights	Market trends from leading industry players
Analysis of the competitive landscape	Analysis of market innovations, trends, and emerging applications
Validation of market data and regional breakdowns	Numerical data for statistical analysis

policy framework is crucial to align climate, biodiversity, and water management goals without conflict at the farm level. This would offer farmers clear guidance and incentives while minimizing contradictory policies and administrative challenges. Despite viable revenue models, legislative barriers may still hinder participation. Future policies should focus on removing these obstacles and fostering a supportive, efficient environment for farmers to adopt carbon farming.

6.7.1 Bridging the Knowledge Gap

A lack of knowledge on soil carbon enhancement is a major barrier for both farmers and policymakers in adopting carbon farming practices. Key recommendations to improve knowledge transfer include:

(i) **Farmer education and collaboration**: Promote knowledge sharing on carbon sequestration techniques, benefits, and tracking progress through peer exchanges and independent advisors.
(ii) **Focused research**: Study the effectiveness of various carbon sequestration methods across different soils and climates, with a focus on monitoring and validation.
(iii) **Capacity and permanence**: Increase understanding of soils' carbon storage capacity and the permanence of stored carbon.
(iv) **Integrated measures**: Explore effective sequestration techniques while ensuring compliance with fertilization regulations and maintaining soil carbon stability.

Accurate models and measurements are crucial for selling carbon storage services. Farmers, businesses, and policymakers need improved systems to predict and monitor carbon storage across different management practices. The EU can help by enhancing data collection and adapting models to regional conditions. Key priorities include developing robust methods for tracking SOC and providing farmers with tools to assess short-term benefits like soil structure and water retention. The carbon farming consortium urges the European Commission to support local research and offer tailored advice. Policies should promote independent advisory services and strengthen agricultural knowledge and innovation systems (AKIS). Long-term soil monitoring and research on nutrient flows in carbon-enriched soils will improve models and maximize carbon farming's benefits, including better nutrient retention and reduced fertilizer use (Kumar et al. 2022, 2023).

6.7.2 Providing Economic Incentives

Farmers must invest in carbon sequestration practices, which have long-term benefits like improved yields and resilience but require upfront costs. Early adopters, who have embraced these practices incrementally, have paved the way for broader adoption. To scale this, short-term incentives such as compensating costs or rewarding farmers for providing climate services are essential. A supportive framework should offer fair compensation, not obligations, and ensure early adopters are eligible for future EU opportunities. Blended funding, combining public payments and private market rewards, can optimize contributions while recognizing co-benefits like biodiversity and climate resilience. Customizable, result-based systems should allow farmers to adapt measures to their conditions, with fair compensation for different soil types. Annual rewards should be given to avoid delays in measuring SOC increases, while administrative burdens are minimized through efficient systems like remote sensing. To promote widespread adoption, policies must recognize and reward early adopters. Research should aim to establish fair pricing for carbon farming and define compensation based on specific techniques and regional conditions. Encouraging local and private carbon farming initiatives, supported by reliable monitoring systems, will ensure project integrity. Clear guidance for farmers and buyers on carbon credit quality is essential to build trust in the system.

6.7.3 Eliminating Contradictory and Restrictive Policy

Carbon farming's full potential is hindered by restrictive and contradictory legislation. Policymakers must adopt supportive policies to increase soil carbon and integrate environmental goals such as CO_2 reduction, biodiversity, and water management. Current policies often lack cohesion and overlook the broader impacts of soil carbon management, potentially conflicting with other objectives.

A clear policy framework is needed to make soil carbon enhancement feasible and administratively manageable, balancing carbon farming ambitions with the constraints of existing regulations. The European Commission's initiatives, like the new CAP and the Climate Pact's carbon farming initiative, recognize soil carbon's importance, with a focus on motivational policies rather than new obligations. Integrating direct payments and market measures can support carbon farming.

The success of carbon farming depends on adequate incentives to drive demand for carbon credits, with the EU setting ambitious climate targets, including a 55% emissions reduction by 2030 and climate neutrality by 2050. Agriculture's role in addressing both CO_2 and non-CO_2 emissions is crucial, with land use, land-use change and forestry (LULUCF) regulations aiming for a "no-debit" rule on land use emissions, though concerns remain about its impact on agricultural emissions reductions (Fern 2018).

6.8 Future Thrust

The future of carbon credits and markets is shaped by regulations, market trends, and technology. Brinkman et al. (2009) emphasized the need for detailed abatement data and financial flows in carbon trading, focusing on real-world market conditions and policies over theoretical potential. This approach underscores key drivers of trading volumes and carbon prices in evolving markets. Recent studies highlight the rise of environmental markets driven by emission regulations and consumer demand for sustainable products. Byrareddy et al. (2023) show that premium pricing for eco-friendly goods encourages carbon market growth and community benefits from trading initiatives. Han et al. (2024) analyze carbon pricing in China and Europe, emphasizing its role in a low-carbon economy and exploring market volatility and investor behavior. Together, these studies underscore the growing complexity of carbon markets, shaped by regulations, economic factors, and community engagement, offering valuable insights into future challenges and opportunities. Shen et al. (2020) reviewed global carbon pricing schemes, highlighting diverse approaches like carbon taxes and cap-and-trade systems tailored to varying economic and political contexts. Diniz Oliveira et al. (2021) explored linking carbon markets internationally, noting benefits like improved liquidity but also challenges in harmonizing regulations and ensuring environmental integrity. Adigun et al. (2024) emphasized blockchain's potential to enhance carbon market transparency and efficiency, addressing issues in credit verification and tracking. Together, these studies outline the complexities and innovations shaping carbon markets' future.

Carbon markets' social and ethical dimensions highlight the need for fairness and equity in distributing benefits and burdens. Sreekar et al. (2022) emphasize the role of nature-based solutions, like conservation and land management, in expanding the carbon markets' impact. Michaelowa et al. (2019) explore Article 6 of the Paris Agreement, noting its potential to enhance climate action while addressing challenges like double counting. Together, these insights underscore the complexity of carbon markets, integrating policy, technology, ethics, and ecosystems to drive emissions reductions and sustainable development. Carbon credits and carbon markets are emerging as important tools in the global effort to mitigate climate change, with agriculture playing a crucial role in this landscape. As the world grapples with the urgent need to reduce GHG emissions, the agricultural sector is increasingly recognized for its potential to sequester carbon and contribute to climate change mitigation efforts. The concept of carbon credits in agriculture revolves around incentivizing farmers and landowners to adopt practices that reduce GHG emissions or enhance carbon sequestration in soil and biomass. These practices can include no-till farming, cover cropping, agroforestry, and improved grazing management. By implementing these methods, farmers can generate carbon credits, which can then be sold in carbon markets to companies or organizations looking to offset their emissions. The future thrust of carbon credits and carbon markets in agriculture is likely to be shaped by several key factors:

- Rising demand for carbon offsets is accelerating as countries and companies chase net-zero goals; yet, Probst et al. (2023) found that only 12% of existing credits reflect genuine emissions reductions, and Trencher et al. (2024) underscored the risks of ineffective offsets, calling for stricter oversight and enforcement.
- Technological advancements like remote sensing, satellite imagery, and AI are improving carbon measurement and verification in agriculture, boosting the reliability of carbon markets.
- Policy and standards set by governments and international bodies are essential for ensuring market integrity and building trust among stakeholders.
- Sustainable agriculture integration is increasing, with carbon credit schemes aligning more closely with regenerative agriculture to enhance soil health, biodiversity, and climate resilience.
- Focus on smallholder farmers is growing through simplified verification methods, bundled credits, and community-led projects, promoting greater equity in market participation.
- Soil carbon sequestration practices that enhance soil organic matter are gaining momentum due to their potential for substantial carbon storage.
- New credit methodologies are emerging, including credits for methane reduction, precision agriculture, and climate-resilient crop varieties.
- Private sector growth is fueling innovation with new financing models that support the transition to low-carbon farming practices.
- Enhanced monitoring systems using blockchain and data technologies are strengthening the transparency and credibility of carbon markets.
- Integrated ecosystem markets that combine carbon credits with water and biodiversity conservation efforts are encouraging more holistic sustainability approaches.
- Permanence and additionality remain critical concerns, with emphasis on ensuring carbon storage is long-term and actions are driven by carbon finance.
- Capacity building through education and training is enabling farmers to adopt carbon-smart practices and engage effectively in markets.
- Local adaptation of carbon markets to specific ecosystems and regional contexts is increasing participation and enhancing impact.

- Climate adaptation focus is growing, with practices that both sequester carbon and improve resilience offering dual environmental and economic benefits.
- Equity and justice considerations are vital to ensure that smallholders and marginalized groups benefit fairly from carbon markets.
- Carbon-neutral value chains are emerging in response to consumer demand for low-carbon products, potentially reshaping agricultural supply chains.
- Lifecycle assessments are becoming more comprehensive, offering improved accounting of emissions throughout agricultural production systems.
- Alignment with climate goals is necessary, ensuring agricultural carbon credits contribute meaningfully to global commitments like the Paris Agreement.
- Expanded credit programs and increased funding for agroforestry, cover cropping, and other practices are expected to drive further market growth.
- Digital integration through platforms leveraging AI and blockchain is streamlining market access, reducing transaction costs, and expanding participation.

The future of agricultural carbon credits and markets holds immense potential for growth and innovation as agriculture's role in climate change mitigation gains recognition. Expanding these markets will require tackling challenges like accurate measurement, equitable farmer access, and integration with broader sustainability goals. Success depends on collaboration among farmers, policymakers, researchers, and the private sector to establish fair, effective, and sustainable carbon market mechanisms that drive meaningful climate action.

6.9 Conclusion

Agricultural carbon credits provide a dual advantage: they help mitigate climate change while fostering sustainable development. Practices such as conservation tillage, cover cropping, and agroforestry can sequester between 4 and 12 gigatons of CO_2 each year, thereby contributing to the objectives of the Paris Agreement. Additionally, these methods enhance soil health, improve water retention, and increase resilience to climate stressors, all while supporting biodiversity. However, challenges persist, including the complexities of carbon measurement, high implementation costs, and limited access for smallholder farmers. Fortunately, emerging technologies like remote sensing, blockchain, and machine learning are enhancing precision and lowering barriers to entry. Policy support and corporate demand are propelling market growth, yet experts emphasize that these credits should complement, rather than replace, emissions reductions in other sectors. With strong governance and innovative approaches, agricultural carbon credits have the potential to align environmental, economic, and social objectives, positioning agriculture as a pivotal player in climate action.

References

Acharya P, Ghimire R, Acosta-Martínez V (2024) Cover crop-mediated soil carbon storage and soil health in semi-arid irrigated cropping systems. Agric Ecosyst Environ 361:108813. https://doi.org/10.1016/j.agee.2023.108813

Adigun OA, Falola BO, Esebre SD, Wada I, Adewuyi AT, Olajide TD, Oladokun P (2024) Enhancing carbon markets with fintech innovations: the role of artificial intelligence and blockchain. World J Adv Res Rev 23(2):579–586. https://doi.org/10.30574/wjarr.2024.23.2.2387

Angwaomaodoko EA (2024) The impact of climate change on global economic stability. J Environ Econ Manag 10(1):102–120. https://doi.org/10.5296/ieb.v10i1.22277

Babu S, Mohapatra KP, Yadav GS, Lal R, Singh R, Avasthe RK, Das A, Chandra P, Gudade BA, Kumar A (2020) Soil carbon dynamics in diverse organic land use systems in north eastern Himalayan ecosystem of India. Catena 194:104785. https://doi.org/10.1016/j.catena.2020.104785

Baker JM, Ochsner TE, Venterea RT, Griffis TJ (2007) Tillage and soil carbon sequestration-what do we really know? Agric Ecosys Environ 118(4):1–5. https://doi.org/10.1016/j.agee.2006.05.014

Bijarniya D, Parihar CM, Jat RK, Kalvania K, Kakraliya SK, Jat ML (2020) Portfolios of climate smart agriculture practices in smallholder rice-wheat system of eastern indo-Gangetic plains—crop productivity, resource use efficiency and environmental foot prints. Agronomy 10(10). https://doi.org/10.3390/AGRONOMY10101561

Brinkman M, Fankhauser S, Irons B, Weyers S (2009) The carbon market in 2020 : volumes, prices and gains from trade, vol 13, pp 1–26

Bruce V, Henderson Z, Greenwood K, Hancock PJB, Burton AM, Miller P (1999) Verification of face identities from images captured on video. J Exp Psychol Appl 5(4):339–360. https://doi.org/10.1037/1076-898X.5.4.339

Burney JA, Davis SJ, Lobell DB (2010) GHG mitigation by agricultural intensification. Proc Natl Acad Sci USA 107(26):12052–12057. https://doi.org/10.1073/pnas.0914216107

Button ES, Pett-Ridge J, Murphy DV, Kuzyakov Y, Chadwick DR, Jones DL (2022) Deep-C storage: biological, chemical and physical strategies to enhance carbon stocks in agricultural subsoils. Soil Biol Biochem 170:108697. https://doi.org/10.1016/j.soilbio.2022.108697

Byrareddy MV, Islam MA, Nguyen-Huy T, Slaughter G (2023) A systematic review of emerging environmental markets: potential pathways to creating shared value for communities. Heliyon 9(9):e19754. https://doi.org/10.1016/j.heliyon.2023.e19754

Chachar M, Ahmed S, Murtaza G, Shah Jillani PSAN, Baloch H, Ali Hakro R (2023) The impact of climate change on horticulture: A global perspective and adaptation strategies. Ecofemin Climate Change 4(1):41–44. https://doi.org/10.26480/efcc.01.2023.41.44

Chakraborty D, Ladha JK, Rana DS, Jat ML, Gathala MK, Yadav S, Rao AN, Ramesha MS, Raman A (2017) A global analysis of alternative tillage and crop establishment practices for economically and environmentally efficient rice production. Sci Rep 7(1):1–11. https://doi.org/10.1038/s41598-017-09742-9

Chen Y, Guo W, Ngo HH, Wei W, Ding A, Ni B, Hoang NB, Zhang H (2024) Ways to mitigate GHG production from rice cultivation. J Environ Manag 368:122139. https://doi.org/10.1016/j.jenvman.2024.122139

Choudhary KM, Jat HS, Nandal DP, Bishnoi DK, Sutaliya JM, Choudhary M, Singh Y, Sharma PC, Jat ML (2018) Evaluating alternatives to rice-wheat system in western indo-Gangetic Plains: crop yields, water productivity and economic profitability. Field Crop Res 218:1–10. https://doi.org/10.1016/j.fcr.2017.12.023

Das TK, Saharawat YS, Bhattacharyya R, Sudhishri S, Bandyopadhyay KK, Sharma AR, Jat ML (2018) Conservation agriculture effects on crop and water productivity, profitability and soil organic carbon accumulation under a maize-wheat cropping system in the north-western indo-Gangetic Plains. Field Crop Res 215:222–231. https://doi.org/10.1016/j.fcr.2017.10.021

Devi AS, Singh KS (2021) Carbon storage and sequestration potential in aboveground biomass of bamboos in north East India. Sci Rep 11(1):837. https://doi.org/10.1038/s41598-020-80887-w

Diniz Oliveira T, Costa Gurgel A, Tonry S (2021) Potential trading partners of a Brazilian emissions trading scheme: the effects of linking with a developed region (Europe) and two developing regions (Latin America and China). Technol Forecast Soc Chang 171:120947. https://doi.org/10.1016/j.techfore.2021.120947

Dondini M, Abdalla M, Aini FK, Albanito F, Beckert MR, Begum K, Brand A, Cheng K, Comeau LP, Jones EO, Farmer JA, Feliciano DMS, Fitton N, Hastings A, Henner DN, Kuhnert M, Nayak DR, Oyesikublakemore J, Phillips L, Smith P (2018) In: Singh BKBT-SCS (ed) Chapter 9 - projecting soil C under future climate and land-use scenarios (modeling). Academic Press, pp 281–309. https://doi.org/10.1016/B978-0-12-812766-7.00009-3

Farage PK, Ardö J, Olsson L, Rienzi EA, Ball AS, Pretty JN (2007) The potential for soil carbon sequestration in three tropical dryland farming systems of Africa and Latin America: A modelling approach. Soil Tillage Res 94(2):457–472. https://doi.org/10.1016/j.still.2006.09.006

Fern (2018) Fern' s analysis of the EU'S LULUCF regulation. April

Filonchyk M, Peterson MP, Zhang L, Hurynovich V, He Y (2024) GHGes emissions and global climate change: examining the influence of CO_2, CH_4, and N_2O. Sci Total Environ 935:173359. https://doi.org/10.1016/j.scitotenv.2024.173359

Gathala MK, Kumar V, Sharma PC, Saharawat YS, Jat HS, Singh M, Kumar A, Jat ML, Humphreys E, Sharma DK, Sharma S, Ladha JK (2014) Reprint of "optimizing intensive cereal-based cropping systems addressing current and future drivers of agricultural change in the Northwestern Indo-Gangetic Plains of India". Agric Ecosyst Environ 187:33–46. https://doi.org/10.1016/j.agee.2013.12.011

Griscom BW, Adams J, Ellis PW, Houghton RA, Lomax G, Miteva DA, Schlesinger WH, Shoch D, Siikamäki JV, Smith P, Woodbury P, Zganjar C, Blackman A, Campari J, Conant RT, Delgado C, Elias P, Gopalakrishna T, Hamsik MR, Fargione J (2017) Natural climate solutions. Proc Natl Acad Sci USA 114(44):11645–11650. https://doi.org/10.1073/pnas.1710465114

Han A, Yu T, Ke Y, Liu C, Liu Y (2024) Study on the effect of carbon trading on the carbon emission intensity of enterprises—a mechanism test based on ESG performance. Front Environ Sci 12(June):1–23. https://doi.org/10.3389/fenvs.2024.1406577

Hulme M (2007) Climate conflict. New Sci 196(2629):26. https://doi.org/10.1016/S0262-4079(07)62834-6

Intergovernmental Panel on Climate Change (IPCC) (2023) Health, wellbeing and the changing structure of communities. In: Climate Change 2022—impacts, adaptation and vulnerability, pp 1041–1170. https://doi.org/10.1017/9781009325844.009

IPCC (2023) Summary for policymakers of the intergovernmental panel on climate change

Jangir CK, Sangwan PS, Panghaal D, Kumar S, Meena RS, Bharti J, Jat RD, Singh N (2024) Spatial variability and statistical analysis of soil properties in the rice wheat-based systems of north-West India. Commun Soil Sci Plant Anal 55(8):1205–1223

Jat M, Maize I, Singh Y (2015) Innovations in soil and water management/conservation research through integrated approaches of nuclear and isotopic techniques and precision agriculture. In: Soil-specific farming, pp 262–297. https://doi.org/10.1201/b18759-14

Jat HS, Sharma PC, Datta A, Choudhary M, Kakraliya SK, Singh Y, Sidhu HS, Gerard B, Jat ML (2019) Re-designing irrigated intensive cereal systems through bundling precision agronomic innovations for transitioning towards agricultural sustainability in north-West India. Sci Rep 9(1):1–14. https://doi.org/10.1038/s41598-019-54086-1

Jhariya MK, Raj A, Banerjee A, Meena RS, Bargali SS, Kumar S, Nema S, Poonam, Oraon PR (2022) Plan and policies for soil organic carbon management under agroforestry system. In: Plans and policies for soil organic carbon management in agriculture. Springer, Singapore, pp 191–219

Kantwa SR, Choudhary M, Agrawal RK, Dixit AK, Kumar S, Chary GR (2024) Reducing energy and carbon footprint through diversified rainfed cropping systems. Energy Nexus 14:100306. https://doi.org/10.1016/j.nexus.2024.100306

Kumar R, Punia H, Kumar S (2020) Nanomaterials for agriculture input use efficiency. In: Resources use efficiency in agriculture. Springer, Singapore. https://doi.org/10.1007/978-981-15-6953-1_5

Kumar S, Bamboriya SD, Rani K, Meena RS, Sheoran S, Loyal A, Kumawat A, Jhariya MK (2022) Grain legumes: a diversified diet for sustainable livelihood, food, and nutritional security. In: Advances in legumes for sustainable intensification. Academic Press, pp 157–178

Kumar S, Sharma SK, Dhaka AK, Bedwal S, Sheoran S, Meena RS, Jangir CK, Kumar D, Kumar R, Jat RD, Meena AK (2023) Efficient nutrient management for enhancing crop productivity, quality and nutrient dynamics in lentil (*Lens culinaris* Medik.) in the semi-arid region of northern India. PLoS One 18(2):e0280636

Kumar S, Meena RS, Kumar S, Pradhan G, Jangir CK, Ghosh S, Punia H, Sheoran P, Meena R, Ahmad MA, Goyal SK (2025) Designing a diversified Indian mustard production system for energy-carbon-cum-heat use efficiency and sowing dates assessment. GCB Bioenergy 17(6): e70044

Lal R (2004) Soil carbon sequestration impacts on global climate change and food security. Science 304(5677):1623–1627. https://doi.org/10.1126/science.1097396

Lal R (2010) Enhancing eco-efficiency in agro-ecosystems through soil carbon sequestration. Crop Sci 50(April):120–131. https://doi.org/10.2135/cropsci2010.01.0012

Leharwan M, Kumar Y, Kumar R, Kumar Saraswat P, Kumar R, Veetil KT, A., Bhattacharjee, S., Kumar, A. and Kumar, S. (2023) Assessing the effects of conservation tillage and in-situ crop residue management on crop yield and soil properties in rice–wheat cropping system. Sustainability 15(17):12736

Mandal B, Majumder B, Bandyopadhyay PK, Hazra GC, Gangopadhyay A, Samantaray RN, Mishra AK, Chaudhury J, Saha MN, Kundu S (2007) The potential of cropping systems and soil amendments for carbon sequestration in soils under long-term experiments in subtropical India. Glob Chang Biol 13(2):357–369. https://doi.org/10.1111/j.1365-2486.2006.01309.x

Meena RS, Pradhan G (2023) Industrial garbage-derived biocompost enhances soil organic carbon fractions, CO_2 biosequestration, potential carbon credits, and sustainability index in a rice-wheat ecosystem. Environ Res 116525. https://doi.org/10.1016/j.envres.2023.116525

Meena RS, Pradhan G, Kumar S, Lal R (2023) Using industrial wastes for rice-wheat cropping and food-energy-carbon-water-economic nexus to the sustainable food system. Renew Sust Energ Rev 187:113756. https://doi.org/10.1016/j.rser.2023.113756

Meena RS, Pradhan G, Singh K, Kumar S, Singh AK, Shashidhar KS, Mina KK, Rao CS (2024a) Agriculture models for restoring degraded land to enhance CO_2 biosequestration and carbon credits in the Vindhyan region of India. Sci Total Environ 929:172661. https://doi.org/10.1016/j.scitotenv.2024.172661

Meena RS, Lal R, Kumar S, Pradhan G, Rao CS, Singh AK, Pathak H, Abhilash PC, Kumar A, Sharma SK, Jat ML, Singh S (2024b) Potential of Indian agriculture for capturing atmospheric CO_2 and monetizing carbon credits to the farmers: an overview and policy framework. Adv Agron 188:101–206. https://doi.org/10.1016/bs.agron.2024.06.001

Meena RS, Pradhan G, Nalani, Singh K, Singh AK, Verma SK, Mina KK, Kumar S, Chaturvedi RK (2024c) Diversified cropping modules designed for soil restoration, CO_2 sequestration, and generating carbon credits. Land Degrad Dev. https://doi.org/10.1002/ldr.5224

Meena RS, Pradhan G, Kumar S (2025) Energy flow, eco-efficiency, and economic circulation with recycled industrial waste compost application in wheat and subsequent rice farming. Sci Total Environ 967:178779. https://doi.org/10.1016/j.scitotenv.2025.178779

Michaelowa A, Shishlov I, Brescia D (2019) Evolution of international carbon markets: lessons for the Paris agreement. Wiley Interdiscip Rev Clim Chang 10(6):1–24. https://doi.org/10.1002/wcc.613

Nair PKR (2012) Carbon sequestration studies in agroforestry systems: A reality-check. Agrofor Syst 86(2):243–253. https://doi.org/10.1007/s10457-011-9434-z

Ogle SM, Breidt FJ, Paustian K (2005) Agricultural management impacts on soil organic carbon storage under moist and dry climatic conditions of temperate and tropical regions. Biogeochemistry 72(1):87–121. https://doi.org/10.1007/s10533-004-0360-2

Palinkas LA, Wong M (2020) Global climate change and mental health. Curr Opin Psychol 32:12–16. https://doi.org/10.1016/j.copsyc.2019.06.023

Pan SL, Carter L, Tim Y, Sandeep MS (2022) Digital sustainability, climate change, and information systems solutions: opportunities for future research. Int J Inf Manag 63:102444. https://doi.org/10.1016/j.ijinfomgt.2021.102444

Pathak H, Srinivasarao CH, Jat ML (2021) Conservation agriculture for climate change adaptation and mitigation in India. J Agric Phys 21(1):182–196. http://agrophysics.in/admin/adminjournalpdf/20220315121810163007592l/journal-780307173.pdf

Paustian K, Lehmann J, Ogle S, Reay D, Robertson GP, Smith P (2016) Climate-smart soils. Nature 532(7597):49–57. https://doi.org/10.1038/nature17174

Pradhan G, Meena RS (2023) Interaction impacts of biocompost on nutrient dynamics and relations with soil biota, carbon fractions index, societal value of CO_2 equivalent, and ecosystem services in the wheat-rice farming. Chemosphere 139695. https://doi.org/10.1016/j.chemosphere.2023.139695

Pradhan G, Meena RS (2023a) Utilizing waste compost to improve the atmospheric CO_2 capturing in the rice-wheat cropping system and energy-cum-carbon credit audit for a circular economy. Sci Total Environ 164572. https://doi.org/10.1016/j.scitotenv.2023.164572

Pradhan G, Meena RS, Kumar S, Lal R (2023) Utilizing industrial wastes as compost in wheat-rice production to improve the above and below-ground ecosystem services. Agric Ecosyst Environ 358:108704. https://doi.org/10.1016/j.agee.2023.108704

Probst B, Toetzke M, Anadon LD, Kontoleon A, Hoffmann V (2023) Systematic review of the actual emissions reductions of carbon offset projects across all major sectors. Nature Portfolio 10:1–31

Punia H, Tokas J, Malik A, Satpal, Rani A, Gupta P, Kumari A, Mor VS, Bhuker A, Kumar S (2020) Solar radiation and nitrogen use efficiency for sustainable agriculture. In: Resources use efficiency in agriculture, pp 177–212

Raihan A, Muhtasim DA, Pavel MI, Faruk O, Rahman M (2022) An econometric analysis of the potential emission reduction components in Indonesia. Cleaner Prod Lett 3:100008. https://doi.org/10.1016/j.clpl.2022.100008

Raina N, Zavalloni M, Viaggi D (2024) Incentive mechanisms of carbon farming contracts: A systematic mapping study. J Environ Manag 352:120126. https://doi.org/10.1016/j.jenvman.2024.120126

Reichle DE (2020) In: D.E.B.T.-T.G.C.C, Reichle CC (eds) Chapter 1 – an introduction to ecological energetics and the global carbon cycle. Elsevier, pp 1–3. https://doi.org/10.1016/B978-0-12-820244-9.00001-9

Roy S, Tandukar S, Bhattarai U (2022) Gender, climate change adaptation, and cultural sustainability: insights from Bangladesh. Front Climate 4(May). https://doi.org/10.3389/fclim.2022.841488

Sahoo A, Rath N (2023) Assessment of climate change, water poverty and risk communities: some insights from Western Odisha. GeoJournal 88(4):3665–3678. https://doi.org/10.1007/s10708-022-10819-w

Saimun MSR, Karim MR, Sultana F, Arfin-Khan MAS (2021) Multiple drivers of tree and soil carbon stock in the tropical forest ecosystems of Bangladesh. Trees Forests People 5:100108. https://doi.org/10.1016/j.tfp.2021.100108

Sapkota TB, Majumdar K, Jat ML, Kumar A, Bishnoi DK, McDonald AJ, Pampolino M (2014) Precision nutrient management in conservation agriculture based wheat production of North-west India: profitability, nutrient use efficiency and environmental footprint. Field Crop Res 155:233–244. https://doi.org/10.1016/j.fcr.2013.09.001

Shen N, Zhao Y, Deng R (2020) A review of carbon trading based on an evolutionary perspective. Int J Climate Change Strat Manag 12(5):739–756. https://doi.org/10.1108/IJCCSM-11-2019-0066
Sheoran S, Gupta M, Kumari S, Kumar S, Rakshit S (2022) Meta-QTL analysis and candidate genes identification for various abiotic stresses in maize (*Zea mays* L.) and their implications in breeding programs. Mol Breed 42(5):26
Sidhu HS, Jat ML, Singh Y, Sidhu RK, Gupta N, Singh P, Singh P, Jat HS, Gerard B (2019) Sub-surface drip fertigation with conservation agriculture in a rice-wheat system: A breakthrough for addressing water and nitrogen use efficiency. Agric Water Manag 216:273–283. https://doi.org/10.1016/j.agwat.2019.02.019
Singh B, Singh V, Mohamed A (2020) Site-specific fertilizer nitrogen Management in Cereals in South Asia, vol 39, pp 137–178. https://doi.org/10.1007/978-3-030-38881-2_6
Smith P, Soussana JF, Angers D, Schipper L, Chenu C, Rasse DP, Batjes NH, van Egmond F, McNeill S, Kuhnert M, Arias-Navarro C, Olesen JE, Chirinda N, Fornara D, Wollenberg E, Álvaro-Fuentes J, Sanz-Cobena A, Klumpp K (2020) How to measure, report and verify soil carbon change to realize the potential of soil carbon sequestration for atmospheric GHG removal. Glob Chang Biol 26(1):219–241. https://doi.org/10.1111/gcb.14815
Sreekar R, Zeng Y, Zheng Q, Lamba A, Teo HC, Sarira TV, Koh LP (2022) Nature-based climate solutions for expanding the global protected area network. Biol Conserv 269:109529. https://doi.org/10.1016/j.biocon.2022.109529
Thind H, Sharma S, Singh YS, Sidhu H (2019) Rice–wheat productivity and profitability with residue, tillage and green manure management. Nutr Cycl Agroecosyst 113. https://doi.org/10.1007/s10705-018-09967-8
Trencher G, Nick S, Carlson J, Johnson M (2024) Demand for low-quality offsets by major companies undermines climate integrity of the voluntary carbon market. Nat Commun 15(1): 1–15. https://doi.org/10.1038/s41467-024-51151-w
Weiskopf SR, Rubenstein MA, Crozier LG, Gaichas S, Griffis R, Halofsky JE, Hyde KJW, Morelli TL, Morisette JT, Muñoz RC, Pershing AJ, Peterson DL, Poudel R, Staudinger MD, Sutton-Grier AE, Thompson L, Vose J, Weltzin JF, Whyte KP (2020) Climate change effects on biodiversity, ecosystems, ecosystem services, and natural resource management in the United States. Sci Total Environ 733:137782. https://doi.org/10.1016/j.scitotenv.2020.137782
Yadvinder-Singh SM, Sidhu HS, Humphreys E, Thind HS, Jat ML, Blackwell J, Singh V (2014) Nitrogen management for zero till wheat with surface retention of rice residues in north-West India. Field Crop Res 184(April):183–191. https://doi.org/10.1016/j.fcr.2015.03.025
Yang Z, Cui X, Fan X, Ruan Y, Xiang Z, Ji L, Gao H, Zhang M, Shan S, Liu W (2024) "Active carbon" is more advantageous to the bacterial community in the rice rhizosphere than "stable carbon". Comput Struct Biotechnol J 23:1288–1297. https://doi.org/10.1016/j.csbj.2024.03.012
Zhou W, Guan K, Peng B, Margenot A, Lee D, Tang J, Jin Z, Grant R, DeLucia E, Qin Z, Wander MM, Wang S (2023) How does uncertainty of soil organic carbon stock affect the calculation of carbon budgets and soil carbon credits for croplands in the U.S. Midwest? Geoderma 429: 116254. https://doi.org/10.1016/j.geoderma.2022.116254
Zomer RJ, Bossio DA, Sommer R, Verchot LV (2017) Global sequestration potential of increased organic carbon in cropland soils. Sci Rep 7(1):1–8. https://doi.org/10.1038/s41598-017-15794-8

Chapter 7
Agroforestry Management Plans for Carbon-Negative Emissions

Sharanjit Kaur Brar and Ayushi Singh

Abstract Agroforestry, a climate-smart approach of integrating woody perennials with arable crops, is gaining significant importance for its potential to capture and store carbon from the atmosphere in the soil. Agroforestry appears to be a viable way to address the issue of climate change by promoting carbon sequestration through soil carbon sequestration, thereby ensuring enhanced benefits to the farmers in terms of gains at the field and economic level. This system has high potential to accomplish the goals of food security and climate change while simultaneously generating carbon credits for the local economy through sequestering carbon in both above- and belowground biomass. The carbon negation under agroforestry is governed through various drivers and processes, including soil organic matter, plant density regulation, species diversity, litter addition, rhizodeposition, and biomass production. By 2050, it is anticipated that the conversion of 630 million hectares of unproductive croplands to agroforestry systems will sequester carbon up to the range of 1.13–2.24 Pg C year^{-1} globally. Additionally, these systems contribute towards higher sequestration rates (600 Mt C year^{-1}) as compared to forests and croplands. The conversion of forest land use to agricultural systems resulted in a 50%–61% loss in soil organic carbon (SOC) stocks. Furthermore, the cost-benefit analysis of carbon-negative solutions emphasizes the importance of considering multiple factors such as setup costs, maintenance, and potential income from carbon credits. Although agroforestry systems have an immense potential in providing various mitigation benefits, widespread adoption is hindered due to challenges like policy barriers and a lack of awareness. Research and development organizations should collaborate to promote agroforestry by identifying future areas of research and thereby designing effective systems so as to move towards sustainable and effective agroforestry systems. Moreover, there is a need to address climate change and promote sustainable development through various schemes and policies. Policies that integrate agroforestry systems with rural development, climate change mitigation, and economic incentives are required by governments and international organizations for wider-scale adoption to achieve carbon-negative goals.

S. K. Brar · A. Singh (✉)
Department of Soil Science, Punjab Agricultural University, Ludhiana, Punjab, India

R. S. Meena et al. (eds.), *Carbon-Negative Agriculture*, Sustainability Sciences in Asia and Africa, https://doi.org/10.1007/978-981-95-1483-0_7

Keywords Agroforestry · Carbon negative · Climate change · Carbon credits · Policy framework

7.1 Introduction

Agroforestry (AF) integrates woody perennials with crops or livestock on the same land, promoting efficient utilization over monocropping through diversification of structural and functional components (Fahad et al. 2022). This system harnesses the benefits of both crops and trees to provide essential environmental services. The design and management of agroforestry systems involve complex interactions between trees and crops, representing a potential pathway for intensifying sustainable land management. Unlike other systems, agroforestry can sequester greater amounts of both above and belowground biomass compared to sole crops and forestlands. It creates a win-win scenario for the farming community and environmental sustainability by delivering numerous benefits, including the maintenance of atmospheric carbon dioxide (CO_2) levels, enhanced biodiversity, and ensured food security (Quandt et al. 2017; Meena et al. 2025). Trees play a crucial role in regulating local microclimates, improving moisture retention, reducing temperature variability, and storing atmospheric carbon long-term due to their deeper roots and functional traits that promote resource partitioning and increased carbon uptake (Singh et al. 2022).

Agroforestry is a climate-smart practice that has garnered significant global attention for its potential to mitigate climate change and reduce greenhouse gas emissions (Raj and Lal 2014) through atmospheric carbon sequestration (Possu et al. 2016). Carbon sequestration refers to the removal of CO_2 from the atmosphere and its storage in green plants. Carbon exists in various forms within the biosphere, and the carbon cycle, a vital component of soil and vegetation systems, is influenced by numerous geochemical and biological processes. Atmospheric CO_2 is converted into organic compounds through photosynthesis, and increased levels of CO_2 stimulate plant productivity. Additionally, soil organic carbon (SOC) serves as a key indicator of soil quality, regulating its physicochemical and biological properties (Pradhan et al. 2023). As a "storehouse of soil nutrients," SOC significantly impacts soil fertility and productivity. Furthermore, SOC plays a critical role in maintaining soil health, enhancing food production, and mitigating climate change (Francaviglia et al. 2019; Meena et al. 2024). Moreover, carbon sequestration in soil contributes to the stabilization of soil aggregates, the gradual release of nutrients, and improved water-holding capacity (Kane 2015; Kumar et al. 2022a).

Agroforestry serves as a powerful environmental conservation strategy, primarily due to its significant capacity to sequester carbon in both soil and biomass. This is achieved through the implementation of diverse management practices, including windbreaks, shelterbelts, silvopastoral systems, and agri-silviculture, among others. Moreover, agroforestry systems not only enhance soil health but also boost farmers' income through carbon sequestration and the establishment of carbon credit markets. For the effective execution of any agroforestry system, it is essential to involve local communities alongside planning and management experts to maximize the benefits. Engaging farmers, local specialists, academics, and policymakers will yield valuable insights and foster a comprehensive approach to developing agroforestry systems that are tailored to local conditions and sustainable over the long term (Kumar et al. 2022).

7.2 Nexus Between Agroforestry and Carbon Negativity

The carbon in the Earth is present and continually cycled in various pools like soil, atmosphere, vegetation, oceans, and geological depositions. The carbon cycle operates in balanced conditions, where carbon moving into a given pool is roughly equivalent to carbon moving out of other pools (Fig. 7.1). Carbon storage in soil accounts for more than 2000 billion tons, and it is more than carbon stored in vegetation (550 billion tons) and the atmosphere (800 billion tons). Carbon sequestration is successful when carbon storage from a given management option exceeds carbon losses. Global estimates indicated that agricultural lands contain approximately 12% of the world's terrestrial carbon, which presents a significant opportunity for carbon sequestration (Raj and Lal 2014). Additionally, the atmospheric CO_2 levels have shown a 40% increase, from 280 ppm (in 1750) to 422 ppm (in 2025), which highlights the need for management plans that prioritize carbon negative rather than carbon neutrality. The active removal of CO_2 from the atmosphere is

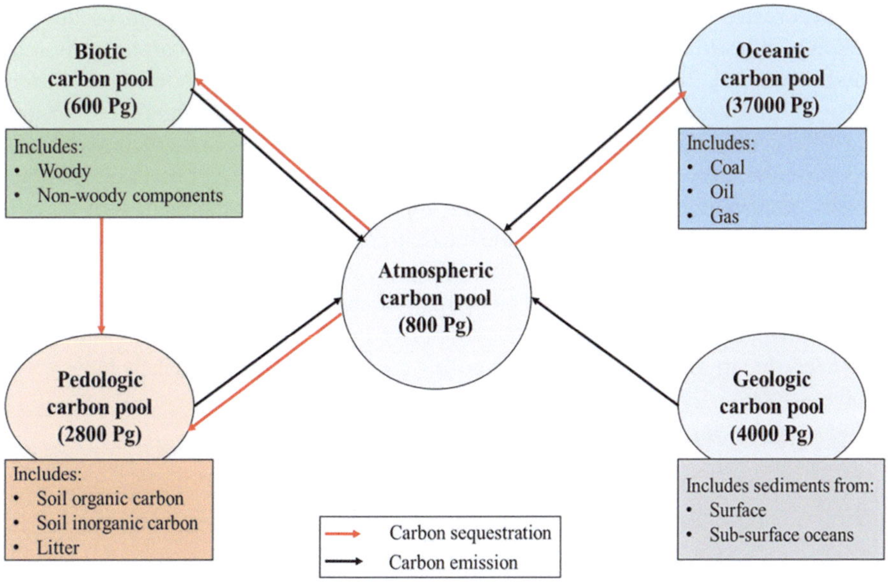

Fig. 7.1 Movement of carbon between land, atmosphere, and oceans (in Petagram C)

referred to as "carbon negative," and it is essential to preserving equilibrium in the global carbon cycle.

The concept of ecological engineering for the restoration of soil health can be applied to designing appropriate agroforestry land use systems. Agroforestry is a very complex system with high plant diversity and functions that are heavily influenced by the environmental and socio-economic factors. Through the integration of trees and crops, AF systems are effective in promoting carbon negation by sequestering carbon in soil and biomass, thereby contributing towards climate change mitigation and healthier ecosystems. Furthermore, AF systems would aid in providing favorable microclimate through regulation of soil moisture and sunlight penetration, biological nitrogen fixation through leguminous trees, and creation of more diverse and stable ecosystems (Bana et al. 2022). The ability of a system to sequester carbon depends on various factors, like physical, biological, and socio-economic features of the given land-use system. Agroforestry operates by sequestering carbon in two primary segments: aboveground and belowground carbon pools. Plant parts, including leaves, branches, and stems, are included in the aboveground carbon pools. In agroforestry systems, CO_2 is captured and stored by trees and other permanent plants, increasing the amount of biomass aboveground. The agroforestry model that is used, together with the type of trees chosen and their growth traits, affects how much carbon is stored in this pool. Soil microbes and plant roots, and their exudates, contribute to the belowground carbon pools, which are present in resistant form due to their slower decomposition rate and stronger interactions with soil particles. The great plant diversity and complexity of agroforestry systems improve the soil's capacity to store carbon, which makes it easier to create essential subsurface carbon pools.

Agroforestry is actually a form of modified forest system to sustain production and soil health while ensuring climate regulation by bringing down carbon emissions (Bamboriya et al. 2022). It serves as a valuable source of soil carbon in two forms: (1) for climate change mitigation, recalcitrant pools must reside in soils to sustain long-term carbon storage, and (2) above- and belowground biomass carbon contributed through multiple cropping cycles sustain short-term labile carbon pools. Among various contributions, biodiversity is the most important component that significantly helps in the build-up of both the recalcitrant and labile pools of carbon. These types of pools of carbon ensure that CO_2 is not immediately liberated to the atmosphere after decomposition. The storage of carbon in soils is the ultimate goal to promote carbon carbon-negative economy and provide incentives to the farmers. However, due to many legislative and policy constraints, carbon markets have yet to evolve.

7.3 Carbon Currency Through the Agroforestry Approach

Carbon currency is a tradable asset aimed at bringing down carbon emissions by superimposing climate costs on the everyday economic decision-making of consumers and businesses (Liu et al. 2022). Carbon currency paves the way for establishing standardized carbon-backed securities that represent the right to emit or remove one unit of CO_2. Farmers of the developing countries (especially in the tropics) have a great opportunity to earn carbon credits, in the form of points, earned for each ton of carbon sequestered by agroforestry, which helps to remove carbon emissions from the atmosphere. Under the carbon market, farmers receive payment from the polluters (industries) as approved by the Kyoto protocol's clean development mechanism (Atangana et al. 2014).

Carbon-backed securities function similarly to traditional asset-backed securities, but they focus on the carbon credits derived from agroforestry practices. These carbon credits represent the amount of CO_2 sequestered, where their validity and permanence are evaluated and verified by third-party certification bodies, thereby ensuring the accuracy of the amount of CO_2 sequestered. In agroforestry systems, the validation of these credits includes two steps: (A) creation of carbon credits and (B) carbon finance organization. In a nutshell, carbon credit creation under various AF systems relies on the sequestration of carbon in soil and biomass, which are further aggregated and securitized to ensure several investment choices for individuals and businesses to participate in projects related to climate change mitigation and potentially earn returns at the same time. Furthermore, carbon financing would particularly boost the involvement of the private sector to assist in projects aimed at agroforestry initiatives in addressing climate change.

Agroforestry is a promising approach to address climate change, achieve food security, and support local economies through the generation of carbon credits (Walden et al. 2020; Meena et al. 2024a). Therefore, it serves not just as a primary income source but also as an additional revenue from carbon markets to strengthen the financial status of the farmer under the changing climate scenario. When farmers and landowners can directly profit from the sale of carbon credits, they are more likely to adopt sustainable land management practices. The investors interested in supporting projects related to climate change are intrigued by carbon-backed securities that not only address environmental impact but also offer financial revenue, thereby offering a pathway for more funding for various agroforestry-related projects.

7.4 Potential of Agroforestry for Achieving Carbon-Negative Outcomes

7.4.1 Key Drivers and Processes of Carbon Negation

Agroforestry systems have the potential to significantly reduce CO_2 levels in the atmosphere by (1) storing carbon in biomass (above- and belowground) and soil, and (2) producing biomass for biopower and biofuels to replace fossil fuels. Agroforestry may conserve current carbon stocks by improving fallows and providing food and fuelwood, lowering deforestation rates. Various efforts have been made to estimate carbon sequestration under these systems. The rate at which carbon is sequestered varies across geographical areas as well as with the plant species used in the agroforestry systems (Newaj and Dhyani 2008; Meena et al. 2024b). Carbon sequestration under agroforestry systems varies with soil, climate, and species composition. Jose (2009) and Dixon (1995) predicted carbon sequestration rates in 2050 up to the range of 1.13–2.24 Pg C year^{-1} globally by the conversion of unproductive croplands (630 million ha) to agroforestry systems. Researchers have reported that carbon sequestration serves to be an economically viable opportunity for farmers practicing agroforestry in low-income countries (Nair et al. 2010). Furthermore, among various land use systems, agroforestry is documented to have the greatest carbon sequestration capability (IPCC 2000) as described in Table 7.1.

Agroforestry strategies may boost direct carbon inputs to the soil by: (a) retention/incorporation of pruned woody material as mulch and allowing abundant tree litter to decompose on site, (b) addition of surface and belowground litter during crop fallow phases, (c) allowing livestock to graze and adding dung to the soil, (d) integration of trees and their litter input in animal production systems, and (e) addition of litter inputs from shade-tolerant species growing under trees (Fig. 7.2).

7.4.2 Carbon Negation via *Above- and Belowground Biomass*

There is considerable variation in carbon sequestered in above- and belowground biomass, which is usually higher than in treeless land use systems (Fialho and Zinn 2014). Furthermore, even after achieving the targeted carbon sequestration by tree

Table 7.1 Carbon sequestration potential by 2040

Sources	Carbon sequestration (Mt C year^{-1})
Agroforestry	600
Forest management	250
Grazing management	375
Cropland management	150

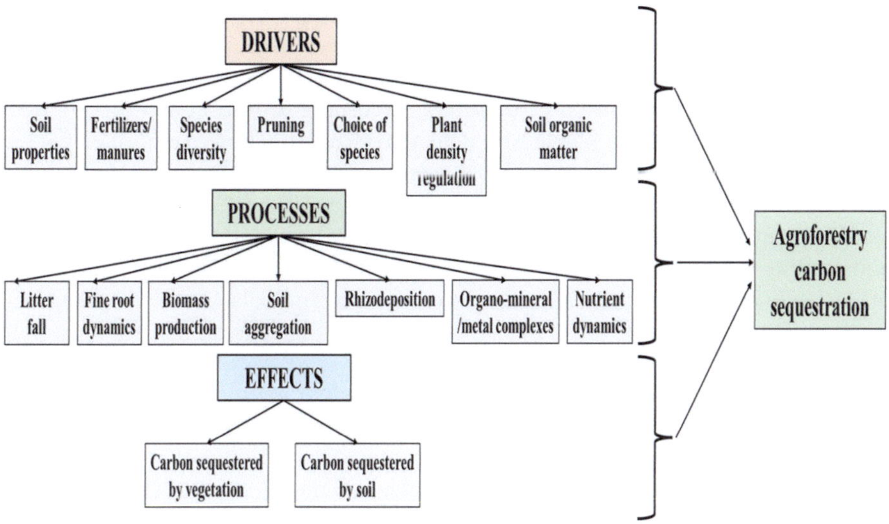

Fig. 7.2 Carbon sequestration under agroforestry

biomass, the efforts to sequester previously emitted carbon will be necessary to achieve safe levels of carbon in the atmosphere for climate change mitigation (Smith et al. 2014; Dawar et al. 2024). Higher SOC can be sequestered in agroforestry systems through various management options like avoiding burning, increasing soil organic matter, minimizing soil disturbance through adopting zero or minimum tillage practices, and erosion control practices (Lal 2004; Soto-Pinto et al. 2010; Meena et al. 2023). No tillage systems have a large potential to contribute towards increased soil carbon sequestration rates as compared to conventional tillage practices due to minimum disturbance in the fields (Sá et al. 2015). The amount of carbon sequestrated can be known through simple mathematical calculations, where aboveground biomass is assessed through general equations, and belowground biomass is 30% of aboveground biomass. The carbon sequestered varies extensively depending on the agroforestry system. Oelbermann et al. (2006) and Nair et al. (2009b) have reported that SOC pools range from 1.3 to 170 Mg C ha^{-1} in the top 40 to 100 cm of soil layer with the continuous 13-year alley cropping practice in southern Canada and 16-year silvopastoral systems in the Atlantic Coast of Costa Rica, respectively.

The practice of agroforestry led to higher SOC, which could be attributed to higher quantity as well as quality of litter inputs (Kumar et al. 2020). Higher sequestration rates resulted from increased aggregation, humification, and biomass translocation into soil by roots at deeper depths. Likewise, Nair et al. (2009a) categorized higher carbon sequestration potential in agroforestry, followed by tree plantations and arable crops. The extensive root systems of trees may reach deep into mineral soil and help in recovering nutrients from below the rooting zone. Root-derived carbon inputs are important for the SOC pool in deeper soil layers because it is more likely to be stabilized in soil through physicochemical interactions with particles than shoot-derived carbon (Kell 2012). European beech (*Fagus sylvatica* L.) roots contributed 1.55 times more to SOC than shoots (Scheu and Schauermann 1994). In croplands, root-derived carbon contributed 1.5–3 times more SOC than shoot-derived carbon (Johnson et al. 2006). According to Nair et al. (2009b), trees help in storing carbon in stable silt- and clay-sized fractions (less than 53 μm diameter) in deeper soil profiles. Also, trees in intensively managed agricultural settings can minimize SOC losses from soil erosion. The effect of four land use systems: natural forest, agricultural, and two agroforestry (30–60 and > 60 years) on soil carbon stocks was studied by Hombegowda et al. (2016). They reported a 50–61% loss in SOC stocks after the conversion of forest land use to agricultural systems. On the contrary, forest plantation resulted in increasing SOC in topsoil up to 862 Mg ha^{-1} (Muñoz-Rojas et al. 2015).

Higher quality and quantity of belowground carbon inputs, microclimatic modification, slower decomposition of root litter, and increased nutrient cycling help in higher carbon sequestration under agroforestry systems (Kumar et al. 2023). Litter with higher molecular weight biopolymers, which are less accessible to enzymes and more hydrophobic, has slower breakdown rates. These biopolymers include condensed cutin, tannins, lignin, and modified lignin. Lignin is often the most resistant part and does not get decomposed like other components. Gama-Rodrigues et al. (2010) reported the occlusion of carbon in soil macroaggregates to be a key mechanism for protecting carbon in cocoa (*Theobroma cacao*) agroforest soils in Brazil.

Agroforestry systems can act as a source as well as a sink of carbon. This is because crops and trees in agri-silvicultural systems act as a sink of CO_2, while agri-silvopastoral systems act as a source of greenhouse gases. Likewise, silvopastoral systems having trees and grasses as component crops sequestered higher SOC in comparison to systems with sole trees or pastures (Ajayi et al. 2011; Meena and Pradhan 2023). De Stefano and Jacobson (2018) concluded that the silvopastoral system resulted in higher levels of aboveground and belowground tree carbon, as well as SOC, compared to other systems studied. The conversion of agricultural land to agroforestry led to significant changes in SOC stocks, with a 13% increase in SOC stocks. The inclusion of perennials in pasture/grassland systems like silvopasture and agro-silvipastoral systems significantly improved soil organic carbon levels due to their dense foliage and highly developed root systems that helped in the translocation of belowground biomass carbon. There were no significant changes in SOC stocks with a shift from pasture/grassland or uncultivated/other to agroforestry. On the contrary, the shift from agriculture to agroforestry increased SOC stocks by 25%. Furthermore, the shift from forest to agroforestry led to a 25% decrease in SOC stocks. Likewise, specific agroforestry systems led to significant differences in SOC stocks, where the shift from forest to agri-silviculture and silvopastoral decreased SOC stocks by 12 and 44%, respectively.

To quantify carbon sequestration under agroforestry systems precisely, long-term data needs to be studied and evaluated (Table 7.2). Udawatta and Jose (2011) estimated carbon sequestration under four different agroforestry systems in the United States. They estimated carbon sequestration at a rate of 530 Tg $yr.^{-1}$,

Table 7.2 Soil carbon sequestration under different agroforestry systems or species

Agroforestry system/species	Description	Soil C (Mg ha^{-1})	References
Home gardens	Soil depth: 0–100 cm, Age: >35 years Location: Kerala, India	101–126	Saha et al. (2010)
Shaded cacao systems	Soil depth: 0–100 cm, Age: 30 years Location: Bahia, Brazil	302	Gama-Rodrigues et al. (2010)
Silvopastoral	Age: >25 years Soil depth: 0–100 cm	30–50	Nair et al. (2009b)
Different tree species (*Populus deltoides*, *Eucalyptus tereticornis*, *Melia azedarach*)	Age: 7 years Soil depth: 0–45 cm Location: Punjab, India	159–201	Singh and Gill (2014)
Poplar-based agroforestry	Age: 9 years	9.24 per year	Chauhan et al. (2015)
Eucalyptus	Age: 7–10 years	3.71	Gupta et al. (2013)
Agri-silviculture	Age: 8 years Location: Pantnagar, India	7.69 per year	Adhikari et al. (2020)

accounting for 33% of the US emission rate of 1600 Tg C $yr.^{-1}$ from fossil fuel combustion (coal, oil, and gas). An additional 4.7 Tg C might be sequestered annually by riparian buffers. In AF, functionally important tree species with broader and deeper root systems can enhance soil carbon sequestration (Kell 2012; Lorenz and Lal 2010). Compared to coniferous tree species, broadleaf trees have larger and deeper root systems, which leads to higher ratios of root biomass to aboveground biomass, and thereby higher SOC inputs from roots at soil depth (Laganière et al. 2010). The amount of carbon sequestered in aboveground biomass varied between 0.29 and 15.21 Mg C ha^{-1} $yr.^{-1}$ in major agroforestry systems (Nair et al. 2009a).

7.4.3 Stabilization of Soil Organic Carbon in Agroforestry Systems

The stability of carbon in soil is influenced by both biotic and abiotic factors (Sheoran et al. 2022). Plant inputs and organic matter have a secondary role in determining SOC retention, which can last decades to millennia. Microbially generated materials may play a significant role in SOC stabilization (Kleber et al. 2011; Pradhan and Meena 2023). SOC is protected and stabilized for the long term through physical and chemical protection, where the former involves the prevention of decomposers from accessing organic matter (OM), while the latter renders stabilization through the formation of various organo-mineral complexes. The soil organic matter interacts with minerals, metal ions, and other organic substances through various physical and chemical forces and thereby preserves the carbon sequestered in soil for the long term in comparison to physical protection. Furthermore, biochemical recalcitrance and physical protection can prolong the presence of SOC in soil, allowing for the formation of organo-mineral complexes.

The management practices like integrated application of inorganic fertilizers and organic manures along with crop residues, shifting towards conservation tillage, cover cropping, mulching, windbreaks, shelterbelts and reducing cultivation intensity, grazing, and burning can significantly enhance organic carbon levels in soil (Choudhary et al. 2019; Leharwan et al. 2023). The adoption of these conservation-effective practices can increase biomass input to soil in agroforestry systems, reduce nutrient and water losses, increase biomass production, and protect SOC from losses by influencing physical and chemical stabilization mechanisms. The quantity of carbon preserved under different management techniques in agroforestry is also influenced by aggregates and their size fractions. The carbon stored in aggregates is prevented from being lost through (1) establishing a physical barrier between aggregate stored carbon and microbes and enzymes, and (2) controlled interactions among various soil microorganisms. Additionally, compared to microaggregates, macroaggregates retain more carbon, indicating the contribution of plant roots and fungal hyphae to the elevation of carbon concentrations. In light of this, several studies have emphasized the critical function of aggregates in elucidating the complexities of soil carbon sequestration in agroforestry systems.

7.5 Economic Assessment in Carbon-Negative Agroforestry

Understanding the financial viability and environmental impact of agroforestry for carbon-negative goals requires an economic assessment of agricultural practices to sequester carbon and generate income. Carbon-negative energy systems, such as the pyrolysis-biochar platform, produce bio-oil and biochar while removing CO_2 from the atmosphere (Kumar et al. 2025). These systems not only produce energy but also evaluate techno-economic performance to promote carbon-negative emissions (Shahane et al. 2022; Scheibenreif 2023).

Venance-Pâques Gniayou et al. (2021) reported an improvement in carbon storage as well as the generation of additional revenue due to the practice of cocoa-agroforestry systems. In Indonesia, the potential of cocoa-based agroforestry systems was evaluated by Roziaty (2023), focusing on the trade-off between carbon accumulation and agricultural productivity. Four types of cocoa plantations over a

20-year financial period under two scenarios: one without carbon payments and another with participation in the voluntary carbon market (VCM). The practice of cocoa-coconut AF systems resulted in the generation of the highest net present value (NPV—6647 USD) and benefit-to-cost ratio (5.8). Although the fastest payback period (2 years) was reported under monoculture systems, the practice of cocoa-coconut AF system increased NPV by 15%–25% at verified emission reduction (VER—8 USD Mt CO_2 e^{-1}). In a nutshell, these studies suggest agroforestry to be a sustainable and profitable method through the sequestration of carbon and generating financial revenue.

The cost-benefit analysis of carbon-negative solutions emphasizes the importance of considering multiple factors such as setup costs, maintenance, and potential income from carbon credits (Flugge and Abadi 2006). A study by Giannitsopoulos et al. (2020) in the UK suggested that agroforestry provided societal benefits equivalent to arable farming when the value of sequestered carbon was at least €16 per ton of CO_2, comparable to the €18.9 per ton used in India's rice production. On the other hand, the practice of agroforestry in Switzerland provided higher values of €35 per ton, while lower values in the range of €109–€137 per ton were reported in Spain. The expected carbon price is to rise up to €91 per ton by 2030, thereby creating a strong financial incentive for implementing AF systems. Furthermore, the oak-based tree planting in Spain took more than 60 years to recover original carbon losses.

7.6 Prospects of Agroforestry for Enhancing Carbon-Negativity Under a Changing Climate

Agroforestry presents a wide scope towards carbon negative under a changing climate without compromising on farmers' existing production systems. It plays a significant role in mitigating atmospheric CO_2 levels and has been recognized under the Kyoto Protocol's Article 3.3, which emphasizes afforestation and reforestation as options for greenhouse gas mitigation. The growing awareness about the potential of AF for carbon sequestration has paved the way for initiatives like the National Climate Change Action Plan to transform 1.5 Mha of fallows and degraded agricultural land into AF systems. The terrestrial carbon sinks account for about one-third of anthropogenic CO_2 emissions, especially agricultural ecosystems. Carbon sequestration under agroforestry mainly deals with long-term C retention in soil through stabilization mechanisms that lower the rate of SOC breakdown. Agroforestry systems can help mitigate climate change by lowering SOC losses and stabilizing SOC pools, which include organic materials with radiocarbon dates ranging from 1000 to more than 10,000 years, particularly in subsurface layers (Schmidt et al. 2011). Three primary mechanisms can help to reduce atmospheric CO_2: carbon sequestration, carbon conservation, and carbon substitution.

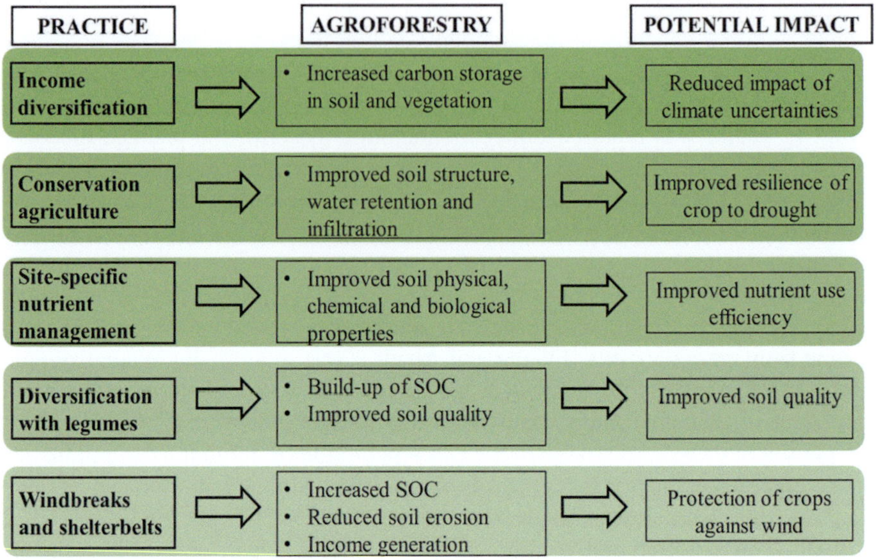

Fig. 7.3 Summary of practices with their potential impact on carbon-negative under agroforestry

Tree age, location, and management are some of the factors that affect the capacity to trap carbon. Kaul et al. (2010) observed the variation of carbon storage in India from 101 to 156 Mg C ha^{-1}, where the maximum carbon storage (82 Mg C ha^{-1}) was observed under Sal forests. The highest rate of carbon sequestration was recorded under fast-growing and short rotation eucalyptus plantation (6 Mg C ha^{-1} year^{-1}) and poplar (8 Mg C ha^{-1} year^{-1}) compared to sal (1 Mg C ha^{-1} year^{-1}) and teak (2 Mg C ha^{-1} year^{-1}) plantations with lower sequestration rates. Owing to their rapid growth and adaptability, short-rotation plantations not only sequester carbon efficiently but also provide biomass for energy, helping to reduce greenhouse gas emissions. Under the carbon substitution mechanism through agroforestry, fossil fuel-based products are replaced with renewable biomass alternatives. Similarly, the National Mission on Biodiesel was launched by the Government of India (GOI), carried out over 4,00,000 hectares, to include biofuels into gasoline mix with a 5% ethanol doping program. The emphasis is laid on AF with hydrocarbon-producing trees containing higher oil content, such as *Madhuca indica*, *Jatropha curcas,* and *Pongamia pinnata*. Agroforestry with hydrocarbon-producing trees reduces dependency on fossil fuels by augmenting biodiesel production. However, their full potential is still underutilized. Studies by Lal (2001) suggest that biofuels could offset 0.3–0.7 Pg C year^{-1} from fossil fuel emissions. Additionally, the biomass from agroforestry provides food, construction materials, shade, and medicines, etc., that decreases the pressure on the natural forests to let them thrive. Furthermore, it may allow land to be taken out of fallow under the shifting cultivation system. The potential impacts of agroforestry on carbon sequestration are summarized in Fig. 7.3.

7.7 R&D Priorities in Agroforestry for Generating Carbon Currency

Agroforestry combines climate change adaptation and mitigation objectives through carbon accretion in plants and soil (FAO 2009), providing significant benefits to local biodiversity and ecosystems. However, it needs to be promoted in both voluntary and compliance markets, operating on the cap-and-trade system, providing wide opportunities for farmers to actively participate in offsetting carbon emissions and earning monetary benefits (Stavi and Lal 2013; Pradhan and Meena 2023a). R&D in agroforestry should focus on innovative practices that enhance productivity, sustainability, and resilience in farming systems. This includes developing techniques for integrating trees and crops, optimizing land use, and improving soil health and water conservation, especially considering region specificity and farmer goals (Jangir et al. 2024). Important inferences could be derived from not only successful cases but also negative cases, especially when the same agroforestry system is practiced over a widespread area, highlighting any significant factors. Additionally, R&D should be towards methodologies to quantify carbon sequestration potential in agroforestry systems, for establishing the groundwork required to generate carbon credits or carbon currency. A reliable measurement, reporting, and verification systems for agroforestry need to be established to contribute significantly to climate change mitigation efforts, enabling farmers to participate in carbon markets and derive financial benefits from sustainable practices. Conjugated efforts on agroforestry R&D and carbon currency generation can accelerate the adoption of agroforestry, strengthen rural economies, and advance national climate goals.

To provide insights on the long-term carbon sequestration potential under agroforestry systems ecosystem models could be employed to assess the plant and soil carbon stocks. Schelhaas et al. (2004) utilized the CO2FIX model to assess carbon stock and flow patterns across various ecosystems worldwide. However, it has not been tested with real data from agroforestry systems. Kaonga et al. (Kaonga and Bayliss-Smith 2012) used empirical data from 7-year-old *Leucaena* woodlots in Msekera, Zambia, to assess the CO2FIX model's ability to estimate aboveground and soil carbon stocks. The aboveground C stock in *Leucaena* woodlots varied from 24.5 to 55.9 Mg ha^{-1}. Likewise, SOC (0–200 cm depth) varied from 106.9 (in *L. diversifolia*) to 186.0 Mg/ha (in *L. leucocephala*). Despite broad awareness of agroforestry's potential for carbon sequestration and biomass production, our understanding of these areas remains restricted. Furthermore, inaccuracy in the prediction of long-term empirical data by these models could be attributed to inadequate availability of data on climate, tree growth, litter quantity, and quality. Biased outcomes result from a lack of information and standardized methods. The inconsistency in results can be dealt with (1) higher sample size (2) detailed information about previous land-use system (3) inclusion of explanatory variables like age of systems and C sequestration rates, and (4) soil characteristics like bulk density, pH, and texture (Kim et al. 2016; Kumar et al. 2021).

The current approaches for determining C vary greatly and are based on various assumptions. The current measurements and estimations in large-scale global models can lead to significant under- or overestimation of carbon. Therefore, practical guidelines for researchers, such as accurate descriptions of methodologies and processes, would enable other academics to incorporate them into bigger databases. Also, the errors in estimation result from flaws in the current literature, including incorrect assumptions, operational shortcomings, and inaccurate information, could be further avoided. Thus, regular and consistent methods for estimating carbon sequestration under agroforestry systems would help in positioning agroforestry systems as a credible option for carbon credit generation (Jhariya et al. 2022).

7.8 Policy Framework for Agroforestry and Carbon Incentive Initiatives

While conducting a literature survey for this chapter, very few investigations were reported that talked about the policies or examined their effects on the adoption and impact of agroforestry on carbon incentives. More analysis needs to be done on the policies affecting or related to agroforestry systems in various countries. These could include (1) Providing better evidence through field research on the impact of agroforestry on carbon currency, (2) Examining various examples to better understand in-depth knowledge about the role and success that the policy played, (3) Reviewing documents of major policy to identify major policies that could enhance the adoption of agroforestry practices at national level, and (4) participation of key players and institutions at government and non-government levels to strengthen this aspect of agriculture.

About climate change adaptation and mitigation efforts, agroforestry is becoming more widely recognized, which might further lead to improved agroforestry policies. Various problems like resolving policy restrictions in land and tree tenure, availability of germplasm, distribution of technical knowledge, and markets for tree products might lead to increasing agroforestry use and climate change adaptation benefits for communities. Agroforestry has the potential to deliver mitigation benefits, but recognition and reward are hindered by many obstacles. Various strategies include (1) expansion of mitigation possibilities to recognize the importance of agroforestry for carbon sequestration and (2) raising awareness of the link between agroforestry and carbon accretion. To better promote agroforestry, research and development organizations should collaborate to design areas of inquiry and learn from previous efforts.

Policies that integrate agroforestry with rural development, climate change mitigation, and economic incentives are required by governments and international organizations. Clear land tenure rights, which guarantee farmers and communities long-term ownership or access to land for agroforestry investments, should be a part of these programs. Simplified registration and verification procedures should be implemented for carbon credit schemes like the clean development mechanism (CDM) and voluntary carbon markets like Verra and Gold Standard to promote farmer involvement. Financial incentives like tax breaks, subsidies, and concessional loans can also promote the widespread use of agroforestry. Governments should also integrate agroforestry into national climate action plans (NAPs), afforestation programs, and emissions trading schemes, ensuring that farmers benefit from both environmental and economic gains. Strategic coordination and collaboration among high-level decision makers across multiple sectors are required, particularly in agriculture, environmental management, and forestry. Furthermore, capacity-building programs, extension services, and digital tools must be provided to farmers with technical knowledge, access to carbon markets, and support for sustainable agroforestry practices.

7.9 Future Directions

Agroforestry research should expand research for capacity building at the regional and national levels to allow policymakers and practitioners to promote agroforestry. A lot of improvement is still required to build an appropriate design of planting, tree age, and low-input-high-yielding systems with higher resource use efficiency (Punia et al. 2020; Kumar et al. 2025). Optimal agroforestry design for higher carbon sequestration is desired, which amounts to increased returns and adoption. Furthermore, researchers should explore innovative platforms that empower smallholders to improve the quality of agroforestry products by focusing on key enabling factors, such as access to stable and reliable markets. Soil carbon is the greatest carbon sink, which can help with problems arising from carbon emissions. The regulatory frameworks should therefore acknowledge AF to sequester carbon back into the soil. The biggest loopholes arise because of carbon price variations. Thus, to attain maximum benefits of AF, there is a dire need to properly set carbon prices to bring down the opportunity cost to landowners, even after the carbon earnings provide additional financial incentives to farmers. The practice of AF is competitive yet attractive because alternative land uses, either for growing crops or infrastructure which should be outweighed by the benefits and economic returns provided by the agroforestry system, which is possible with appropriate carbon credit pricing. The government must provide market and technical assistance to overcome hurdles in adopting agroforestry. The government should also secure grounds for public-private partnerships to ascertain market accessibility and funding sources. The lack of infrastructure should also be met by strengthening institutional mechanisms, such as creating specialized agencies specifically for AF, creating and setting forth a national program on agroforestry, and financing mechanisms. Regular monitoring of these indicators and fixing their problems at the earliest can speed up the regional agroforestry progress. Moreover, the government should recognize climate change as a threat and actively promote sustainable development through schemes and policies.

7.10 Conclusion

Agroforestry holds immense potential in the era of climate change in both tropical and temperate areas, deeply rooted in strong ecological principles, offering immense potential to address multiple regional developmental goals by providing essential ecosystem services. Researchers have highlighted the capacity of these systems as sustainable land-use options, particularly their carbon sequestration potential. AF offers incredible opportunities for climate change mitigation and storage of carbon in soil while simultaneously sustaining farm income. Besides, AF provides several goods such as timber, fodder, fuelwood, and food, all while monitoring the atmospheric CO_2 levels within an acceptable range. Significant amounts of carbon can be stored by AF systems like boundary plantations, silvipastoral, agri-horticulture, home gardens, windbreaks, shelterbelts, and multistoried systems. Among different plantation types fast-growing trees have shown higher sequestration potential than the slow-growing species. Moreover, management strategies practiced under the AF system should focus on optimizing carbon storage in soil and biomass to lower carbon emissions and generate carbon credits significantly. Additionally, pairing AF with techniques such as minimum or no-till, residue management, and building soil health through practices that promote organic matter retention boosts the sequestration of carbon in the soil.

References

Adhikari B, Lodhiyal N, Lodhiyal LS (2020) Assessment of crop yield, productivity and carbon sequestration in agroforestry systems in Central Himalaya, India. Agrofor Syst 94(1):281–296

Ajayi OC, Place F, Akinnifesi FK, Sileshi GW (2011) Agricultural success from Africa: the case of fertilizer tree systems in southern Africa (Malawi, Tanzania, Mozambique, Zambia and Zimbabwe). Inter J Agri Sustain 9:130–136

Atangana A, Khasa D, Chang S, Degrande A, Atangana A, Khasa D, Chang S, Degrande A (2014) Agroforestry and the carbon market in the tropics. Trop Agrofor:353–365

Bamboriya SD, Bana RS, Kuri BR, Kumar V, Bamboriya S, Meena RP (2022) Achieving higher production from low inputs using synergistic crop interactions under maize-based polyculture systems. Environ Sustain 5:145–159

Bana RC, Gupta AK, Bana RS, Shivay YS, Bamboriya SD, Thakur NP, Puniya R, Gupta M, Jakhar SR, Kailash, Choudhary RS, Bochalya RS, Bajaya T, Kumar V, Kumar P, Choudhary AK (2022) Zinc-coated urea for enhanced zinc biofortification, nitrogen use efficiency and yield of basmati rice under typic fluvents. Sustainability 14:104. https://doi.org/10.3390/su14010104

Chauhan SK, Sharma R, Singh B, Sharma SC (2015) Biomass production, carbon sequestration and economics of on-farm poplar plantations in Punjab. India J Appl Nat Sci 7(1):452–458

Choudhary M, Rana KS, Meena MC, Bana RS, Jakhar P, Ghasal PC, Verma RK (2019) Changes in physico-chemical and biological properties of soil under conservation agriculture based pearl millet –mustard cropping system in rainfed semi-arid region. Arch Agron Soil Sci 65(7): 911–927. https://doi.org/10.1080/03650340.2018.1538556

Dawar R, Yalamalle VR, Bana RS, Kaur R, Shivay YS, Choudhary AK, Singh T, Meena SL, Vijay D, Vijayakumar HP, Kumar V, Yadav AL (2024) Seed dormancy dynamics and

germination characteristics of *Malva parviflora* L. Agriculture 14:266. https://doi.org/10.3390/agriculture14020266
De Stefano A, Jacobson MG (2018) Soil carbon sequestration in agroforestry systems: a meta-analysis. Agrofor Syst 92:285–299
Dixon RK (1995) Agroforestry systems: sources or sinks of greenhouse gases? Agrofor Sys 31:99–116
Fahad S, Chavan SB, Chichaghare AR, Uthappa AR, Kumar M, Kakade V, Pradhan A, Jinger D, Rawale G, Yadav DK, Kumar V (2022) Agroforestry systems for soil health improvement and maintenance. Sustainability 14(22):14877
FAO (2009) Enabling agriculture to contribute to climate change mitigation. The Food and Agriculture Organization of the United Nations, Rome, FAO submission to the UNFCCC
Fialho RC, Zinn YL (2014) Changes in soil organic carbon under eucalyptus plantations in Brazil: a comparative analysis. Land Degrad Develop 25:428–437. https://doi.org/10.1002/ldr.2158
Flugge F, Abadi A (2006) Farming carbon: an economic analysis of agroforestry for carbon sequestration and dryland salinity reduction in Western Australia. Agrofor Syst 68(3):181–192
Francaviglia R, Di Bene C, Farina R, Salvati L, Vicente-Vicente JL (2019) Assessing "4 Per 1000" soil organic carbon storage rates under Mediterranean climate: a comprehensive data analysis. Mitig Adapt Strateg Glob Chang 24:795–818
Gama-Rodrigues EF, Nair PKR, Nair VD, Gama-Rodrigues AC, Baligar VC, Machado RCR (2010) Carbon storage in soil size fractions under two cacao agroforestry systems in Bahia, Brazil. Environ Manag 45:274–283. https://doi.org/10.1007/s00267-009-9420-7
Giannitsopoulos ML, Graves AR, Burgess PJ, Crous-Duran J, Moreno G, Herzog F, Palma JH, Kay S, de Jalón SG (2020 Oct) Whole system valuation of arable, agroforestry and tree-only systems at three case study sites in Europe. J Clean Prod 1(269):122283
Gupta A, Dhyani SK, Ramnewaj Handa AK, Prasad R, Alam B, Rizvi RH, Gupta G, Pandey KK, Jain A (2013) Modeling analysis of potential carbon sequestration under existing agroforestry systems in three districts of Indo-Gangetic plains in India. Agrofor Syst 87:1129–1146
Hombegowda HC, Van-Straaten O, Köhler M, Hölscher D (2016) On the rebound: soil organic carbon stocks can bounce back to near forest levels when agroforests replace agriculture in southern India. Soil 2:13–23
IPCC (2000) Land use, land-use change, and forestry. A special report of the IPCC. Cambridge University Press, Cambridge
Jangir CK, Sangwan PS, Panghaal D, Kumar S, Meena RS, Bharti J, R.D. and Singh, N. (2024) Spatial variability and statistical analysis of soil properties in the rice wheat-based systems of North-West India. Commun Soil Sci Plant Anal 55(8):1205–1223
Jhariya MK, Raj A, Banerjee A, Meena RS, Bargali SS, Kumar S, Nema S, Poonam, Oraon PR (2022) Plan and policies for soil organic carbon management under agroforestry system. In: Plans and policies for soil organic carbon management in agriculture. Springer, Singapore, pp 191–219
Johnson JMF, Allmaras RR, Reicosky DC (2006) Estimating source carbon from crop residues, roots and rhizodeposits using the national grain-yield database. Agron J 98:622–636. https://doi.org/10.2134/agronj2005.0179
Jose S (2009) Agroforestry for ecosystem services and environmental benefits: an overview. Agrofor Syst 76:1–10
Kane D (2015) Carbon sequestration potential on agricultural lands a review of current science and available practices. Nat Sustain Agri Coalition:1–35
Kaonga ML, Bayliss-Smith TP (2012) Simulation of carbon pool changes in woodlots in eastern Zambia using the CO2FIX model. Agrofor Syst 86:213–223
Kaul M, Mohren GMJ, Dadhwal VK (2010) Carbon storage and sequestration potential of selected tree species in India. Mitig Adapt Strateg Glob Chang 15(5):489–510
Kell DB (2012) Large-scale sequestration of atmospheric carbon via plant roots in natural and agricultural ecosystems: why and how. Philos Trans R Soc B 367:1589–1597. https://doi.org/10.1098/rstb.2011.0244

Kim D-G, Kirschbaum MUF, Beedy TL (2016) Carbon sequestration and net emissions of CH4 and N2O under agroforestry: synthesizing available data and suggestions for future studies. Agric Ecosyst Environ 226:65–78. https://doi.org/10.1016/j.agee.2016.04.011

Kleber M, Nico PS, Plante A, Filley T, Kramer M, Swanston C, Sollins P (2011) Old and stable soil organic matter is not necessarily chemically recalcitrant: implications for modeling concepts and temperature sensitivity. Glob Chang Biol 17:1097–1107. https://doi.org/10.1111/j.1365-2486.2010.02278.x

Kumar R, Punia H, Kumar S (2020) Nanomaterials for agriculture input use efficiency. Resources use efficiency in agriculture. Springer, Singapore. https://doi.org/10.1007/978-981-15-6953-1_5

Kumar V, Bana RS, Singh T, Louhar G (2021) Ecological weed management approaches for wheat under rice–wheat cropping system. Environ Sustain 4(1):51–61. https://doi.org/10.1007/s42398-020-00157-3

Kumar S, Bamboriya SD, Rani K, Meena RS, Sheoran S, Loyal A, Kumawat A, Jhariya MK (2022) Grain legumes: a diversified diet for sustainable livelihood, food, and nutritional security. In: Advances in legumes for sustainable intensification. Academic Press, pp 157–178

Kumar S, Sharma SK, Thakral SK, Bhardwaj KK, Jhariya MK, Meena RS, Jangir CK, Bedwal S, Jat RD, Gaber A, Atta AA (2022a) Integrated nutrient management improves the productivity and nutrient use efficiency of *Lens culinaris* Medik. Sustainability 14(3):1284

Kumar S, Sharma SK, Dhaka AK, Bedwal S, Sheoran S, Meena RS, Jangir CK, Kumar D, Kumar R, Jat RD, Meena AK (2023) Efficient nutrient management for enhancing crop productivity, quality and nutrient dynamics in lentil (*Lens culinaris* Medik.) in the semi-arid region of northern India. PLoS One 18(2):e0280636

Kumar S, Meena RS, Kumar S, Pradhan G, Jangir CK, Ghosh S, Punia H, Sheoran P, Meena R, Ahmad MA, Goyal SK (2025) Designing a diversified Indian mustard production system for energy-carbon-cum-heat use efficiency and sowing dates assessment. GCB Bioenergy 17(6): e70044

Laganière J, Angers D, Paré D (2010) Carbon accumulation in agricultural soils after afforestation: a meta-analysis. Glob Chang Biol 16:439–453. https://doi.org/10.1111/j.1365-2486.2009.01930.x

Lal R (ed) (2001) Soil carbon sequestration and the greenhouse effect. Soil Science Society of America, Madison, pp 1–26

Lal R (2004) Soil carbon sequestration impacts on global climate change and food security. Science 304(5677):1623–1627

Leharwan M, Kumar Y, Kumar R, Kumar Saraswat P, Kumar R, Veetil AKT, Bhattacharjee S, Kumar A, Kumar S (2023) Assessing the effects of conservation tillage and in-situ crop residue management on crop yield and soil properties in rice–wheat cropping system. Sustainability 15(17):12736

Liu Q, Chen Z, Xiao SX (2022) A theory of carbon currency. Fundam Res 2(3):375–383

Lorenz K, Lal R (2010) Carbon sequestration in forest ecosystems. Springer, Dordrecht

Meena RS, Pradhan G (2023) Industrial garbage-derived biocompost enhances soil organic carbon fractions, CO_2 biosequestration, potential carbon credits, and sustainability index in a rice-wheat ecosystem. Environ Res 116525. https://doi.org/10.1016/j.envres.2023.116525

Meena RS, Pradhan G, Kumar S, Lal R (2023) Using industrial wastes for rice-wheat cropping and food-energy-carbon-water-economic nexus to the sustainable food system. Renew Sust Energ Rev 187:113756. https://doi.org/10.1016/j.rser.2023.113756

Meena RS, Pradhan G, Nalani Singh K, Singh AK, Verma SK, Mina KK, Kumar S, Chaturvedi RK (2024) Diversified cropping modules designed for soil restoration, CO_2 sequestration, and generating carbon credits. Land Degrad Dev 35. https://doi.org/10.1002/ldr.5224

Meena RS, Pradhan G, Singh K, Kumar S, Singh AK, Shashidhar KS, Mina KK, Rao CS (2024a) Agriculture models for restoring degraded land to enhance CO_2 biosequestration and carbon credits in the Vindhyan region of India. Sci Total Environ 929:172661. https://doi.org/10.1016/j.scitotenv.2024.172661

Meena RS, Lal R, Kumar S, Pradhan G, Rao CS, Singh AK, Pathak H, Abhilash PC, Kumar A, Sharma SK, Jat ML, Singh S (2024b) Potential of Indian agriculture for capturing atmospheric CO_2 and monetizing carbon credits to the farmers: an overview and policy framework. Adv Agron 188:101–206. https://doi.org/10.1016/bs.agron.2024.06.001

Meena RS, Pradhan G, Kumar S (2025) Energy flow, eco-efficiency, and economic circulation with recycled industrial waste compost application in wheat and subsequent rice farming. Sci Total Environ 967:178779. https://doi.org/10.1016/j.scitotenv.2025.178779

Muñoz-Rojas M, Jordán A, Zavala LM, De la Rosa D, Abd-Elmabod SK, Anaya-Romero M (2015) Impact of land use and land cover changes on organic carbon stocks in Mediterranean soils (1956–2007). Land Degrad Develop 26:168–179

Nair PKR, Kumar BM, Nair VD (2009a) Agroforestry as a strategy for carbon sequestration. J Plant Nutri Soil Sci 172:10–23

Nair PKR, Nair V, Gama-Rodrigues E, Garcia R, Haile S, Howlett D, Kumar BM, Mosquera-Losada MR, Saha S, Takimoto A, Tonucci R (2009b) Soil carbon in agroforestry systems: an unexplored treasure? Available from Nature Proceedings

Nair PKR, Nair VD, Kumar BM, Showalter JM (2010) Carbon sequestration in agroforestry systems. Adv Agron 108:237–307

Newaj R, Dhyani SK (2008) Agroforestry systems for carbon sequestration: present status and scope. Ind J Agrofor 10(1):1–9

Oelbermann M, Voroney RP, Gordon AM, Kass DCL, Schlönvoigt AM, Thevathasan NV (2006) Soil carbon dynamics and residue stabilization in a Costa Rican and southern Canadian alley cropping system. Agrofor Sys 68:27–36

Possu WB, Brandle JR, Domke GM, Schoeneberger M, Blankenship E (2016) Estimating carbon storage in windbreak trees on U.S. agricultural lands. Agrofor Syst 90:889. https://doi.org/10.1007/s10457-016-9896-0

Pradhan G, Meena RS (2023) Interaction impacts of biocompost on nutrient dynamics and relations with soil biota, carbon fractions index, societal value of CO_2 equivalent, and ecosystem services in the wheat-rice farming. Chemosphere 139695. https://doi.org/10.1016/j.chemosphere.2023.139695

Pradhan G, Meena RS (2023a) Utilizing waste compost to improve the atmospheric CO_2 capturing in the rice-wheat cropping system and energy-cum-carbon credit audit for a circular economy. Sci Total Environ 164572. https://doi.org/10.1016/j.scitotenv.2023.164572

Pradhan G, Meena RS, Kumar S, Lal R (2023) Utilizing industrial wastes as compost in wheat-rice production to improve the above and below-ground ecosystem services. Agric Ecosyst Environ 358:108704. https://doi.org/10.1016/j.agee.2023.108704

Punia H, Tokas J, Malik A, Satpal, Rani A, Gupta P, Kumari A, Mor VS, Bhuker A, Kumar S (2020) Solar radiation and nitrogen use efficiency for sustainable agriculture. In: Resources use efficiency in agriculture, pp 177–212

Quandt A, Neufeldt H, McCabe JT (2017) The role of agroforestry in building livelihood resilience to floods and drought in semiarid Kenya. Ecol Soc 22(3)

Raj AJ, Lal SB (2014) Agroforestry theory and practices. Scientific publishers

Roziaty E (2023) Trade-off between carbon offset and economic benefit: potential of cocoa-based agroforestry system implemented in voluntary carbon market. J Agric Rural Dev Trop Subtrop 124(2)

Sá JCD, Séguy L, Tivet F, Lal R, Bouzinac S, Borszowskei PR, Briedis C, dos Santos JB, Hartman DC, Bertoloni CG, Rosa J, Friedrich T (2015) Carbon depletion by plowing and its restoration by no-till cropping systems in oxisols of subtropical and tropical agro ecoregions in Brazil. Land Degrad Develop 26:531–543

Saha SK, Nair PKR, Nair VD, Kumar BM (2010) Carbon storage in relation to soil size-fractions under tropical tree-based land-use systems. Plant Soil 328:433–446. https://doi.org/10.1007/s11104-009-0123-x

Scheibenreif M (2023) Assessing the potential for negative emission technologies in Austria. Master's thesis

Schelhaas MJ, van Esch PW, Groen TA, de Jong BHJ, Kanninen M, Liski J, Masera O, Mohren GMJ, Nabuurs GJ, Palosuo T, Pedroni L, Vallejo A, Vilén T (2004) CO_2FIX V 3.1– A modelling framework for quantifying carbon sequestration in forest ecosystems. Wageningen, Alterra, Alterra-rapport 1068, 122 p

Scheu S, Schauermann J (1994) Decomposition of roots and twigs: effects of wood type (beech and ash), diameter, site of exposure and macrofauna exclusion. Plant Soil 241:155–176. https://doi.org/10.1007/BF00033936

Schmidt MWI, Torn MS, Abiven S, Dittmar T, Guggenberger G, Janssens IA, Kleber M, Kögel-Knabner I, Lehmann J, Manning DAC, Nannipieri P, Rasse DP, Weiner S, Trumbore SE (2011) Persistence of soil organic matter as an ecosystem property. Nature 478:49–56. https://doi.org/10.1038/nature10386

Shahane AA, Shivay YS, Prasanna R, Kumar D, Bana RS (2022) Effect of crop establishment methods and microbial inoculations on augmenting the energy efficiency and nutritional status of rice and wheat in cropping system mode. Sustainability 2022(14):5986. https://doi.org/10.3390/su14105986

Sheoran S, Gupta M, Kumari S, Kumar S, Rakshit S (2022) Meta-QTL analysis and candidate genes identification for various abiotic stresses in maize (*Zea mays* L.) and their implications in breeding programs. Mol Breed 42(5):26

Singh B, Gill RIS (2014) Carbon sequestration and nutrient removal by some tree species in an agrisilviculture system in Punjab, India. Range Manage Agrofor 35(1):107–114

Singh T, Bana RS, Satapathy BS, Lal B, Yogi AK, Singh R (2022) Energy balance, productivity and resource-use efficiency of diverse sustainable intensification options of rainfed-lowland rice systems under different fertility scenarios. Sustainability 2022(14):3657. https://doi.org/10.3390/su14063657

Smith P, Edenhofer O, Pichs-Madruga R, Sokona Y, Farahani E et al (2014) Agriculture, forestry and other land use (AFOLU). In: Climate change 2014: mitigation of climate change. Contribution of working group III to the fifth assessment report of the intergovernmental panel on climate change. Cambridge University Press, Cambridge/New York

Soto-Pinto L, Anzueto M, Mendoza J, Ferrer GJ, de Jong B (2010) Carbon sequestration through agroforestry in indigenous communities of Chiapas, Mexico. Agrofor Sys 78:39–51

Stavi I, Lal R (2013) Agroforestry and biochar to offset climate change: a review. Agron Sustain Dev 33:81–96. https://doi.org/10.1007/s13593-0120081-1

Udawatta RP, Jose S (2011) Carbon sequestration potential of agroforestry practices in temperate North America. Carbon sequestration potential of agroforestry systems: opportunities and challenges. pp 17–42

Venance-Pâques Gniayou K, Affia Sonmia Francia K, N'guessan Lucien D, Kouamé Kra Modeste A, Constant Yves AY (2021) Socio-economic assessment of different cocoa agroforestry systems in the forest-Savannah transition zone in central Côte d'Ivoire. For Trees Livelihoods 30(3):195–212

Walden P, Ollikainen M, Kahiluoto H (2020) Carbon revenue in the profitability of agroforestry relative to monocultures. Agrofor Syst 94:15–28. https://doi.org/10.1007/s10457-019-00355-x

Chapter 8
Management of Agrochemicals and Policy Planning for Carbon Credits and Ecosystem Services

Heena, P. Chaitra, K. R. Anusha, and N. J. Gowthamchand

Abstract The increased use of agrochemicals has improved agricultural productivity and economic returns, but it has also resulted in serious environmental problems, such as soil degradation, water contamination, air pollution and biodiversity loss. This chapter examines the multifaceted impacts of agrochemicals on agroecosystem services (AES), which are essential for sustaining agricultural productivity and environmental health. In order to maintain agricultural production and environmental health, agrochemicals have a multifaceted impact on agroecosystem services. Agroecosystem services (AES) are categorized into provisioning (such as food and raw materials), regulatory (such as carbon sequestration), supporting (such as nutrient cycling) and cultural (such as recreation) services, all of which are essential to preserving ecological balance. However, agroecosystem disservices (AEDS), such as soil erosion, pest resurgence and habitat destruction, threaten these beneficial functions. The study examines the benefits and downsides of several cropping systems, such as monocropping, multi-cropping and agroforestry, with regard to ecosystem services. Sustainable techniques, including cover crops, conservation agriculture and regenerative farming, are suggested as ways to mitigate the negative environmental effects of agrochemicals and make it easier to generate carbon credits. By strengthening soil health, increasing carbon absorption and lowering greenhouse gas emissions, these techniques increase agriculture's resilience and sustainability. To encourage sustainable farming and ecosystem service management, policy

Heena (✉)
Department of Chemistry, Baba Farid College of Engineering and Technology, Bathinda, Punjab, India

P. Chaitra
Department of Soil Science, VC Farm, University of Agricultural Sciences, Bangalore, Karnataka, India

K. R. Anusha (✉)
Department of Vegetable Science, Punjab Agricultural University, Ludhiana, Punjab, India

N. J. Gowthamchand
Department of Soil Science, Dr. Rajendra Prasad Central Agricultural University, Pusa, Bihar, India

R. S. Meena et al. (eds.), *Carbon-Negative Agriculture*, Sustainability Sciences in Asia and Africa, https://doi.org/10.1007/978-981-95-1483-0_8

proposals also place a strong emphasis on integrated pest management (IPM), regulatory changes and financial incentives. Farmers are encouraged to implement methods that lower emissions and improve soil carbon storage by including carbon credit systems, which eventually promote a balanced approach to environmental stewardship and food production. In order to achieve carbon-neutral agriculture while preserving vital agroecosystem functions, this chapter presents a thorough understanding of how agrochemicals affect ecosystem services.

Keywords Agriculture · Carbon credit · Economy · Fertilizers · Over-usage · Production

Abbreviations

AEDS	Agroecosystem disservices
AES	Agroecosystem services
CSA	Climate-smart agriculture
GHG	Greenhouse gas
IPM	Integrated pest management
PES	Payments for ecosystem services
SHC	Soil Health Card
SOC	Soil organic carbon

8.1 Introduction

Agriculture plays a vital role in the global economy, serving as the backbone of nations by providing food, raw materials for industries and job opportunities for a large population. As of 2021, agricultural land, which encompasses croplands and permanent pastures, accounts for approximately 37% of the Earth's terrestrial surface, equating to around 4.79 billion hectares (Hyde 2023; FAO 2023). Agricultural systems are intricately linked with natural ecosystems, which offer essential services that support and enhance agricultural productivity. For these systems to operate effectively, they rely heavily on the services provided by natural ecosystems, including the maintenance of soil structure, nutrient cycling, biological pest control, genetic diversity for crop breeding and hydrological services (Power 2010). Recent advancements in agriculture aimed at achieving high yields and introducing desirable traits in crops have seen significant success; however, they have also placed considerable stress on the environment and disrupted the entire agroecosystem, which includes soil, water, climate and livestock. In response to growing food demands, farmers are enhancing productivity through various agricultural intensification methods, with the widespread use of agrochemicals being a prominent factor. Agrochemicals, in the form of fertilizers and pesticides, contribute to agricultural economies by mitigating pests and diseases, thereby lowering production costs and increasing yields (Bhandari 2014). Nonetheless, these practices have detrimental effects on food security, leading to soil and water pollution, elevated greenhouse gas (GHG) emissions and biodiversity loss (Taylor and Rising 2021; Kumar et al. 2022). The increased toxicity in crops has led to harmful diseases and adverse effects on soil microorganisms, which in turn indirectly impact humanity as a whole. Consequently, reducing the reliance on agrochemicals raises concerns regarding agricultural production and environmental protection, both of which require significant attention. Carbon credit generation represents a market-based strategy that allows farmers to earn credits by adopting practices that capture carbon or reduce greenhouse gas (GHG) emissions. These credits can be traded in carbon markets, providing financial incentives for sustainable land management (Cariappa et al. 2024). The agroecosystem services (AES) are crucial for enhancing carbon sequestration; however, the imprudent use of agrochemicals for yield enhancement can hinder these services. Therefore, it is essential to cultivate a comprehensive understanding of agroecosystem services and disservices across various cropping systems to design effective management plans that facilitate sustainable future planning.

8.2 Overview of Usages of Agrochemicals in Agriculture

The usage of agrochemicals in the agriculture sector has successfully improved economic efficiency and enhanced food security concerns (Kumar et al. 2022a). Agriculture has been associated with agrochemicals since the 1930s to date for their promising role in improving agricultural production. Modern agriculture has been associated with the use of chemical-based fertilizers and pesticides. The strong emergence of huge types of agrochemicals was triggered after the discovery of the first pesticide in 1939, dichloro-diphenyl-trichloroethane (DDT), which further initiated the discovery of numerous fertilizers and pesticides. Agrochemicals are broadly classified into fertilizers, chemical pesticides, plant growth regulators and some other allied chemical substances (FICCI 2019). The most commonly used fertilizers constitute zinc, potassium, nitrate, methomyl, dimethoate, oxamyl, cymoxanil and pyrimethanil (Oller et al. 2007). Common pesticides are DDT, fluvalinate, coumaphos, chlorothalonil and chlorpyrifos (Zhu et al. 2014). However, each of the agrochemicals has a severe impact on living beings and the environment in different ways, which makes it a challenging task for researchers to primarily focus on millions of chemicals to study their side effects and their control issues. Recent research reports have suggested that the global agrochemicals and pesticides market, which was estimated at 278,428.78 USD million in the year 2023, is expected to reach USD 320,916.55 million by 2028, with a Compound Annual Growth Rate (CAGR) of 2.4% in the subsequent years (Mishra et al. 2024). Presently, the global usage of pesticides is approximately 2 million tonnes, of which 47.5% are herbicides, followed by insecticides (29.5%), fungicides (17.5%) and the rest of the pesticides (5.5%) (FAO 2024; Quandahor et al. 2024). The total

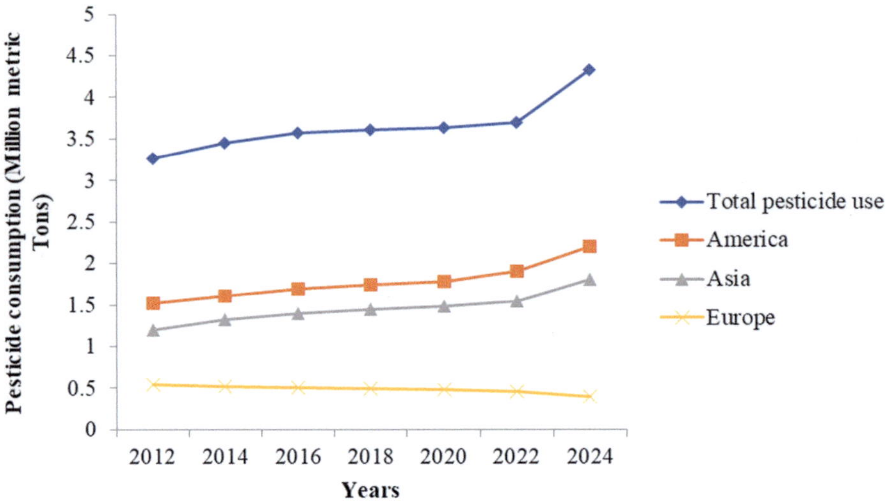

Fig. 8.1 Global usage of agrochemicals from 2012 to 2024

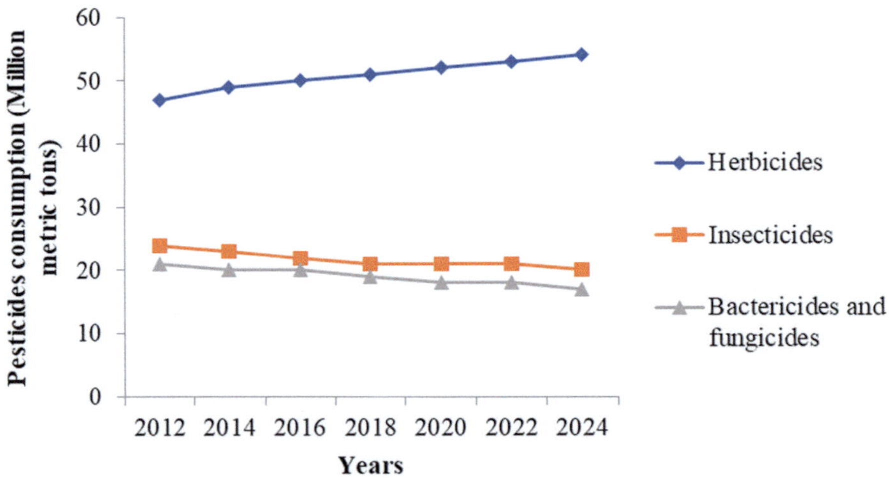

Fig. 8.2 Trend of different types of agrochemicals from 2012 to 2024

pesticide consumption has been estimated as 3.39 million tonnes; however, in India, 61,702 tonnes have been reported in the year 2024 (FAOSTAT 2024). The leading ten countries in pesticide consumption worldwide are China, the USA, Argentina, Thailand, Brazil, Italy, France, Canada, Japan and India (World Atlas 2018). In addition to agrochemicals used in the form of pesticides, fertilizer consumption is highest in the Asia region, with 58.5% consumption in East and South Asia. It showed nitrogen, phosphate and potash consumption as 62.1%, 57.6% and 46.4% respectively (FAO 2015) (Figs. 8.1 and 8.2).

The current global population is about 7.2 billion, which is projected to reach 9.3 billion by 2050 (FICCI 2016). To meet their rising food demands, modern agriculture is highly dependent on various agrochemicals that significantly contribute to enhancing crop production and quality to meet the fast-growing demands. However, the overuse of agrochemicals has been seen worldwide, by consuming pesticide concentrations above the maximum residual limit (MRL) that affects different crops and their food components (Sarkar et al. 2021; Meena et al. 2025). The injudicious application of agrochemicals and their improper handling and unawareness of the regulatory guidelines while using them have resulted in the accumulation of harmful chemicals in treated crops, thereby leading to environmental degradation, affecting living being's health and further posing challenges and risks to agricultural ecosystems and soil health (Zhang et al. 2011; Shammi et al. 2020; Meena et al. 2024).

8.3 Impacts of Agrochemicals on Agroecosystem Services

Agrochemicals are widely used for enriching plant nutrition and against a wide range of pests infesting agricultural produce (Zhang et al. 2011). The persistent use of agrochemicals worldwide has disturbed the whole agroecosystem comprises of soil, water, air and biodiversity health briefly described in Fig. 8.3. This has led to the increased rate of soil degradation, pest resurgence, destruction of pollinators and

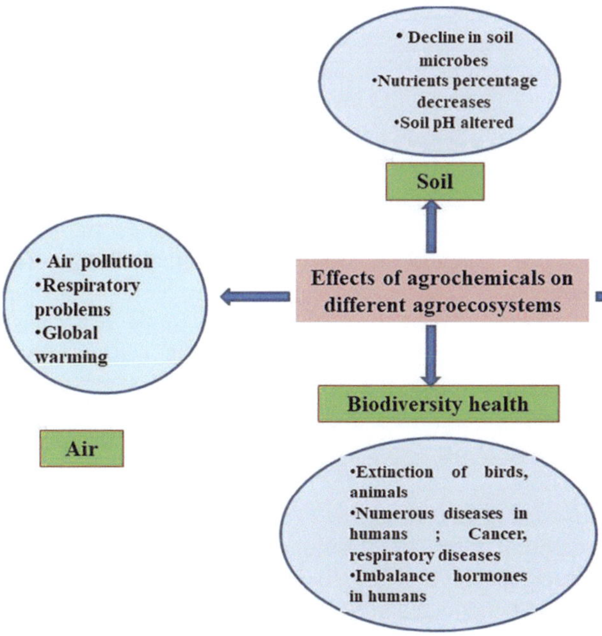

Fig. 8.3 Effects of agrochemicals on different agroecosystems

natural enemies, pesticide resistance, drug-resistant phytopathogens, human and animal health hazard, numerous diseases in human limbs, devastation of native flora and fauna, also resulted in bioaccumulation and biomagnification, diverse changes in climate change, environmental pollution etc., therefore, posing serious threats to life and health.

8.3.1 Effects of Agrochemicals on Soil

Agrochemicals have long persistence in the soil systems, which are severely affecting the soil health (Prashar and Shah 2016; Meena et al. 2024a). The use of agrochemicals mainly aims to enhance the quantity and quality of food supply for humans, but it is now hurting soil biological activity and diversity. They have substantial impact on soils nutrients profile and further affecting their roles in various metabolic pathways in plants, physicochemical properties of soil (pH, moisture, soil organic carbon [SOC], amount of nutrients) and also declining organic matter, microbial enzymes and affecting many microorganisms' life that are beneficial for soil (Kalia and Gosal 2011; Meena et al. 2024b). Soil microflora, constituting bacteria, fungi, algae and protozoa, have a significant contribution in maintaining soil fertility through their catabolic role by degrading plants and animal residues, which helps in the cycling of the organic, inorganic nutrient content of soil. Pesticide that disrupts the activities of the soil microorganisms could be expected to affect the nutritional quality of soils and would therefore have serious ecological consequences The amount of applied agrochemicals in the form of pesticides is only 0.1% that has significant effect on the target organism; however, the remaining chemicals applied deteriorate the soil's profile (Carriger et al. 2006). The use of agrochemicals affects the soil microflora growth, activity and enzyme levels, which further decrease the soil fertility and overall soil health. Soil gets polluted due to the presence of harmful chemicals or other alterations in the soil environment by industrial wastes and overuse of pesticides rather than recommended doses (Meena et al. 2023). Researchers showed that pesticide use decreased the soil microbial biomass and imbalance respiration rate in soil (Chen et al. 2001; Haney et al. 2000). The continuous application has modified the soil properties. The study on the impact of agrochemicals in Abeokuta, Nigeria, in which different plantations were used (maize, plantain, rice, vegetables and control) (Sheoran et al. 2022; Osinuga et al. 2023) showed the changes in morphological characteristics such as colour, texture, structure and consistency. The results showed declining trends in soil chemical indicators like pH, OM, TN and K in the agrochemical fields as compared to the fields with no agrochemicals. Moreover, the total trace element contents in the soil were highly affected and declined, and their levels have reached the permissible limit of WHO standards for agricultural purposes (Osinuga et al. 2023). Therefore, indiscriminate use of agrochemicals might work for a few years, but in future, there will be a decreasing trend of beneficial soil organisms to cling to the nutrients (Biswas et al. 2014).

8.3.2 Effects of Agrochemicals on Underground Water

Similar to soil, agrochemical contamination in water has become a serious worldwide concern, which is posing a risk to human health. Among agrochemicals, pesticides are the major moieties affecting the water bodies, mainly those which are adjacent to agricultural areas (Sjerps et al. 2019; Meena and Pradhan 2023). Their residues are commonly found in surface water systems, irrigation drains, which pollute and harm the aquatic life when they enter into pond and river water, and can harm the aquatic environment (Biswas et al. 2014). Farmers use agrochemicals to increase soil fertility, eliminating weeds or harmful pests and diseases to enhance agricultural productivity and meet human demands for food; however, they create problems when agrochemicals accumulate as residue in soils and percolate into the groundwater (Adeoye et al. 2013). Thus, the relative rate of percolation and degradation within the soil profile, which processes are regulated by climate, soil characteristics, chemical characteristics, application rate, aquifer depth and farming practice, determines the risk of contamination. It has been surveyed that the regions with high rainfall and irrigation practices showed high runoff of pesticide residues into the water bodies (Larsen et al. 2017). Several investigations stated that the traces of pesticide residues in the atmosphere resulted in contamination of the surface water system via wet precipitation and increased moisture conditions in the atmosphere (National Pesticide Information Centre –NPIC 2022). Moreover, the residents living near the agricultural lands and affected water bodies are more exposed to pesticide side effects and are further creating higher carcinogenic risks to life (Tang et al. 2019; Dereumeaux et al. 2020). Also, these harmful chemicals can cause severe health problems such as infertility and chromosomal changes as a consequence of DNA mutation, and oxidative stress, which are further responsible for ageing and diseases like Parkinson's and Alzheimer's disease (Bolognesi and Merlo 2011). Studies have reported the presence of huge amounts of pesticides in the water that are affecting human as well as aquatic life. Large amounts of agrochemicals (mostly pesticides/herbicides) are discharged into the water, endangering human health, aquatic ecosystems and plants (Anonymous 2024). In a study investigated by Zardosht et al. (2023), the impact of insecticides (diazinon and deltamethrin; 86 and 11.62 μg/L) on the quality of surface and groundwater was examined using Gas chromatography and Mass spectrometry (GC-MS) and the results showed that the quality of the surface water affected gets decreased using agrochemicals in terms of dissolved oxygen, nitrate, turbidity, Total organic carbon, Biochemical oxygen demand and Chemical oxygen demand, resulting in 15.2, 189.6, 00.0, 53, 176 and 57.5 respectively.

8.3.3 Effects of Agrochemicals on the Environment

Global agrochemicals usage has increased dramatically in recent years, resulting in soil and water deterioration, but also affecting air quality. Generally, pesticides in the air get accumulated via different means (Mueller 2015) such as evaporation of pesticide molecules before they hit their intended target, and further get mixed with fine spray particles, drifting of vapours post pesticide application and the movement caused by wind and wind-carried soil particles. In addition, some chemicals are volatile and evaporate easily depending on the vapor pressure, temperature and relative humidity of the air (Bish et al. 2021). Some agrochemical ingredients, due to their toxic chemical nature, leave residues in the atmosphere for a short time, while some last longer and harm the environment. It has been found that organochlorine pesticides were banned after the 1960s due to their long persistence and bioaccumulative nature, and therefore banned in many countries. Similarly, due to the usage of other chemicals such as carbamates, phosphates, chloro compounds and many other synthetic derivatives in the urban and rural sites, it has resulted in persistent accumulation of harmful chemical moieties due to their volatile nature that imparts the atmospheric pollution (Zhang et al. 2013).

8.3.4 Effects of Agrochemicals on Biodiversity

Agrochemicals enter the ecosystem via soil, water and air, which ultimately affect the ecosystem, thus entering in food chain and affecting human health. These are used to remove or kill the unwanted plants and microorganisms, pests, but unintentionally, they are harming the non-target plants and desirable living species. Many species of birds, soil microbes, beneficial insects and other small wild animals are in danger because of the overuse of pesticides (Biswas et al. 2014; Khan et al. 2023; Ahmad et al. 2024; Pradhan and Meena 2023). Herbicide contamination of water has also shown distressing effects on aquatic plants. It has been reported that one of the agrochemicals, Oxadiazon, reported to significantly lower algal growth in one research. Various chemicals used for multipurpose have been reported to affect the thyroid function in amphibians, rodents, fish and birds, including insecticides, herbicides and fungicides, which ultimately impact the immune system (Galloway and Handy 2003). The commonly used pesticides 2, 4-D, diazinon, atrazine, simazine dicamba have been reported to affect the mortality of various species because of toxic chemicals (Raimondo et al. 2007). Therefore, their regular use has initiated problems for human health and multiple environmental issues (Geiger et al. 2010). Studies investigating food toxicity have shown that contaminated food results in higher toxicity than contaminated water and air (Margni et al. 2002). The pesticide intake further leads to disturbances in natural hormones, and their long-term exposure causes severe imbalances, reduces intelligence and generates reproduction-related problems and cancer (Yadav et al. 2015). Notably, farmers are at high risk of exposure to the harmful chemicals. Moreover, specifically, its use of chlorinated pesticides and methyl bromide and their derivatives has shown more cases of prostate cancer in farmers (Alavanja et al. 2003), and also numerous cases of allergic and non-allergic asthma in farmers have been seen (Hoppin et al. 2009). Studies have shown that there are numerous commercial pesticides affecting biodiversity such as chlorpyrifos which is an organophosphate insecticide used for controlling number of pests that has reported severe risks to human lives with the main target on their neurodevelopment and also due to its safety issues, it has banned nowadays in several countries (Altaf et al. 2022). Similarly, pesticides like paraquat, malathion, which are commercially used in agricultural farms, are extremely poisonous and can cause serious poisoning when consumed (Kim et al. 2022). These types of pesticides are severely affecting the lungs, leading to their damage and failure. In this way, agrochemicals are disturbing the whole ecosystem, starting from soil, water, air and whole living beings.

8.4 Agroecosystems Services and Disservices Under Different Cropping Systems

Agriculture not only delivers commodity-based services like food, bioenergy and fodder, but it also provides non-commodity-based services such as pollution control, biodiversity conservation and carbon sequestration (Renting et al. 2009). For instance, coffee plantations in agroforestry systems not only yield coffee cherries, which fulfil human needs, but the shade trees themselves can also provide timber for building houses or making furniture, further meeting human requirements (Cerda et al. 2014; Pinoargote et al 2017). Both trees and coffee plants absorb CO_2 from the atmosphere and store it in their biomass, playing a role in climate regulation. This type of perennial plantation can also support biodiversity conservation (DeClerck et al. 2010; Pradhan and Meena 2023a). Both approaches involve two distinct types of goods and services: marketable commodities, such as coffee cherries and timber, and non-market commodities, like climate regulation and flood mitigation, which provide broader benefits to human societies (Rapidel et al. 2015). Whether any agroecosystem contributes to such services relies on the management, which in turn relies on the parity between short-term (immediate crop yields) and long-term gains (maintaining soil health and biodiversity) (Power 2010).

The AES are not solely the result of agroecosystem functioning; their provision is largely influenced by the intensity and nature of human activity within each agroecosystem (Mach et al. 2015). This human intervention can result in the emergence of agroecosystem disservices (AEDS), which refer to 'functions, processes, and attributes that produce perceived or actual negative impacts on human well-being', highlighting that agroecosystem contributions can sometimes be detrimental (Shackleton et al. 2016). They are loss of biodiversity, nutrient runoff, loss of wildlife habitat, pesticide poisoning, greenhouse gas emissions, sedimentation of waterways etc.

8.4.1 Agroecosystem Services

The AES across various agricultural landscapes are considered agroecosystem services (Liu et al. 2020). The AES comprises provisioning, regulating, supporting and cultural services. Provisioning services include food (crops), fodder, fibre, raw materials (cash crops like hemp, cotton, ginseng and rape), medicinal materials, energy and fuel (Rana et al. 2018; Zabala et al. 2021). Soil conservation, waste decomposition, carbon sequestration and local climate regulation come under regulatory services. Maintaining the crop cover shields the soil from water and wind, preventing soil erosion. Soil microbes break down the dead plant materials and recycle the nutrients into the soil, maintaining fertility (Rana et al. 2018; Zabala et al. 2021; Huang et al. 2015). Through photosynthesis, plants fix carbon dioxide and store it in their biomass, which later decays and forms soil organic matter (SOM), further restoring the soil carbon (Zhang et al. 2007). This process is known as carbon sequestration. Climate regulation can be achieved by the reduction in the heat island effect. Transpiration of crop plants and evaporation from irrigated farmland cool down the surrounding environment, reducing the temperatures (Zabala et al. 2021; Kumar et al. 2017). Supporting services encompass nutrient cycling, pollination, soil structure and fertility, biological control and weed control. Nutrient cycling forms a basis for maintaining soil fertility, further sustaining agricultural productivity over time. Pollination by various pollinators such as bees, butterflies, bats and birds plays a major role in enhancing crop yields. Soil structure and aggregation are vital for nutrient absorption by crops and retaining water in the soil, both essential for optimal plant growth (Rana et al. 2018; Zabala et al. 2021). Natural enemies of pests exist within the ecosystem, which helps to keep the pest population at bay, and seed predators like birds, ants and beetles help prevent weed seeds from infestation in crop fields (Zhang et al. 2007; Huang et al. 2015). Cultural services include aesthetic value, spiritual and religious values, science and education, recreation and tourism and cultural heritage. The agricultural landscapes are aesthetically pleasing, and entertainment based on them creates an opportunity for recreational activities like harvesting, fishing and experiencing the essence of rural life (O'Brien et al. 2020; Rana et al. 2018). Agricultural practices give an insight into the science behind farming and help people gain knowledge through scientific research. Agricultural folklores are connected to the beliefs and traditions that worship the spiritual aspects of farming cultures (Hanacek and Rodriguez-Labajos 2018) (Fig. 8.2).

8.4.2 Agroecosystem Disservices

Agroecosystem disservices (AEDS) reduce the productivity of crop plants by negatively impacting crop health and soil quality, thereby affecting the ecosystem balance (Fig. 8.4). Competition for resources can adversely affect the yield of crop plants. Weeds compete with crop plants for water, nutrients, sunlight and pollination, thereby decreasing productivity (Zhang et al. 2007; Gaba et al. 2017; Welbank 1963. They also exhibit allelopathic effects (chemical inhibition by root exudates), limiting the growth of plants (Weston and Duke 2003). Water consumption for irrigating crops in arid and semi-arid regions causes scarcity, leading to the depletion of groundwater reserves, to some extent, can cause soil salinization, rendering the land less fertile over time (Hardaker et al. 2020; Zabala et al. 2021). Plastic films used to enhance the productivity of crops when not properly recycled hinder the growth of the plants, pollute the soil and water, accumulate in food chains and affect human health (Zhang and Zhou 2018). Excessive usage of chemical fertilizers and insecticides to enhance the yield of crops causes eutrophication in the water bodies, hindering aquatic life and human health (Hardaker et al. 2020; Zabala et al. 2021). Soil pollution resulting from the use of sludge, sewage irrigation, pesticides and microplastics causes chronic health issues due to heavy metal poisoning (Zhang and Zhou 2018). The use of pesticides has led to an increase in pest outbreaks, as well as the emergence of new strains of microbes. This is due to the overuse of agricultural chemicals, which has weakened the effectiveness of previous pest resistance mechanisms (Huang et al. 2015). This can increase the cost of chemical control and lead to unintended negative health effects on non-target organisms, including humans.

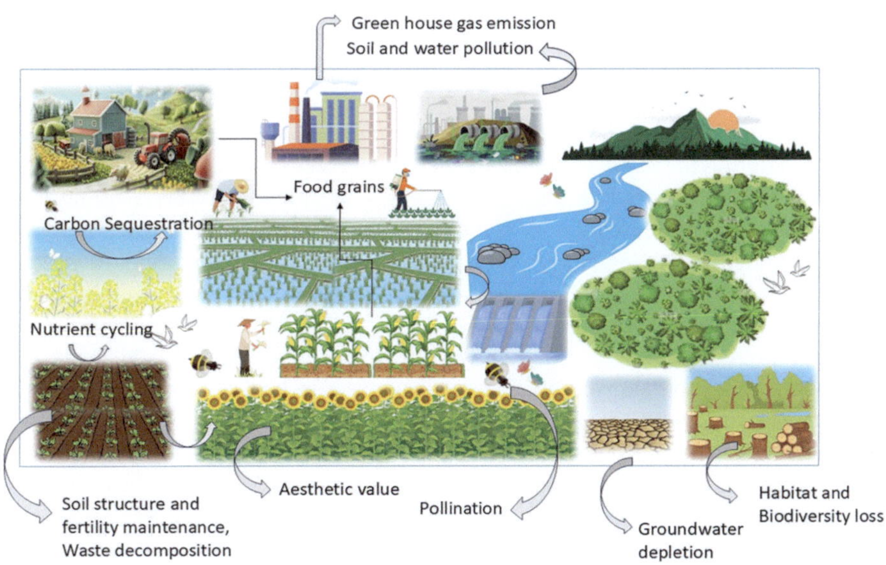

Fig. 8.4 Agroecosystem services and disservices

Greenhouse gas emissions from agricultural activities such as crop cultivation, livestock breeding and use of machinery significantly contribute to climate change (Ma et al. 2015). Habitat and biodiversity loss due to deforestation, monoculture farming and the killing of non-target organisms by weedicides disrupts the balance of the ecosystem (Zhang et al. 2007; Power 2010).

8.4.3 Understanding Agroecosystem Services and Disservices in Diverse Cropping Systems

Agroecosystems are dynamic, where agricultural activities interact with the environment, creating benefits and shortcomings for ecological sustainability (Power 2010). Different cropping systems, such as monocropping, mixed cropping, cover crop, crop rotation and agroforestry, offer varying levels of services and disservices Table 8.1.

Monocropping aims to achieve maximum yield from a single crop in a single growing season with the aid of fertilizers, fossil fuels and farming machinery (Rahaman et al. 2020). While it is necessary for deriving provisional services (food, fodder and fibre) for the growing population, in the long run, it will deplete the soil of its nutrients and render it unfit for further cultivation. They deplete the soil of beneficial microbes, which are imperative for the functioning of the ecosystem and for maintaining human microbiome diversity (Blum et al. 2019). Continuous monocropping impedes the balance in the edaphic ecosystem, making it vulnerable to soil-borne diseases and the accumulation of harmful chemicals (Mo et al. 2016). Monocropping leads to the loss of biodiversity, making agriculture the strongest driver of biodiversity loss (Scholes et al. 2018; Yahya et al. 2017). AEDS of monocropping outweigh the AES and are unsustainable in the long run.

Gallaher (2009) described multi-cropping as 'the cultivation of two or more crops on the same plot of land each year'. If two or more crops are grown simultaneously in the field during the same season, it's called intercropping (Mousavi and Eskandari 2011). Intercropping offers a wide range of agroecosystem services, starting from provisional, supporting, regulatory and cultural services. Raseduzzaman and Jensen (2017) conducted a meta-analysis for yield enhancement in intercropping systems (ICS). They concluded that intercropping systems provide yield stability under diverse environmental conditions. Replacement design of intercropping in cereal–legume intercropping systems has shown a steady increase in the yield. Also, in a study fenugreek crop showed a substantial increase in land on intercropping with equivalent ratio of 19%–38% as compared to sole crop (Wasaya et al. 2013). This increase in yield is attributed to the resource use efficiency (of water, light and nutrients) in ICS provided there is no competition among the crops and has different growing periods (Chen et al. 2018; Hu et al. 2016; Ghanbari et al. 2010; Reddy and Reddi 2007). Pathogen expansion is restricted due to the functional diversity in the ICS, limiting the spread of diseases (Finckh et al. 2000). A pea–cereal ICS favoured

Table 8.1 Agroecosystems services and disservices under different cropping systems

Cropping system	AES	AEDS	Economic impact	Potential for sustainability	Climate resilience	References
Monocropping	Provisioning services (high productivity)	Nutrient depletion Increased incidence of new pests and diseases Loss of biodiversity	Increase in short-term gains Long-term degradation of soil structure and fertility	Low	Low	Blum et al. (2019) and Scholes et al. (2018)
Multi-cropping (intercropping, relay cropping)	Provisioning services Supporting services (nutrient cycling, soil fertility conservation, reduction in pests and diseases by push-pull dilution effect, water conservation by protection against water and wind erosion, regulation of microclimate) Regulating services (lower emissions of greenhouse gases from the soil)	Groundwater depletion Pollution of air and waterbodies due to agrochemical industries	Stable, but an increase in management costs	High	High	Balandaite et al. (2024), Gaba et al. (2015), Fargione and Tilman (2005), Bedoussac and Justes (2010a, b), Ratnadass et al. (2012), Thobatsi (2009), Morris and Garitty (1993) and Lupwayi and Kennedy (2007)
Crop rotation	Provisioning services Supporting services (increase in soil organic matter, soil microbial biomass Regulating services		Stable, with strategic planning	High	High	West and Post (2002), McDaniel et al. (2014a, b) and Suzuki et al. (2012)
Agroforestry	Provisioning services Supporting services (reduce erosion and enhance soil fertility Regulating services (carbon sequestration, local climate		Long-term gains	Very high	Very high	Jose (2009), Williams-Guille'n et al. (2008), Nair and Graetz (2004), Kirby and Potvin (2007), Vansynghel et al. (2022) and Kuyah et al. (2017)

(continued)

Table 8.1 (continued)

Cropping system	AES	AEDS	Economic impact	Potential for sustainability	Climate resilience	References
	regulation, improved air and water quality) Cultural services (increase aesthetics)					

lower severity of Ascochyta blight (*Mycosphaerella pinodes*) than mono-cropped pea (Schoeny et al. 2010). The modification of the microclimate around the canopy of the intercrop combinations was attributed to this reduction. An increase in crop diversity in ICS enhances nutrient cycling and thus soil fertility, especially in the case of cereal–legume ICS (Russel 2002). The ICS have higher below-ground activity, increasing soil organic matter (SOM) and soil carbon sequestration (Cong et al. 2015). Maize–soybean relay strip cropping by Yang et al. (2025) over the period of 10 years revealed that intercropped maize crops not only enhanced the light interception of the crops and enhanced C fixation but also increased the below-ground allocation of carbon and microbial immobilization of carbon. Overall, the ICS increased 13.9% of soil organic carbon (SOC) after 10 years.

Crop rotation refers to the cultivation of multiple crops in the same piece of land in a recurring manner (Malik et al 2024). Crop rotation has been known to boost crop yields by enhancing plant biodiversity, soil health improvement and decreasing weed, pathogen and insect populations (Bowles et al. 2020; Yu et al. 2022; Zhao et al. 2020). Zhao et al. (2020) conducted a meta-analysis to analyse the yield variability under different environments and field management practices. Crop rotation significantly increased the yields by 20% in comparison to monocropping; especially in legume-based rotations, there was an increase of 14% compared to non-legume. The yield benefits of rotation lasted for 2–3 years. Crop rotations have been known to control the spread of disease and pests because of the non-availability of compatible hosts and thus prevent severe outbreaks (Volsi et al. 2022; He et al. 2019; Balloux and van Dorp 2017). Incorporation of pulses, cereals and oil seeds under crop rotation significantly reduced the occurrence of diseases. Among them, cereals and other broadleaved crops showed minimal disease overlap, while the lentils and oil seeds showed a minimal number of common diseases (He et al. 2019). Resilience to climate change is a prime factor to be prioritized in today's world. Crop rotation improves the water retention in the soil by reducing the run-off and decreasing transpiration and soil water evaporation (Pala et al. 2007; Hatfield and Dold 2019). A 6-year maize–potato rotation experiment by Wang et al. (2021) reported an increase in soil water storage capacity by 11.29 mm in the 0–280 cm layer and decreased water consumption by 16.81%–24.83% in 3 years. Crop rotation coupled with conservation tillage, fallowing and cover crops paves the way for sustainable practices which can maximize agroecosystem services (Gebremedhin and Schwab 1998; Leharwan et al. 2023).

Agroforestry is a multi-functional approach of combining trees and shrubs with crops or livestock (Nair et al. 2012). It offers various agroecosystem services like providing food and timber, reducing nutrient runoff and carbon sequestration, increasing the population of natural enemies and pollinators, controlling soil erosion and improving local climate conditions (Davis et al. 2012; Caudill et al. 2015; Winans et al. 2015; Yirga 2019; Pradhan et al. 2023). Nair et al. (2012) illustrated that the carbon sequestration potential of vegetation, above-ground and below-ground, varied from 0.29 Mg ha^{-1} $year^{-1}$ within a fodder bank agroforestry system located in West African Sahel to 15.21 Mg ha^{-1} $year^{-1}$ found in mixed-species stands present in Puerto Rico. Estimates for soil carbon ranged from 1.25 Mg ha^{-1} in a Canadian-based alley cropping system to 173 Mg ha^{-1} observed in an Atlantic Coast silvopastoral system situated in Costa Rica. Enhanced soil aggregate stability, increased soil carbon and nitrogen levels, as well as improved soil enzyme activity in soils under agroforestry, were observed with a comparative study of row crops by Udawatta et al. (2008). Agroforestry conserves biodiversity by preserving the sensitive species germplasm, reduces the rate of conversion of natural habitats and provides natural habitats for those species that can withstand certain level of disturbances and provide a wide range of ecosystem services such as control of soil erosion and loss of habitat (Jose 2009).

The prevalence of varied agroecosystems facilitates the increase in ecosystem services and the promotion of biodiversity, which cover a range of farming techniques like agroforestry and mixed cropping. These systems increase soil health, pest management and overall agricultural resilience in addition to accelerating the carbon sequestration process (Maskell et al 2023 and Yousefi et al 2024).

8.5 Carbon Credit Generation Through Agroecosystem Services

GHG emissions are being identified as the potential contributors to climate change and global warming, causing hazardous consequences on the livelihood of humans (Poolen and Ryszka 2021). In order to address this issue, the Paris Agreement was formed in 2016, aiming to achieve a climate-neutral world by limiting global warming to 1.5 °C by reducing the emission of greenhouse gases. Agriculture and livestock contribute significantly to GHGs (carbon dioxide, methane, nitrous oxide, ozone and perfluorocarbons) at all stages from crop production to distribution, consumption and waste disposal (Uppala et al. 2016; FAO 2024; Kumar et al. 2025). Among them, primary contributions arise from the manufacturing units of fertilizers, feed imports, operation of farm equipment, net CO_2 emissions from soil (determined by the balance between the organic matter input and leaching) (Dyer et al. 2010; Gingrich et al. 2007; Aguilera et al. 2013). Adapting to new methods in response to climate change and lowering GHG emissions while still producing enough food to meet the world's expanding population are the challenges that agriculture faces today (Lokuge and Anders 2022). Agricultural activities produce comparatively less GHG emissions when compared to industrial emissions, particularly CO_2. Thus, carbon credit schemes in the agriculture sector have drawn much interest (Shockley and Snell 2021). Carbon credit systems enable farmers to earn additional income by trading their surplus carbon credits to companies/industries that release GHGs (Lokuge and Anders 2022). Agroecosystem services contribute to carbon sequestration and thereby reduce the carbon footprint. The carbon footprint of agriculture can be divided into three hypothetical tiers. Tier 1 includes all direct GHG emissions from agriculture machineries and soil. Tier 2 encompasses the indirect emissions, such as the GHGs during electricity generation, and Tier

3 pertains to the production and transport of agri-based chemicals and machinery (Jaiswal and Agarwal 2019).

A prime factor in mitigating these emissions is soil organic carbon (SOC), the carbon (C) which is linked to soil organic matter and a component of the natural carbon cycle, that is, a main reservoir of carbon in the terrestrial ecosystem (Brenna et al. 2014). Plants fix the atmospheric CO_2 as biomass through the process of photosynthesis. Eventually, they complete their lifecycle and return to the soil, and a slow decomposition of their remains leads to the formation of SOC (Juan 2021). During the process of mineralization, some organic carbon resistant to decomposition is released into the soil, while some organic matter releases carbon dioxide back into the atmosphere (Chan 2008). Soil organic carbon that resists decomposition serves as a major carbon reservoir, accounting for approximately 80% of terrestrial carbon reservoirs. It is vital to life on Earth (Ontl and Schulte 2012). As a result, promoting agricultural practices that promote increasing C stock will help to mitigate the emission levels. Several strategies can be adopted to sequester higher amounts of carbon in the soil, namely increasing crop residue inputs, reducing tillage intensity, optimizing nutrient management, applying manure, restoring permanent forests and grasslands and eliminating summer fallows (Abbas et al. 2023). Implementing these sustainable practices would make agroecosystem services a potential driver for the development of the carbon credit market.

8.6 Carbon Credit Generation Through Reducing Agrochemical Usage in Agriculture

A carbon credit is a tradable permit that allows its holder to emit a specified amount of carbon dioxide or other GHGs, typically measured as one metric tonne of CO_2 equivalent (Lokuge and Andres 2022). This term was introduced under the Kyoto Protocol and reinforced by the Paris Agreement. Carbon credits are a market-driven mechanism designed to cap and reduce global emissions, incentivizing businesses and nations to adopt sustainable practices (Aggarwal 2024). To improve agricultural sustainability and lower carbon emissions, which can result in carbon credit generation, agrochemical management techniques are crucial (Kamyab et al. 2024). Farmers can help to sequester and reduce carbon emissions and earn carbon credits by implementing sustainable agrochemical management techniques into practice. Below are several key strategies that can contribute to carbon credit generation by reducing agrochemical usage in agriculture.

8.6.1 Integrated Pest Management

The comprehensive ecological approach to managing pests in agricultural systems known as integrated pest management (IPM) entails the strategic integration of several control techniques, such as chemical, biological and cultural approaches, to keep pest populations below the economic threshold level (ETL) while lowering environmental and public health risks (Deguine et al. 2021). IPM is a key component of sustainable agriculture, widely adopted in both developed and developing countries for safe, high-quality food production (Angon et al. 2023). Rather than following rigid protocols, IPM is a dynamic, science-driven approach that integrates local resources, cutting-edge research and advanced technology. By enhancing resilience, preserving biodiversity and improving soil health, IPM ensures long-term agricultural sustainability. Its adaptability makes it an essential strategy for farmers seeking both productivity and environmental stewardship (Srivastav et al. 2024). Approximately 2.7 million tonnes of CO2 equivalent have been reduced by farmers adopting IPM on their farms (Mulungu et al. 2024). A non-market worldwide public good created by farmers using environmentally conscious pest management methods, this mitigation has a value of US$12.2 million. Since 2006, the total amount of carbon trapped by these techniques is equivalent to protecting roughly 44,000 hectares of forest (Deutz et al. 2020). Reducing the use of agrochemicals directly lowers greenhouse gas emissions, as their production, transportation and application are highly energy-intensive (Cech et al. 2022). By minimizing pesticide use, farmers can significantly reduce their carbon footprint, cutting emissions throughout the entire pesticide lifecycle (Kantwa et al. 2024).

8.6.2 Precision Farming

Conventional farming often uses blanket treatments that are applied across wide areas, which leads to inefficient use of resources and suboptimal yield. On the other hand, precision farming revolutionizes existing conventional techniques by increasing crop accuracy, decreasing waste, maximizing resource utilization and minimizing environmental impact (Jayaraman et al. 2023). Precision farming is a pioneer in sustainable agriculture, emphasizing the maximization of crop production while reducing ecological footprints. The fundamental idea behind precision farming is data-driven farming that recognizes each field as a unique entity and adjusts agricultural methods to suit certain conditions (Vinod Chandra et al. 2023). It accounts for variations in soil properties, topography and crop health, ensuring that each field receives the precise care it needs for optimal results.

The improper use of agrochemicals in agriculture is a major driver of greenhouse gas emissions (Elumalai et al. 2025). Precision agriculture optimizes the fertilizer use by preventing its excess application; improper application of fertilizers contributes to the release of greenhouse gases, especially nitrous oxide (Shanmugavel et al.

2023). By maximizing resource usage, reducing chemical inputs and enhancing soil health, precision farming techniques can dramatically lower these emissions and promote sustainable agriculture practices while reducing their negative effects on the environment (Getahun et al. CR2024).

8.6.3 Reduced or Optimized Use of Fertilizers

One of the main causes of emissions of the highly potent greenhouse gases is the usage of nitrogen-based fertilizers (Ferland et al. CR2312024). Optimizing fertilizer use through precise timing, targeted application methods and selection of suitable fertilizers can help to reduce these emissions while maintaining crop productivity (Xing and Wang 2024). In addition to improving soil health, effective fertilizer management increases the soil's capacity to store carbon and promote long-term agricultural sustainability. Furthermore, a lot of carbon credit programmes provide financial incentives to farmers who utilize sustainable fertilizing techniques by rewarding actions that reduce greenhouse gas emissions (Raina et al. 2024). Farmers may improve soil fertility, lower input costs and increase overall farm resilience in addition to lowering environmental impacts by eliminating excessive fertilizer use and putting best management practices into place. Adopting optimal fertilizer techniques eventually results in a more sustainable agricultural system that balances environmental responsibility and productivity while creating chances for financial incentives through sustainable farming initiatives and carbon credit markets (Shanmugavel et al. 2023). Organic fertilizers are more effective than chemical fertilizers in reducing greenhouse gas emissions. Additionally, a dry, cold climate and soils with low silt and high sand content are more favourable for minimizing CO_2 and nitrous oxide (N_2O) emissions (Zeng et al. 2025).

8.6.4 Use of Bio-Based Agrochemicals

Bio-based agrochemicals, which come from renewable sources like plants, algae and fungi, are crucial for carbon credit generation and sustainable agriculture. They improve soil health, increase carbon absorption and drastically lower greenhouse gas emissions as environmentally acceptable substitutes for synthetic agrochemicals (Ammar et al. 2023). These bio-based agrochemicals, including bio-fertilizers, bio-pesticides and plant growth regulators, support regenerative farming by fostering beneficial soil microbial activity, increasing organic matter and reducing reliance on chemical-based inputs, essential factors in carbon credit generation.

By improving soil structure and increasing its ability to store carbon, bio-based agrochemicals contribute to lowering the overall carbon footprint of agricultural practices. Because they are biodegradable, they also reduce pollutants in the environment, maintaining the long-term equilibrium of the ecosystem (Ozlu et al. 2022).

By using climate-smart techniques that absorb and hold onto atmospheric carbon, farmers and agribusinesses that use these sustainable inputs can produce carbon credits. After that, carbon credits can be traded on carbon markets, generating monetary rewards for cutting emissions and encouraging eco-friendly farming. With the growing need for carbon offsets and sustainable food production worldwide, bio-based agrochemicals offer a twofold benefit: they boost agricultural output while reducing the effects of climate change (Jayaraman et al. 2023). Their incorporation into contemporary agricultural systems guarantees enhanced crop yields, environmental sustainability and a profitable pathway for farmers participating in carbon trading initiatives.

Effective agrochemical management plays a crucial role in reducing the carbon footprint of agricultural systems while enhancing soil health and carbon sequestration (Jat et al. 2022). Farmers can reduce greenhouse gas emissions and increase total land productivity by implementing sustainable methods like conservation tillage, integrated pest management, precision agriculture and optimum fertilizer use (Muhie 2022). These sustainable approaches align with carbon credit programmes that reward climate-smart farming. By implementing agrochemical management strategies that lower emissions and sequester carbon, farmers can generate carbon credits, providing both financial incentives and environmental benefits (Cariyappa and Krishna 2024). As carbon markets expand, agriculture's role in climate change mitigation strengthens, making responsible agrochemical management a key driver of sustainable and profitable farming.

8.7 Technological Practices in Agriculture to Gain More Carbon Credits

As climate change concerns intensify, agriculture plays a vital role in reducing GHG emissions (Xing and Wang 2024). By adopting innovative technologies and sustainable agriculture practices, farmers and agribusinesses can not only lower their carbon footprint but also earn carbon credits (Thamarai et al. 2024). Optimizing the use of agrochemicals and implementing advanced agricultural practices can enhance carbon sequestration, allowing farmers to benefit from carbon markets while promoting environmental sustainability (Saikanth et al. 2023).

8.7.1 Precision Agriculture for Optimized Agrochemical Use

Precision agriculture optimizes the use of agrochemicals like pesticides and fertilizers by utilizing the global positioning system (GPS), sensors and drones (Guebsi et al. 2024). By reducing the excessive application of agrochemicals, farmers can cut the emission of greenhouse gases from synthetic fertilizers (Tripathi et al. 2020). By applying the proper amount of agrochemicals where and when they are needed, farmers can reduce their environmental impact and increase productivity with the aid of remote sensing (Sishodia et al. 2020). Key technologies of remote sensing include

GPS-Guided Application It ensures the precise placement of agrochemicals, that is, fertilizers and pesticides, minimizing their excess use, reducing the runoff of agrochemicals and enhancing their efficiency for better crop health and environmental sustainability (Roberts et al. 2021).

Remote Sensing It utilizes satellites, drones and multispectral cameras to monitor crop health and detect nutrient deficiencies, enabling precise and targeted agrochemical applications (Wang et al. 2024).

Variable Rate Technology It dynamically adjusts the fertilizer and pesticide application rates in real-time based on soil and crop conditions for maximizing the efficiency of agrochemicals and promoting sustainable resource management in agriculture (Pawase et al. 2024).

Soil and Crop Sensors It help to measure soil moisture, nutrient levels and plant health, which help in determining the exact amount of agrochemicals required for application (Rajak et al. 2023).

Automated Machinery and Robotics Autonomous sprayers and robotic weeders apply agrochemicals with very high precision (Upadhyay et al. 2024).

Use of Internet of Things (IoT) in Smart Farming It connects different sensors and devices for real-time monitoring and automated decision making (Morchid et al. 2023).

By implementing these technologies, farmers can cut the emission of greenhouse gases, improve soil health and help in carbon credit generation.

8.7.2 Practicing Regenerative Farming

This method of farming will limit greenhouse gas emissions, enhance farmers' financial returns, prevent soil erosion and depletion, actively build soil, provide appropriate crop nutrients with minimal external inputs, produce high-yielding, healthy crops with few weeds and pests and improve human health. It mainly focuses on improving soil health, the water cycle and the storage of carbon in soil (Khangura et al. 2023).

Regenerative farming employs a variety of techniques that integrate ecological and biological systems. The main objective is to profit from natural processes, which include:

- Utilizing high-biomass plants' photosynthesis to capture soil carbon.
- Improving soil structure and water retention through the use of biological systems.
- Include cattle, with a predicted improvement in ecosystem services.

8.7.3 Introducing Cover Crops

Cover crops like legumes, rye or clover can improve the soil's ability to sequester carbon by adding organic matter (Koudahe et al. 2024). In many respects, cover crops improve biodiversity on the farm and support a more robust ecosystem. Beneficial insects and crucial pollinators can find food and habitat in flowering cover crops (Van Eerd et al. 2023). By preventing soil erosion and lowering nitrogen losses by an average of 48%, cover crops preserve water quality. Large-scale cover crops can trap enormous amounts of atmospheric carbon by promoting biological activity in the soil (Punia et al. 2020; Quintarelli et al. 2022).

8.7.4 Biological Alternatives to Synthetic Agrochemicals

A sustainable way to lower greenhouse gas (GHG) emissions while improving soil health and crop yield is to utilize bio-based fertilizers and bio-pesticides. Biological substitutes for synthetic agrochemicals reduce environmental damage and promote long-term agricultural sustainability (Avasthe et al. 2023). Biological nitrogen fixation (BNF), in which nitrogen-fixing bacteria, like Rhizobium, establish symbiotic associations with leguminous plants, is one of the most successful biological treatments (Lindström and Mousavi 2020). By converting atmospheric nitrogen into forms that plants can use, these bacteria greatly lessen reliance on manufactured nitrogen fertilizers, which are a major source of greenhouse gas emissions. Over time, this organic way of nutrient fixation improves soil fertility while reducing carbon footprints (Guo et al. 2023).

In the same manner, natural biopesticides offer environmentally beneficial substitutes for chemical pesticides. The natural insect repellent properties of neem oil, which are extracted from the neem tree, interfere with the lifecycles of pests without endangering beneficial creatures (Lengai et al. 2020). *Bacillus thuringiensis (Bt)*, another popular biopesticide, releases proteins that are harmful to particular insect pests but safe for humans, animals and non-target organisms. The environmental effects of the manufacture, use and breakdown of synthetic agrochemicals are lessened by these biological remedies (Ragasruthi et al. 2024). By integrating bio-based fertilizers and biopesticides into modern farming, agriculture can enhance productivity while fostering biodiversity, soil health and climate resilience. The shift towards biological alternatives supports global sustainability efforts and enables farmers to participate in carbon markets and climate-smart farming incentives (Shanmugavel et al. 2023).

8.7.5 Use of Slow-Release Fertilizers

Slow-release fertilizers offer an advanced solution for minimizing nitrogen losses and reducing GHG emissions. These fertilizers regulate the release of nutrients, ensuring a steady supply to crops while preventing excessive leaching and volatilization losses (Govil et al. 2024). The significant advancement is the application of nitrification inhibitors, which slow down the conversion of ammonium into nitrate, reducing nitrous oxide emissions. These inhibitors enhance fertilizer efficiency by stabilizing nitrogen in the soil, which improves crop uptake and lessens environmental impact (Tariq et al. 2022). Farmers who apply slow-release fertilizers not only optimize nutrient use but also significantly lower their carbon footprint. These sustainable fertilizer application practices contribute to climate-smart agriculture (CSA) and qualify farmers for carbon credit programmes, offering financial incentives while promoting long-term soil health and sustainability (Asadu et al. 2024).

8.7.6 Agrochemical Waste Management and Recycling

Effective agrochemical waste management is crucial for preventing soil and water contamination while reducing emissions from hazardous waste decomposition. Agrochemical containers and leftovers should be disposed of and recycled properly to protect the environment and promote sustainable farming methods (Koul et al. 2022). Reusing old agrochemical containers, cutting down on landfill waste and lowering the carbon footprint of producing new materials are all made possible by closed-loop recycling systems (Hajam et al. 2023). Additionally, safe incineration technologies ensure the responsible disposal of agrochemical residues, preventing harmful emissions and toxic buildup in ecosystems.

Farmers can reduce environmental concerns, adhere to legal requirements and support circular economy objectives by implementing these waste management strategies. Additionally, carbon credit systems, which offer financial incentives for appropriate waste processing, are in line with sustainable pesticide disposal practices (Yang et al. 2023). Incorporating effective recycling and disposal practices into farming operations enhances the sector's potential to fight climate change while also promoting long-term sustainability (Islam et al. 2024).

The integration of precision agriculture, use of bio-based agrochemicals, slow-release fertilizers and efficient agrochemical waste management is crucial for reducing GHG emissions and promoting sustainable farming (Sanyaolu and Sadowski 2024). These innovations enable carbon credit opportunities, increase resource efficiency and improve soil health, all of which have positive effects on the environment and the economy (Raina et al. 2024). Farmers may reduce climate change, boost production and support global sustainability initiatives by using climate-smart practices, guaranteeing a resilient and environmentally friendly agricultural future.

8.8 Research and Development Towards Carbon Credit in Various Agroecosystems

Research and development (R&D) initiatives and methods about carbon credits in agroecosystems include the following:

Conservation Agriculture
Crop rotation, regular soil cover and minimal soil disturbance are all part of this technique. It can increase crop yields, sequester carbon and lower greenhouse gas emissions (Tiefenbacher et al. 2021). Researchers evaluated conservation agriculture's capacity to provide carbon credits and extra revenue for farmers.

Low-Carbon Farming
This strategy incentivizes farmers to implement methods that enhance soil carbon content and minimize the usage of synthetic fertilizers (Lal 2013). Reduced tillage, mulching, intercropping and the use of organic fertilizers are some of these techniques.

Agroforestry
An agroforestry system based on gamhar (*Gmelina arborea* Roxb.) was evaluated for its capacity to sequester carbon and provide credit. The study discovered that treatments had a considerable impact on the price of carbon, carbon credit, carbon sequestration and total carbon stock (Kumar et al. 2024).

Schemes for Carbon Farming
These programmes provide farmers with cash rewards for implementing carbon farming techniques. Farmers are able to sell their carbon credits (Everhart 2023).

Voluntary Carbon Markets
Businesses and individuals can buy carbon credits in these markets to make up for their carbon emissions. Voluntary carbon markets (VCMs) have expanded quickly in the last few years. The difficulty and cost of measuring SOC restrict the size of data that can be gathered (De Pinto et al. 2010). Research is being conducted to understand how soil microbes can convert plant organic matter into stable carbon, how soil amendments such as biochar can increase greenhouse gas sequestration and the ability of agroforestry systems can reduce GHG emissions.

8.9 Policy Perspective for Reducing Agrochemicals and Agroecosystem Services Management Plans

The development of AES management plans is crucial to promoting sustainable agricultural practices while ensuring the continued provision of these essential ecosystem functions (Dale and Polasky 2007). These plans aim to integrate agricultural production with the maintenance and enhancement of ecosystem services, providing a framework for balancing food security with environmental sustainability. The reduction in agrochemicals usage while improving agroecosystem services is a significant aim for a sustainable agriculture strategy. Agrochemicals, such as pesticides and fertilizers, have made major contributions to food security, but their excessive usage has resulted in serious environmental degradation, such as biodiversity loss, soil depletion and water pollution (Foley et al. 2011).

Policies aiming at resolving these difficulties must strike a balance between the need to preserve agricultural output and the conservation of ecosystems and their functions, such as pollination, water filtering and nutrient cycling (Tilman et al. 2002). One effective policy approach is promoting IPM strategies, which reduce reliance on chemical pesticides by combining biological controls, crop diversification and precise application of chemicals only when necessary (Naranjo et al. 2015). Governments may encourage farmers to use sustainable pest management techniques by offering incentives and by providing the required training and information (Pretty et al. 2018). Policies should support the use of biological control agents and biopesticides in addition to IPM since they are less damaging to non-target species and promote the long-term health of ecosystems (Geiger et al. 2010). Regulatory frameworks play a crucial role in reducing agrochemical use. Governments must impose more stringent laws on dangerous agrochemicals, outlawing compounds that seriously threaten public and environmental health (Wilson and Tisdell 2001). Simultaneously, economic instruments such as taxes on harmful chemicals and subsidies for organic and sustainable farming practices can motivate farmers to transition away from conventional chemical-intensive methods (Pretty et al. 2018). According to encouraging organic farming in this situation provides a strategy to dramatically cut pesticide inputs while enhancing soil fertility and biodiversity.

Investing in research and innovation is also crucial for lowering agrochemical use. Precision agricultural technology, such as GPS-guided pesticide applications and sensor-based soil health monitoring, enables farmers to hand out pesticides more efficiently, decreasing waste and environmental pollution (Gebbers and Adamchuk 2010). Furthermore, governments can subsidize research into safer and more ecologically friendly agrochemicals (Foley et al. 2011). In addition to technical advancements, policymakers should encourage agroecological methods, which emphasize agricultural systems that function in harmony with nature by improving natural soil fertility and pest control mechanisms (Altieri et al. 2015). Education and extension services are essential in promoting the adoption of sustainable practices. Governments and non-governmental organizations should engage in farmer education initiatives that emphasize the environmental and economic benefits of lowering pesticide usage (Pretty et al. 2018). Extension services may also supply farmers with up-to-date knowledge on new techniques and technology, facilitating the transition to sustainable farming systems (Naranjo et al. 2015). In this sense, measures that encourage peer-to-peer learning among farmers can be especially beneficial, as farmers frequently trust their peers' advice more than outsider specialists (Larsen et al. 2017).

The AES management plans should be incorporated into agricultural policy to guarantee that critical services like pollination, water control and soil fertility are maintained alongside food production (Tilman et al. 2002). Payments for ecosystem services (PES) programmes, which reward farmers for implementing techniques that improve ecosystem services, might provide financial incentives for sustainable land management (Foley et al. 2011). Such regulations not only encourage farmers to use less agrochemicals but also help restore and safeguard key ecosystems, ensuring their long-term viability (Seufert and Ramankutty 2017). Lastly, effective policy implementation depends on stakeholder participation. To develop comprehensive plans that address the intricate relationships between agriculture and ecosystem services, policymakers, scientists, farmers and environmental organizations must collaborate (Altieri et al. 2015). According to Pretty et al. (2018), collaborative governance can promote the creation of locally relevant solutions and a sense of shared responsibility. Furthermore, regulations must have comprehensive monitoring and evaluation mechanisms to monitor the effectiveness of pesticide reduction strategies and make necessary modifications (Foley et al. 2011). In conclusion, reducing agrochemical use while managing agroecosystem services requires a holistic, multi-faceted policy approach. Integrated pest management, regulatory frameworks, research, education and ecosystem service management plans are all critical components of a successful strategy.

The expansion of agricultural production has relied significantly on agrochemicals, such as synthetic fertilizers and insecticides, to increase crop productivity. However, indiscriminate use has had negative environmental and economic implications. To minimize these effects, several national and international policy frameworks have been developed, with an emphasis on sustainable agriculture practices, cost reduction and soil health enhancement. Furthermore, carbon credit programmes have evolved as a feasible way to reward farmers for practicing climate-smart agriculture.

8.9.1 Policy Planning Schemes and Programmes

8.9.1.1 Integrated Nutrient and Pest Management

The integrated nutrient management (INM) and IPM programmes promote balanced fertilization and biological pest management strategies in order to reduce chemical dependency while preserving soil fertility and ecosystem health. Governments offer financial incentives, training and subsidies to encourage these activities (FAO 2017; Lal 2020).

8.9.1.2 Organic Farming and Natural Farming Initiatives

Organic and natural farming suggests making use of fewer agrochemicals by substituting synthetic fertilizers and pesticides with biofertilizers, compost and natural pest control methods. These measures promote soil health, increase microbial activity and reduce chemical residue building, resulting in long-term sustainability. Programmes such as the Organic Certification Cost-Share Programme (OCCSP) in the United States and the Paramparagat Krishi Vikas Yojana (PKVY) in India assist farmers in switching to natural and organic farming, lowering their dependency on artificial agrochemicals and increasing soil organic matter.

8.9.1.3 Climate-Smart Agriculture Programmes

Climate-smart agriculture (CSA) programmes encourage sustainable farming by lowering pesticide use via organic farming, biofertilizers and integrated pest management (Kumar and Kumar 2023). They promote precision agriculture, which optimizes fertilizer and pesticide use while minimizing environmental effects. Crop rotation, agroforestry and conservation tillage are examples of CSA techniques that improve soil quality and biodiversity. By incorporating climate-resilient crops and smart irrigation, CSA promotes long-term sustainability and food security. The National Mission for Sustainable Agriculture (NMSA) in India and the EU's Green Deal are two instances of such initiatives.

8.9.1.4 Soil Health Card Scheme

Soil Health Card (SHC) is a Government of India scheme promoted by the Department of Agriculture & Cooperation under the Ministry of Agriculture and Farmers' Welfare. A SHC is used to assess the current state of soil health and, when used over time, to identify changes in soil health caused by land management. An SHC provides soil health indicators and descriptive phrases. Farmers can apply fertilizer with knowledge by utilizing programmes like the Soil Health Card Scheme, which give them reports on the nutrient status of their soil. As a result, less pesticide is used excessively and soil production is increased economically.

8.9.1.5 Carbon Credit and Payment to Ecosystem Services Mechanisms

Using biochar, conservation tillage and agroforestry, governments and international organizations encourage carbon sequestration, allowing farmers to take part in carbon credit markets. In India, PES, sometimes known as payments for environmental services (or benefits), are incentives given to farmers or landowners in exchange for managing their property to provide an ecological service. In the United Nations financial returns for sustainable activities are facilitated by initiatives like the Clean Development Mechanism (CDM) and Verified Carbon Standard (VCS).

8.9.1.6 The Pesticide Management Bill and Regulatory Frameworks

In India, the Pesticide Management Bill (PMB), 2020, aims to control the production, importation, packaging and distribution of pesticides while ensuring their judicious application. The Sustainable Use Directive (SUD) of the European Union and other similar regulations seek to reduce the dangers of pesticides to the environment and human health (Fetting 2020).

8.9.1.7 Fertilizer Subsidy Reforms and Direct Benefit Transfer

To rationalize the use of fertilizers, the Government of India implemented Direct Benefit Transfer (DBT) in fertilizer subsidies in 2013, ensuring that benefits reach farmers directly. This promotes balanced fertilizing and reduces overuse.

8.9.2 Economic and Environmental Benefits

The use of sustainable agrochemical management practices not only reduces environmental risks but also minimizes input costs, increasing farmer profitability. Additionally, increased soil organic carbon improves soil health, resulting in sustained productivity. Participation in carbon credit programmes provides an alternative revenue stream, fostering resilience against market fluctuations (Lal 2011; FAO 2019). The integration of government policies and incentive-based programmes plays a crucial role in transitioning towards sustainable agriculture. These initiatives not only reduce dependence on chemical inputs but also create opportunities for carbon credit generation, fostering economic stability for farmers. A robust policy framework, coupled with technological advancements, will further drive the global agenda for sustainable and climate-resilient agriculture.

8.10 Future Thrust

The usage of agrochemicals in the future will primarily concentrate on resolving the mounting environmental issues related to their use. Agrochemicals, such as fertilizers and pesticides, have increased agricultural output, but they also have a negative influence on ecosystems. This is especially true when it comes to soil health, water pollution and biodiversity loss. In the future, minimizing negative environmental consequences will be crucial to the sustainable use of agrochemicals to maintain food security for an expanding population.

The use of agrochemicals raises serious concerns about biodiversity loss. Numerous pesticides damage non-target organisms, including helpful insects like pollinators and naturally occurring predators of pests. Pollination and naturally occurring pest control are two ecosystem services that are vital to agriculture and are threatened by the reduction in these species. According to Geiger et al. (2010), future research and development will probably focus on biopesticides and more focused chemical agents that target pests without endangering beneficial species. The focus will also be on generating ecologically benign agrochemicals and boosting biological control strategies through Integrated Pest Management (IPM) systems (Naranjo et al. 2015). Degradation of the soil as a result of the overuse of fertilizers is another emerging issue. The consistent use of synthetic fertilizers may result in an imbalance of nutrients and a reduction in soil organic matter, both of which are essential for soil fertility and carbon sequestration. It is anticipated that in the future, there will be less synthetic fertilizer used and more focus on sustainable soil management techniques, such as adding organic and biofertilizers that gradually enhance soil health (Foley et al. 2011). This will enhance long-term agricultural sustainability and encourage a more balanced nutrient cycle.

Water pollution from agrochemical runoff is already a significant environmental issue, causing eutrophication in freshwater and marine ecosystems. Nutrient leaching, particularly nitrogen and phosphorus from fertilizers, leads to the excessive growth of algae and creates hypoxic zones that disrupt aquatic ecosystems (Smith et al. 1999). Precision agriculture technologies, which allow for the precise application of fertilizers and pesticides based on real-time data, are expected to reduce over-application and minimize runoff into water bodies (Gebbers and Adamchuk 2010).

The impacts of agrochemicals will worsen due to climate change, as changes in temperature and precipitation patterns could render pest control and soil amendments more necessary. Concerns about a growing reliance on agrochemicals are raised by this. Future methods must incorporate climate-resilient agricultural techniques that improve ecosystem services while lowering the demand for chemical inputs, including agroforestry (Altieri et al. 2015). The usage of agrochemicals in the future will be greatly influenced by regulatory rules. Governments are anticipated to encourage research towards safer, more environmentally friendly pesticide alternatives while simultaneously enforcing stronger limits on the most hazardous agrochemicals. Promotion of organic farming, financial incentives for implementing sustainable practices and international collaboration to control transboundary pollution are a few examples of potential policy frameworks (Pretty 2008; Pretty et al. 2018). There will probably be a shift in governance towards a multi-level strategy that strikes a balance between maintaining the environment and agricultural output. Furthermore, the future of agrochemical use will be focused on reducing their environmental impact through breakthroughs in biopesticides, sustainable soil and water management, precision agriculture and adaptable climate strategies.

Governments and the agriculture industry will need to work together on policies and technology that lower the environmental dangers associated with agrochemicals while maintaining global food security.

8.11 Conclusion

The extensive use of agrochemicals to boost crop yields has significantly disrupted agroecosystem functions, harming soil health and degrading water quality, which ultimately leads to a loss of biodiversity. These consequences collectively undermine the long-term sustainability of farming systems and threaten the existence of various species. It is crucial to understand the detrimental effects of agrochemicals to develop effective strategies for their mitigation and to restore ecological balance. By integrating sustainable management practices such as agroforestry, biofertilizers, regenerative farming, crop rotation and reduced tillage, we can greatly enhance carbon sequestration while simultaneously improving essential ecosystem functions. These advancements also present opportunities for farmers to participate in carbon credit markets, yielding not only environmental benefits but also potential financial rewards. Therefore, aligning sustainable agricultural practices with carbon credit generation presents a promising pathway towards climate-smart agriculture and the economic empowerment of rural communities.

References

Abbas A, Zhang Z, Zheng H, Alami MM, Alrefaei AF, Abbas Q, Naqvi SAH, Rao MJ, Mosa WF, Abbas Q, Hussain A (2023) Drones in plant disease assessment, efficient monitoring, and detection: a way forward to smart agriculture. Agronomy 13(6):1524

Adeoye PA, Abubakar SK, Adeolu AR, El-Mohsen AAA, Sabbour M, Farhadi A, Hamidi R, Pirasteh Anosheh H (2013) Effect of agrochemicals on groundwater quality: a review. Sci Agric 1:1–7

Aggarwal R (2024) Carbon offsets compatible with the Paris Agreement to limit global warming: Call for a direct action. Environmental Challenges 17,101034

Aguilera E, Lassaletta L, Sanz-Cobena A, Garnier J, Vallejo A (2013) The potential of organic fertilizers and water management to reduce N2O emissions in Mediterranean climate cropping systems: a review. Agric Ecosyst Environ 164:32–52

Ahmad MF, Ahmad FA, Alsayegh AA, Zeyaullah M, AlShahrani AM, Muzammil K, Saati AA, Wahab S, Elbendary EY, Kambal N, Abdelrahman MH, Hussain S (2024) Pesticides impacts on human health and the environment with their mechanisms of action and possible countermeasures. Heliyon 10(7):e29128. https://doi.org/10.1016/j.heliyon.2024.e29128

Alavanja MC, Samanic C, Dosemeci M, Lubin J, Tarone R, Lynch CF, Knott C, Thomas K, Hoppin JA, Barker J (2003) Use of agricultural pesticides and prostate cancer risk in the agricultural health study cohort. Am J Epidemiol 157(9):800–814

Alonso-Serra J (2021) Carbon sequestration: Counterintuitive feedback of plant growth. Quantitative Plant Biology 2:e11. https://doi.org/10.1017/qpb.2021.11

Altaf R, Ullah Z, Darko DA, Iqbal A, Khan MS, Asif M (2022) Molecularly imprinted polymers for the detection of chlorpyrifos (an organophosphate pesticide). ASEAN Journal of Science and Engineering 2(3):257–266. https://doi.org/10.17509/ajse.v2i3.44993

Altieri MA, Funes-Monzote FR, Petersen P (2015) Agroecologically efficient agricultural systems for smallholder farmers: contributions to food sovereignty. Agron Sustain Dev 32(1):1–13

Ammar EE, Rady HA, Khattab AM, Amer MH, Mohamed SA, Elodamy NI, Al-Farga A, Aioub, AA. (2023) A comprehensive overview of eco-friendly bio fertilizers extracted from living organisms. Environmental Science and Pollution Research 30(53):113119–113137

Angon PB, Mondal S, Jahan I, Datto M, Antu UB, Ayshi FJ, Islam MS (2023) Integrated pest management (IPM) in agriculture and its role in maintaining ecological balance and biodiversity. Advances in Agriculture 2023,5546373

Anonymous (2024). FAOSTAT. Available online: https://www.fao.org/faostat/en/#data/RP (accessed on 1 January 2024).

Asadu CO, Ezema CA, Ekwueme BN, Onu CE, Onoh IM, Adejoh T, Ezeorba TPC, Ogbonna CC, Otuh PI, Okoye JO, Emmanuel UO (2024) Enhanced efficiency fertilizers: Overview of production methods, materials used, nutrients release mechanisms, benefits and considerations. Environmental Pollution and Management

Avasthe RK, Devi SH, Bhupenchandra I, Kumar A, Chongtham SK, Babu S, Singh R, Das A, Gudade BA, Bora SS (2023) Agro-tactics for reducing carbon footprint in agricultural production systems: A review. Indian Journal of Agronomy 68(2):115–125

Balandaitė J, Romaneckas K, Kimbirauskienė R, Sinkevičienė A (2024) Comprehensive assessment of the effect of multi-cropping on agroecosystems. Plants 13(10):1372. https://doi.org/10.3390/plants13101372

Balloux F, van Dorp L (2017) Q&A: what are pathogens, and what have they done to and for us? BMC Biol 15(1):91. https://doi.org/10.1186/s12915-017-0433-z

Bedoussac L, Justes E (2010a) Dynamic analysis of competition and complementarity for light and nitrogen use to understand the yield and the protein content of a durum wheat-winter pea intercrop. Plant Soil 330:37–54. https://doi.org/10.1007/s11104-010-0303-8

Bedoussac L, Justes E (2010b) The efficiency of a durum wheat-winter pea intercrop to improve yield and wheat grain protein concentration depends on nitrogen availability during early growth. Plant Soil 330:19–35. https://doi.org/10.1007/s11104-009-0082-2

Bhandari G (2014) An overview of agrochemicals and their effects on the environment in Nepal. Appl Ecol Environ Sci 2:66–73

Bish M, Oseland E, Bradley K (2021) Off-target pesticide movement: a review of our current understanding of drift due to inversions and secondary movement. Weed Technol 35:345–356

Biswas SK, Rahman S, Kobir SMA, Ferdous T, Banu NA (2014) A review on impact of agrochemicals on human health and environment: Bangladesh perspective. Plant Environ Dev 3(2):31–35

Blum ZB, Zechmeister-Boltenstern S, Keiblinger KM (2019) Does soil contribute to the human gut microbiome? Microorganisms 7:287. https://doi.org/10.3390/microorganisms7090287

Bolognesi C, Merlo FD (2011) Pesticides: human health effects. In: Nriagu JO (ed) Encyclopedia of environmental health. Elsevier, pp 438–453

Bowles TM, Mooshammer M, Socolar Y, Calderón F, Cavigelli MA, Culman SW, Deen W, Drury CF, Garcia AG, Gaudin AC (2020) Long-term evidence shows that crop-rotation diversification increases agricultural resilience to adverse growing conditions in North America. One Earth 2: 284–293

Brenna S, Acutis M, Tabaglio V, Grandi M (2014) Conservation agriculture as a driver for carbon credit market. In European conference on green carbon: making sustainable agriculture real. Brussels, pp 1–9

Cariappa AA, Krishna VV (2024) Carbon farming in India: Are the existing projects inclusive, additional, and permanent? Climate Policy 1–16

Carriger JF, Rand GM, Gardinali PR, Perry WB, Tompkins MS, Fernandez AM (2006) Pesticides of potential ecological concern in sediment from South Florida canals: an ecological risk prioritization for aquatic arthropods. Soil Sediment Contam 15:214–245

Caudill SA, DeClerck FJA, Husband TP (2015) Connecting sustainable agriculture and wildlife conservation: does shade coffee provide habitat for mammals? Agric Ecosyst Environ 199:85–93

Cech R, Leisch F, Zaller JG (2022) Pesticide use and associated greenhouse gas emissions in sugar beet, apples, and viticulture in Austria from 2000 to 2019. Agriculture 12(6):879

Cerda R, Deheuvels O, Calvache D, Niehaus L, Sáenz Y, Kent J, Vílchez S, Villota A, Martínez C, Somarriba E (2014) Contribution of cocoa agroforestry systems to family income and domestic consumption: Looking toward intensification. Agroforestry Systems 88(6):957–981. https://doi.org/10.1007/s10457-014-9691-8

Chan Y (2008) Increasing soil organic carbon of agricultural land. Primefact 735:1–5

Chen SK, Edwards CA, Subler S (2001) A microcosm approach for evaluating the effects of the fungicides benomyl and captan on soil ecological processes and plant growth. Appl Soil Ecol 18:69–82

Chen G, Kong X, Gan Y, Zhang R, Feng F, Yu A, Zhao C, Wan S, Chai Q (2018) Enhancing the systems productivity and water use efficiency through coordinated soil water sharing and compensation in strip-intercropping. Sci Rep 8:10494

Cong WF, Hoffland E, Li L, Six J, Sun JH, Bao XG, Van Der Werf W (2015) Intercropping enhances soil carbon and nitrogen. Glob Chang Biol 21(4):1715–1726

Dale VH, Polasky S (2007) Measures of the effects of agricultural practices on ecosystem services. Ecol Econ 64(2):286–296

Davis AS, Hill JD, Chase CA, Johanns AM, Liebman M (2012) Increasing cropping system diversity balances productivity, profitability, and environmental health. PLoS One 7:e47149

Deguine JP, Aubertot JN, Flor RJ, Lescourret F, Wyckhuys KA, Ratnadass A (2021) Integrated pest management: Good intentions, hard realities. A review. Agronomy for Sustainable Development 41(3):38

De Pinto A, Maghalaes M, Ringler C (2010) Potential of carbon markets for small farmers. International Food Policy Research Institute

DeClerck FA, Chazdon R, Holl KD, Milder JC, Finegan B et al (2010) Biodiversity conservation in human-modified landscapes of Mesoamerica: past, present and future. Biol Conserv 143:2301–2313

Dereumeaux C, Fillol C, Quenel P, Denys S (2020) Pesticide exposures for residents living close to agricultural lands: a review. Environ Int 134:105210. https://doi.org/10.1016/j.envint.2019.105210

Deutz A, Heal GM, Niu R, Swanson E, Townshend T, Zhu L, Delmar A, Meghji A, Sethi SA, Tobin-de la Puente J (2020) Financing nature: Closing the global biodiversity financing gap. The Paulson Institute, The Nature Conservancy, and the Cornell Atkinson Center for Sustainability, 256

Dyer JA, Kulshreshtha SN, McConkey BG, Desjardins RL (2010) An assessment of fossil fuel energy use and CO2 emissions from farm field operations using a regional level crop and land use database for Canada. Energy 35(5):2261–2269

Elumalai P, Gao X, Parthipan P, Luo J, Cui J (2025) Agrochemical pollution: A serious threat to environmental health. Current Opinion in Environmental Science & Health 100597

Everhart S (2023) Growing carbon credits: strengthening the agricultural sector's participation in voluntary carbon markets through law and policy. N Y Univ Environ Law J 31:65

FAO (2017) The future of food and agriculture: trends and challenges. Food and Agriculture Organization of the United Nations

FAO (2024) FAOSTAT: pesticides use. FAO, Rome

Fargione J, Tilman D (2005) Niche differences in phenology and rooting depth promote coexistence with a dominant C4 bunchgrass. Oecologia 143:598–606. https://doi.org/10.1007/s00442-005-0010-y

Federation of Indian Chambers of Commerce and Industry (2019) Report on use of agrochemicals for sustainable farming: 8th agrochemicals conference 2019. New Delhi, India. http://ficci.in/past-event-page.asp?evid=24230

Federation of Indian Chambers of Commerce and Industry (FICCI) (2016) A report on Indian agrochemical industry

Fetting C (2020) The European green deal. ESDN Rep (December) 2(9)

Ferland D, Wagner-Riddle C, Brown SE, Bourgault M, Helgason W, Farrell RE, Congreves KA (2024) Improved nitrogen fertilizer management reduces nitrous oxide emissions in a northern Prairie cropland. Science of The Total Environment 956:177211

Finckh MR, Gacek ES, Goyeau H, Lannou C, Merz U, Mundt CC, Munk L, Nadziak J, Newton AC, de Vallavieille-Pope C, Wolfe MS (2000) Cereal variety and species mixtures in practice, with emphasis on disease resistance. Agronomie 20:813–837

Foley JA, Ramankutty N, Brauman KA, Cassidy ES, Gerber JS, Johnston M, Zaks DP (2011) Solutions for a cultivated planet. Nature 478(7369):337–342

Food and Agriculture Organization (2015) World fertilizer trends and outlook to 2018. Food and Agriculture Organization of the United Nations

Food and Agriculture Organization of the United Nations. (2024, November 14). Greenhouse gas emissions from agrifood systems: Global, regional and country trends, 2000–2022. FAOSTAT Highlights

Gaba S, Perronne R, Fried G, Gardarin A, Bretagnolle F, Biju-Duval L, Colbach N, Cordeau S, Fernández-Aparicio M, Gauvrit C, Gibot-Leclerc S, Guillemin JP, Moreau D, Munier-Jolain N, Strbik F (2017) Response and effect traits of arable weeds in agro-ecosystems: a review of current knowledge. Weed Res 57(3):123–147

Galloway T, Handy R (2003) Immunotoxicity of organophosphorous pesticides. Ecotoxicology 12: 345–363

Gallaher RN (2009) Multiple cropping systems. In Management of Agricultural, Forestry and Fisheries Enterprises. Vol. I: Multiple Cropping Systems (Encyclopedia of Life Support Systems)

Gebbers R, Adamchuk VI (2010) Precision agriculture and food security. Science 327(5967): 828–831

Gebremedhin B, Schwab G (1998) The economic importance of crop rotation systems: evidence from the literature

Geiger F, Bengtsson J, Berendse F, Weisser WW, Emmerson M, Morales MB, Inchausti P (2010) Persistent negative effects of pesticides on biodiversity and biological control potential on European farmland. Basic Appl Ecol 11(2):97–105. https://doi.org/10.1016/j.baae.2009.12.001

Ghanbari A, Dahmardeh M, Siahsar BA, Ramroudi M (2010) Effect of maize (*Zea mays* L.) – cowpea (*Vigna unguiculata* L.) intercropping on light distribution, soil temperature, and soil moisture in an environment. J Food Agric Environ 8:102–108

Getahun S, Kefale, H., Gelaye, Y (2024) Application of precision agriculture technologies for sustainable crop production and environmental sustainability: A systematic review. The Scientific World Journal 2024,2126734

Gingrich S, Erb KH, Krausmann F, Gaube V, Haberl H (2007) Long-term dynamics of terrestrial carbon stocks in Austria: A comprehensive assessment of the time period from 1830 to 2000. Reg Environ Chang 7(1):37–47

Govil S, Long NV D, Escribà-Gelonch M., Hessel, V. (2024) Controlled-release fertiliser: Recent developments and perspectives. Industrial Crops and Products 219:119160.

Guebsi, R., Mami, S., Chokmani, K. (2024). Drones in precision agriculture: A comprehensive review of applications, technologies, and challenges. Drones, 8(11), 686.

Guo, K., Yang, J., Yu, N., Luo, L., Wang, E. (2023). Biological nitrogen fixation in cereal crops: Progress, strategies, and perspectives. Plant Communications, 4(2).

Hajam YA, Kumar R, Kumar A (2023) Environmental waste management strategies and vermi transformation for sustainable development. Environmental Challenges, 100747

Hanaček K, Rodríguez-Labajos B (2018) Impacts of land-use and management changes on cultural agroecosystem services and environmental conflicts—A global review. Global Environmental Change, 50:41–59. https://doi.org/10.1016/j.gloenvcha.2018.02.016

Haney RL, Senseman SA, Hons FM, Zuberer DA (2000) Effect of glyphosate on soil microbial activity and biomass. Weed Sci 48:89–93

Hardaker A, Pagella T, Rayment M (2020) Integrated assessment, valuation and mapping of ecosystem services and dis-services from upland land use in Wales. Ecosyst Serv 43:101098

Hatfield JL, Dold C (2019) Water-use efficiency: advances and challenges in a changing climate. Front Plant Sci 10:103

He HM, Liu LN, Munir S, Bashir NH, Yi W, Jing Y, Li CY (2019) Crop diversity and pest management in sustainable agriculture. J Integr Agric 18(9):1945–1952. https://doi.org/10.1016/S2095-3119(19)62689-4

Hoppin JA, Umbach DM, London SJ, Henneberger PK, Kullman GJ, Coble J, Alavanja MC, Beane Freeman LE, Sandler DP (2009) Pesticide use and adult-onset asthma among male farmers in the agricultural health study. Eur Respir J 34(6):1296–1303

Hu F, Gan Y, Chai Q, Feng F, Zhao C, Yu A, Mu Y, Zhang Y (2016) Boosting system productivity through the improved coordination of interspecific competition in maize/pea strip intercropping. Field Crop Res 198:50–60

Huang J, Tichit M, Poulot M, Darly S, Li SC, Petit C, Aubry C (2015) Comparative review of multifunctionality and ecosystem services in sustainable agriculture. J Environ Manag 149:138–147

HYDE (2023) Land use: cropland PBL Netherlands environmental assessment agency, History satabase of the global environment 3.3. Available at: https://ourworldindata.org/land-use

Islam NF, Gogoi B, Saikia R, Yousaf B, Narayan M, Sarma H (2024) Encouraging circular economy and sustainable environmental practices by addressing waste management and biomass energy production. Regional Sustainability 5(4):100174

Jaiswal B, Agrawal M (2019) Carbon footprints of agriculture sector. In S. S. Muthu (Ed.), Carbon footprints, environmental footprints and eco-design of products and processes (pp. 81–99). Singapore: Springer. https://doi.org/10.1007/978-981-13-7916-1_4

Jat ML, Singh VK, Jat HS, Parihar CM, Sutaliya JM, Poonia S, Jat RK, Jat MK (2022) Resilient and diversified Agri-food systems for nutritional security and climate resilience. Indian J Agron 67 (Special issue):S218–S226

Jat ML, Chakraborty D, Ladha, JK, Parihar, CM, Datta, A., Mandal, B., Nayak HS, Maity P, Rana DS, Chaudhari SK, Gerard B (2022) Carbon sequestration potential, challenges, and strategies towards climate action in smallholder agricultural systems of South Asia. Crop and Environment 1(1):86–101

Jayaraman, T., Aparajay, K., Devshali, C. C., Jaiswal, S (2023) Greenhouse gas mitigation and carbon markets in Indian agriculture. Economic & Political Weekly 58(43),71

Jose S (2009) Agroforestry for ecosystem services and environmental benefits: an overview. Agrofor Syst 76:1–10. https://doi.org/10.1007/s10457-009-9229-7

Kalia A, Gosal SK (2011) Effect of pesticide application on soil microorganisms. Arch Agron Soil Sci 57:569–596

Kantwa SR, Choudhary M, Agrawal RK, Dixit AK, Kumar S, Chary GR (2024) Reducing energy and carbon footprint through diversified rainfed cropping systems. Energy Nexus 14:100306

Kamyab H, SaberiKamarposhti M, Hashim H, Yusuf M (2024). Carbon dynamics in agricultural greenhouse gas emissions and removals: A comprehensive review. Carbon Letters 34(1): 265–289

Khan BA, Nadeem MA, Nawaz H, Amin MM, Abbasi GH, Nadeem M, Ali M, Ameen M, Javaid MM, Maqbool R (2023) Pesticides: Impacts on agriculture productivity, environment, and management strategies. In Emerging contaminants and plants: Interactions, adaptations and remediation technologies (pp. 109–134). Springer. https://doi.org/10.1007/978-3-031-22680-4_5

Khangura R, Ferris D, Wagg C, Bowyer J (2023) Regenerative agriculture—A literature review on the practices and mechanisms used to improve soil health. Sustainability 15(3):2338

Kim D, Burkett-Cadena ND, Reeves LE (2022) Pollinator biological traits and ecological interactions mediate the impacts of mosquito-targeting malathion application. Scientific Reports 12(1): 17039. https://doi.org/10.1038/s41598-022-21094-x

Kirby KR, Potvin C (2007) Variation in carbon storage among tree species: implications for the management of a small scale carbon sink project. For Ecol Manag 246:208–221

Koul B, Yakoob M, Shah MP (2022) Agricultural waste management strategies for environmental sustainability. Environmental Research, 206, 112285

Koudahe K, Allen SC, Djaman K (2024) Critical review of the impact of cover crops on soil properties. International Soil and Water Conservation Research 10(3):343–354

Kumar A, Kumar S (2023) Prospects of carbon farming in India: an innovative climate-resilient strategy. Indian J Agron 68(1):1–12

Kumar R, Mishra V, Buzan J, Kumar R, Shindell D, Huber M (2017) Dominant control of agriculture and irrigation on urban heat islands in India. Sci Rep 7:1405

Kumar S, Bamboriya SD, Rani K, Meena RS, Sheoran S, Loyal A, Kumawat A, Jhariya MK (2022) Grain legumes: a diversified diet for sustainable livelihood, food, and nutritional security. In: Advances in legumes for sustainable intensification. Academic Press, pp 157–178

Kumar S, Sharma SK, Thakral SK, Bhardwaj KK, Jhariya MK, Meena RS, Jangir CK, Bedwal S, Jat RD, Gaber A, Atta AA (2022a) Integrated nutrient management improves the productivity and nutrient use efficiency of *Lens culinaris* Medik. Sustainability 14(3):1284

Kumar S, Sharma SK, Dhaka AK, Bedwal S, Sheoran S, Meena RS, Jangir CK, Kumar D, Kumar R, Jat RD, Meena AK (2023) Efficient nutrient management for enhancing crop productivity, quality and nutrient dynamics in lentil (*Lens culinaris* Medik.) in the semi-arid region of northern India. PLoS One 18(2):e0280636

Kumar A, Malik MS, Shabnam S, Kumar R, Karmakar S, Das SS, Lakra K, Singh I, Kumar R, Sinha AK, Barla S (2024) Carbon sequestration and credit potential of gamhar (*Gmelina arborea* Roxb.) based agroforestry system for zero carbon emission of India. Sci Rep 14(1): 4828

Kumar S, Meena RS, Kumar S, Pradhan G, Jangir CK, Ghosh S, Punia H, Sheoran P, Meena R, Ahmad MA, Goyal SK (2025) Designing a diversified Indian mustard production system for energy-carbon-cum-heat use efficiency and sowing dates assessment. GCB Bioenergy 17(6): e70044

Kuyah S, Öborn I, Jonsson M (2017) Regulating ecosystem services delivered in agroforestry systems. In agroforestry: anecdotal to modern science. pp 797–815

Lal R (2011) Sequestering carbon in soils of agro-ecosystems. Food Policy 36:S33–S39

Lal R (2013) Intensive agriculture and the soil carbon pool. J Crop Improv 27(6):735–751

Lal R (2020) Regenerative agriculture for food and climate. J Soil Water Conserv 75(5):123A–124A

Larsen AE, Gaines SD, Deschenes O (2017) Agricultural pesticide use and adverse birth outcomes in the San Joaquin Valley of California. Nat Commun 8:302. https://doi.org/10.1038/s41467-017-00349-2

Leharwan M, Kumar Y, Kumar R, Kumar Saraswat P, Kumar R, Veetil AKT, Bhattacharjee S, Kumar A, Kumar S (2023) Assessing the effects of conservation tillage and in-situ crop residue management on crop yield and soil properties in rice–wheat cropping system. Sustainability 15(17):12736

Lengai GM, Muthomi JW, Mbega ER (2020) Phytochemical activity and role of botanical pesticides in pest management for sustainable agricultural crop production. Scientific African, 7, e00239

Lindström K, Mousavi SA (2020) Effectiveness of nitrogen fixation in rhizobia. Microbial Biotechnology, 13(5):1314–1335

Liu X, Elgowainy A, Wang M (2020) Life cycle energy use and greenhouse gas emissions of ammonia production from renewable resources and industrial by-products. Green Chem 22: 5751–5761

Lokuge N, Anders S (2022) Carbon-credit systems in agriculture: A review of literature. School Pub Policy Publ 15. https://doi.org/10.11575/sppp.v15i1.7459

Lokuge N, Anders S (2022) Carbon-credit systems in agriculture: A review of literature. The School of Public Policy Publications, 15

Lupwayi N, Kennedy A (2007) Grain legumes in Northern Great Plains: impact on selected biological soil processes. J Agron 99:1700–1709

Ma F, Eneji AE, Liu J (2015) Assessment of ecosystem services and dis-services of an agro-ecosystem based on extended emergy framework: a case study of Luancheng county, North China. Ecol Eng 82:241–251

Mach ME, Martone RG, Chan KMA (2015) Human impacts and ecosystem services: insufficient research for trade-off evaluation. Ecosyst Serv 16:112–120. https://doi.org/10.1016/j.ecoser.2015.10.018

Malik AI, Bell R, Zang H et al (2024) Exploring the plant and soil mechanisms by which crop rotations benefit farming systems. Plant and Soil 507:1–9. https://doi.org/10.1007/s11104-024-06994-z

Margni M, Rossier D, Crettaz P, Jolliet O (2002) Life cycle impact assessment of pesticides on human health and ecosystems. Agric Ecosyst Environ 93(1–3):379–392

Maskell LC, Radbourne A, Norton LR, Reinsch S, Alison J, Bowles L, Geudens K, Robinson DA (2023) Functional agro-biodiversity: An evaluation of current approaches and outcomes. Land 12(11):2078. https://doi.org/10.3390/land12112078

McDaniel MD, Grandy AS, Tiemann LK, Weintraub MN (2014a) Crop rotation complexity regulates the decomposition of high and low-quality residues. Soil Biol Biochem 78:243–254

McDaniel MD, Tiemann LK, Grandy AS (2014b) Does agricultural crop diversity enhance soil microbial biomass and organic matter dynamics? A meta-analysis. Ecol Appl 24:560–570

Meena RS, Pradhan G (2023) Industrial garbage-derived biocompost enhances soil organic carbon fractions, CO2 biosequestration, potential carbon credits, and sustainability index in a rice-wheat ecosystem. Environ Res 116525. https://doi.org/10.1016/j.envres.2023.116525

Meena RS, Pradhan G, Kumar S, Lal R (2023) Using industrial wastes for rice-wheat cropping and food-energy-carbon-water-economic nexus to the sustainable food system. Renew Sust Energ Rev 187:113756. https://doi.org/10.1016/j.rser.2023.113756

Meena RS, Pradhan G, Nalani Singh K, Singh AK, Verma SK, Mina KK, Kumar S, Chaturvedi RK (2024) Diversified cropping modules designed for soil restoration, CO2 sequestration, and generating carbon credits. Land Degrad Dev 35. https://doi.org/10.1002/ldr.5224

Meena RS, Pradhan G, Singh K, Kumar S, Singh AK, Shashidhar KS, Mina KK, Rao CS (2024a) Agriculture models for restoring degraded land to enhance CO2 biosequestration and carbon credits in the Vindhyan region of India. Sci Total Environ 929:172661. https://doi.org/10.1016/j.scitotenv.2024.172661

Meena RS, Lal R, Kumar S, Pradhan G, Rao CS, Singh AK, Pathak H, Abhilash PC, Kumar A, Sharma SK, Jat ML, Singh S (2024b) Potential of Indian agriculture for capturing atmospheric CO2 and monetizing carbon credits to the farmers: an overview and policy framework. Adv Agron 188:101–206. https://doi.org/10.1016/bs.agron.2024.06.001

Meena RS, Pradhan G, Kumar S (2025) Energy flow, eco-efficiency, and economic circulation with recycled industrial waste compost application in wheat and subsequent rice farming. Sci Total Environ 967:178779. https://doi.org/10.1016/j.scitotenv.2025.178779

Mishra MK, Mishra SK, Panday L, Pandey R (2024) Agrochemicals and their impact on the environment. In book: Recent trends in plant protection

Mo AS, Qiu ZQ, He Q, Wu HY, Zhou XB (2016) Effect of continuous monocropping of tomato on soil microorganism and microbial biomass carbon. Commun Soil Sci Plant Anal 47(9): 1069–1077. https://doi.org/10.1080/00103624.2016.1165832

Morchid A (2023) Applications of internet of things (IoT) and sensors technology to increase food security and agricultural Sustainability: Benefits and challenges. Ain Shams Engineering Journal [Internet]. Sep 30;15(3):102509. https://doi.org/10.1016/j.asej.2023.102509
Morris RA, Garrity DP (1993) Resource capture and utilization in intercropping. Field Crop Res 34: 303–317
Mousavi SR, Eskandari H (2011) A general overview on intercropping and its advantages in sustainable agriculture. J Appl Environ Biol Sci 1(11):482–486
Muhie SH (2022) Novel approaches and practices to sustainable agriculture. Journal of Agriculture and Food Research 10, p.100446
Mueller T (2015) Methods to measure herbicide volatility. Weed Sci 63:116–120. https://doi.org/10.1614/WS-D-13-00127.1
Mulungu K, Abro Z, Niassy S, Muriithi B, Picthar J, Kidoido M, Subramanian S, Mohamed S, Khan Z, Hailu G, Kassie M (2024) The economic, social, and environmental impact of ecologically centered integrated pest management practices in East Africa. Journal of Environmental Management 371, p.123241
Nair VD, Graetz DA (2004) Agroforestry as an approach to minimizing nutrient loss from heavily fertilized soils: the Florida experience. Agrofor Syst 61:269–279
Nair PKR, Kumar BM, Nair VD (2012) Agroforestry as a strategy for carbon sequestration. Journal of Plant Nutrition and Soil Science 175(1):10–23. https://doi.org/10.1002/jpln.201100047
Nair PKR, Kumar BM, Nair VD (2009) Agroforestry as a strategy for carbon sequestration [Review of carbon sequestration potentials in agroforestry systems]. Journal of Plant Nutrition and Soil Science 172:10–23
Naranjo SE, Ellsworth PC, Frisvold GB (2015) Economic and environmental impacts of IPM practices. Annu Rev Entomol 60:271–289
National Pesticide Information Center (NPIC) (2022) Pesticide information profiles—Chlorpyrifos and Diazinon. Corvallis
O'Brien PL, Hatfield JL, Dold C, Kistner-Thomas EJ, Wacha KM (2020) Cropping pattern changes diminish agroecosystem services in North and South Dakota, USA. Agron J 112:1–24
Oller I, Malato S, Sánchez-Pérez JA, Maldonado MI, Gasso R (2007) Detoxification of wastewater containing five common pesticides by solar AOPs–biological coupled system. Catal Today 129(1):69–78
Ontl TA, Schulte LA (2012) Soil carbon storage. Nat Educ Know 3:35
Osinuga OA, Aduloju AB, Oyegoke CO (2023) Impact of agrochemicals application on soil quality indicators and trace elements level of wetlands under different uses. Journal of Trace Elements and Minerals 5:100090. https://doi.org/10.1016/j.jtemin.2023.100090
Ozlu E, Arriaga FJ, Bilen, S, Gozukara G, Babur E (2022) Carbon footprint management by agricultural practices. Biology, 11(10), p.1453
Pala M, Ryan J, Zhang H, Singh M, Harris HC (2007) Water-use efficiency of wheat-based rotation systems in a Mediterranean environment. Agric Water Manag 93:136–144
Pawase PP, Nalawade SM, Walunj AA, Bhanage GB, Kadam PB, Durgude AG, Patil MR (2024) Comprehensive study of on-the-go sensing and variable rate application of liquid nitrogenous fertilizer. Computers and Electronics in Agriculture 216, p.108482
Pinoargote M, Cerda R, Mercado L, et al (2017) Carbon stocks, net cash flow and family benefits from four small coffee plantation types in Nicaragua. Forest, Trees and Livelihoods 26(3), 183–198. https://doi.org/10.1080/14728028.2016.1268544
Poolen D, Ryszka K (2021) Can voluntary carbon markets change the game for climate change? RaboResearch—Economic Research
Power AG (2010) Ecosystem services and agriculture: tradeoffs and synergies. Philos Transac R Soc B Biol Sci 365(1554):2959–2971
Pradhan G, Meena RS (2023) Interaction impacts of biocompost on nutrient dynamics and relations with soil biota, carbon fractions index, societal value of CO2 equivalent, and ecosystem services in the wheat-rice farming. Chemosphere 139695. https://doi.org/10.1016/j.chemosphere.2023.139695

Pradhan G, Meena RS (2023a) Utilizing waste compost to improve the atmospheric CO2 capturing in the rice-wheat cropping system and energy-cum-carbon credit audit for a circular economy. Sci Total Environ 164572. https://doi.org/10.1016/j.scitotenv.2023.164572

Pradhan G, Meena RS, Kumar S, Lal R (2023) Utilizing industrial wastes as compost in wheat-rice production to improve the above and below-ground ecosystem services. Agric Ecosyst Environ 358:108704. https://doi.org/10.1016/j.agee.2023.108704

Prashar P, Shah S (2016) Impact of fertilizers and pesticides on soil microflora in agriculture. In: Sustainable agriculture reviews. Springer, Cham, pp 331–361

Pretty J (2008) Agricultural sustainability: concepts, principles and evidence. Philos Transac R Soc B Biol Sci 363(1491):447–465

Pretty J, Benton TG, Bharucha ZP, Dicks LV, Flora CB, Godfray HCJ, Wratten S (2018) Global assessment of agricultural system redesign for sustainable intensification. Nat Sustain 1(8): 441–446

Punia H, Tokas J, Malik A, Satpal, Rani A, Gupta P, Kumari A, Mor VS, Bhuker A, Kumar S (2020) Solar radiation and nitrogen use efficiency for sustainable agriculture. Resources use efficiency in agriculture, pp 177–212

Quandahor P, Kim L, Kim M, Lee K, Kusi F, Jeong I-h (2024) Effects of agricultural pesticides on decline in insect species and individual numbers. Environments 11(8):182. https://doi.org/10.3390/environments11080182

Quintarelli V, Radicetti E, Allevato E, Stazi SR, Haider G, Abideen Z, Bibi S, Jamal A Mancinelli R (2022) Cover crops for sustainable cropping systems: a review. Agriculture, 12(12), p.2076

Ragasruthi M, Balakrishnan N, Murugan M, Swarnakumari N, Harish S, Sharmila DJS (2024) Bacillus thuringiensis (Bt)-based biopesticide: Navigating success, challenges, and future horizons in sustainable pest control. Science of The Total Environment, p.176594.

Rahaman MT, Gurung DB, Pitol MNS (2020) Comparative study of understory between exotic monoculture plantation (Acacia sp.) and adjacent natural sal (Shorea robusta) forest. Eur J Agric Food Sci 2

Raimondo S, Mineau P, Barron MG (2007) Estimation of chemical toxicity to wildlife species using interspecies correlation models. Environ Sci Technol 41(16):5888–5894

Raina N, Zavalloni M, Viaggi D (2024) Incentive mechanisms of carbon farming contracts: a systematic mapping study. J Environ Manag 352:120126

Rajak P, Ganguly A, Adhikary S, Bhattacharya S (2023) Internet of Things and smart sensors in agriculture: Scopes and challenges. Journal of Agriculture and Food Research 14, p.100776

Rana K, Goyal N, Sharma GP (2018) Staging stewards of agro-ecosystems in the ecosystem services framework. Ecosyst Serv 33:89–101

Rapidel B, Ripoche A, Allinne C, Metay A, Deheuvels O, Lamanda N, Blazy JM, Valdés-Gómez H, Gary C (2015) Analysis of ecosystem services trade-offs to design agroecosystems with perennial crops. Agron Sustain Dev 35:1373–1390. https://doi.org/10.1007/s13593-015-0317-y

Raseduzzaman MD, Jensen ES (2017) Does intercropping enhance yield stability in arable crop production? A meta-analysis. Eur J Agron 91:25–33

Ratnadass A, Fernandes P, Avelino J, Habib R (2012) Plant species diversity for sustainable management of crop pests and diseases in agroecosystems: a review. Agron Sustain Dev 32: 273–303. https://doi.org/10.1007/s13593-011-0022-4

Reddy TY, Reddi GHS (2007) Principles of agronomy. Kalyani Publishers, New Delhi, pp 468–489

Renting H, Rossing WAH, Groot JCJ, Van der Ploeg JD, Laurent C et al (2009) Exploring multifunctional agriculture. A review of conceptual approaches and prospects for an integrative transitional framework. J Environ Manag 90:S112–S123. https://doi.org/10.1016/j.jenvman.2008.11.014

Roberts DP, Short Jr NM., Sill J, Lakshman DK, Hu X, Buser M (2021) Precision agriculture and geospatial techniques for sustainable disease control. Indian Phytopathology 74(2), pp.287–305

Russel DF (2002) Intercropping – Prospects and limitations. In Management of Agricultural, Forestry and Fisheries Enterprises. Vol. I: Multiple Cropping Systems (Encyclopedia of Life Support Systems, EOLSS). Oxford, UK: UNESCO-EOLSS Publishers

Saikanth K, Singh BV, Sachan DS, Singh B (2023) Advancing sustainable agriculture: a comprehensive review for optimizing food production and environmental conservation. International Journal of Plant & Soil Science 35(16), pp.417-425

Sanyaolu M, Sadowski A (2024) The role of precision agriculture technologies in enhancing sustainable agriculture. Sustainability 16(15), p.6668

Sarkar S, Gil JDB, Keeley J, Jansen K (2021) The use of pesticides in developing countries and their impact on health and the right to food. European Union

Schoeny A, Jumel S, Rouault F, Lemarchand E, Tivoli B (2010) Effect and underlying mechanisms of pea-cereal intercropping on the epidemic development of ascochyta blight. Eur J Plant Pathol 126:317–331

Scholes R, Huang Y, Roué M, Saw LG, Mketeni FG (2018). IPBES (2018): summary for policymakers of the assessment report on land degradation and restoration of the intergovernmental science-policy platform on biodiversity and ecosystem services

Seufert V, Ramankutty N (2017) Many shades of gray – the context-dependent performance of organic agriculture. Sci Adv 3(3):e1602638

Shackleton CM, Ruwanza S, Sanni GKS, Bennett S, De Lacy P, Modipa R, Mtati N, Sachikonye M, Thondhlana G (2016) Unpacking Pandora's box: understanding and categorising ecosystem disservices for environmental management and human wellbeing. Ecosystems 19:587–600. https://doi.org/10.1007/s10021-015-9952-z

Shanmugavel D, Rusyn I, Solorza-Feria O, Kamaraj S.-K (2023) Sustainable SMART fertilizers in agriculture systems: A review on fundamentals to in-field applications. Science of the Total Environment, 904, Article 166729. https://doi.org/10.1016/j.scitotenv.2023.166729

Shammi M, Sultana A, Hasan N, Rahman MM, Islam MS, Bodrud-Doza M, Uddin MK (2020) Pesticide exposures towards health and environmental hazard in Bangladesh: a case study on farmers' perception. J Saudi Soc Agric Sci 19(2):161–173

Sheoran S, Gupta M, Kumari S, Kumar S, Rakshit S (2022) Meta-QTL analysis and candidate genes identification for various abiotic stresses in maize (*Zea mays* L.) and their implications in breeding programs. Mol Breed 42(5):26

Shockley J, Snell W (2021, April 29) Carbon markets 101. Econ Policy Update 21(4). https://agecon.ca.uky.edu/files/carbon_markets_101.pdf

Sishodia RP, Ray RL, Singh SK (2020) Applications of remote sensing in precision agriculture: A review. Remote sensing, 12(19), p.3136

Sjerps RMA, Kooij PJF, van Loon A, van Wezel AP (2019) Occurrence of pesticides in Dutch drinking water sources. Chemosphere 235:510–518. https://doi.org/10.1016/j.chemosphere.2019.06.207

Smith VH, Tilman GD, Nekola JC (1999) Eutrophication: impacts of excess nutrient inputs on freshwater, marine, and terrestrial ecosystems. Environ Pollut 100(1–3):179–196

Srivastav AK, Das P, Srivastava AK (2024) Precision Agriculture and Environmental Monitoring. InBiotech and IoT: An Introduction Using Cloud-Driven Labs. 151–203

Suzuki C, Takenaka M, Oka N, Nagaoka K, Karasawa T (2012) A DGGE analysis shows that crop rotation systems influence the bacterial and fungal communities in soils. Soil Sci Plant Nutr 58: 288–296

Tariq A, Larsen KS, Hansen LV, Jensen LS, Bruun S (2022) Effect of nitrification inhibitor (DMPP) on nitrous oxide emissions from agricultural fields: Automated and manual measurements. Science of the Total Environment 847, p.157650

Tang XY, Yang Y, McBride MB, Tao R, Dai YN, Zhang XM (2019) Removal of chlorpyrifos in recirculating vertical flow constructed wetlands with five wetland plant species. Chemosphere 216:195–202

Taylor CA, Rising J (2021) Tipping point dynamics in global land use. Environ Res Lett 16:125012

Thamarai P, Deivayanai VC, Saravanan A, Vickram AS, Yaashikaa PR (2024) Carbon mitigation in agriculture: Pioneering technologies for a sustainable food system. Trends in Food Science & Technology, p.104477

Thobatsi TH (2009) Growth and yield of maize (*Zea mays* L.) and cowpea (*Vigna unguiculata*) in an intercropping system. University of Natural and Agricultural Sciences, Department of Plant Production and Soil Science, University of Pretoria

Tiefenbacher A, Sandén T, Haslmayr HP, Miloczki J, Wenzel W, Spiegel H (2021) Optimizing carbon sequestration in croplands: a synthesis. Agronomy 11(5):882

Tilman D, Cassman KG, Matson PA, Naylor R, Polasky S (2002) Agricultural sustainability and intensive production practices. Nature 418(6898):671–677

Tripathi S, Srivastava P, Devi RS, Bhadouria R (2020) Influence of synthetic fertilizers and pesticides on soil health and soil microbiology. In Agrochemicals detection, treatment and remediation (pp. 25-54). Butterworth-Heinemann

Udawatta RP, Kremer RJ, Anderson SH, Garrett HE (2008) Variations in soil aggregate stability and enzyme activities in a temperate agroforestry practice. Applied Soil Ecology 39:153–160

Upadhyay A, Sunil GC, Zhang Y, Koparan C, Sun X (2024) Development and evaluation of a machine vision and deep learning-based smart sprayer system for site-specific weed management in row crops: An edge computing approach. Journal of Agriculture and Food Research 18, p.101331

Uppala S, Chapala MM, Kumar KVK, Pavuluri K, Chandra KJ (2016) Climate change, carbon offsets and low carbon technologies in agriculture: a review. Int J Hortic Agric 2(1):1–8

Van Eerd LL, Chahal I, Peng Y, Awrey J.C (2023) Influence of cover crops at the four spheres: A review of ecosystem services, potential barriers, and future directions for North America. Science of the Total Environment 858,p.159990

Vansynghel J, Ocampo-Ariza C, Maas B, Martin EA, Thomas E, Hanf-Dressler T et al (2022) Quantifying services and disservices provided by insects and vertebrates in cacao agroforestry landscapes. Proc R Soc B 289(1982):20221309

Vinod Chandra SS, Anand H, Ghassan FA. Precision farming for sustainability: An agricultural intelligence model. Computers and Electronics in Agriculture 226:109386

Volsi B, Higashi GE, Bordin I, Telles TS (2022) The diversification of species in crop rotation increases the profitability of grain production systems. Sci Rep 12(1):19849. https://doi.org/10.1038/s41598-022-23718-4

Wang H, Zhang X, Yu X, Hou H, Fang Y, Ma Y, Zhang G (2021) Maize-potato rotation maintains soil water balance and improves productivity. Agron J 113:645–656

Wang J, Wang Y, Li G, Qi Z (2024) Integration of remote sensing and machine learning for precision agriculture: a comprehensive perspective on applications. Agronomy 14(9),p.1975

Wasaya A, Ahmad R, Hassan FU, Ansar M, Manaf A, Sher A (2013) Enhancing crop productivity through wheat (Triticum *aestivum L.*)-fenugreek intercropping system. J Anim Plant Sci 23(1): 210–215

Welbank P (1963) A comparison of competitive effects of some common weed species. Ann Appl Biol 51:107

West TO, Post WM (2002) Soil organic carbon sequestration rates by tillage and crop rotation. Soil Sci Soc Am J 66:1930–1946

Weston LA, Duke SO (2003) Weed and crop allelopathy. Critical Reviews in Plant Sciences, 22(3–4):367–389. https://doi.org/10.1080/713610861

Williams-Guillén K, Perfecto I, Vandermeer J (2008) Bats limit insects in a tropical agroforestry system. Science 320:70

Winans KS, Tardif A-S, Lteif AE, Whalen JK (2015) Carbon sequestration potential and cost-benefit analysis of hybrid poplar, grain corn, and hay cultivation in southern Quebec, Canada. Agrofor Syst 89:421–433

World Atlas (2018) Top pesticide-consuming countries of the world. https://www.worldatlas.com/articles/top-pesticide-consuming-countries-of-the-world.html

Xing Y, Wang X (2024) Impact of agricultural activities on climate change: a review of greenhouse gas emission patterns in field crop systems. Plants, 13(16), p.2285

Yadav IC, Devi NL, Syed JH, Cheng Z, Li J, Zhang G, Jones KC (2015) Current status of persistent organic pesticides residues in air, water, and soil, and their possible effect on neighboring countries: a comprehensive review of India. Sci Total Environ 511:123–137

Yahya MS, Syafiq M, Ashton-Butt A, Ghazali A, Asmah S, Azhar B (2017) Switching from monoculture to polyculture farming benefits birds in oil palm production landscapes: evidence from mist netting data. Ecol Evol 7:6314–6325

Yang M, Chen L, Wang J, Msigwa G, Osman AI, Fawzy S, Rooney DW and Yap P (2023) Circular economy strategies for combating climate change and other environmental issues. Environmental Chemistry Letters 21(1),pp.55-80

Yang H, Su Y, Wang L, Whalen JK, Pu T, Wang X, Wu Y (2025) Strip intercropped maize with more light interception during post-silking promotes photosynthesized carbon sequestration in the soil. Agric Ecosyst Environ 378:109301

Yirga SA (2019) Agroforestry for sustainable agriculture and climate change: a review. Int J Environ Sci Nat Resour 19(5):127–137

Yousefi M, Marja R, Barmettler E et al (2024) The effectiveness of intercropping and agri-environmental schemes on ecosystem service of biological pest control: A meta-analysis. Agronomy for Sustainable Development 44:15. https://doi.org/10.1007/s13593-024-00947-7

Yu T, Mahe L, Li Y, Wei X, Deng X, Zhang D (2022) Benefits of crop rotation on climate resilience and its prospects in China. Agronomy 12(2):436

Zabala JA, Martínez-Paz JM, Alcon F (2021) Comprehensive approach for agroecosystem services and disservices valuation. Sci Tot Environ 768:1–14

Zardosht Z, Khosravani F, Rezaei S, Ghaderi S, Hassani G (2023) The impact of two insecticides on the pollutant cycle and quality of surface and groundwater resources in the irrigated lands of Yasuj, Iran. Heliyon 9(6):e16872. https://doi.org/10.1016/j.heliyon.2023.e16872

Zeng J, Wang X, Li Y, Xing, L., Li X, Zhang Q, Liu H, Liu Y, Liu W, Feng Y, Wang X (2025) Optimizing fertilization rate can maintain the greenhouse gas mitigation effect of no-tillage: A global-scale meta-analysis. Field Crops Research, 322, p.109720

Zhang N, Zhou Z (2018) Positive and negative agro-ecosystem services evaluation in urbanizing area: a case study of Chang'an District. Arid Land Geo 41(2):410–418

Zhang W, Ricketts TH, Kremen C, Carney K, Swinton SM (2007) Ecosystem services and dis-services to agriculture. Ecol Econ 64:253–260

Zhang WJ, Jiang FB, Ou JF (2011) Global pesticide consumption and pollution: with China as a focus. Proc Int Acad Ecol Environ Sci 1(2):125–144

Zhang X, Brown TN, Ansari A, Yeun B, Kitaoka K, Kondo A (2013) Effect of wind on the chemical uptake kinetics of a passive air sampler. Environ Sci Technol 47:7868–7875

Zhao J, Yang Y, Zhang K, Jeong J, Zeng Z, Zang H (2020) Does crop rotation yield more in China? A meta-analysis. Field Crop Res 245:107659

Zhu W, Schmehl DR, Mullin CA, Frazier JL (2014) Four common pesticides, their mixtures, and a formulation solvent in the hive environment have high oral toxicity to honey bee larvae. PLoS One 9(1):e77547

Chapter 9
Farm Waste Management and Recycling for Negative Carbon Sink in Agriculture

Md Basit Raza, Saheed Garnaik, Amit Kumar Dash, Arkaprava Roy, Priyanka Meena, Soura Shuvra Gupta, Tanmaya Kumar Bhoi, and Debasis Golui

Abstract Agriculture plays a dual role in global climate change, contributing to greenhouse gas (GHG) emissions while also offering the potential for carbon sequestration. The management of agricultural waste is a critical component in the global effort to enhance environmental quality and combat climate change. While essential for food security, agricultural activities significantly contribute to GHG emissions and environmental degradation, primarily due to improper waste management practices. This chapter explores various strategies for recycling farm wastes, such as composting (vermicomposting and anaerobic digestion), integration of agroforestry, biochar production and application to soil for soil carbon stabilization, and improved agronomic practices to enhance soil carbon sequestration and reduce the environmental footprint of agriculture. Various best management practices that have the potential to enhance soil carbon storage have been emphasized. Adopting these methods not only improves soil health and increases soil organic carbon (SOC) but also offers a pathway to transform agriculture into a net-negative carbon sink. The integration of farm waste management into the circular bioeconomy has been put forth as a strategy to mitigate environmental pollution and enhance soil fertility, while showcasing as approaches by which farmers could earn carbon credits to monetize their efforts. The future of negative carbon sinks in

M. B. Raza
ICAR-Directorate of Floricultural Research, Pune, Maharashtra, India

S. Garnaik
ICAR-National Academy of Agricultural Research Management, Hyderabad, Telangana, India

A. K. Dash
ICAR-National Institute of Seed Science and Technology, Mau, Uttar Pradesh, India

A. Roy · P. Meena
ICAR-National Institute of Biotic Stress Management, Baronda, Raipur, Chhattisgarh, India

S. S. Gupta · D. Golui (✉)
ICAR-Indian Agricultural Research Institute, New Delhi, India

T. K. Bhoi
ICFRE-Arid Forest Research Institute, Jodhpur, Rajasthan, India

R. S. Meena et al. (eds.), *Carbon-Negative Agriculture*, Sustainability Sciences in Asia and Africa, https://doi.org/10.1007/978-981-95-1483-0_9

Indian agriculture is envisioned through the adoption of sustainable practices, technological advancements, and supportive policies. This holistic approach is essential for transforming agricultural systems into effective carbon sinks, thereby contributing to global climate change mitigation.

Keywords Carbon sequestration · Circular bioeconomy · Climate change · Negative carbon sink · Sustainable agriculture

Abbreviations

CO_2	Carbondioxide
GHG	Greenhouse gas
GWP	Global warming potential
INR	Indian Rupees
NP	Nonylphenols
PCA	Polychlorinated alkane
PCN	Polychlorinated naphthalene
Pg	Petagram
POP	Persistent organic pollutant
SOC	Soil organic carbon
SOM	Soil organic matter
US$	US Dollar

9.1 Introduction

Farm waste management is a formidable challenge worldwide. Since environmental quality enhancement is given greater priority these days, more concerted efforts are being put forth towards better waste management in agriculture. While regional studies have demonstrated the substantial impact of agricultural activities and the waste generated therein on environmental health, the effects at the global scale remain largely unknown. However, agricultural activities are regarded as crucial for achieving global food and nutritional security, so it is imperative to intensify efforts to normalize their environmental footprint and mitigate the associated pollution effectively. The environmental quality problems due to agricultural operations are more dependent on the on-farm waste management techniques adopted and the production practices followed (Loehr 2012). The growing global population has made it necessary to adopt improved agricultural practices and enhance crop and livestock production, which further contributes to agricultural waste. Recent reports show that agricultural wastes produced globally quantify to around 1 billion tons annually, significantly contributing to greenhouse gas (GHG) emissions (Karić et al. 2022; Peng et al. 2023; Meena et al. 2025).

India is among the top producers of agricultural waste, standing next to China (Singh and Sidhu 2014), generating around 350–990 million ton (Mt) of waste annually (Koul et al. 2022). Improper disposal of farm wastes poses a threat to human health and environmental quality (Kaab et al. 2019; Meena et al. 2024). The major categories of farm wastes that could be of potential concern when disposed improperly include crop residues, livestock wastes, poultry wastes, and aquaculture wastes (Duque-Acevedo et al. 2020). A major proportion of farm waste is organic (80%) (Obi et al. 2016), thus bearing a significant potential as an input for strengthening the bioeconomy. It is common knowledge that there are losses at every stage of the food system, which could in turn be used as input in farm operations to improve/sustain productivity. A study conducted by Alexander (2017) reports that as high as 44% of harvested crops (on a dry weight basis) can get lost (accounting for 36.9% of energy) before human consumption, with the highest loss rate being recorded in livestock production. Such losses are greater when proper waste management techniques are absent in both on-farm and off-farm, resulting in a detrimental effect on the environment due to greater GHG emissions. Almost all the organic wastes generated on-farm can be of great value economically or in terms of enhancers of soil health, turning into valuable resources contributing towards sustainable agriculture. Thus, reducing, reusing, and recycling (3Rs) of farm wastes, either individually or collectively, is a fertile avenue for depressurizing the dependence on natural resources, biodiversity, and soil quality (Koul et al. 2022; Meena et al. 2024a).

Agriculture, being one of the highest producers of biomass, can act as a huge input for enhancing soil carbon (C) stock, if the wastes generated at different steps are managed strategically. Farm waste management is a cost-effective, environment-friendly, and practical option that could be largely exploited to reduce GHG emissions from the agricultural sector. Since agriculture is both a source and sink of C, it is imperative to devise a strategy to entrap more C from the atmosphere than it emits, thus making soil a net C-negative system. Various methods, including composting (including vermicomposting), biochar production, anaerobic digestion of organic farm wastes (biogas production), agroforestry, and adoption of improved agronomic practices, can potentially help to mitigate GHG emissions and promote C sequestration, enabling farmers to monetize their efforts through C credits. On proper adoption of management methods, it is envisaged that as high as 50%–66% of historically lost C could be replenished over time (Lal 2004). It is estimated that under recommended management methods, soils offer a sequestration potential of 0.6–2.0 Pg C year^{-1} (Lal 2000). Reports from long-term studies have shown that C sequestration could be increased to the tune of 70–551 kg C ha^{-1} when sole organic manure is applied, when compared to sole inorganic fertilizer use (Mandal et al. 2008). Anaerobic digestion of organic farm wastes (including livestock wastes) for the generation of biogas can, on one hand, reduce the demand for fossil fuels, while on the other hand, it can help to capture uncontrolled methane (CH_4)emissions. A case study by Bist et al. (2012) implies the potential of biogas plants in utilizing farm wastes to generate C credits worth 134 lakh Indian Rupees (INR). Further, a case study from Taiwan on biochar production using Poplar tree wood reports the greater value of biochar in soil (estimated at US$ 64.28 t^{-1}) than its combustion value (US$ 60.34 t^{-1}, when its energy value is compared relative to coal), which was mainly because of its other shadow benefits on soil health and crop yield (Kung et al. 2013; Meena et al. 2024b).

9.2 Status of Farm Waste Generation

Farm wastes can be categorized into crop residues (like leaf litter, seed pods, stalks, stems, straws, husks, and weeds), horticultural wastes, animal (livestock) wastes (including urine, dung, wash water, residual milk, and waste feed), aquaculture wastes, and hazardous wastes (includes plastics, agro-chemicals, packaging materials). Broadly, farm wastes can also be categorized into biodegradable and non-biodegradable wastes depending on their decomposability. The biodegradable farm wastes primarily consist of crop stalks, leaves, roots, fruit peels, and seed/nut shells, which are often discarded or burned. Crop stalks, leaves, and roots are collectively called crop residues, which play a significant role in C sequestration (Dash et al. 2021, 2023). Crop residues have considerable potential to produce C-neutral energy. They can also serve as a C sink, helping to maintain soil quality and aiding in nutrient management and conservation (Lal 1997; Amit 2020; Kumar et al. 2023b). Annual crop residue production is currently projected to be 2802 Mt globally. A report released by the Ministry of New and Renewable Energy, Government of India (GoI), estimates that India generates approximately 500 Mt of crop residue annually, with 92 Mt being burned (Tables 9.1 and 9.2) (MNRE 2009). Cereals, fibers, oilseeds, pulses, and sugarcane are the top contributors to crop residue generation. Among cereal crops, rice, wheat, maize, and millet together account for 70% of the crop residue, followed by fiber crops (13%), oilseeds (6%), pulses (3%), sugarcane (2%), and others (6%) (MNRE 2009). Burning paddy straw

Table 9.1 Generation of crop residues worldwide

Country	Crop residue generated (Mt $year^{-1}$)
India	501
Indonesia	55
Bangladesh	72
Myanmar	19
Afghanistan	7.7
Bhutan	0.4
Pakistan	123
Sri Lanka	5
Nepal	23
USA	488
China	1039
Total	3420

Sen et al. (2024)

Table 9.2 Generation of crop residues in India

States	Crop residue generated (Mt $year^{-1}$)
Punjab	51
Uttar Pradesh	60
Maharashtra	46
Andhra Pradesh	44
West Bengal	36
Madhya Pradesh	33
Karnataka	34
Gujarat	29
Rajasthan	29
Haryana	28
Other states	111
Total	501

MNRE (2009)

Table 9.3 Annual production of excreta by farm animals

Animals	Excreta production (Mt)
Cattle	210
Buffaloes	147
Sheep and goats	30
Pigs	2
Poultry	1
Others	171

Annual Report, Ministry of Agriculture, Govt. of India, 2012

produced from one acre of land (equivalent to 2.5–3.0 metric tons) results in the loss of nutrients equivalent to 32 kg of urea, 5.5 kg of di-ammonium phosphate, and 51 kg of potash, which are already present in the residues (Mandal et al. 2004).

In central and western India, approximately 61.6 Mt of vegetable and fruit waste is produced annually (Annual Report 2014). The highest amount of vegetable and fruit waste is generated in Maharashtra, followed by Gujarat. Tomato (*Solanum lycopersicum*), potato (*Solanum tuberosum*), onion (*Allium cepa*), brinjal (*Solanum melongena*), okra (*Abelmoschus esculentus*), cabbage (*Brassica oleracea* var. *capitata*), and cauliflower (*Brassica oleracea* var. *botrytis*) are the main contributors of vegetable waste, whereas mango (*Mangifera indica*), banana (*Musa acuminata*), citrus (*Citrus* sps), guava (*Psidium guajava*), papaya (*Carica papaya*), sapota (*Manilkara zapota*), grapes (*Vitis vinifera*), and pomegranate (*Punica granatum*) constitute the majority of fruit plantation wastes. Another part of farm waste is animal excreta. It is estimated that India generates approximately 471 Mt of animal excreta from various livestock (Annual Report 2012) (Table 9.3). Aquaculture is also one of the rapidly growing fields of agriculture. Notably, global aquaculture production grew sixfold within 26 years after 1990 (FAO 2018). With the growing scale of production, the wastes generated from this field have also been increasing. Feed waste shares a major share of the total aquaculture waste, which is followed by fecal waste from fish. When left unmanaged, these wastes could cause water pollution. Therefore, aquaculture has also gained attention from researchers to prevent emerging soil and water pollution.

9.3 Farm Waste Recycling and Negative Carbon Sink

To achieve a negative C sink in agriculture, farm wastes such as crop residues, horticultural wastes, and animal wastes should be recycled in an integrated manner (Fig. 9.1). Some of the strategies to recycle these farm wastes include composting (aerobic and anaerobic), vermicomposting, anaerobic digestion, and biochar production.

9.3.1 Agronomic Practices to Enhance the Carbon Sink

Agronomic waste, including crop residues, plant discards, compost, and organic mulches, plays a crucial role in transforming agricultural systems into carbon sinks. When these organic materials are returned to the soil instead of being burned or discarded, they become valuable inputs for building soil organic carbon (SOC). Practices such as composting, vermicomposting, and residue incorporation enrich the soil with stable organic matter, enhancing soil fertility while also increasing its capacity to store C over the long term. Carbon sequestration through agronomic waste management provides a dual benefit, that is, it mitigates GHG emissions and creates opportunities for farmers to participate in C credit schemes. For instance,

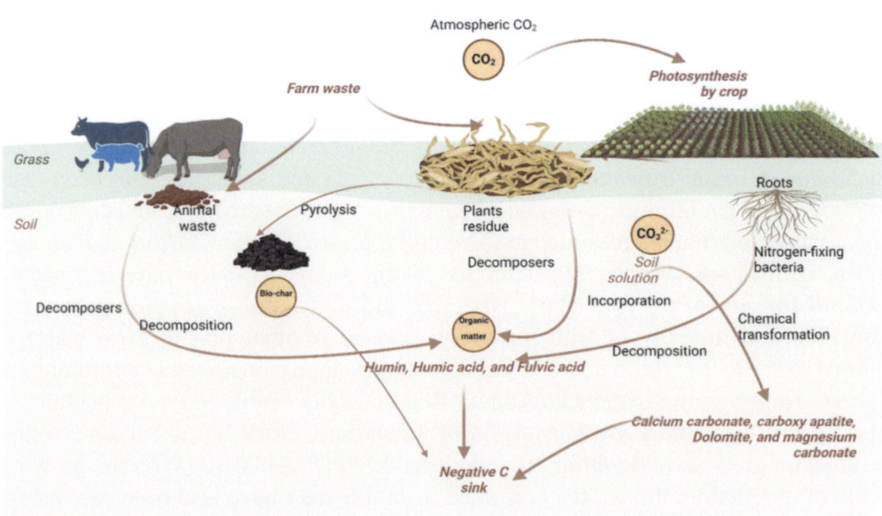

Fig. 9.1 Farm waste recycling and negative carbon sink

incorporating crop residues into the soil promotes microbial activity, humus formation, and aggregate stability, all of which contribute to SOC accumulation (Meena et al. 2023). Research shows that straw incorporation, compared to removal, increases SOC sequestration by an average of 0.35 Mg C ha^{-1} $year^{-1}$ at 0–20 cm depth (Han et al. 2018). These measurable gains in soil C can be verified using established C accounting frameworks, enabling farmers to claim credits in voluntary or compliance-based C markets.

The use of enriched organic amendments such as compost mixed with nutrients or beneficial microbes accelerates the decomposition of wastes while enhancing soil C and nutrient value. This not only improves soil productivity but also leads to more efficient C stabilization. Similarly, biochar, produced via pyrolysis of crop residues, is a particularly effective strategy for long-term C storage. Unlike decomposable organic matter, biochar is highly stable in soil and can persist for decades to centuries, acting as a durable C sink. Its application improves soil health, nutrient retention, and microbial activity, making it a win-win solution for residue management and climate action (Song et al. 2022).

Organic mulching and conservation tillage are additional practices that reduce soil disturbance and encourage residue retention on the soil surface, especially in tropical systems where organic matter decomposition is rapid. These techniques help maintain and increase SOC stocks, reduce erosion, and enhance soil water retention. While intensive farming systems may struggle to manage large volumes of residue due to short cropping cycles, integrating residue recycling with biochar production or cover crop cultivation provides alternative C-positive solutions. Importantly, the transition from open residue burning, a major source of C emissions to sustainable practices like composting, biochar application, and organic mulching significantly lowers the C footprint of agriculture. Burning not only emits CO_2, CH_4, and nitrous oxide (N_2O), but also degrades soil structure and microbial life, thereby reducing SOC sequestration potential (Panneerselvam et al. 2024; Meena and Pradhan 2023). In contrast, sustainable management practices turn agronomic wastes into a climate asset.

As these practices build up soil C stock, the additional C can be monitored, reported, and verified under recognized C credit frameworks. This allows farmers and land managers to monetize their climate-positive actions through C trading mechanisms. By adopting improved waste management practices, agriculture can shift from being a net emitter to a net sink of C, delivering both environmental and economic benefits. With growing interest in climate-smart agriculture, agronomic waste management offers a scalable pathway for C credit generation and long-term sustainability (Singh and Chaturvedi 2023; Paul et al. 2023; Pradhan and Meena 2023).

9.3.2 Biochar

Biochar is a C-rich material produced by the pyrolysis of biomass and transforming 10%–50% of the initial biomass C into biochar C, which remains in the soil for extended periods, ranging from hundreds to thousands of years (Lehmann and Joseph 2024). The thermochemical process converts biomass into biochar, resulting in a more stable biological and chemical form of sequestered C compared to its original state (Bhoi et al. 2024; Dash et al. 2024; Pradhan and Meena 2023a). Its extended half-life makes it a promising residue management alternative for mitigating climate change by sequestering C in soil and curbing emissions of GHGs (Fig. 9.2). The potential for global C-sequestration through the utilization of biochar production from sources such as forest residues, mill residues, field crop residues, and urban waste has been estimated to reach approximately 0.16 Gt year^{-1} (Lehmann et al. 2006). By slowing the release of CO_2 from soil, it effectively traps C in a long-lasting soil C pool, thus preventing its return to the atmosphere. Zimmerman (2010) suggests that applying biochar (produced under anaerobic conditions) to soil is a viable method for boosting SOC. This process sequesters about 50% of the initial biomass C, significantly more than the roughly 3% retained after burning and less than 10%–20% remaining after 5–10 years of biological decomposition (Lehmann et al. 2006). In a study conducted by Creamer et al. (2014), biochar produced from sugarcane bagasse and hickory wood was evaluated for CO_2 adsorption capabilities. They found that biochar derived from bagasse at 600 °C exhibited the highest CO_2 adsorption capacity, reaching 73.55 mg g^{-1} at 25 °C. At the global level, agricultural wastes hold the capacity to generate approximately 373 Mt of biochar each year, offering the potential to sequester around 0.55 Pg of CO_2 annually (Ganguly et al. 2022). Biochar prepared from feedstocks with abundant lignocellulosic material, such as wood biomass processed at higher

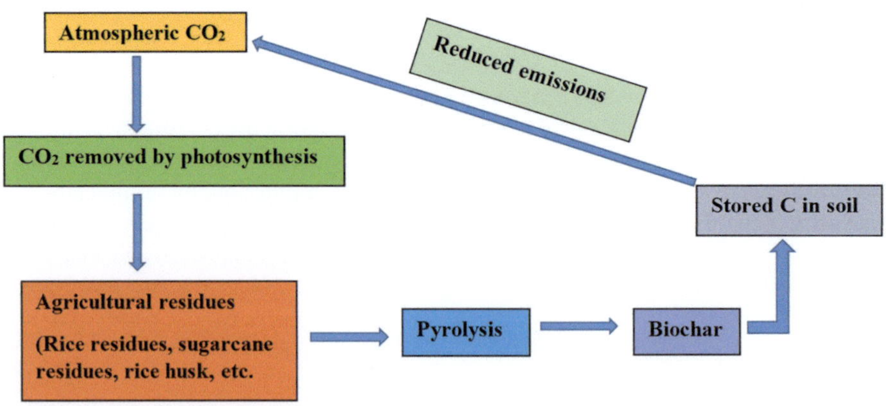

Fig. 9.2 Biochar for negative C sink in agriculture

pyrolysis temperatures, exhibit greater C-sequestration potential compared to those prepared from lower lignocellulosic sources like straw biomass or manure (Chen and Pilla 2022).

The use of biochar for C sequestration represents not only a technological innovation but also an economically viable opportunity within the expanding C sink economy (Yadav and Ramakrishna 2023). Voluntary C markets allow entities to offset their emissions on a non-mandatory basis (Honegger et al. 2021), enabling them to strengthen C neutrality claims, lower net emissions, or contribute to national climate goals. In this emerging market, obtaining certifications for biochar production and its C removal capabilities has become essential for producers aiming to access these C credit systems (Yadav and Ramakrishna 2023; Pradhan et al. 2023). Presently, only a limited number of market platforms support C credit generation through biochar. A significant milestone was achieved when Pacific Biochar became the first US company to earn C credits for biochar-based C removal (Sage et al. 2023; Kumar et al. 2025). A notable example is the Cool Terra C project in the United States, where the use of biochar demonstrated measurable benefits for soil health, crop yield, and C sequestration. Through the adoption of enhanced agricultural practices involving biochar, the project effectively quantified the C sequestered in soils, along with associated reductions in greenhouse gas emissions and improvements in SOC levels (Graf et al. 2023).

However, the effectiveness of biochar is contingent on factors like its physical and chemical composition, source material, pyrolysis temperature, and soil type. It can reduce GHG emissions by improving soil water holding capacity, which can reduce the need for irrigation and can also reduce CH_4 emissions from livestock manure and N_2O emissions from fertilizer. Interestingly, biochar application to soil can also return a significant amount of the nutrients that are removed from soil when biomass is harvested. Biochar, when applied to soil, helps to improve the soil's physical properties, thus improving the overall soil health status.

9.3.3 Composting

Traditionally, farm wastes have been utilized to produce composts like farm yard manure (FYM), straw manure, and other compost types. Composting waste biomass stabilizes organic matter, improves soil moisture content, enhances nutrient levels, and minimizes GHG emissions (Badawi 2023; Darjee et al. 2023; Dash et al. 2025). A C-negative farming system strictly avoids the use of chemical fertilizers and any agricultural practices that elevate the emissions of CO_2, CH_4, N_2O, and other GHGs. As a result, compost, with its high nutrient content, serves as an effective fertilizer in such systems, reducing the reliance on fossil-based fertilizers (Reimer et al. 2023; Raza et al. 2024). Composting rice crop residues from one hectare of land yields approximately 3 tons of nutrient-rich manure, comparable in quality to FYM. Field experiments demonstrated that applying 200 Mt ha^{-1} year^{-1} of this manure significantly increases organic matter levels compared to control (Blair et al. 2006).

According to Powlson et al. (2011), long-term studies spanning over 49 years indicated annual rates of SOC sequestration ranging from 10 to 22 kg C ha^{-1} $year^{-1}$ t^{-1} of dry solids with manure applications. Another study compared the effects of FYM and nitrogen, phosphorus, and potassium (NPK) fertilizers on a rice-wheat cropping system and found a notable rise in C sequestration in plots treated with FYM compared to those treated with only NPK fertilizers (Naresh et al. 2017). Additionally, applying rice straw compost at a rate of 2 t ha^{-1} alongside inorganic fertilizer significantly boosted C and N concentrations in sandy loam soil (Sodhi et al. 2009; Punia et al. 2020). Fortuna et al. (2003) reported that substituting compost for nitrogen fertilizer increased particulate organic matter carbon (POM-C) by 45% and SOC by 16%. When farm waste like crop residues, leaves, and manure is left untreated, it decomposes quickly, releasing CO_2 and CH_4 into the atmosphere (Obame et al. 2024; Dash et al. 2025). During composting, much of the easily degradable C is lost as CO_2, but the remaining C is transformed into stable, complex compounds like humic substances. These stable forms of C resist further decomposition once added to soil, thus locking C into the soil for long periods (Obame et al. 2024; Pradhan et al. 2025; Dash et al. 2025). Carbon sequestration by compost is measured scientifically, verified independently, and rewarded financially through C credits (Desai 2021). By composting farm waste and enhancing SOC under the Kenya Agricultural Carbon Project, smallholder farmers have generated over 150,000 C credits (Atela 2012). California's Healthy Soils Program provides financial incentives for farmers to apply compost to fields to build soil C and reduce emissions, partially funded through C credit mechanisms (Desai 2021). Therefore, composting offers a sustainable method to recycle substantial amounts of cow dung, urine, and other organic farm wastes, effectively utilizing them to support crop growth while serving as a C sink through the production of nutrient-rich organic inputs for soil health enhancement.

9.3.4 Vermicomposting

Vermicomposting is another environmental friendly method involving the decomposition of organic waste into nutrient-rich vermicast, facilitated by earthworms (Huang and Xia 2018). Earthworms contribute vermicast to the soil, improving its physicochemical and biological properties facilitating better plant growth and greater soil C accumulation (Urmi et al. 2022). An increase in organic C in the soil supports sustainable agriculture and acts as an effective sink for atmospheric CO_2 (Gnanavelrajah et al. 2008; Tripathy et al. 2023; Bamel et al. 2025; Dash et al., 2025). The extent of C sequestration largely relies on the size of the active C pool and its role in forming stabilized and mineralized C. Earthworms influence the C cycle in several ways, depending on SOC content. While they do not significantly contribute to CO_2 emissions, they enhance net C sequestration. This is because they stabilize large amounts of C, known as earthworm-activated C (Zhang et al. 2013). Farooqi et al. (2024) reported that vermicompost-amended soil showed notable improvement in SOC sequestration compared to soils amended with biochar and compost. The application of vermicompost enhanced soil aggregation by promoting the formation of microaggregates and macroaggregates, which were subsequently cemented or coagulated together (Farooqi et al. 2023). This process also enhanced aggregate-associated C, which acted as a sink for atmospheric CO_2 (Dash et al. 2023; Yadav et al. 2024; Sharma et al. 2024). Vermicomposting plays a critical role in C sequestration and the mitigation of GHG emissions, particularly CH_4 and CO_2, positioning it as a viable strategy for earning C credits under established climate mitigation frameworks. According to the IPCC (2006), diverting one ton of organic waste from open dumping to vermicomposting can prevent the emission of approximately 0.25–1 ton of CO_2-equivalent, depending on the type of waste involved. According to Dey Chowdhury et al. (2023), the vermicomposting of maize residues blended with pig manure, cow dung, and biochar resulted in relatively low GH Gemissions, ranging between 0.003 and 0.081 g CO_2-eq kg^{-1} for CO_2, 0 and 0.17 g CO_2-eq kg^{-1} for CH_4, and 130.40 and 189.10 g CO_2-eq kg^{-1} for N_2O. Similarly, the vermicomposting of tomato stems and cow dung ensured 2.28 and 5.76 g CO_2-eq kg^{-1} CO_2 emissions of CH_4 and N_2O, respectively. The superior GHG reduction potential of vermicomposting compared to traditional composting can be attributed to two key factors. First, earthworm activity enhances aeration within the composting substrate, maintaining aerobic conditions that significantly reduce CH_4 generation. Second, vermicomposting occurs at ambient temperatures, in contrast to thermophilic (high-temperature) composting. The GHG emissions, particularly CH_4 and CO_2, increase proportionally with temperature, making vermicomposting a more climate-friendly alternative (Berdier et al. 2025).

9.3.5 Anaerobic Digestion of Farm Waste

Anaerobic digestion is a process involving biological anaerobic decomposition of organic matter, producing biogas (primarily CH_4 and CO_2) and digestate. This process is commonly used for managing farm wastes such as animal excreta, crop residues, and food waste (Feng and Rosa 2024). Farm wastes are placed in an anaerobic digester, where microbes break down the organic matter in a series of stages (hydrolysis, acidogenesis, acetogenesis, and methanogenesis). Biogas containing CH_4 and CO_2 is collected and used for electricity or heating purposes (Ankathi et al. 2024). This biogas is a renewable energy source that represents an alternative to fossil energies, helping in negative CO_2 emission (Möller 2015). The remaining material is called digestate, which when applied to soils, can improve soil health and increase its organic C content, contributing to long-term soil carbon sequestration. Digestate can be further processed into biochar, a stable form of carbon, which can also greatly enhance C sequestration and soil fertility (Breunig et al. 2019). Recycling these nutrient-rich digestates in agricultural systems reduces mineral fertilizer applications, thus ultimately reducing CO_2 emissions. A study even reports an increase in total organic C content in soil when digestate at a 5% dose is applied to soil, due to enhancement in recalcitrance of entrapped SOC (Cardelli et al. 2018). Incorporating C capture and storage into anaerobic digestion systems is increasingly recognized as a cost-effective strategy for achieving negative GHG emissions. Multiple studies have demonstrated that energy production via anaerobic digestion emits significantly fewer GHG gases compared to conventional fossil fuel-based systems. The global warming potential (GWP) associated with such systems has been reported to reach as low as −167.52 kg CO_2-equivalent, highlighting their strong capacity for C offsetting (Zhang et al. 2025). Additionally, the economic value of this C reduction has been estimated at approximately 13.22 EUR per megawatt-hour of electricity produced, based on a C pricing factor of 0.08 EUR per kg of CO_2 avoided (Zhang et al. 2025).

9.4 Negative Carbon Sink Strategy in Agriculture

Soils have been significantly contributing to atmospheric CO_2. Agricultural soils have lost 50%–75% of their soil C pool, equivalent to 30–60 Mg C $ha^{-1,}$ due to practices like ploughing and biomass burning. To enhance SOC, recommended practices include no-till farming, crop residue mulch, and wetland restoration. The SOC sequestration rate is 300–500 Kg C ha^{-1} $year^{-1}$ under intensive agriculture and 0.8–1.0 Mg ha^{-1} $year^{-1}$ through wetland restoration. Soil C sequestration mechanisms include aggregation, biosolid humification, and bicarbonate leaching. The global potential of soil C sequestration is 0.6–1.2 Pg C $year^{-1}$, potentially offsetting 15% of fossil fuel emissions (Busari et al. 2015). The following plans are presented hereunder, which may comprehensively be put forth to achieve the goal of making the agriculture sector a net negative C sink in the future.

9.4.1 Soil Health Improvement

Conservation tillage, coupled with practices like soil cover and crop diversity, has proven effective for sustainable food production and environmental preservation (Leharwan et al. 2023). Methods such as no-tillage, reduced tillage, stubble mulch, and ridge tillage help maintain soil structure and enhance C storage. These practices boost organic matter retention, reduce erosion, and improve soil biological activity (McClelland et al. 2021). Enhancing the quantity and quality of plant biomass over time and space can also help boost the ability of agroecosystems to capture and store atmospheric CO_2 in the soil. Cover cropping is essential for increasing net primary productivity (NPP) and providing high-quality plant residues that contribute to soil organic matter (SOM) (Jarecki and Lal 2003). However, the changes in soil C stock is affected by cover crop management and local environmental conditions.

Reducing GHG emissions from agriculture involves increasing and protecting SOM. Agricultural soils can become significant sinks of atmospheric C by boosting SOM levels. Natural ecosystems lose much of their C pool when put to agricultural use. Implementing good management practices can enhance the SOC pool, utilizing the large C sink capacity in agricultural soils. Cover crops planted either continuously or during autumn season and later terminated, could increase total soil C stocks by 20–30% compared to other cover crop growing windows (McClelland et al. 2021). High annual cover crop biomass production (>7 Mg ha^{-1} $year^{-1}$) leads to 30% higher soil C stocks (McClelland et al. 2021). Enhancing NPP through better synchronization of cover crop growing time and climate will improve CO_2 drawdown in agroecosystems. Integrating growing windows, climate, and soil factors in decision-support tools is crucial for accurately quantifying soil C stock changes under cover crops, particularly as terrestrial soil C markets expand (Handa et al. 2020).

9.4.2 Agroforestry Inclusion

Agroforestry has deep roots in the Indian subcontinent, where socio-religious traditions emphasize the importance of trees. Integrated into crop and livestock systems, agroforestry adapts to local agroclimatic conditions to optimize positive interactions, leading to more productive, sustainable, and diverse land use. Research in India dates back over a century, with early studies on tree-crop interactions and silvopastoralism. Diagnostic surveys way back in the 1980s revealed diverse agroforestry practices across the country, including multifunctional fallows, home gardens, mixed-species systems, alley cropping, and windbreaks. Recently, agroforestry has gained popularity for enhancing livelihood and environmental security amidst extreme weather events and global warming (Kumar et al. 2022a). It plays a vital role in providing products and services to both rural and urban populations. Approximately 1.2 billion people worldwide practice agroforestry, covering an estimated 823 million hectares. Remote sensing data from 2010 showed that 43% of global agricultural land had at least 10% tree cover, showing an increase of 2% compared to that 10 years before (Lal 2007). The estimated C sequestration potential of agroforestry systems worldwide ranges from 1.1 to 2.2 Pg C per year for a half-century period (Dixon 1995). Generally, increased C sequestration rates can be expected upon inclusion of agroforestry systems in fertile, humid sites than those on arid, semiarid, and degraded sites. Additionally, tropical agroforestry systems exhibit higher vegetation C sequestration rates than temperate agroforestry systems. Agroforestry systems are more efficient than mono-culturing in facilitating higher C inputs to the soil (Lorenz and Lal 2014).

9.4.3 Carbon Monitoring and Verification

Around half of the habitable land is presently occupied by agricultural activities. With the increasing global population, agricultural expansion is required, which is possible if natural forest areas are cut down, which will lead to the depletion of major C sinks on a global scale. Moreover, with widespread agricultural activities, more soils (storehouses of C) will be disrupted, which will release CO_2 into the atmosphere. To avoid these, farmers need to adopt preventive measures, which may include minimal soil disturbance, maximum on-farm residue retention, and innovative technologies for entrapping and enhancing SOC, and livestock management. Now adoption of such activities greatly varies from the practice presently followed by farmers worldwide. This being said, the farmers need to be incentivized to facilitate a large-scale transition into farm activities promoting C sequestration. In the voluntary market, agriculture C credit projects have shown exponential growth from 3 lakh tons transacted to 3.4 million (Donofrio et al. 2021). The C market is a way of earning revenue for farmers. For this, project developers tie up with the farmers, spread awareness about the benefits of the C market, and help them adopt new management practices that could help improve on-farm soil C sequestration and decarbonize the food chain. Carbon offset certifiers are non-profit organizations for reporting and tracking C credit project information, including credit generated, ownership, sale, and retirement (Oldfield et al. 2021). Emerging certifiers focused on agriculture include the Regen Network and the soil C programs from the Australian and Canadian governments (Liu 2022). Private soil C certificates aim to offset these costs by certifying SOC increases, which farmers can sell as voluntary emission offsets. The process examines SOC enrichment processes, measurement methods, and management strategies. It also considers governance challenges like ensuring permanence, additionality, and preventing leakage. Despite benefits, uncertainties remain about SOC permanence and governance effectiveness. Future research should focus on developing robust SOC metrics and monitoring standards to clarify their impact on atmospheric GHG concentrations and climate stability (Paul et al. 2023).

9.5 Best Management Practices for Achieving Negative Carbon Sinks in Agriculture

Best management practices (BMPs) in agriculture are a set of practices designed to optimize agricultural productivity while minimizing negative environmental impacts (Lal 2015). These practices are scientifically tested and proven, focusing on improving soil health, water management, nutrient cycling, and pest control. The aim is to achieve sustainable farming systems that balance economic viability with ecological health. BMPs are region-specific which take various factors into account, including local climate, soil types, and crop choices. In the agricultural context, negative C sinks are achieved through practices that enhance the C storage capacity of soils and plants. Waste management plays a crucial role here by promoting the recycling of organic residues, such as crop residues, livestock manure, and agro-industrial by-products, back into the soil. Through practices like composting, biochar application, and residue incorporation, agricultural waste is transformed into a resource that enriches SOM, boosts microbial activity, and enhances C sequestration (Roy et al. 2024). Enhancing SOM levels, integrating cover crops and perennial trees, and adopting improved soil management practices can transform agriculture into a substantial C sink (Roy et al. 2024). This not only helps to mitigate climate change by reducing atmospheric CO_2 levels but also improves soil fertility and water retention, making agricultural systems more resilient to environmental stressors (Sheoran et al. 2022). The adoption of BMPs is critical in transforming agricultural lands into effective negative C sinks, contributing to global efforts to combat climate change. Recently, artificial intelligence (AI) has emerged as a powerful tool for valorizing agricultural waste by enabling innovative recycling into biofuels, fertilizers, and feed, thus guiding BMPs for negative C sinks in agriculture (Sen et al. 2025). Table 9.4 summarizes various agricultural practices aimed at reducing C emissions and enhancing C sequestration.

9.6 Policy Perspectives on Farm Waste Management Through Best Management Practices

Effective farm waste management is vital for achieving sustainable agriculture and environmental protection, especially in the context of rising global food demands and escalating GHG emissions. The expansion of agricultural activities has dramatically increased farm waste generation, posing significant challenges for its disposal and environmentally sound utilization. Strategic policy interventions focusing on BMPs and the integration of agricultural waste into a circular bioeconomy are necessary to address these issues. A circular bioeconomy in agriculture emphasizes the sustainable use of biomass generated as farm waste by converting it into bioenergy, bioproducts, and bio-based materials such as compost, biochar, and biogas, thereby mitigating C emissions and enhancing soil health (Sen et al. 2025). Practices like biochar production through pyrolysis of crop residues, anaerobic digestion of animal manure to produce biogas, and composting of organic farm wastes are promising solutions to transform waste into resources while sequestering C and promoting climate resilience. In India, several government initiatives already support such transitions. Programs like the National Bio-Energy Mission (NBEM) aim to promote bioenergy production from agricultural residues; the Pradhan Mantri Kisan Urja Suraksha evam Utthaan Mahabhiyan (PM-KUSUM) supports decentralized renewable energy generation, including biomass-based plants; the Paramparagat Krishi Vikas Yojana (PKVY) promotes organic farming practices using on-farm biomass, while the National Mission on Sustainable Agriculture

Table 9.4 Agricultural management practices aimed at reducing carbon emissions and enhancing carbon sequestration and the role of waste management

Management practice	Description	Benefits	Scope/relevance of waste management	References
Conservation tillage	Minimal soil disturbance, preserving soil structure, and organic matter	Reduces soil erosion, enhances water retention, and increases carbon sequestration	Stubbles help maintain permanent soil cover	Lal (2015) and Roy et al. (2024)
Cover cropping	Planting cover crops like legumes, grasses, and brassicas during off-season periods	Improves soil health, prevents erosion, and enhances carbon storage in the soil	Residues of cover crops serve as organic amendments, recycling biomass back into the soil system.	Sharma et al. (2018)
Agroforestry	Integrating trees and shrubs into agricultural landscapes	Provides habitat diversity, improves soil structure, and increases carbon sequestration through tree biomass	Litterfall and pruned biomass act as organic inputs, enhancing soil carbon and preventing waste	Dollinger and Jose (2018)
Crop diversification	Growing different crops including legumes in the same field over time	Enhances carbon sequestration, biodiversity, soil health, and reduces pest risks	Diversified cropping generates varied biomass waste, which can be reused to maintain soil carbon	Ghosh et al. (2023)
Crop residue management	Managing the remains of crops left in the field after harvest	Enhances SOM, reduces erosion, and improves water retention	Residues are incorporated into soil instead of being burnt, improving soil health and carbon storage	Bag et al. (2022) and Sen et al. (2024)
Organic farming practices	Utilizing organic fertilizers, composting, and avoiding synthetic chemicals	Enhances SOM, reduces greenhouse gas emissions, and promotes biodiversity	Farm waste is composted or directly applied as organic amendments, minimizing nutrient loss	Reeve et al. (2016)
Natural farming	Farming method using natural inputs and minimal external chemicals	Enhances soil health, promotes biodiversity, and reduces carbon emissions	Utilizes on-farm wastes like cow dung, plant residues, and biopesticides, closing nutrient loops	Das et al. (2024)
Precision agriculture	Using technology to optimize field-level	Reduces inputs (fertilizers, pesticides), minimizes waste, and enhances	Prevents overapplication of agrochemicals, minimizing the	Ackerson (2022) and Roy et al. (2022a)

(continued)

Table 9.4 (continued)

Management practice	Description	Benefits	Scope/relevance of waste management	References
	management regarding crop farming	carbon efficiency, better monitoring of C and other nutrient status	generation of chemical waste	
Manure management	Proper collection, storage, and application of animal manure	Reduces methane emissions, recycles nutrients, and improves soil carbon content	Converts animal waste into a valuable resource rather than a pollution source	Rayne and Aula (2020)
Integrated pest management (IPM)	Combining biological, cultural, physical, and chemical tools to manage pests	Reduces the need for chemical pesticides, lowers environmental impact, and supports carbon sequestration	Promotes organic and biological pest control using plant or microbial waste-derived biopesticides	Daelemans et al. (2022)
Composting	Converting organic waste into nutrient-rich compost	Enhances soil fertility, reduces landfill waste, and sequesters carbon in the soil	Directly transforms agricultural and food waste into valuable organic amendments	Prakash et al. (2022) and Wright et al. (2022)
Biochar application	Adding biochar (charred organic material) to the soil	Improves soil fertility, increases water retention, and sequesters carbon for long periods	Converts biomass waste into biochar, preventing CO_2 emissions from open decomposition	Yadav and Ramakrishna (2023)
Efficient water management	Using irrigation methods like drip or sprinkler systems to conserve water	Reduces water usage, enhances crop yield, and maintains soil carbon levels	Avoids wastage of nutrient-rich runoff that would otherwise carry nutrients into water bodies	Lal (2016) and Roy et al. (2022b)
Inorganic nutrient management	Using synthetic fertilizers efficiently to meet crop nutrient requirements	Optimizes nutrient use efficiency, reduces nutrient runoff, and enhances crop yield	Reduces chemical fertilizer wastage and complements organic nutrient recycling efforts	Antil and Raj (2020)
Application of PGPR	Plant growth-promoting rhizobacteria (PGPR) are beneficial soil bacteria	Boosts plant growth, improves nutrient uptake, and enhances soil health	Reduces need for synthetic fertilizers, encouraging recycling of organic materials	Rani et al. (2022)
Pasture management	Rotational grazing and proper livestock management to maintain healthy pastures	Enhances soil health, increases forage productivity, and sequesters carbon in soil and plant biomass	Manages livestock waste naturally through rotational grazing, improving soil organic carbon	Amorim et al. (2020)

(NMSA) encourages residue management to prevent stubble burning. Additionally, initiatives like the Sub-Mission on Agricultural Mechanization (SMAM) provide subsidies for equipment that can aid in sustainable waste management practices. These government schemes aim not only to manage waste but also to enable farmers to generate additional income through C credit trading, biogas production, and organic certification. Policymakers can strengthen farm waste management by mandating nutrient recycling plans, incentivizing sustainable waste utilization through grants, low-interest loans, or tax credits, and promoting farmer education through extension services, workshops, and demonstration projects. However, significant challenges persist, including the high initial cost of infrastructure for biogas or biochar production, lack of access to modern technologies for small and marginal farmers, and logistic difficulties in waste collection and processing at a large scale. Tailoring BMPs and circular economy practices to local soil types, climatic conditions, and agricultural systems is crucial, as one-size-fits-all solutions are often ineffective. Technological and logistical barriers, as noted by Kumar et al. (2023a), must be addressed through innovative approaches such as decentralized waste processing units, farmer cooperatives, and public–private partnerships. Despite these hurdles, the opportunity to convert farm waste from an environmental liability into an economic and ecological asset remains immense. Promoting circular bioeconomy practices within the agricultural sector can lead to significant reductions in C footprints, improve soil fertility and water retention, and support rural livelihoods. By integrating waste management into sustainable agricultural frameworks through enabling policies, financial support, education, and infrastructure development, it is possible to transform agricultural systems into net contributors to environmental sustainability and climate change mitigation.

9.7 Adoption Challenges in Promoting Farm Waste Recycling

With ongoing urbanization and population expansion, annual farm waste generation is anticipated to increase at a greater rate. Among these, agricultural waste/farm wastes hold particular significance for the circular economy. By redirecting farm waste, which includes crop residues and animal manure, farmers can enhance soil fertility, mitigate environmental pollution, and promote a circular economy within the agricultural sector (Sharma et al. 2019). However, despite the potential benefits, the adoption of farm waste recycling practices is very limited and varies widely across regions and farming systems. There are a few constraints that restrict the adoption of farm wastes, hindering the C credit generation. Hence, these bottlenecks need to be overcome to facilitate the widespread adoption of farm waste recycling and contribute to enhancing soil C stock.

9.7.1 Low Nutrient Content

Agronomic evaluation of compost involves assessing its impact on plant growth. Research has shown that compost can enhance crop yields due to its nutrient content and stimulatory effect on the soil's physicochemical properties (Sharma et al. 2017). However, some composts with low nutrient levels may fail to promote plant growth, underscoring the importance of evaluating composts for their nutritional value. Compost with low nutrient content doesn't boost the SOC significantly, which also affects the C credit generation, as the projects depend on a measurable increase within a stipulated period. However, adding nutrient-rich substrates can further improve the compost's nutritive and agronomic properties (Lawal and Babalola 2014). Compost quality is evaluated through chemical analyses that quantify the concentrations of essential elements. Organic fertilizers should contain at least 1% nitrogen, and 1.5% each of phosphorus and potassium. Additionally, compost should supply essential micronutrients such as calcium, zinc, and copper in concentrations ranging from 0.01 to 0.05%.

9.7.2 Contamination with Emerging Pollutants

Modern society relies heavily on a diverse array of organic chemicals, many of which become sequestered in solid materials during waste treatment processes (Saraswat et al. 2024). Over the past few decades, extensive research has been conducted specifically focusing on certain groups of persistent organic pollutants (POP) (Saha et al. 2024). More recently, attention has expanded to include other hazardous chemicals, such as PCN (polychlorinated naphthalenes), PCA (polychlorinated alkanes), synthetic musks, estrogens, organotin compounds, and NP (nonylphenols) (Sjöström et al. 2008; Clarke and Smith 2011). The POPs are particularly concerning due to their direct toxicity to living organisms and their tendency to bioaccumulate at higher trophic levels. Chronic exposure of lower organisms to small amounts of these pollutants can lead to potentially harmful concentrations in humans and wildlife over time (Anonymous 2011). Presence of these POPs in farm wastes can have a serious impact on the C market, as compost contaminated with POPs could disqualify projects from certification or lead to reputation and compliance risks.

9.7.3 Longer Preparation Period

Organic farm wastes include leaves, plant remnants, and dead vegetation, some of which contain recalcitrant compounds and have low nutrient content, making them difficult to compost. When such wastes are added to a compost pile, they can slow down the overall composting process. Co-composting, which involves mixing different types of materials, can either accelerate or hinder the composting rate depending on the combination. The addition of nutrient-rich substrates enhances microbial activity, speeding up decomposition. Conversely, when materials with low nutrient levels, high cellulose, or high lignin content are included, the composting process slows down. Plant taxa differ in toughness (sclerophylly) due to the presence of various chemical compounds such as lignin, suberin, cellulose, phenolics, tannins, irritants, and allergens (Lewis and Brown 2010). High-quality leaves, like nutrient-rich alder leaves, decompose more rapidly than low-quality leaves, such as nutrient-poor conifer needles. Additionally, the presence of high nicotine content in leaves, such as those from tobacco, extends their decomposition time (Zhao et al. 2017). Plant wastes rich in lignin, polyphenols, and suberin significantly prolong the composting process. Simultaneously, the longer duration of composting affects the C credit generation. Most of the C credit projects often have a scheduled reporting period (e.g., yearly or biannually). If composting is too slow, it can delay achieving C benefits, postponing the issuance.

9.7.4 Introduction of a Foreign Organism

The source of degradable materials and activators, such as animal dung or sewage, significantly influences the diversity and population of microorganisms, as well as the presence of potential pathogens in the compost (Alvarenga et al. 2015). Pathogenic organisms like *Escherichia coli* and *Salmonella* sp. have been reported in compost (Epelde et al. 2018). The presence of these organisms reduces the quality of compost and questions the sustainability and standards of the C credit projects. Therefore, it is essential to thoroughly assess compost for both microbial and chemical components to ensure the safety of plants, soil organisms, animals, and humans.

9.7.5 Financial Constraints

Processes such as composting, vermicomposting, and anaerobic digestion are effective in degrading a substantial portion of organic solid waste, converting it into valuable products for organic agriculture and other applications. Economic analysis of composting systems typically considers various factors, including capital investment, labor, biomass, processing energy, maintenance and repair costs, as well as revenue from compost production and disposal credits (Blumenstein et al. 2012). Generating C credits requires upfront investment for compost infrastructure and documentation systems. Cukjati et al. (2012) reported that an anaerobic sludge composting facility with a capacity of 7.12 million kg required an initial investment of approximately €462,646, with annual operating costs ranging from €250,000 to

€360,000. Without financial capacity, farmers or project developers cannot participate in or scale up the C markets.

9.7.6 Knowledge and Awareness Gap

Farmers often face challenges in implementing effective recycling practices due to a lack of knowledge and technical expertise. Poorly implemented practices lead to low-quality projects that fail validation or do not generate significant C credits. This gap is further exacerbated by limited access to extension services or training programs that could offer essential guidance. Bridging these knowledge gaps is crucial for encouraging the widespread adoption of farm waste recycling practices.

9.8 Future Prospects of Negative Carbon Sinks

The future of farm waste management is set to be driven by climate-smart strategies that transform agricultural residues into tools for C sequestration and environmental restoration. The future lies in developing integrated, circular approaches that not only mitigate emissions but actively sequester C, transforming agriculture into a negative C sink. Emerging technologies like integrated vermicomposting, anaerobic digestion, and biochar production offer scalable solutions to recycle organic waste while locking C into soil systems (Kumar et al. 2022b). Innovations such as decentralized biorefineries and nutrient recovery units can convert waste into clean energy, biofertilizers, and soil amendments, significantly reducing GHG emissions. Furthermore, digital tools powered by Internet of Things (IoT) and AI will enable precision waste tracking, optimize composting processes, and support data-driven C accounting. Linking these sustainable practices with verified C credit markets can provide economic incentives for farmers to adopt them. As policies shift toward regenerative agriculture and net-zero targets, farm waste will increasingly be recognized not as a burden but as a valuable resource for building resilient, low-emission agricultural systems.

9.9 Conclusions

Agricultural waste management presents both challenges and opportunities in the global effort to improve environmental quality and sustainability. Agricultural activities could potentially help in climate change mitigation through the strategic implementation of improved farm management practices that enhance carbon sequestration and reduce GHG emissions. Proper management and recycling of farm wastes, by composting (including vermicomposting), biochar production, anaerobic digestion of organic farm wastes (biogas production), agroforestry, and adoption of improved agronomic practices, offer substantial potential to enhance soil health, add carbon to soil, and support sustainable agriculture. Techniques such as conservation tillage, composting, vermicomposting, biochar production, and anaerobic digestion are effective strategies that contribute to soil health improvement, carbon sequestration, and climate change mitigation. The BMPs, including agroforestry, cover cropping, conservation tillage, and biochar application, are critical to transforming agricultural lands into effective carbon sinks. These practices contribute to the increase in SOM, improved soil structure, and enhanced water retention, all of which are crucial for sustainable agricultural productivity. By integrating these practices, agriculture can transition towards a net-negative carbon sink, thereby playing a vital role in combating climate change while ensuring long-term soil fertility and productivity.

Farm waste management can also be integrated with circular bioeconomy, offering a promising pathway to reduce the environmental impact of agriculture. Biochar, anaerobic digestion, and composting are highlighted as key processes in this regard. However, several challenges hinder the widespread adoption of these practices; some of which include poor nutrient status of compost, contamination with pollutants, extended composting periods due to their recalcitrant nature, financial constraints, and a lack of knowledge and awareness among farmers are significant barriers that need to be addressed. Policy support from the government, alongside international funding and carbon credit incentives, will be crucial in driving the successful implementation of these sustainable practices. Therefore, transforming Indian agriculture into a significant negative carbon sink requires a concerted effort involving multiple stakeholders, including farmers, policymakers, researchers, and the private sector. By adopting a holistic approach that integrates BMPs, circular bioeconomy practices, and innovative technologies, Indian agriculture can make a substantial contribution to global climate change mitigation efforts.

References

Ackerson JP (2022) Precision agriculture and soil health. Crop Soil 55:35–39. https://doi.org/10.1002/crso.20214

Alexander P (2017) Losses, inefficiencies and waste in global food system. Retrieved from www.elsevier.com/locate/agsy

Alvarenga P, Mourinha C, Farto M, Santos T, Palma P, Sengo J, Morais MC, Cunha-Queda C (2015) Sewage sludge, compost and other representative organic wastes as agricultural soil amendments: Benefits versus limiting factors. Waste Manag 40:44–52. https://doi.org/10.1016/j.wasman.2015.01.027. PubMed: 25708406

Amit KD (2020) Stability of soil organic carbon as affected by crop residue and nutrient management options under conservation agriculture. Doctoral Dissertation, Division of Soil Science and Agricultural Chemistry, ICAR-Indian Agricultural Research Institute, New Delhi

Amorim HC, Ashworth AJ, Moore Jr PA, Wienhold BJ, Savin MC, Owens PR, Jagadamma S, Carvalho TS, Xu S (2020) Soil quality indices following long-term conservation pasture management practices. Agric Ecosyst Environ 301:107060

Ankathi SK, Chaudhari US, Handler RM, Shonnard DR (2024) Sustainability of biogas production from anaerobic digestion of food waste and animal manure. Appl Microbiol 4:418–438. https://doi.org/10.3390/applmicrobiol4010029

Annual report (2012) Department of Agriculture & Cooperation Ministry of Agriculture Government of India. https://agriwelfare.gov.in/Documents/Annual-Report%202011-12.pdf. Accessed 30 Aug 2024

Annual report (2014) Department of Agriculture & Cooperation Ministry of Agriculture Government of India. https://agriwelfare.gov.in/Documents/Annual-Report-2013-14.pdf. Accessed 30 Aug 2024

Anonymous (2011) Persistent organic pollutant (POP) emissions (APE 006). European Environment Agency. www.eea.europa.eu/data-and-maps/indicators/eea32-persistent-organic-pollutant-pop-missions-1/assessment-1

Antil RS, Raj D (2020) Integrated nutrient management for sustainable crop production and improving soil health. Nutrient dynamics for sustainable crop production, pp 67–101

Atela JO (2012) The politics of agricultural carbon finance: the case of the Kenya agricultural carbon project, STEPS Working Paper 49. STEPS Centre, Brighton. https://www.arin-africa.org/wp-content/uploads/2024/10/Carbon-Agriculture.pdf. Accessed 24 Apr 2025

Badawi MA (2023) Composting farm waste for production of high quality organic fertilizers and sustainable development. Integrated waste management: the circular economy. UK, p 196

Bag K, Roy A, Tigga P, Didawat RK, Kumar P, Dey A (2022) Real-time monitoring of crop residue burning through satellite remote sensing. In: Nemichandrappa M, Halidoddi RR (eds) Advanced innovative technologies in agricultural engineering for sustainable agriculture. AkiNik Publications, Delhi, pp 55–79. https://doi.org/10.22271/ed.book.1810

Bamel D, Dey A, Mondal BK, Bhattacharyya R, Das D, Das S, Das TK, Dash AK (2025) Non-abundance of Cellulose and Lignin in Crop Residue Biomass Increases Decomposability and Lowers Temperature Sensitivity of the Process and Helps in Labile Soil Organic Carbon build-up. Journal of Soil Science and Plant Nutrition 1–18

Berdier C, Maillefert M, Girault M (2025) Carbon footprint of composting and vermicomposting of household biowaste: a decision-making factor for regional biowaste recovery policies? Recycling 10(2):44

Bhoi TK, Samal I, Saraswat A, Hombegowda HC, Samal SK, Dash AK, Sharma S, Lawate P, Vyas V, Raza MB (2024) Biochar as a soil amendment: effects on microbial communities and soil health. In: Biochar production for green economy. Academic Press, Cambridge, MA, pp 137–159

Bist AS, Bisht K, Pargaien S (2012) Sustainability a step forward to land grant pattern. Int J Eng Sci Res Technol 1(7)

Blair N, Faulkner RD, Till AR, Korschens M, Schulz E (2006) Long-term management impacts on soil C, N and physical fertility. Soil Tillage Res 91:39–47. https://doi.org/10.1016/j.still.2005.11.001

Blumenstein B, Bühle L, Wachendorf M, Möller D (2012) Economic assessment of the integrated generation of solid fuel and biogas from biomass (IFBB) in comparison to different energy recovery, animal-based and non-refining management systems. Bioresour Technol 119:312–323. https://doi.org/10.1016/j.biortech.2012.05.077. PubMed: 22750498

Breunig HM, Amirebrahimi J, Smith S, Scown CD (2019) Role of digestate and biochar in carbon-negative bioenergy. Environ Sci Technol 53:12989–12998. https://doi.org/10.1021/acs.est.9b03763. PubMed: 31626735

Busari MA, Kukal SS, Kaur A, Bhatt R, Dulazi AA (2015) Conservation tillage impacts on soil, crop and the environment. Int Soil Water Conserv Res 3:119–129. https://doi.org/10.1016/j.iswcr.2015.05.002

Cardelli R, Giussani G, Marchini F, Saviozzi A (2018) Short-term effects on soil of biogas digestate, biochar and their combinations. Soil Res 56:623–631. https://doi.org/10.1071/SR18017

Chen N, Pilla S (2022) A comprehensive review on transforming lignocellulosic materials into biocarbon and its utilization for composites applications. Compos Part C 7:100225. https://doi.org/10.1016/j.jcomc.2021.100225

Clarke BO, Smith SR (2011) Review of emerging organic contaminants in biosolids and assessment of international research priorities for the agricultural use of biosolids. Environ Int 37:226–247. https://doi.org/10.1016/j.envint.2010.06.004. PubMed: 20797791

Creamer AE, Gao B, Zhang M (2014) Carbon dioxide capture using biochar produced from sugarcane bagasse and hickory wood. Chem Eng J 249:174–179. https://doi.org/10.1016/j.cej.2014.03.105

Cukjati N, Zupančič GD, Roš M, Grilc V (2012) Composting of anaerobic sludge: an economically feasible element of a sustainable sewage sludge management. J Environ Manag 106:48–55. https://doi.org/10.1016/j.jenvman.2012.04.001. PubMed: 22562011

Daelemans R, Hulsmans E, Honnay O (2022) Both organic and integrated pest management of apple orchards maintain soil health as compared to a semi-natural reference system. J Environ Manag 303:114191. https://doi.org/10.1016/j.jenvman.2021.114191. PubMed: 34861508

Darjee S, Alekhya G, Sudarshan S, Padhan SR, Rajareddy G, Baishya M, Dash AK (2023) Enhancing crop productivity with biofertilizers: exploring multifunctional benefits. Microbiol Res J Int 34:101–112

Das A, Singh S, Rani K, Gouda HS, Mahakuda B, Roy A, Thakur JK, Chakraborty R (2024) Natural farming: status, challenges and prospects. In: Mohapatra B, Singh AK, Gaikwad SB, Singh K, Ghosh A (eds) Futuristic trends in agriculture engineering & food sciences, vol 3, Book 23. IIP Iterative International Publishers, Chikkamagaluru, pp 225–236

Dash AK, Singh C, Varma S (2021) Chemistry and recent advances in management of acid sulphate soils. Latest Trend Soil Sci:97–116

Dash AK, Dwivedi BS, Dey A, Meena MC, Chakraborty D (2023) Temperature sensitivity of soil organic carbon as affected by crop residue and nutrient management options under conservation agriculture. J Soil Sci Plant Nutr 23:4183–4197. https://doi.org/10.1007/s42729-023-01335-y

Dash AK, Tripathy S, Naveenkumar A, Bhoi TK, Kumari A, Latare AM, Nandipamu TMK, Singh V, Raza MB, Saraswat A, Bhada J (2024) Biochar for pollutants bioremediation from soil and water ecosystem. In: Biochar production for green economy. Academic Press, pp 433–452

Dash AK, Meena MC, Das S, Dey A, Raza MB, Tripathy S, Kumar A, Panda D, Divyadarshan A (2025) Temporal effects of conservation agriculture-based rice–wheat cropping system on soil aggregation and organic carbon dynamics in northwestern Indo-Gangetic Plains. J Soil Sci Plant Nutr 25:1–17

Dash AK, Meena MC, Das S, Dey A, Sangwan S, Bana RS, Aditya K, Raza MB, Padhan D, Tripathy S, Bardhan T, Jat HS, Kumar A, Al-Ansari N, Salim A, Mattar MA (2025) Long-term conservation agriculture to enhance soil properties and quality in rice–wheat cropping system. Scientific Reports 15(1):28393

Dash AK, Dey A, Bhattacharyya R, Das TK, Bhatia A, Das S (2025) Conservation agriculture strategies for enhancing soil health. Soil Health and Sustainability in India 423

Dash AK, Sannagoudar MS, Kumari K, Vinesh B, Kumar A (2025) Sustainable soil health management practices for quality seed production. Recent innovations in seed production and quality control system. ICAR-National Institute of Seed Science and Technology

Desai D (2021) The healthy soil program: an analysis of California's effort to build soil and mitigate climate change. Int Yearbook Soil Law Policy 2019:261–293

Dey Chowdhury S, Suhaib KH, Bhunia P, Surampalli RY (2023) A critical review on the vermicomposting of organic wastes as a strategy in circular bioeconomy: mechanism, performance, and future perspectives. Environ Technol, pp 1–38

Dixon RK (1995) Agroforestry systems: Sources of sinks of greenhouse gases? Agrofor Syst 31: 99–116. https://doi.org/10.1007/BF00711719

Dollinger J, Jose S (2018) Agroforestry for soil health. Agrofor Syst 92:213–219. https://doi.org/10.1007/s10457-018-0223-9

Donofrio SD, Maguire PM, Myers KM, Daley CD, Lin KL (2021) State of the voluntary carbon markets 2021. For Trend. https://www.forest-trends.org/publications/state-of-the-voluntary-carbon-markets-2021/

Duque-Acevedo M, Belmonte-Ureña LJ, Cortés-García FJ, Camacho-Ferre F (2020) Agricultural waste: review of the evolution, approaches and perspectives on alternative uses. Glob Ecol Conserv 22:e00902. https://doi.org/10.1016/j.gecco.2020.e00902

Epelde L, Jauregi L, Urra J, Ibarretxe L, Romo J, Goikoetxea I, Garbisu C (2018) Characterization of composted organic amendments for agricultural use. Front Sustain Food Syst 2:44. https://doi.org/10.3389/fsufs.2018.00044

FAO (2018) The state of world fisheries and aquaculture: meeting the sustainable development goals. Food and Agriculture Organization of the United Nations, Rome

Farooqi ZUR, Sabir M, Ahmad HR, Shahbaz M, Smith J (2023) Reclaimed salt-affected soils can effectively contribute to carbon sequestration and food grain production: evidence from Pakistan. Appl Sci 13:1436. https://doi.org/10.3390/app13031436

Farooqi ZUR, Qadir AA, Khalid S, Murtaza G, Ashraf MN, Shafeeq-Ur-Rahman W, Javed W, Waqas MA, Xu M (2024) Greenhouse gas emissions, carbon stocks and wheat productivity following biochar, compost and vermicompost amendments: comparison of non-saline and salt-affected soils. Sci Rep 14:7752. https://doi.org/10.1038/s41598-024-56381-y. PubMed: 38565858

Feng Y, Rosa L (2024) Global biomethane and carbon dioxide removal potential through anaerobic digestion of waste biomass. Environ Res Lett 19:024024. https://doi.org/10.1088/1748-9326/ad1e81

Fortuna A, Harwood RR, Paul EA (2003) The effects of compost and crop rotations on carbon turnover and the particulate organic matter fraction. Soil Sci 168:434–444. https://doi.org/10.1097/00010694-200306000-00005

Ganguly A, Brown RC, Wright MM (2022) Techno-economic and greenhouse gas emission assessment of carbon negative pyrolysis technology. Green Chem 24:9290–9302

Ghosh PK, Hazra KK, Roy A (2023) Contributions of diversified cropping systems to plant nutrition, soil health, and environmental services. Indian J Fertil 19:1108–1117

Gnanavelrajah N, Shrestha RP, Schmidt-Vogt D, Samarakoon L (2008) Carbon stock assessment and soil carbon management in agricultural land-uses in Thailand. Land Degrad Dev 19:242–256. https://doi.org/10.1002/ldr.838

Graf A, Wohlfahrt G, Aranda-Barranco S, Arriga N, Brümmer C, Ceschia E, Ciais P, Desai AR, Di Lonardo S, Gharun M et al (2023) Joint optimization of land carbon uptake and albedo can help achieve moderate instantaneous and long-term cooling effects. Commun Earth Environ 4:298

Han X, Xu C, Dungait JA, Bol R, Wang X, Wu W, Meng F (2018) Straw incorporation increases crop yield and soil organic carbon sequestration but varies under different natural conditions and farming practices in China: a system analysis. Biogeosciences 15(7):1933–1946

Handa AK, Chavan SB, Sirohi C, Rizvi RH (2020) Importance of agroforestry systems in carbon sequestration. In: Proceedings of the National Agroforestry Symposium

Honegger M, Poralla M, Michaelowa A, Ahonen H-M (2021) Who is paying for carbon dioxide removal? Designing policy instruments for mobilizing negative emissions technologies. Front Clim 3:672996

Huang K, Xia H (2018) Role of earthworms' mucus in vermicomposting system: Biodegradation tests based on humification and microbial activity. Sci Total Environ 610–611:703–708. https://doi.org/10.1016/j.scitotenv.2017.08.104. PubMed: 28822937

IPCC (2006) Guidelines for national greenhouse gas inventories. Prepared by the National Greenhouse Gas Inventories Programme, Intergovernmental Panel on Climate Change

Jarecki MK, Lal R (2003) Crop management for soil carbon sequestration. Crit Rev Plant Sci 22: 471–502. https://doi.org/10.1080/713608318

Kaab A, Sharifi M, Mobli H, Nabavi-Pelesaraei A, Chau KW (2019) Combined life cycle assessment and artificial intelligence for prediction of output energy and environmental impacts of sugarcane production. Sci Total Environ 664:1005–1019. https://doi.org/10.1016/j.scitotenv.2019.02.004. PubMed: 30769303

Karić N, Maia AS, Teodorović A, Atanasova N, Langergraber G, Crini G, Ribeiro ARL, Đolić M (2022) Bio-waste valorisation: agricultural wastes as biosorbents for removal of (in)organic pollutants in wastewater treatment. Chem Eng J Adv 9:100239. https://doi.org/10.1016/j.ceja.2021.100239

Koul B, Yakoob M, Shah MP (2022) Agricultural waste management strategies for environmental sustainability. Environ Res 206:112285. https://doi.org/10.1016/j.envres.2021.112285. PubMed: 34710442

Kumar S, Bamboriya SD, Rani K, Meena RS, Sheoran S, Loyal A, Kumawat A, Jhariya MK (2022a) Grain legumes: a diversified diet for sustainable livelihood, food, and nutritional security. In: Advances in legumes for sustainable intensification. Academic Press, pp 157–178

Kumar S, Sharma SK, Thakral SK, Bhardwaj KK, Jhariya MK, Meena RS, Jangir CK, Bedwal S, Jat RD, Gaber A, Atta AA (2022b) Integrated nutrient management improves the productivity and nutrient use efficiency of *Lens culinaris* Medik. Sustainability 14(3):1284

Kumar SP, Subudhi S, Bhatia L, Saha K, Mudgil D, Prasad Shadangi K, Srivastava RK, Pattnaik B, Arya RK (2023a) Utilization of agricultural waste biomass and recycling toward circular bioeconomy. Environ Sci Pollut Res Int 30:8526–8539. https://doi.org/10.1007/s11356-022-20669-1. PubMed: 35554831

Kumar S, Sharma SK, Dhaka AK, Bedwal S, Sheoran S, Meena RS, Jangir CK, Kumar D, Kumar R, Jat RD, Meena AK (2023b) Efficient nutrient management for enhancing crop productivity, quality and nutrient dynamics in lentil (*Lens culinaris* Medik.) in the semi-arid region of northern India. PLoS One 18(2):e0280636

Kumar S, Meena RS, Kumar S, Pradhan G, Jangir CK, Ghosh S, Punia H, Sheoran P, Meena R, Ahmad MA, Goyal SK (2025) Designing a diversified Indian mustard production system for energy-carbon-cum-heat use efficiency and sowing dates assessment. GCB Bioenergy 17(6): e70044

Kung CC, McCarl BA, Cao X (2013) Economics of pyrolysis-based energy production and biochar utilization: a case study in Taiwan'. Energy Policy 60:317–323

Lal R (1997) Residue management, conservation tillage and soil restoration for mitigating greenhouse effect by CO_2-enrichment. Soil Tillage Res 43:81–107. https://doi.org/10.1016/S0167-1987(97)00036-6

Lal R (2000) Soil management in the developing countries. Soil Sci 165:57–72. https://doi.org/10.1097/00010694-200001000-00008

Lal R (2004) Soil carbon sequestration impacts on global climate change and food security. Science 304:1623–1627. https://doi.org/10.1126/science.1097396. PubMed: 15192216

Lal R (2007) Carbon management in agricultural soils. Mitig Adapt Strateg Glob Chang 12:303–322. https://doi.org/10.1007/s11027-006-9036-7

Lal R (2015) Restoring soil quality to mitigate soil degradation. Sustainability 7:5875–5895. https://doi.org/10.3390/su7055875

Lal R (2016) Soil health and carbon management. Food Energy Sec 5:212–222. https://doi.org/10.1002/fes3.96

Lawal TE, Babalola OO (2014) Assessing the associated challenges in the use of animal manure in plant growth. J Hum Ecol 48:285–297. https://doi.org/10.1080/09709274.2014.11906798

Leharwan M, Kumar Y, Kumar R, Kumar Saraswat P, Kumar R, Veetil AKT, Bhattacharjee S, Kumar A, Kumar S (2023) Assessing the effects of conservation tillage and in-situ crop residue management on crop yield and soil properties in rice–wheat cropping system. Sustainability 15(17):12736

Lehmann J, Joseph S (eds) (2024) Biochar for environmental management: science, technology and implementation. Taylor & Francis

Lehmann J, Gaunt J, Rondon M (2006) Bio-char sequestration in terrestrial ecosystems – a review. Mitig Adapt Strateg Glob Chang 11:403–427. https://doi.org/10.1007/s11027-005-9006-5

Lewis SE, Brown AV (2010) Comparative leaf decomposition rates including a non-native species in an urban Ozark stream. J Arkansas Acad Sci 64:92–96

Liu S (2022) Restoring carbon sinks in agriculture through carbon credits. Sustainable Finance Initiative, Stanford

Loehr R (2012) Agricultural waste management: problems, processes, and approaches. Elsevier, Amsterdam

Lorenz K, Lal R (2014) Soil organic carbon sequestration in agroforestry systems. A review. Agron Sustain Dev 34:443–454. https://doi.org/10.1007/s13593-014-0212-y

Mandal KG, Misra AK, Hati KM, Bandyopadhyay KK, Ghosh PK, Mohanty M (2004) Rice residue management options and effects on soil properties and crop productivity. J Food Agric Environ 2:224–231

Mandal B, Majumder B, Adhya TK, Bandyopadhyay PK, Gangopadhyay A, Sarkar D, Kundu MC, Choudhury SG, Hazra GC, Kundu S, Samantaray RN, Misra AK (2008) Potential of double-cropped rice ecology to conserve organic carbon under subtropical climate. Glob Chang Biol 14: 2139–2151. https://doi.org/10.1111/j.1365-2486.2008.01627.x

McClelland SC, Paustian K, Schipanski ME (2021) Management of cover crops in temperate climates influences soil organic carbon stocks: a meta-analysis. Ecol Appl 31:e02278. https://doi.org/10.1002/eap.2278. PubMed: 33320994

Meena RS, Pradhan G (2023) Industrial garbage-derived biocompost enhances soil organic carbon fractions, CO_2 biosequestration, potential carbon credits, and sustainability index in a rice-wheat ecosystem. Environ Res 116525. https://doi.org/10.1016/j.envres.2023.116525

Meena RS, Pradhan G, Kumar S, Lal R (2023) Using industrial wastes for rice-wheat cropping and food-energy-carbon-water-economic nexus to the sustainable food system. Renew Sust Energ Rev 187:113756. https://doi.org/10.1016/j.rser.2023.113756

Meena, R.S, Pradhan. G., Nalani, Singh, K., Singh, A.K., Verma, S.K., Mina, K.K., Kumar, S., and Chaturvedi, R.K. (2024). Diversified cropping modules designed for soil restoration, CO_2 sequestration, and generating carbon credits. Land Degrad Dev, doi:https://doi.org/10.1002/ldr.5224, 35

Meena RS, Pradhan G, Singh K, Kumar S, Singh AK, Shashidhar KS, Mina KK, Rao CS (2024a) Agriculture models for restoring degraded land to enhance CO_2 biosequestration and carbon credits in the Vindhyan region of India. Sci Total Environ 929:172661. https://doi.org/10.1016/j.scitotenv.2024.172661

Meena RS, Lal R, Kumar S, Pradhana G, Rao CS, Singh AK, Pathak H, Abhilash PC, Kumar A, Sharma SK, Jat ML, Singh S (2024b) Potential of Indian agriculture for capturing atmospheric CO_2 and monetizing carbon credits to the farmers: an overview and policy framework. Adv Agron 188:101–206. https://doi.org/10.1016/bs.agron.2024.06.001

Meena RS, Pradhan G, Kumar S (2025) Energy flow, eco-efficiency, and economic circulation with recycled industrial waste compost application in wheat and subsequent rice farming. Sci Total Environ 967:178779. https://doi.org/10.1016/j.scitotenv.2025.178779

MNRE (Ministry of New, Renewable Energy Resources) (2009) Government of India, New Delhi. www.mnre.gov.in/biomassrsources

Möller K (2015) Effects of anaerobic digestion on soil carbon and nitrogen turnover, N emissions, and soil biological activity. A review. Agron Sustain Develop 35:1021–1041. https://doi.org/10.1007/s13593-015-0284-3

Naresh RK, Timsina J, Bhaskar S, Gupta RK, Singh AK, Dhaliwal SS et al (2017) Effects of tillage, residue and nutrient management on soil organic carbon dynamics and its fractions, soil aggregate stability and soil carbon sequestration: a review. EC Nutr 12:53–80

Obame RM, Musadji NY, Menken EF, Ntoma RO, Koutika LS, Mouaromba W, Sereme BKZ, Mounguengui MM (2024) Farm waste management and recycling for negative carbon sink in agriculture: a review. J Agric Sci 17(1):128–128

Obi FO, Ugwuishiwu BO, Nwakaire JN (2016) Agricultural waste concept, generation, utilization and management. Niger J Technol 35:957–964. https://doi.org/10.4314/njt.v35i4.34

Oldfield EE, Eagle AJ, Rubin RL, Rudek J, Sanderman J, Gordon DR (2021) Agricultural soil carbon credits: making sense of protocols for carbon sequestration and net greenhouse gas removals. Environmental Defense Fund, New York. https://www.edf.org/sites/default/files/content/agricultural-soil-carbon-credits-protocol-synthesis.pdf. Accessed on 24 Apr 2025

Panneerselvam P, Senapati A, Mitra D, Priyadarshini A, Shadangi S, Behera S, Kumar U, Kumar A, Shahid M, Sharma S, Garhwal RS, Mani I, Kumar Nayak AK (2024) Enhancing soil quality and yield through microbial assisted in-situ residue management in rice-rice cropping system in Odisha, Eastern India. J Environ Manag 358:120916. https://doi.org/10.1016/j.jenvman.2024.120916. PubMed: 38642486

Paul C, Bartkowski B, Dönmez C, Don A, Mayer S, Steffens M, Weigl S, Wiesmeier M, Wolf A, Helming K (2023) Carbon farming: are soil carbon certificates a suitable tool for climate change mitigation? J Environ Manag 330:117142. https://doi.org/10.1016/j.jenvman.2022.117142. PubMed: 36608610

Peng X, Jiang Y, Chen Z, Osman AI, Farghali M, Rooney DW, Yap PS (2023) Recycling municipal, agricultural and industrial waste into energy, fertilizers, food and construction materials, and economic feasibility: a review. Environ Chem Lett 21:765–801. https://doi.org/10.1007/s10311-022-01551-5

Powlson DS, Whitmore AP, Goulding KWT (2011) Soil carbon sequestration to mitigate climate change: a critical re-examination to identify the true and the false. Eur J Soil Sci 62:42–55. https://doi.org/10.1111/j.1365-2389.2010.01342.x

Pradhan G, Meena RS (2023) Interaction impacts of biocompost on nutrient dynamics and relations with soil biota, carbon fractions index, societal value of CO_2 equivalent, and ecosystem services in the wheat-rice farming. Chemosphere 139695. https://doi.org/10.1016/j.chemosphere.2023.139695

Pradhan G, Meena RS (2023a) Utilizing waste compost to improve the atmospheric CO_2 capturing in the rice-wheat cropping system and energy-cum-carbon credit audit for a circular economy. Sci Total Environ 164572. https://doi.org/10.1016/j.scitotenv.2023.164572

Pradhan G, Meena RS, Kumar S, Lal R (2023) Utilizing industrial wastes as compost in wheat-rice production to improve the above and below-ground ecosystem services. Agric Ecosyst Environ 358:108704. https://doi.org/10.1016/j.agee.2023.108704

Pradhan A, Rana KS, Choudhary AK, Bana RS, Thapa S, Dash AK et al (2025) Dual-crop basis residue-retained bed-planting and zinc fertilization lead to improved food-energy-water-carbon nexus in pearl millet–wheat cropping system in semi-arid agro-ecologies. Renew Sust Energ Rev 215:115579

Prakash C, Roy A, Gawdiya S (2022) Utilization of municipal wastes in agriculture. Food Scient Repor 3:13–16

Punia H, Tokas J, Malik A, Satpal, Rani A, Gupta P, Kumari A, Mor VS, Bhuker A, Kumar S (2020) Solar radiation and nitrogen use efficiency for sustainable agriculture. Resources use efficiency in agriculture, pp 177–212

Rani K, Tigga P, Roy A, Das A, Trivedi A (2022) Nutrient availability and plant productivity through PGPR: mechanisms, potential and constraints. In: Mishra S, Kumar S (eds) Advances in Agricultural Biotechnology. AkiNik Publications, Delhi, pp 140–153. https://doi.org/10.22271/ed.book.1747

Rayne N, Aula L (2020) Livestock manure and the impacts on soil health: a review. Soil Syst 4:64. https://doi.org/10.3390/soilsystems4040064

Raza MB, Meena BP, Behera SK, Rani K, Wanjari RH, Biswas AK et al (2024) Long-term influence of addition of organic and inorganic sources of nutrients on soil Zn fractions, yield and Zn uptake by maize (Zea mays L.). J Soil Sci Plant Nutr:1–19

Reeve JR, Hoagland LA, Villalba JJ, Carr PM, Atucha A, Cambardella C, Davis DR, Delate K (2016) Organic farming, soil health, and food quality: considering possible links. Adv Agron 137:319–367. https://doi.org/10.1016/bs.agron.2015.12.003

Reimer M, Kopp C, Hartmann T, Zimmermann H, Ruser R, Schulz R, Müller T, Möller K (2023) Assessing long term effects of compost fertilization on soil fertility and nitrogen mineralization rate. J Plant Nutr Soil Sci 186:217–233. https://doi.org/10.1002/jpln.202200270

Roy A, Bag K, Tigga P, Kumar P, Didawat RK (2022a) Estimation of soil organic carbon by remote sensing. In: Kumar N (ed) Current research in soil science. AkiNik Publications, Delhi, pp 77–92. https://doi.org/10.22271/ed.book.1736

Roy D, Ghosh T, Roy A, Gogoi M, Alam K, Mitra R (2022b) Enhancing agricultural water productivity of crops at field scale. In: Ghosh A, Mahajan A, Langangmeilu G, Ghosh A, Soujanya K (eds) Futuristic trends in agriculture engineering & food sciences. IIP Iterative International Publishers, Chikkamagaluru, pp 376–390

Roy, A., Jadhav, K.P., Ash, M., Dash, A.K., Raza, M.B., Reddy, D.D. & Srinivasa Rao, Ch. (2024) Carbon farming strategies in Indian agriculture: perspective and challenges. In: Research and technology advancements in agriculture (edited by Ch Srinivasa Rao, A. Dhandapani & S. Kumar). ICAR-National Academy of Agricultural Research Management: Hyderabad, pp 379–410

Sage RF, Gilman IS, Smith JAC, Silvera K, Edwards EJ (2023) Atmospheric CO_2 decline and the timing of CAM plant evolution. Ann Bot 132:753–770

Saha B, Ateia M, Tolaymat T, Fernando S, Varghese JR, Golui D, Bezbaruah AN, Xu J, Aich N, Briest J, Iskander SM (2024) The unique distribution pattern of PFAS in landfill organics. J Hazard Mater 479:135678. https://doi.org/10.1016/j.jhazmat.2024.135678

Saraswat A, Ram S, Sharma S, Chawla R, Khardia N, Chauhan D, Vishwakarma DK, Raza MB, Rajput VD, Dey P, Meena RS, Behera B (2024) EDCs exposure-induced alteration in the germination, growth, and physiological trait of the plant. In: Emerging contaminants. Woodhead Publishing, pp 159–177. https://doi.org/10.1016/B978-0-443-18985-2.00001-8

Sen M, Roy A, Rani K, Nalia A, Das T, Tigga P, Rakshit D, Atta K, Mandal S, Das A (2024) Crop residue: status, distribution, management, and agricultural sustainability. In: Meena VS, Rakshit A, Meena MD, Baslam M, Fattah IMR, Lam SS, Kaba JS (eds) Waste management for sustainable and restored agricultural soil. Academic Press, Cambridge, MA, pp 167–201. https://doi.org/10.1016/B978-0-443-18486-4.00017-8

Sen M, Garai S, Dutta S, Rakshit D, Roy A, Das R, Das S (2025) Harnessing artificial intelligence for circular economy in agriculture: reducing waste and maximising efficiency. In: Garwi J, Dzingirai M, Masengu R (eds) Integrating agriculture, green marketing strategies, and artificial intelligence. IGI Global Scientific Publishing, Pennsylvania, pp 317–362. https://doi.org/10.4018/979-8-3693-6468-0.ch011

Sharma A, Saha TN, Arora A, Shah R, Nain L (2017) Efficient microorganism compost benefits plant growth and improves soil health in Calendula and Marigold. Hortic Plant Jo 3:67–72. https://doi.org/10.1016/j.hpj.2017.07.003

Sharma P, Singh A, Kahlon CS, Brar AS, Grover KK, Dia M, Steiner RL (2018) The role of cover crops towards sustainable soil health and agriculture—a review paper. Am J Plant Sci 09:1935–1951. https://doi.org/10.4236/ajps.2018.99140

Sharma B, Vaish B, Monika, Singh UK, Singh P, Singh RP (2019) Recycling of organic wastes in agriculture: an environmental perspective. Int J Environ Res 13:409–429

Sharma S, Saraswat A, Khardia N, Chawla R, Ram S, Kumar R, Raza MB, Kumar G, Behera B, Dash AK, Al Tawaha AR (2024) Soil-based implication approach for environmental nexus. Environmental Nexus for Resource Management. CRC Press, pp 67–80

Sheoran S, Gupta M, Kumari S, Kumar S, Rakshit S (2022) Meta-QTL analysis and candidate genes identification for various abiotic stresses in maize (*Zea mays* L.) and their implications in breeding programs. Mol Breed 42(5):26

Singh N, Chaturvedi V (2023) Understanding carbon markets: prospects for India and stakeholder perspectives. https://ercst.org/document/understanding-carbon-markets-prospects-for-India-and-stakeholder-perspectives/. Accessed on 2 Mar 2024

Singh Y, Sidhu HS (2014) Management of cereal crop residues for sustainable rice-wheat production system in the Indo-Gangetic Plains of India. Proc Indian Nat Sci Acad 80:95–114. https://doi.org/10.16943/ptinsa/2014/v80i1/55089

Sjöström AE, Collins CD, Smith SR, Shaw G (2008) Degradation and plant uptake of nonylphenol (NP) and nonylphenol-12-ethoxylate (NP12EO) in four contrasting agricultural soils. Environ Pollut 156:1284–1289. https://doi.org/10.1016/j.envpol.2008.03.005. PubMed: 18433956

Sodhi GPS, Beri V, Benbi DK (2009) Soil aggregation and distribution of carbon and nitrogen in different fractions under long-term application of compost in rice–wheat system. Soil Tillage Res 103:412–418. https://doi.org/10.1016/j.still.2008.12.005

Song B, Almatrafi E, Tan X, Luo S, Xiong W, Zhou C, Qin M, Liu Y, Cheng M, Zeng G, Gong J (2022) Biochar-based agricultural soil management: an application-dependent strategy for contributing to carbon neutrality. Renew Sust Energ Rev 164:112529. https://doi.org/10.1016/j.rser.2022.112529

Tripathy S, Biswas S, Singh P, Ghose TJ, Purakayastha TJ, Ahmed N, Pandey RN, Das TK, Ramakrishnan B (2023) Soil quality, resilience, and crop productivity under 32-year-old long-term rice–rice system in acidic alfisol of Assam. J Soil Sci Plant Nutr 23(4):5333–5344

Urmi TA, Rahman MM, Islam MM, Islam MA, Jahan NA, Mia MAB, Akhter S, Siddiqui MH, Kalaji HM (2022) Integrated nutrient management for rice yield, soil fertility, and carbon sequestration. Plants 11:138. https://doi.org/10.3390/plants11010138. PubMed: 35009141

Wright J, Kenner S, Lingwall B (2022) Utilization of compost as a soil amendment to increase soil health and to improve crop yields. Open J Soil Sci 12:216–224. https://doi.org/10.4236/ojss.2022.126009

Yadav R, Ramakrishna W (2023) Biochar as an environment-friendly alternative for multiple applications. Sustainability 15:13421

Yadav B, Moharana PC, Malav LC, Jangir A, Sharma RP, Dash AK, Mina BL (2024) Geospatial techniques for mapping land degradation vulnerability in a hot arid agroecosystem of India. Sustainable Land Management in India. Springer Singapore pp 201–217

Zhang W, Hendrix PF, Dame LE, Burke RA, Wu J, Neher DA, Li J, Shao Y, Fu S (2013) Earthworms facilitate carbon sequestration through unequal amplification of carbon stabilization compared with mineralization. Nat Commun 4:2576. https://doi.org/10.1038/ncomms3576. PubMed: 24129390

Zhang D, Li D, Bywater A, Short M, Sadhukhan J (2025) Carbon credits monetary value for anaerobic digestion systems and energy policy implication in the UK. Innov Energy 2(1): 100066-1

Zhao GH, Yu YL, Zhou XT, Lu BY, Li ZM, Feng YJ (2017) Effects of drying pretreatment and particle size adjustment on the composting process of discarded flue-cured tobacco leaves. Waste Manag Res 35:534–540. https://doi.org/10.1177/0734242X17690448. PubMed: 28190373

Zimmerman AR (2010) Abiotic and microbial oxidation of laboratory-produced black carbon (biochar). Environ Sci Technol 44:1295–1301. https://doi.org/10.1021/es903140c. PubMed: 20085259

Chapter 10
Solar Energy for Carbon-Negative Agriculture and Carbon-Credit Generation

Neha Thakur, Dixit Chaudhary, Ashna Acharya, and Priyanka Sharma

Abstract The entire world is currently dealing with a range of critical challenges, including unchecked population growth, excessive resource exploitation, a lack of clean water, and climate change. Among the many problems, climate change can be considered the elephant in the room. Numerous man-made and natural activities are to blame for it, but one of the largest and most inevitable sources is greenhouse gas (GHG) emissions from agriculture. Recognizing this, governments across the globe have undertaken steps to lessen agriculture's carbon footprint. A crucial part of both national and international emission trading programs is carbon credits. These credits that farmers obtain by lowering their carbon footprint might be used as an extra revenue stream. Utilizing solar energy in agriculture and allied activities, such as greenhouses, irrigation systems, dryers, pasteurizers, and farm machinery, can help farmers reduce their carbon footprint by significantly reducing their dependence on fossil fuels. This reduction in GHG emissions can be quantified and certified, enabling farmers to earn carbon credits. With an estimated potential of 6500 TW, which is sufficient to meet a major portion of global energy needs, solar energy is not only environmentally friendly but economical as well. Additionally, many governments throughout the world have issued regulations and subsidies to promote the use of solar energy in agriculture. Aside from that, using solar energy is the most straightforward and accessible of all renewable energy sources, making it an effective tool for reducing carbon footprint as well as generating additional income.

N. Thakur (✉)
University School of Agriculture Sciences, Rayat Bahra University, Mohali, Punjab, India

D. Chaudhary
Department of Floriculture and Landscape Architecture, Dr. Y S Parmar University of Horticulture and Forestry, Nauni, Solan, Himachal Pradesh, India

A. Acharya
Department of Floriculture and Landscaping, Punjab Agriculture University, Ludhiana, India

P. Sharma
NICE School of Business Studies, Shobhit University, Meerut, India

R. S. Meena et al. (eds.), *Carbon-Negative Agriculture*, Sustainability Sciences in Asia and Africa, https://doi.org/10.1007/978-981-95-1483-0_10

Keywords Agrivoltaics · Carbon credit · Carbon footprint · Greenhouse gases · Solar energy · Solar energy policies

Abbreviations

CBAM	Carbon Border Adjustment Mechanism
CER	Certified Emission Reductions
CFA	Central Financial Assistance
CH_4	Methane
CO_2	Carbon dioxide
DISCOMs	Distribution companies
DOER	Department of Energy Resources
DWPs	Diesel water pumps
ETs	Electric tractors
EUETS	European Union's Emission Trading System
FiT	Feed-in-tariff
FPC	Flat plate collector
FPOs	Farmer producer organizations
GDP	Gross domestic product
GHGs	Greenhouse gases
HSAT	Horizontal single-axis tracking
IPCC	Intergovernmental Panel on Climate Change
LCA	Life cycle assessment
LED	Light-emitting diode
MIIT	Ministry of Industry and Information Technology
N_2O	Nitrous oxide
NABARD	National Bank for Agriculture and Rural Development
NEA	National Energy Administration
PFC	Power Finance Corporation
PV	Photovoltaic
REC	Rural Electrification Corporation
REPPs	Renewable energy power plants
SERC	State Electricity Regulatory Commission
SMART	Solar Massachusetts Renewable Target
SPG	Solar power generators
SPVWPS	Solar photovoltaic water pumping system
SPWP	Solar photovoltaic water pump
TW	Terawatt
WUA	Water user associations

10.1 Introduction

Numerous nations are increasingly compelled to investigate and switch to environmentally friendly sources of energy. This shift is driven by the growing demand for energy and the detrimental effects that fossil fuels have on the environment and other related issues (Solangi et al. 2011). In terms of revenue generation, energy is the foundation of the contemporary world, and solar energy has been placed at the forefront, essentially for its dual role in mitigating carbon emissions and generating carbon credits in agroecosystems. To lower greenhouse gas (GHG) emissions and heating expenses, green energy sources like solar energy are required due to uncertainties about the security of energy. Solar energy is a clean form of energy, which is also non-exhaustible, making it more reliable than traditional sources of energy (Hassanien et al. 2016). By 2050, solar energy is projected to be the main source of electricity (Krasner et al. 2025). Further, the use of solar energy in agriculture and allied activities offers a great solution to the problem of elevated levels of carbon emissions from the agriculture sector and reduces the overall carbon footprint (Jhariya et al. 2022). The term "carbon footprint" was first used to refer to a subset of Wackernagel and Rees' (1996) "ecological footprint." The ecological footprint, measured in global hectares, is the amount of ecologically productive land and ocean needed to support a specific human population. This idea defines carbon footprint as the amount of land needed to absorb carbon dioxide (CO_2) emitted throughout an individual's life or the lifetime of the entire world's population.

Although solar energy can be used to reduce the carbon footprint in varied ways, its integration in agroecosystems offers the most sustainable approach as it offers environmental benefits through carbon sequestration (Nordberg et al. 2021; Meena et al. 2025) and monetary benefits via carbon credits and governmental incentives. Solar energy can be used in a wide array of farm operations such as greenhouses, irrigation systems, solar dryers, pasteurization units, farm machinery, and robotics. The photovoltaic (PV) system, in particular, is in trend and is being installed by farmers on farms. Further, agrivoltaics enable farmers to go for concurrent production of crop and energy (Cuppari et al. 2024). By using solar energy in farm operations, farmers can mitigate the amount of CO_2. This mitigation can be quantified, certified, and monetized by using carbon-credit schemes, offering both environmental and economic leverage. Farmers can sell these tradeable carbon credits to producers with higher GHG emissions, thus generating additional income. Further, the excess electricity generated by installed solar systems can be sold to local buyers or distribution companies (DISCOMs).

To attract farmers to incorporate solar power in agriculture, various government schemes providing incentives for using solar energy in agroecosystems have been announced worldwide, such as India's PM KUSUM (*Kisan Urja Suraksha evam Uthaan Mahabhiyan*), *Magel Tyala Saur Krushi Pump Yojana, Atal Solar Krishi Pump Yojana* (ASKP), China's PV poverty alleviation projects, and the United States' Solar Massachusetts Renewable Target (SMART), which can be leveraged by farmers. A global revenue of $104 billion collected from carbon taxes and emissions trading schemes in 2023 (World Bank 2024) further underscores the significance of carbon credits and trading. Solar energy is an effective tool for converting agroecosystems into carbon-neutral, sustainable businesses. Its integration into agriculture creates prospects for economic growth through the trading of renewable energy and carbon credits, in addition to helping to replace traditional energy sources. In this chapter, we have made an attempt to discuss the role of solar energy in reducing the overall carbon footprint of agroecosystems using solar-assisted farming systems and the generation of carbon credits. In addition, we have explored the various measures implemented by governments worldwide to enhance the adoption of solar energy in the agricultural sector through policies and financial incentives.

10.2 Solar Energy and Carbon Credits in Agriculture

The agriculture sector plays a major role in global greenhouse gas (GHG) emissions, particularly methane (CH_4), nitrous oxide (N_2O), and carbon dioxide (CO_2) (Kabir et al. 2022; Meena et al. 2023). Every day, farming practices, including irrigation and the use of machinery, rely on fossil fuels, which further contribute to GHG emissions (Lokuge and Anders 2022). Solar energy captured from the sun is a clean and renewable alternative to fossil fuels. Technologies like solar panels can transform this energy into practical electricity for diverse applications, such as lighting, heating, water pumping, and cooling systems like refrigeration. Since the solar system emits no GHG when in use, it is considered an important way to lessen agriculture's environmental footprint. Solar panels are durable, operating efficiently for up to 25 years with minimal maintenance after installation (Anusuya et al. 2023).

Using solar energy helps farmers lower electricity costs while also reducing reliance on traditional energy sources. Various solar systems are available for use today to meet different needs. One common type is the photovoltaic (PV) systems, which use semiconductor materials (silicon) to convert sunlight into electricity through the photovoltaic effect (Fthenakis and Leccisi 2021). When sunlight hits the PV cells, the electrons are energized and create an electric current that can be used directly or stored in batteries for later use. Another type is the solar thermal systems, which capture heat from sunlight and use it for heating water in solar water heaters, regulating greenhouse temperatures, powering solar dryers, pasteurizing milk, and driving turbines in concentrated solar power (CSP) plants.

In addition to reducing emissions, farmers can also benefit from carbon credits. A carbon credit is a tradable permit that represents the right to emit one metric ton of CO_2 or its equivalent in other greenhouse gases (Sellars et al. 2021; Mezzanotte 2019; Meena et al. 2024; Pradhan and Meena 2023). If a farmer adopts practices that cut emissions or increase carbon capture, they can generate carbon credits and sell them, creating a potential income stream.

Combining solar technology with carbon offset programs provides a dual benefit for farmers. Not only can they save on energy costs and possibly sell surplus electricity back to the grid, but they can also earn extra revenue through carbon credit markets. This approach supports sustainable farming, contributes to environmental conservation, and improves profitability (Leharwan et al. 2023). For the carbon-credit system to work, regulatory bodies must establish emission caps, setting limits on how much GHG different industries can emit. Companies that exceed their allowed emissions must purchase credits from those who emit less than their cap. This trading mechanism encourages organizations to invest in cleaner technologies and overall helps reduce emissions on a broader scale.

10.3 Types of Carbon Credits

10.3.1 Compliance/Mandatory Credits

These credits are used within government-regulated carbon markets, such as California's Cap-and-Trade Program or the European Union Emissions Trading System (EU ETS). Companies that fall under these regulations are required to hold a sufficient number of credits to balance out their greenhouse gas emissions (Fig. 10.1).

10.3.2 Voluntary Credits

These credits are not mandatory, but many individuals and companies voluntarily choose to purchase carbon credits to help balance out the emissions produced (Castagné et al. 2020). These credits support efforts like methane capturing,

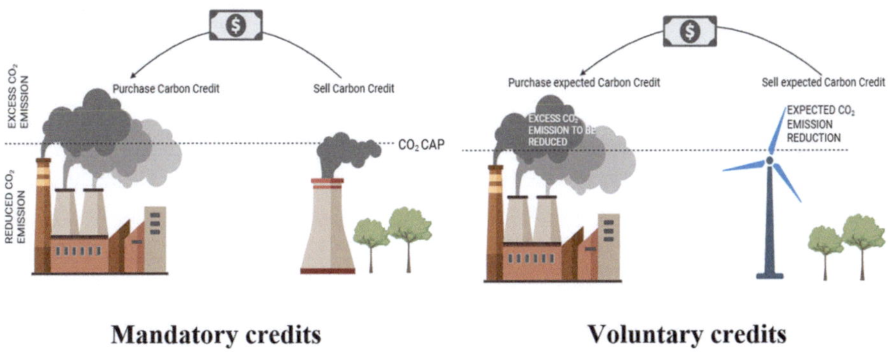

Fig. 10.1 An illustration of the mandatory and voluntary credits market

expanding renewable energy and forest restoration and are usually traded through voluntary carbon markets (Fig. 10.1).

Carbon credits help global collaboration by allowing companies and organizations, regardless of their local regulations, to contribute to emission reduction. Money from these credit systems often supports funding for environmental projects in developing nations, like reforestation, clean energy, and eco-friendly farming practices. These projects not only contribute to lowering carbon output but also help build long-term sustainability and economic growth (Sellars et al. 2021; Pradhan and Meena 2023a).

10.4 Scope and Importance of Solar Energy for Reducing Carbon Footprint and Generating Credits in an Agroecosystem

Use of solar energy in agriculture offers a promising opportunity for realizing SDGs (Sustainable Development Goals) 2015. Solar power can be used effectively to reduce GHG emissions and dependency on traditional energy resources. Additionally, by using agrivoltaics, farmers can go for the simultaneous production of food and energy (Kumar et al. 2022; Cuppari et al. 2024; Meena and Pradhan 2023). Using solar-powered water pumps to meet agricultural water needs (Nasir 2019), solar dryers for preserving agricultural produce (Ibrahim et al. 2024), solar energy for milk pasteurization (Sharma et al. 2017), and solar-powered agricultural robots for various on-farm tasks (Gorjian et al. 2021) result in a considerable reduction in the amount of CO_2 emitted. Besides on-farm mitigation of GHGs, solar energy also aids in the reduction in CO_2 emission in greenhouses by effectively regulating the environmental conditions (heating, cooling, irrigation, and lighting) primarily via employing solar collectors and photovoltaic cells. Solar energy serves as a green and sustainable alternative for the limited sources of fossil fuels (Hassanien et al. 2016).

By using solar energy in agriculture and allied activities, farmers can reduce their carbon footprint as well as participate in carbon offset programs. This reduction in carbon emissions can be converted into tradable carbon credits. This strategy not only has a positive environmental impact but also aids in farmers' income. Although there has been an increase in carbon-credit generation and transactions under the Kyoto Protocol, many countries have introduced initiatives to cut carbon emissions following the 2008 Copenhagen Summit on climate change. For example, India has set a target to cut its carbon dioxide emissions by 25% by 2023, while China has set a goal of a 40% reduction (Sarkar and Dash 2011). By adopting solar energy in agriculture and related activities, farmers can take advantage of these opportunities.

10.5 Global Review of Solar Energy Policies to Promote Use of Solar Energy in Agricultural Systems

Agriculture occupies about 35% of the total land area and contributes around 25%, 50%, and 70% of CO_2, CH_4, and N_2O, respectively, and makes up roughly 13.5% of all anthropogenic GHG emissions worldwide (Montzka et al. 2011). To reduce the carbon footprint from agricultural activities and to promote the use of solar energy in the agriculture sector, countries worldwide have taken initiatives in the form of various laws, policies, and subsidies. Some of these governmental plans and initiatives have been summarized below.

10.5.1 Indian Government's Plans and Initiatives

India is an agriculture-based economy. The agriculture sector contributes an enormous part to driving India's economic growth. In 2021, the sector accounted for roughly 45.5% of the labor force, which is a significant portion, and contributed around 20% of India's gross domestic product (GDP) (Raihan et al. 2024). However, in a developing and hugely populated country like India, improving and sustaining farmers' productivity is a huge challenge. India's geographical area receives around 5000 trillion kWh of energy annually, with the majority of regions receiving 4–7 kWh m^{-2} every day. In India, solar PV energy may be used quite successfully and has enormous scalability. Keeping this in mind, the Indian government has launched several schemes that aim to boost agricultural output while reducing the carbon footprint of the agricultural and allied sectors.

The PM KUSUM (*Kisan Urja Suraksha evam Uthaan Mahabhiyan*) is one of the many schemes launched by the Government of India to make it possible for farmers to use renewable energy in areas where grid connectivity is challenging and expensive. The scheme has been divided into three components for its smooth functioning.

(a) *Component A*: Individual farmers, groups of farmers, cooperatives, panchayats, farmer producer organizations (FPOs), and water user associations (WUA) can set up renewable energy power plants (REPPs) with capacities ranging from 500 kW to 2 MW. These plants will be referred to as solar power generators (SPGs). DISCOMs can buy the surplus solar power at a feed-in tariff (FiT) set by the relevant State Electricity Regulatory Commission (SERC).

(b) *Component B*: When grid supply is unavailable, individual farmers will receive assistance in installing standalone solar agriculture pumps in off-grid locations.

A Central Financial Assistance (CFA) of 30% of the stand-alone solar agriculture pump's benchmark cost or the tender cost, whichever is lower, will be given. At most, the farmer will contribute 40% of the remaining amount, with the state government providing at least a 30% subsidy. Farmers can obtain bank financing, which requires them to pay only 10% of the total cost initially and the remaining up to 30% of the cost as a loan.

(c) *Component C*: Under this component, farmers who own grid-connected agriculture pumps will receive assistance in solarizing their pumps. Under the plan, solar PV capacity could be increased up to double the pump capacity in kW.

The extra solar energy will be sold to DISCOMs, after meeting farmers' irrigation demands. Additionally, states can solarize agricultural feeders in place of individual solar pumps. If agricultural feeders are not separated, National Bank for Agriculture and Rural Development (NABARD) or Power Finance Corporation (PFC)/Rural Electrification Corporation (REC) loans may be obtained for feeder separation (Government of India, Ministry of New and Renewable Energy n.d.).

The *Magel Tyala Saur Krushi Pump Yojana* was started by the Maharashtra government to assist farmers in installing solar-powered pumps to irrigate their land. Farmers having up to 2.5 acres of land are eligible for solar agricultural pumps with a maximum capacity of 3 HP; those with between 2.51 and 5 acres are eligible for solar agricultural pumps with a maximum capacity of 5 HP; and those with more than 5 acres of land are eligible for solar agricultural pumps with a capacity of 7.5 HP. Furthermore, it will be acceptable if farmers request solar agricultural pumps with a capacity lower than that which has been approved (*Mahavitaran Magel Tyala Saur Krushi Pump Yojana*). The government has also introduced the *Atal Solar Krishi Pump Yojana* (ASKP), which provides farmers with up to a 95% subsidy on solar pump sets. Additionally, to encourage farmers to adopt solar farming pumps, the Maharashtra government has chosen to give away two light-emitting diode (LED) lamps, a direct current (DC) fan, and a mobile charging connector.

Additionally, the farming community directly benefited from missions like the National Energy Efficiency Enhanced Mission, National Solar Project, and National Green India Mission. The government of Gujarat is planning to sanction the construction of the largest solar power plant in the world, an undertaking valued at US $10 billion. Haryana Renewable Energy Development Agency (HAREDA) has consented to construct three solar-powered initiatives to empower businesses and industries to utilize solar energy around Haryana's Gurgaon. The government's 2017 plan features the solar proposal as the most recent addition (Government of India Power Sector at a Glance ALL INDIA | Ministry of Power n.d.). The program will assist in moving farmers from farming to diversify their revenue streams to increase farmland profitability while reducing unemployment during slow periods during the offseason.

10.5.2 Chinese Government's Plans and Initiatives

The Chinese government has released several measures to support the expansion of PV energy to encourage the use of solar energy in agriculture and related areas. To encourage the sustainable growth of the PV industry, the Ministry of Industry and Information Technology (MIIT) released the Twelfth Five-Year Plan in February 2012. The plan stated that state financial subsidies might be used to boost electricity generation based on solar photovoltaic greenhouses (China's five-year plan). On October 11, 2014, the National Energy Administration (NEA) and State Council Poverty Alleviation Office (SCPAO) released a "Notice on the Working Scheme of the Implementation of the PV Poverty Alleviation Projects." This shows the Chinese government's determination to boost the use of solar energy in agriculture. Solar projects are carried out by constructing the distributed PV system for impoverished families in poor districts of more than 30, by establishing PV power plants on agricultural greenhouses or less fertile hills and slopes. Furthermore, in 6 years, the installed capacity will reach 1.4 GW, which may provide poor households with an annual income of US$ 0.151 billion in electricity. Besides this, China has issued a series of support policies for the PV industry since 2013, many of which directly target the agriculture sector (Xue 2017).

10.5.3 US Government's Plans and Initiatives

The US government has aided in the expansion of solar energy through a variety of market, regulatory, fiscal, and other tools. In the United States, the Massachusetts Department of Energy Resources (DOER) is subsidizing agrivoltaics with compensation rates as high as US$0.06 kWh^{-1}, under the Solar Massachusetts Renewable Target (SMART) program. With this, Massachusetts, the DOER, is spearheading the development of agrivoltaics incentives. An agrivoltaics site must adhere to strict design guidelines to be eligible for the SMART program. These requirements include a minimum panel height of 10 feet (3.0 m) for horizontal single-axis tracking (HSAT) and 8 ft (2.4 m) for fixed-tilt systems, as well as a maximum rated capacity of 5 Megawatt AC and a maximum of 50% shade in any given spot throughout the solar site concerning baseline conditions on the field, which are determined by a "shading-only" tool created for DOER SMART program assessment (Williams et al. 2025). Agrivoltaic-friendly laws have been approved in ten US states; many of these laws stress solar projects that are pollinator-friendly (Tajima et al. 2022; Sheoran et al. 2022).

Governments all over the world have become conscious of the importance of solar energy in reducing their carbon footprint and generating carbon credits. A lot of initiatives have been taken up by many national authorities to promote sustainable energy use in agroecosystems. In its 2019 "Follow-up on the Growth Strategy," Japan declared that "Farming-photovoltaics, where PV equipment is set up above agricultural land, will be developed countrywide" to support the farming sector in a nation experiencing population decline (Cabinet Office Japan, 2019). In South Korea, the government wants to generate 20% of its energy needs from renewable sources, according to the Renewable Energy 3020 Plan, by 2030. For this purpose, a goal of providing 10 GW of agrivoltaics by 2030 has been adopted (Ministry of

Trade, Industry and Energy, "Korea's Renewable Energy 3020 Plan," 2018). Sri Lanka in 2005 launched SPIS (Solar Powered Irrigation System) equipped with a 150 Wp solar panel and a centrifugal pump under the Sustainable Agriculture Water Management Project (SAWMP) (Aheeyar et al. 2012).

The acknowledgement by governments worldwide of the significance of solar energy in reducing carbon footprint is praiseworthy. But to achieve these objectives, a well-defined and encouraging legal and policy framework is vital.

10.6 Strategies for Reducing Carbon Footprint at the Field Level Through Solar Energy

10.6.1 Using a Solar-Powered Water Pumping System

In rural areas of countries, agricultural fields require water delivery systems that are both economical and dependable. Switching from diesel-powered water pumps to solar-powered ones could be an efficient approach to achieve this, in addition to reducing the carbon footprint. In the past few decades, PV systems have been used more often in all types of irrigation, particularly in isolated and arid regions without a consistent electrical supply. A solar water pump system is a comparatively basic apparatus consisting of solar panels, tubing, an electrical control, and a water pump (Thakur et al. 2022). Solar photovoltaic water pump (SPWP), or SPVWP, is a cost-effective and environmentally conscious way to direct water in isolated areas. It is simple to install, has a secure lifespan, and requires little to no maintenance. However, it requires a substantial initial investment and yields nothing when it's cloudy. The different components that make up SPVWPs can be categorized as mechanical, electrical, or electronic parts. Every SPVWP configuration has its corresponding benefits and drawbacks. The best system for a specific application is chosen after taking into account several variables (Sontake and Kalamkar 2016).

A study by Nasir (2019) reported that a 1500 kg reduction in CO_2 emissions was achieved annually by switching to SPVWPs from Diesel Water Pumps (DWPs). In a different experiment by Chazarra-Zapata et al. (2020), it was shown that using three PV installations reduced yearly CO_2 emissions by a total of 341.99 t CO_2. Employing an economic life cycle analysis, Yang et al. (2014) calculated the direct and indirect CO_2 emissions of a solar PV-driven water pumping system. The authors studied Inner Mongolia, China, as a case study for a PV system of 3.4 kW with an estimated 20-year lifespan. The outcomes of their investigation showed that 20,060 kg of CO_2 were emitted overall, where PV panels provide 43.5% of emissions, of which 89% of this has been attributed to the panels' construction phase. According to the publication, the case study showed that solar water pumping reduces CO_2 emissions overall by about 129 Mg CO_2 ha^{-1}.

10.6.2 Solar-Powered Greenhouses

Solar-powered greenhouses are a fantastic alternative to greenhouses that rely on traditional energy sources. Two of the biggest expenses associated with greenhouse cultivation are heating and cooling systems. Protected agriculture can be performed easily and economically by employing solar energy. Additionally, solar energy can be used to power lighting systems in protected agricultural cultivation facilities to lower carbon emissions.

According to research in Spain, integrating PV panels on greenhouses would take roughly 18 years to pay for itself (Ureña-Sánchez et al. 2012). Moreover, the usage of solar energy in specific areas can lower GHG emissions by up to 2,43,252 tons annually. A greenhouse's carbon footprint can decrease to that of open-field horticulture when solar energy fulfils its heating needs (Ntinas et al. 2011; Meena et al. 2024a). In an energy-life cycle assessment on using the solar-powered greenhouse for strawberry cultivation by Fatmeh et al. (2019), it was reported that the use of solar energy instead of diesel fuel significantly reduced the cumulative energy demand by about 50%, thereby reducing the carbon footprint. Liantas et al. (2023) also experimented to compare the energy use efficiency and carbon footprint of greenhouses powered by electricity and PV. The results of the study showed that solar panels could meet 58.0, 83.3, and 9.6% of energy for heating and cooling in spring, summer, and winter, respectively, with a significantly low carbon footprint of 1.6 kg CO_2eq kg^{-1} as compared to 49.9 kg CO_2eq kg^{-1} from the electricity-powered greenhouse.

10.6.3 Solar Dryers

Since ancient times, people have dried food and agricultural commodities outside in the open air utilizing the sun. Yet, there are several drawbacks to this technique, including damaged goods as a result of wind, rain, dampness, and dust; and deterioration of harvested crops as a result of fungi, insects, and decomposition. In addition, the procedure takes a lot of time, labor, and space to lay out the food so it can dry, and the use of artificial mechanical drying, a relatively recent technique, is energy-consuming and expensive (Sharma et al. 2009). Utilizing solar dryers in agriculture to preserve fruits, vegetables, coffee, and other commodities has been shown to be a sensible, cost-effective, and ecologically acceptable practice in numerous nations. For this purpose, two types of solar dryers can be used, namely, passive and active dryers. Passive dryers rely only on the natural circulation of warm air. These can be easily built with affordable, locally sourced materials, which make them suitable for small farms with easy access to raw materials for building, including wood. Whereas, active solar dryers are made with built-in fans or pumps to transfer hot air, which is produced by the sun, from the collector area to the drying beds.

Ibrahim et al. (2024) experimented to analyze the amount of CO_2 emitted while using traditional electric dryers, hybrid solar dryers using traditional electric energy, and PV solar energy dryers. It was reported that using PV solar dryers can mitigate the highest amount of CO_2 and can operate with 0.0% traditional energy consumption. In another study, Getahun et al. (2021) found that the use of a two-stage solar tunnel dryer could mitigate the net CO_2 emitted in a range of 10.9–13 tons year^{-1}. The carbon footprint mitigation capacity of solar dryers shows their immense potential in reducing GHG emissions and generating carbon credits.

10.6.4 Solar Pasteurizers

Pasteurization is one of the heat-treatment techniques employed at the dairy plant, and as a result, it requires a lot of heat, raising the energy requirement in the dairy industry. Since milk pasteurization happens between 65 °C and 150 °C, there are several solar thermal collector options for different types of pasteurization procedures. The dairy industry can use free solar energy by pasteurizing milk using solar thermal collectors. When solar energy is used in milk heat treatments, less fuel and electricity are used, which lower carbon emissions and encourage sustainability.

In a study conducted by Obando-Ulloa et al. (2017), it was reported that by using an LP gas solar hybrid system, the amount of GHGs emitted to the atmosphere was reduced significantly, and the system was also able to supply more than 54% of the energy required for pasteurization of milk. In the dry region of Northern Kenya, where milk marketing is the primary source of revenue for the locals, Wayua et al. (2013) used a flat plate collector (FPC) to harness solar energy for the pasteurization of camel milk. The complete initiative to create a low-cost solar milk pasteurization system was supported by the Kenya Arid and Semi-arid Lands program. The milk was pasteurized using the batch pasteurization process in the designed system, utilizing hot water from FPC to reach a temperature of 63 °C. Sharma et al. (2017) have also predicted that using solar energy for pasteurization of milk in India can mitigate 32–144 thousand tons of CO_2 emissions per year.

10.6.5 Solar-Powered Farm Machinery

Solar PV technology can also be integrated with farm machinery and robotics, which can contribute to a significant reduction of carbon footprint. Electrical tractors (ETs) are charged using fixed PV panels that are placed in designated locations, such as farm building roofs or charging stations, in the off-board charging configuration. Solar-powered agricultural robots have enormous potential for usage in a variety of agricultural tasks, such as sowing, weeding, spraying, harvesting food, and ploughing in both open cultivation and greenhouses as well. Currently, most agricultural robots run on batteries; however, many companies worldwide are investing

Table 10.1 Some noteworthy studies on solar-assisted farming systems

Application	Key findings	References
PV-integrated electric tractor	Emission reduction of 28,000 kg CO_2 eq. per year and fuel saving of 7.6 L ha^{-1}	Mousazadeh et al. (2011)
Water pumping system	Drinking water in rural and urban areas, as well as irrigation, was successfully provided by solar PV water pumping technology	Chandel et al. (2015)
Domestic hybrid solar dryer	12.28 tons CO_2 mitigation by tomato drying in the dryer's lifetime. A payback period of 6 months for drying tomato crops was estimated	Sharma et al. (2023)
Photovoltaic irrigation system (PVIS)	The use of PVIS for irrigation allowed for saving between 69 and 100% of the energy consumed for irrigation, thereby reducing a significant amount of carbon emissions	Todde et al. (2018)
Solar-assisted greenhouse cooling	This study showed that an underground air tunnel solar PV device, with assistance from an EAHE, can be employed for more effective passive cooling in greenhouses	Yildiz et al. (2012)
Pasteurization system	In the primary circuit, water served as the heat transfer fluid to move heat from the solar panel to the heat exchanger or pasteurizer, with a capacity of 200 L hr^{-1}	Nielsen et al. (2002)
Solar dryer	The amount of CO_2 emission was reduced by 6400 kg per month, and 780 US$ per month was saved while using the solar dryer for carrots.	El-Hage et al. (2018)
Solar electric multipurpose vehicle	Emission reduction of 23,000 kg CO_2eq. per year and fuel saving of 3.58 L hr^{-1}	Mousazadeh et al. (2009)
Solar milk pasteurization system	Milk pasteurization at a temperature of 73 °C could be achieved at a capacity of 350 L hr^{-1}	Lazaar et al. (2021)
Solar-powered greenhouses	The article explores the use of solar energy in greenhouse lighting, cooling, and heating systems	Hassanien et al. (2016)

PV photovoltaic, *CO_2* carbon dioxide, *PVIS* photovoltaic irrigation system, *EAHE* Earth-to-air heat exchanger

immensely in the Research and Development (R&D) of solar-powered robotics. A Swiss firm called ecoRobotix makes completely autonomous spraying robots that run on PV power and are used for removing weeds in meadows, row crops, and intercropping operations (Gorjian et al. 2021). The use of a PV-integrated four-wheeled vehicle for herbicide spraying reduced the emission by 8 kg CO_2eq. ha^{-1} and saved the fuel @ 2.6 L ha^{-1} (Azwan et al. 2017). We have compiled some of the noteworthy studies on solar-assisted farming systems in Table 10.1.

10.7 Strategies for the Assessment of Carbon Credit and Carbon Footprint in an Agroecosystem

According to IPCC (2014), the agriculture sector accounts for one-fourth of total anthropogenic GHG emissions. Among all the continents, Asia contributes a significant amount (44%) of total GHG emissions from agriculture, followed by America, Africa, Europe, and Oceania (FAO 2014). However, carbon emissions from agricultural activities can only be reduced and not avoided completely. In such a situation, tradeable carbon credits and the connection of carbon markets to agricultural output are praised as win-win approaches for reducing emissions (World Bank 2023). Shockley and Snell (2021) claim that when compared with industrial emissions, GHG emissions (particularly CO_2) from agricultural operations are quite modest. To achieve climate neutrality, carbon-credit systems in the agricultural sector have drawn significant attention from around the world (Shockley and Snell 2021; Meena et al. 2024b; Pradhan et al. 2023). These systems allow agricultural producers to earn additional revenue by selling their excess carbon credits to producers who generate huge amounts of GHGs. Carbon credits are certificates given to nations that reduce the GHGs they emit into the atmosphere. These are quantified in CER (Certified Emission Reductions), and one ton of reduced CO_2 is equal to one CER. Through the trading of their surplus carbon credits to companies that release more GHGs, carbon-credit systems enable farmers to make additional money. Even so, because agricultural carbon-credit schemes are still in their infancy and there is a dearth of scientific data supporting them, these advantages cannot be assured. These carbon credits can be quantified by measuring the amount of carbon footprint reduced. Since carbon footprints are quantifiable representations of GHG emissions from activities, they are useful for managing emissions and assessing mitigation strategies. Following the quantification of emissions, the most significant sources of emissions can be located, and priority may be given to areas where emissions can be reduced and efficiency can be increased (Kumar et al. 2022a, 2025).

The quantity of GHGs released, withdrawn, or incorporated into a product's life cycle must be assessed and summed to calculate its carbon footprint. The term "life cycle" refers to all the phases that go into making a product, from the procurement of raw materials to final packaging, shipment, consumption, and disposal. When it comes to the production of air pollutants, water usage, and wastewater generation, consumption of energy, GHG emissions, or any additional related criterion significance and cost–benefit measures, life cycle assessment (LCA) provides a comprehensive picture of inputs and outputs (Kumar et al. 2020). Besides LCA, there are a plethora of methods to assess the carbon footprint of agricultural activities, which have been given in Table 10.2.

Table 10.2 A list of tools employed to assess carbon footprint in agriculture and allied activities

Method	Tool/software	References
LCA	SimaPro, GABI, openLCA	Nitschelm et al. (2021) and Goglio et al. (2018)
Carbon footprint calculators	Cool farm tool, HOLOS model, COMET-Farm	Hillier et al. (2013), Kröbel et al. (2012) and Peter et al. (2017)
Remote sensing and GIS	NDVI, Sentinel-2, FLUXNET	Wani et al. (2010), Khan et al. (2020) and Friend et al. (2007)
Carbon sequestration models	RothC, CQESTR, CENTURY	Shirato (2020), Rickman et al. (2001) and Foereid and Høgh-Jensen (2004)
Emission factor approach	IPCC guidelines	Tubiello et al. (2013) and Ioakeimidis et al. (2013)
Farm energy audits	EnSave, RETScreen	Gulkis and Clarke (2010)

IPCC Intergovernmental Panel on Climate Change, *LCA* life cycle assessment, *GIS* Geographic Information System, *NDVI* Normalized Difference Vegetation Index, *GABI* Ganzheitliche Bilanzierung

10.8 Constraints in the Generation of Carbon Credits Through Solar Energy in Agroecosystems

Carbon-credit generation through the use of solar energy in agroecosystems has observed an increasing trend lately. Numerous initiatives have emerged in the past few years, which have led to the generation of carbon credits by altering solar energy use practices in agroecosystems. As things stand, carbon-credit generation through the use of solar energy mainly comprises two methods:

(i) First, the emissions from fossil fuels can be alleviated by switching to more sustainable sources of energy like solar-powered systems or other energy-efficient agroecosystems instead of traditional energy systems.
(ii) Second, these emissions from the use of fossil fuels for generating energy can also be minimized by harnessing solar energy for carrying out various agricultural activities such as utilizing solar-powered dryers for drying the harvest, for powering farm equipment and other systems and use of solar-powered protected structures, consequently, relieving the burden on conventional energy sources.

Although the adoption of solar energy as a renewable alternative to other conventional sources of energy presents numerous benefits for meeting current energy demands, there are still some barriers that hinder the widespread integration into agroecosystems. Some of these include:

(a) A reduction in confidence and trust of customers due to the complex nature and perceived lack of transparency of the processes that are associated with carbon sequestration, further compromising the credibility of carbon-credit generation.

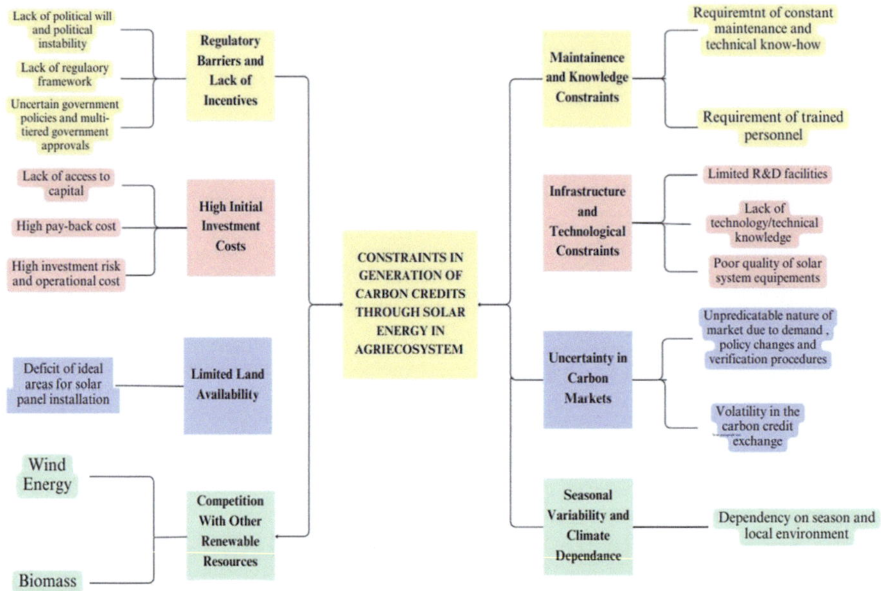

Fig. 10.2 Challenges in the generation of carbon credits through solar energy in the agroecosystem

(b) Overabundance of certification labels such as Verra, Gold Standard, Plan Vivo, and Puro.earth in the carbon markets create confusion and reduce buyer's interest (Khurana et al. 2024).
(c) Wariness regarding the long-term commitments of organizations in reducing carbon emissions (Wongpiyabovorn et al. 2023).

Overall, the market presents lower monetary bids to agricultural carbon credits (Nozaki 2021). On top of that, variations in atmospheric temperature, precipitation, soil characteristics, topographic features, and vegetation together contribute to significant spatial variability in soil carbon storage (Wang et al., 2021; Jangir et al. 2024). The following are the key constraints that are faced by agroecosystems in the generation of carbon credits (Fig. 10.2):

(i) *High initial investment costs:* A significant capital investment is the prerequisite for installing solar energy panels, especially on a scale suitable for agroecosystems. This cost can be prohibitive for many farmers, especially smallholders (Nhamo et al. 2018), while it could be a significant barrier for marginal landholders. Although many incentives and subsidies are offered by the governments, still there is a long path that needs to be maneuverer by the farmers for incorporating solar energy into their agroecosystems.
(ii) *Limited land availability:* Availability of ideal locations suitable for solar panel installation is another major constraint that hinders the adoption of solar energy at the farmer level since agricultural land is constantly in use for crop cultivation or other activities. Making tradeoffs between the

production of food and energy may be necessary when using farmland for solar energy utilization (Dupraz et al. 2011).

(iii) *Infrastructure and technological constraints:* Robust infrastructure, including grids and storage devices, is another requirement in agroecosystems which bridges the gap of distributing and storing the energy produced by solar energy systems effectively. This comprises one of the major obstacles in the rural areas where the infrastructural development is still in the nascent phase, and hence, the adoption and utilization of solar energy as a sustainable source of energy continue to be a challenge (Deichmann et al. 2011).

(iv) *Uncertainty in carbon markets*: Fluctuations in demand and amendments in policies and/or procedures of verification are causes of volatility in carbon markets, which further lead to variations in the value of carbon credits (Williams et al. 2005). This unpredictable nature of the carbon-credit market creates a feeling of distrust among the farmers for investing in the solar-energy projects intended to generate carbon-credits (Lovell et al. 2009).

(v) *Regulatory barriers and lack of incentives*: Many countries across the globe do not have clear regulations and incentives for carbon-credit generation through the use of sustainable energy, which hinders its adoption by farmers. Further, bureaucratic hurdles and other governmental procedures such as registration, tracking, and verification of carbon credits alleviate the challenge and make it very difficult for the farmers to invest in the same (Bauen et al. 2010).

(vi) *Maintenance and knowledge constraints*: Constant maintenance and technical know-how are the prerequisites for the smooth and efficient functioning of solar energy systems. Farmers might not have the expertise to maintain these systems or have access to trained workers (Omer 2008).

(vii) *Competition with other renewable sources*: In certain regions, the appeal of investing in solar energy systems for carbon-credit generation may decline due to competition from other sources of renewable energy, such as wind energy and biomass, which vie for available resources or financial incentives (Upham and Shackley 2006).

(viii) *Seasonal variability and climate dependence*: Solar energy generation relies heavily on sunlight, which can vary significantly with seasonal changes and local environmental conditions. This variability can make it harder to reliably predict the amount of carbon credits that solar energy projects will produce (Bhattacharyya and Timilsina 2010).

10.9 Plan and Policy Framework on the Use of Solar Energy in an Agroecosystem

The practices employed in agroecosystems primarily utilize fossil fuel-based technologies, which release GHGs, add substantially to environmental degradation, and are hazardous. Given these considerations, it is not unexpected that the utilization of solar energy to undertake different agricultural activities such as ploughing, water lifting, and irrigation, along with other operations, is becoming more and more commonplace throughout the world, particularly in low- and middle-income nations, as a means of addressing the energy and climate issues associated with agro-based ecosystems. Many regard solar-powered agro-systems as transformative, enhancing smallholder agricultural output and offering solutions for climate-smart agriculture development (Lefore et al. 2021). However, a major lacuna in the appropriate adoption of solar energy in agroecosystems is the lack of an integrated approach. There needs to be an appropriate plan and policy framework for promoting the use of solar energy in an agroecosystem. This could be accomplished by:

Research and Development The first step in undertaking a policy framework program for the incorporation of solar energy in agroecosystems is to research the existing pedological and agrometeorological conditions of the region. Region-specific research should be conducted to determine optimal solar energy panel geometries, module types, and shading effects on different crops (Aroca-Delgado et al. 2018).

Education and Awareness Campaigns should be conducted to demonstrate the benefits of the use of solar energy in agro-based ecosystems. After disseminating the basic knowledge, training should be provided to the farmers (Cohn 2008). The use of tools like the parametric open-source cold-frame agrivoltaic system (POSCAS), which is in great use in Canada, could be adopted for the large-scale adoption of solar energy by the farmers (Pearce 2021).

Framework of Policy and Regulation The roles of different solar energy systems should be properly defined. The example of Alberta could be taken into consideration for framing a proper policy setup for incorporating solar energy into agroecosystems (Jamil and Pearce 2022). The next step would be implementing a multi-tier system for the application of solar energy to enhance productivity in agricultural systems.

Financial Incentives For the installation of solar energy systems, tax incentives and subsidies should be provided to the end-users or farmers on initial capital expenditure. A policy setup in Canada is a great example where low-interest loans are provided for installing solar energy systems, and these systems are included under carbon-credit schemes to promote adoption (Pearce 2022).

Collaboration Between Ministries A cooperative relationship between the agriculture and energy ministries, along with local governments, should be prioritized for the joint implementation of solar energy-based policies. Municipal land-use planning policies ought to be facilitated at the farmer level (Pascaris 2021).

Standardization and Quality Control Standardization of a suitable protocol for designing, installing, and maintaining solar energy in agroecosystems is an important step for promoting its adoption (Agir et al. 2023). It could be further enhanced by enabling a safe and easy certification process for solar energy installation projects to ensure compliance with technical, environmental, and operational guidelines (Yang 2015).

Socioeconomic Development The installation of solar energy systems could serve the additional purpose of increasing rural employment opportunities in addition to generating revenue from enhanced crop production and the sale of electricity, similar to the positive employment impact seen in Alberta (Ghosh 2023).

Biodiversity Protection These solar energy-based systems encourage pollinator habitats, enhancing ecosystem services provided by agricultural lands (Blaydes et al. 2021). The above-mentioned policy framework could create a synergistic environment for incorporating solar energy into agroecosystems, which would ensure sustainability, increase productivity, and positively reap economic benefits for the end-user.

10.10 Future of Carbon-Credit Program in Agroecosystem Through Harvesting Solar Energy

According to Chandler et al. (2002), developing nations are predicted to surpass the developed nations in terms of greenhouse gas emissions. Integrating solar-powered systems, that is, agrivoltaic systems, which employ PV panels erected over agricultural fields, increases both energy production and agricultural output, thereby serving a dual purpose (Cuppari et al. 2024). The future of carbon-credit generation in agroecosystems through the use of solar energy majorly relies on a combination of government policies, evolving stakeholder perspectives, implementation of technological innovations and international partnerships. For instance, the Carbon Border Adjustment Mechanism (CBAM) proposed by the European Union has the objective of imposing heavy carbon tariffs on imports from nations with lenient and flexible climate policies (Park et al. 2023). Indian exports, including agricultural products, will more likely suffer from this, as additional carbon taxes would have to be paid. Also, this could also possibly lead to stricter legislation among such countries, further resulting in the enactment of cleaner technology and a domestic carbon-pricing system or decarbonization measure (Asian Development Bank 2024; European Parliament 2022). These steps may support investments in low-carbon emitting technologies and conjure governments to intervene in market quality and transparency. India is the third-largest emitter of GHGs, and the agriculture sector could significantly decrease the annual GHG emissions of the country by 84% from

2019 to 2070. Adoption of sustainable agricultural practices, mainly the use of solar energy, can be utilized to achieve net carbon neutrality by 2070 (Khurana et al. 2024). Some of the global initiatives, such as India's PM KUSUM, China's PV poverty alleviation projects and the United States' SMART scheme, aim at incorporating solar energy into agroecosystems and illustrate the potential of generating carbon credits by harvesting solar energy. By providing complete support to farmers, this integration can improve the efficiency and durability of carbon credits in the agriculture sector (World Bank 2011). With the expansion of international carbon markets, solar-powered agroecosystems will become major contributors to carbon-credit trading.

10.11 Conclusion

Increasing emissions of greenhouse gases (GHGs) and elevating levels of carbon footprint is a great challenge to the already bleak global climate scenario. Agriculture and related activities contribute a major portion to the overall world carbon footprint. However, nations all over the world depend largely upon agricultural activities, and carbon emissions from agriculture are inevitable. For this, using solar energy in agriculture as a source of clean energy can have a huge impact on reducing the overall carbon footprint and increasing agriculture's carbon credit. The government have taken many initiatives to promote the usage of solar power in agriculture. This can also help farmers in generating additional income from surplus carbon credits produced. However, uncertain markets, limited land, regulatory barriers, and constraints in technologies and incentives pose a serious challenge to the adoption of these initiatives by farmers. Therefore, a strong policy framework, immaculate regulations, and stringent laws at the national as well as international levels are required to encourage farmers to set up solar-powered agroecosystems.

References

Agir S, Derin-Gure P, Senturk B (2023) Farmers' perspectives on challenges and opportunities of agrivoltaics in Turkiye: an institutional perspective. Renew Energy 212:35–49

Aheeyar M, Bandara MACS, Padmajani (2012) Assessment of solar powered drip irrigation project implemented by Ministry of Agriculture – Phase 1

Anusuya K, Vijayakumar K, Manikandan S (2023) From efficiency to eternity: a holistic review of photovoltaic panel degradation and end-of-life management. Sol Energy 265:112135

Aroca-Delgado R, Pérez-Alonso J, Callejón-Ferre ÁJ, Velázquez-Martí B (2018) Compatibility between crops and solar panels: an overview from shading systems. Sustainability 10(3):743

Asian Development Bank (2024) EU Carbon tariff likely to have limited impact on emissions without global efforts. Available at: Accessed 2 Oct 2024

Azwan MB, Norasikin AL, Sopian K, Abd Rahim S, Norman K, Ramdhan K, Solah D (2017) Assessment of electric vehicle and photovoltaic integration for oil palm mechanisation practise. J Clean Prod 140:1365e1375

Bauen A, Dunnett A, Richter GM, Dailey AG, Aylott M, Casella E, Taylor G (2010) Modelling supply and demand of bioenergy from short rotation coppice and miscanthus in the UK. Bioresour Technol 101(21):8132–8143
Bhattacharyya SC, Timilsina GR (2010) A review of energy system models. Int J Energy Sect Manag 4(4):494–518
Blaydes H, Potts SG, Whyatt JD, Armstrong A (2021) Opportunities to enhance pollinator biodiversity in solar parks. Renew Sust Energ Rev 145:111065
Cabinet Office (2019) Follow-up on the growth strategy. Tokyo, Japan. [Online]
Castagné M, Lickel S, Ritter T, Dufrasne G (2020) Carbon markets and agriculture – why offsetting is putting us on the wrong track. Institute for Agriculture and Trade Policy. https://www.iatp.org/carbon-markets-and-agriculture
Chandel SS, Nagaraju NM, Chandel R (2015) Review of solar photovoltaic water pumping system technology for irrigation and community drinking water supplies. Renew Sust Energ Rev 49: 1084–1099
Chandler W, Secrest TJ, Logan J, Schaeffer R, Szklo AS, Schuler ME, Dadi Z, Kejun Z, Yuezhong Z, Huaqing X, Shukla PR (2002) Climate change mitigation in developing countries: Brazil, China, India, Mexico, South Africa, and Turkey. Available at: https://www.c2es.org/wp-content/uploads/2002/10/dev_mitigation.pdf. Accessed 26 Apr 2025
Chazarra-Zapata J, Molina-Martínez JM, Cruz FJ, Parras-Burgos D, Ruíz- Canales A (2020) How to reduce the carbon footprint of an irrigation community in the South-East of Spain by use of solar energy. Energies 13(11):2848
Cohn A (2008) A revolution fueled by the Sun. Berkeley Review of Latin American Studies. University of California, Berkeley, p 25
Cuppari RI, Branscomb A, Graham M, Negash F, Smith AK, Proctor K, Rupp D, Ayalew AT, Tilaye GG, Higgins CW, Abou Najm M (2024) Agrivoltaics: synergies and trade-offs in achieving the sustainable development goals at the global and local scale. Appl Energy 362: 122970
Deichmann U, Meisner C, Murray S, Wheeler D (2011) The economics of renewable energy expansion in rural Sub-Saharan Africa. Energy Policy 39(1):215–227
Dupraz C, Marrou H, Talbot G, Dufour L, Nogier A, Ferard Y (2011) Combining solar photovoltaic panels and food crops for optimising land use: towards new agrivoltaic schemes. Renew Energy 36(10):2725–2732
El-Hage H, Herez A, Ramadan M, Bazzi H, Khaled M (2018) An investigation on solar drying: a review with economic and environmental assessment. Energy 157:815–829
European Parliament (2022) Opinion of the Committee on Agriculture and Rural Development. Available at: https://www.europarl.europa.eu/doceo/document/AGRI-AD-699239_EN.pdf. Accessed 2 Oct 2024
FAO (2014) Food and agriculture organization of the United Nations: News article assessed on
Fatmeh HF, Ali M, Ashkan NP, Jafar HS, Kwokwing C (2019) Energy-life cycle assessment on applying solar technologies for greenhouse strawberry production. Renew Sust Energ Rev 116: 109411
Foereid B, Høgh-Jensen H (2004) Carbon sequestration potential of organic agriculture in northern Europe–a modelling approach. Nutr Cycl Agroecosyst 68:13–24
Friend AD, Arneth A, Kiang NY, Lomas M, Ogee J, Rödenbeck C, Running SW, Santaren JD, Sitch S, Viovy N, Ian WF (2007) FLUXNET and modelling the global carbon cycle. Glob Chang Biol 13(3):610–633
Fthenakis V, Leccisi E (2021) Updated sustainability status of crystalline silicon-based photovoltaic systems: life-cycle energy and environmental impact reduction trends. Prog Photovolt Res Appl 29(10):068–1077
Getahun E, Gabbiye N, Delele MA, Fanta SW, Vanierschot M (2021) Two-stage solar tunnel chili drying: drying characteristics, performance, product quality, and carbon footprint analysis. Sol Energy 230:73–90

Ghosh A (2023) Nexus between agriculture and photovoltaics (agrivoltaics, agriphotovoltaics) for sustainable development goal: a review. Sol Energy 266:112146

Goglio P, Smith WN, Worth DE, Grant BB, Desjardins RL, Chen W, Tenuta M, McConkey BG, Williams A, Burgess P (2018) Development of crop. LCA, an adaptable screening life cycle assessment tool for agricultural systems: a Canadian scenario assessment. J Clean Prod 172: 3770–3780

Gorjian S, Ebadi H, Trommsdorff M, Sharon H, Demant M, Schindele S (2021) The advent of modern solar-powered electric agricultural machinery: a solution for sustainable farm operations. J Clean Prod 292:126030

Government of India Ministry of New and Renewable Energy (n.d.). https://www.mnre.gov.in/. Accessed 02 Oct 2024

Gulkis A, Clarke A (2010) Agricultural energy management plans: conception to implementation. J Soil Water Conserv 65(1):18A–20A

Hassanien RHE, Li M, Dong LW (2016) Advanced applications of solar energy in agricultural greenhouses. Renew Sust Energ Rev 54:989–1001

Hillier J, Smith P, Bandel T, Daniels S, Malin D, Hamilton H, Walter C (2013) Farm-scale greenhouse gas emissions using the cool farm tool. In: Climate change mitigation and agriculture. Routledge, pp 217–226

Ibrahim A, Amer A, Elsebaee I, Sabahe A, Amer MA (2024) Applied insight: studying reducing the carbon footprint of the drying process and its environmental impact and financial return. Frontiers in Bioengineering and Biotechnology 12:1355133

Ioakeimidis C, Polatidis H, Haralambopoulos D (2013) Use of renewable energy in aquaculture: an energy audit case-study analysis. Global NEST J 15(3):282–294

IPCC (2014) Fifth assessment report of the Intergovernmental Panel on Climate Change. http://www.ipcc.ch/report/ar5/

Jamil U, Pearce JM (2022) Energy policy for agrivoltaics in Alberta Canada. Energies 16(1):53

Jangir CK, Sangwan PS, Panghaal D, Kumar S, Meena RS, Bharti J, R.D. and Singh, N. (2024) Spatial variability and statistical analysis of soil properties in the rice wheat-based systems of North-West India. Commun Soil Sci Plant Anal 55(8):1205–1223

Jhariya MK, Raj A, Banerjee A, Meena RS, Bargali SS, Kumar S, Nema S, Poonam, Oraon PR (2022) Plan and policies for soil organic carbon management under agroforestry system. In: Plans and policies for soil organic carbon management in agriculture. Springer, Singapore, pp 191–219

Kabir M, Kabir Z, Sultana N (2022) Climate change, sustainability, and renewable energy in developing economies. Renew Energy Sustain:377–415

Khan K, Iqbal J, Ali A, Khan SN (2020) Assessment of Sentinel-2-derived vegetation indices for the estimation of above-ground biomass/carbon stock, temporal deforestation and carbon emissions estimation in the moist temperate forests of Pakistan. Appl Ecol Environ Res 18(1): 783

Khurana A, Kajale D, Cariappa AA, Krishna VV (2024) Shaping India's climate future: a perspective on harnessing carbon credits from agriculture. Outlook Agric 113: 00307270241240778

Krasner NZ, Fox J, Armstrong A, Ave K, Carvalho F, Li Y, Walston LJ, Ricketts MP, Jordaan SM, Abou Najm M, Hartmann HM (2025) Impacts of photovoltaic solar energy on soil carbon: a global systematic review and framework. Renew Sust Energ Rev 208:115032

Kröbel R, Janzen HH, Beauchemin KA, Bonesmo H, Little SM, McAllister TA (2012) A proposed approach to estimate and reduce the environmental impact from whole farms. Acta Agric Scandin Sect A Anim Sci 62(4):225–232

Kumar R, Punia H, Kumar S (2020) Nanomaterials for agriculture input use efficiency. Resources use efficiency in agriculture. Springer, Singapore. https://doi.org/10.1007/978-981-15-6953-1_5

Kumar S, Bamboriya SD, Rani K, Meena RS, Sheoran S, Loyal A, Kumawat A, Jhariya MK (2022) Grain legumes: a diversified diet for sustainable livelihood, food, and nutritional security. In: Advances in legumes for sustainable intensification. Academic Press, pp 157–178

Kumar S, Sharma SK, Thakral SK, Bhardwaj KK, Jhariya MK, Meena RS, Jangir CK, Bedwal S, Jat RD, Gaber A, Atta AA (2022a) Integrated nutrient management improves the productivity and nutrient use efficiency of *Lens culinaris* Medik. Sustainability 14(3):1284

Kumar S, Meena RS, Kumar S, Pradhan G, Jangir CK, Ghosh S, Punia H, Sheoran P, Meena R, Ahmad MA, Goyal SK (2025) Designing a diversified Indian mustard production system for energy-carbon-cum-heat use efficiency and sowing dates assessment. GCB Bioenergy 17(6): e70044

Lefore N, Closas A, Schmitter P (2021) Solar for all: a framework to deliver inclusive and environmentally sustainable solar irrigation for smallholder agriculture. Energy Policy 154: 112313

Leharwan M, Kumar Y, Kumar R, Kumar Saraswat P, Kumar R, Veetil AKT, Bhattacharjee S, Kumar A, Kumar S (2023) Assessing the effects of conservation tillage and in-situ crop residue management on crop yield and soil properties in rice–wheat cropping system. Sustainability 15(17):12736

Liantas G, Chatzigeorgiou I, Ravani M, Koukounaras A, Ntinas GK (2023) Energy use efficiency and carbon footprint of greenhouse hydroponic cultivation using public grid and PVs as energy providers. Sustainability 15(2):1024

Lokuge N, Anders S (2022) Carbon-credit systems in agriculture: a review of literature. School Publ Policy Publ 15:12

Lovell H, Bulkeley H, Liverman D (2009) Carbon offsetting: sustaining consumption? Environ Plan A 41(10):2357–2379

Meena RS, Pradhan G (2023) Industrial garbage-derived biocompost enhances soil organic carbon fractions, CO_2 biosequestration, potential carbon credits, and sustainability index in a rice-wheat ecosystem. Environ Res 116525. https://doi.org/10.1016/j.envres.2023.116525

Meena RS, Pradhan G, Kumar S, Lal R (2023) Using industrial wastes for rice-wheat cropping and food-energy-carbon-water-economic nexus to the sustainable food system. Renew Sust Energ Rev 187:113756. https://doi.org/10.1016/j.rser.2023.113756

Meena RS, Pradhan G, Nalani, Singh K, Singh AK, Verma SK, Mina KK, Kumar S, Chaturvedi RK (2024) Diversified cropping modules designed for soil restoration, CO_2 sequestration, and generating carbon credits. Land Degrad Dev 35. https://doi.org/10.1002/ldr.5224

Meena RS, Pradhan G, Singh K, Kumar S, Singh AK, Shashidhar KS, Mina KK, Rao CS (2024a) Agriculture models for restoring degraded land to enhance CO_2 biosequestration and carbon credits in the Vindhyan region of India. Sci Total Environ 929:172661. https://doi.org/10.1016/j.scitotenv.2024.172661

Meena RS, Lal R, Kumar S, Pradhana G, Rao CS, Singh AK, Pathak H, Abhilash PC, Kumar A, Sharma SK, Jat ML, Singh S (2024b) Potential of Indian agriculture for capturing atmospheric CO_2 and monetizing carbon credits to the farmers: an overview and policy framework. Adv Agron 188:101–206. https://doi.org/10.1016/bs.agron.2024.06.001

Meena RS, Pradhan G, Kumar S (2025) Energy flow, eco-efficiency, and economic circulation with recycled industrial waste compost application in wheat and subsequent rice farming. Sci Total Environ 967:178779. https://doi.org/10.1016/j.scitotenv.2025.178779

Mezzanotte M (2019) What are carbon credits? Why we should care

Ministry of Trade Industry and Energy (2018) Korea's renewable energy 3020 plan

Montzka SA, Dlugokencky EJ, Butler JH (2011) Non-CO_2 greenhouse gases and climate change. Nature 476:46–50

Mousazadeh H, Keyhani A, Mobli H, Bardi U, Lombardi G, el Asmar T (2009) Environmental assessment of RAMseS multipurpose electric vehicle compared to a conventional combustion engine vehicle. J Clean Prod 17:781e790

Mousazadeh H, Keyhani A, Javadi A, Mobli H, Abrinia K, Sharifi A (2011) Lifecycle assessment of a solar assist plug-in hybrid electric tractor (SAPHT) in comparison with a conventional tractor. Energy Convers Manag 52:1700e1710

Nasir A (2019) Design and simulation of photo-voltaic water pumping system for irrigation. Adv Appl Sci 4(2):51–63

Nhamo L, Ndlela B, Nhemachena C, Matchaya G, Mabhaudhi T (2018) Adoption and barriers to uptake of climate-smart agriculture technologies in agriculture: a systematic review. Cogent Soc Sci 4(1)

Nielsen K, Mølgaard P, Søndergård T (2002) Solar panel based milk pasteurization. In: Gantenbein R, Shin S (eds) ISCA 2002 -17th international conference computers and their applications. San Francisco, USA, pp 376–379

Nitschelm L, Flipo B, Auberger J, Chambaut H, Dauguet S, Espagnol S, Gac A, Le Gall C, Malnoé C, Perrin A, Ponchant P (2021) Life cycle assessment data of French organic agricultural products. Data Brief 38:107356

Nordberg EJ, Caley MJ, Schwarzkopf L (2021) Designing solar farms for synergistic commercial and conservation outcomes. Sol Energy 228:586–593

Nozaki Y (2021, April) Rise of carbon farming in India. Report, Mitsui and Co Global Strategic Studies Institute, Tokyo

Ntinas G, Kougias P, Nikita-Martzopoulou C (2011) Experimental performance of a hybrid solar energy saving system in greenhouses. Int Agrophys 25(3):257–264

Obando-Ulloa JM, Rivero-Marcos, M., El Sario, E., Sánchez-Ramírez, M. F., Martínez-Mora, E., Castro-Badilla, G., Araya-Rodríguez, F., Guzmán-Hernández, T.J. (2017) Solar energy as an alternative tool in agricultural units in Costa Rica

Omer AM (2008) Green energies and the environment. Renew Sust Energ Rev 12(7):1789–1821

Park, C. Y., Yamamoto, Y., Doong, M. A. L. (2023) European Union carbon border adjustment mechanism: economic impact and implications for Asia

Pascaris AS (2021) Examining existing policy to inform a comprehensive legal framework for agrivoltaics in the US. Energy Policy 159:112620

Pearce JM (2021) Parametric open source cold-frame Agrivoltaic systems. Inventions 6:71

Pearce JM (2022) Agrivoltaics in Ontario Canada: promise and policy. Sustainability 14:3037

Peter C, Helming K, Nendel C (2017) Do greenhouse gas emission calculations from energy crop cultivation reflect actual agricultural management practices? –A review of carbon footprint calculators. Renew Sust Energ Rev 67:461–476

Pradhan G, Meena RS (2023) Interaction impacts of biocompost on nutrient dynamics and relations with soil biota, carbon fractions index, societal value of CO_2 equivalent, and ecosystem services in the wheat-rice farming. Chemosphere 139695. https://doi.org/10.1016/j.chemosphere.2023.139695

Pradhan G, Meena RS (2023a) Utilizing waste compost to improve the atmospheric CO_2 capturing in the rice-wheat cropping system and energy-cum-carbon credit audit for a circular economy. Sci Total Environ 164572. https://doi.org/10.1016/j.scitotenv.2023.164572

Pradhan G, Meena RS, Kumar S, Lal R (2023) Utilizing industrial wastes as compost in wheat-rice production to improve the above and below-ground ecosystem services. Agric Ecosyst Environ 358:108704. https://doi.org/10.1016/j.agee.2023.108704

Raihan A, Tanchangya T, Rahman J, Ridwan M (2024) The influence of agriculture, renewable energy, international trade, and economic growth on India's environmental sustainability. J Environ Energy Econ 21:37–53

Rickman RW, Douglas CL, Albrecht SL, Bundy LG, Berc JL (2001) CQESTR: a model to estimate carbon sequestration in agricultural soils. J Soil Water Conserv 56(3):237–242

Sarkar AN, Dash S (2011) Emissions trading and carbon credit accounting for sustainable energy development with focus on India. Asia Pacific Bus Rev 7(1):50–80

Sellars S, Schnitkey G, Zulauf C, Swanson K, Paulson N (2021) What questions should farmers ask about selling carbon credits? Farmdoc Daily 11:59

Sharma A, Chen CR, Lan NV (2009) Solar-energy drying systems: a review. Renew Sust Energ Rev 13(6–7):1185–1210
Sharma AK, Sharma C, Mullick SC, Kandpal TC (2017) Potential of solar industrial process heating in dairy industry in India and consequent carbon mitigation. J Clean Prod 140:714–724
Sharma M, Atheaya D, Kumar A (2023) Performance evaluation of indirect type domestic hybrid solar dryer for tomato drying: thermal, embodied, economical and quality analysis. Therm Sci Eng Prog 42:101882
Sheoran S, Gupta M, Kumari S, Kumar S, Rakshit S (2022) Meta-QTL analysis and candidate genes identification for various abiotic stresses in maize (*Zea mays* L.) and their implications in breeding programs. Mol Breed 42(5):26
Shirato Y (2020) Use of models to evaluate carbon sequestration in agricultural soils. Soil Sci Plant Nutr 66(1):21–27
Shockley J, Snell W (2021) Carbon markets 101. Econ Policy Update 21(4)
Solangi KH, Islam MR, Saidur R, Rahim NA, Fayaz H (2011) A review on global solar energy policy. Renew Sust Energ Rev 15(4):2149–2163
Sontake VC, Kalamkar VR (2016) Solar photovoltaic water pumping system-A comprehensive review. Renew Sust Energ Rev 59:1038–1067
Tajima M, Doedt C, Iida T (2022) Comparative study on the land-use policy reforms to promote agrivoltaics. In: AIP Conference Proceedings 2635
Thakur AK, Singh R, Gehlot A, Kaviti AK, Aseer R, Suraparaju SK, Natarajan SK, Sikarwar VS (2022) Advancements in solar technologies for sustainable development of agricultural sector in India: A comprehensive review on challenges and opportunities. Environ Sci Pollut Res 29: 43607–43634
The 12th five-year plan for the solar photovoltaic industry. http://www.americansolarmanufacturing.org/news-releases/chinas-five-year-plan-for-solartranslation.pdf
Todde G, Murgia L, Carrelo I, Hogan R, Pazzona A, Ledda L, Narvarte L (2018) Embodied energy and environmental impact of large-power stand-alone photovoltaic irrigation systems. Energies 11(8):2110
Tubiello FN, Salvatore M, Rossi S, Ferrara A, Fitton N, Smith P (2013) The FAOSTAT database of greenhouse gas emissions from agriculture. Environ Res Lett 8(1):015009
Upham P, Shackley S (2006) The case of carbon dioxide capture and storage (CCS) in the UK and The Netherlands. Energy Policy 34(18):3173–3181
Ureña-Sánchez R, Callejón-Ferre ÁJ, Pérez-Alonso J, Carreño-Ortega Á (2012) Greenhouse tomato production with electricity generation by roof-mounted flexible solar panels. Sci Agric 69:233–239
Wackernagel M, Rees WE (1996) Our ecological footprint: reducing human impact on the earth. New Society Publishers, Gabriola Island, p 160
Wang X, Li Y, Duan Y, Wang L, Niu Y, Li X, Yan M (2021) Spatial variability of soil organic carbon and Total nitrogen in desert steppes of China's Hexi corridor. Frontiers in Environmental Science 9:761313
Wani N, Velmurugan A, Dadhwal VK (2010) Assessment of agricultural crop and soil carbon pools in Madhya Pradesh. India Trop Ecol 51(1):11–19
Wayua F, Okoth O, Wangoh MWJ (2013) Design and performance assessment of a flat-plate solar milk pasteurizer for arid pastoral areas of Kenya. J Food Process Preserv 37:120–125
Williams JR, Peterson JM, Mooney S (2005) The value of carbon credits: is there a final answer? J Soil Water Conserv 60(2):36A–40A
Williams HJ, Wang Y, Yuan B, Wang H, Zhang KM (2025) Rethinking agrivoltaic incentive programs: a science-based approach to encourage practical design solutions. Appl Energy 377: 124272
Wongpiyabovorn O, Plastina A, Crespi JM (2023) Challenges to voluntary Ag carbon markets. Appl Econ Perspect Policy 45(2):1154–1167
World Bank (2011) Agricultural innovation systems: an investment sourcebook. World Bank, Washington, DC

World Bank (2023) What is carbon pricing. Available at: https://carbonpricingdashboard.worldbank.org/what-carbon-pricing. Accessed 16 Mar 2023

World Bank (2024) State and trends of carbon pricing 2024. World Bank: Washington DC USA.

Xue J (2017) Photovoltaic agriculture-new opportunity for photovoltaic applications in China. Renew Sust Energ Rev 73:1–9

Yang RJ (2015) Overcoming technical barriers and risks in the application of building integrated photovoltaics (BIPV): hardware and software strategies. Autom Constr 51:92–102

Yang J, Olsson A, Yan J, Chen BA (2014) Hybrid life-cycle assessment of CO_2 emissions of a PV water pumping system in China. Energy Procedia 61:2871–2875

Yildiz A, Ozgener O, Ozgener L (2012) Energetic performance analysis of a solar photovoltaic cell (PV) assisted closed loop earth-to-air heat exchanger for solar greenhouse cooling: an experimental study for low energy architecture in Aegean region. Renew Energy 44:281–287

Chapter 11
Achieving Carbon-Negative Agriculture and Carbon Credits Through Green Renewable Energy

M. B. Fathima Jemziya and M. N. F. Nashath

Abstract Agriculture accounts for about 10%–12% of the world's total emissions of greenhouse gas (GHG), mainly due to changes in the use of land, application of fertilizers, and management of livestock. At the same time, it has enormous potential for mitigating climate change through carbon sequestration and emission reductions. Renewable energy systems, including solar-powered irrigation, biogas digesters for managing manure, wind energy to meet on-farm electricity needs, and biomass-energy conversion, have the potential to reduce the agricultural carbon footprint significantly. For example, biogas systems can reduce subsequent methane emissions up to 50%, or solar irrigation can lower diesel consumption and reduce carbon dioxide (CO_2) emissions between 20% and 30%. The system supports eliminating the need for fossil fuels, increasing carbon credit generation by quantifying emission reductions, and sequestering carbon. Collectively, they are used by farmers to trade on carbon markets. This chapter outlines how such decentralized renewable energy systems can be implemented in agricultural practice to support the reduction targets set under frameworks such as the Paris Agreement and Kyoto Protocol. It further spelt out the economic and operational challenges in adoption, including high upfront costs, technical capacity gaps, and absent policy incentives in the Global South. The focus is on aligning policy instruments, financial instruments, and capacity-building activities for maximum sustainability and farmer participation in carbon credit markets. This is achieved by linking renewable energies to transform the agricultural sector.

Keywords Carbon credits · Greenhouse gas · Renewable energy · Sustainable agriculture

M. B. F. Jemziya (✉) · M. N. F. Nashath
Department of Biosystems Technology, Faculty of Technology, South Eastern University of Sri Lanka, Oluvil, Sri Lanka
e-mail: jemziya@seu.ac.lk; nasath17.mn@seu.ac.lk

R. S. Meena et al. (eds.), *Carbon-Negative Agriculture*, Sustainability Sciences in Asia and Africa, https://doi.org/10.1007/978-981-95-1483-0_11

11.1 Introduction

Of all contemporary environmental problems, climate change may seem to be one of the most pressing problems (Anderegg et al. 2020). Rising atmospheric concentrations of greenhouse gases (GHGs), including carbon dioxide (CO_2), methane (CH_4), and nitrous oxide (N_2O), have significantly changed the Earth's radiative balance, leading to global warming, severe weather events, and host of other adverse impacts (Bellassen et al. 2022; Meena et al. 2025). Agriculture is both the driver and the victim of climate change: it supports global food security and socioeconomic development on a global scale, yet it is a major source of GHG emissions (Kumar et al. 2022, 2025). Accommodating for the entire world, it has been estimated that agriculture contributes to about 10%–12% of the total GHG emissions, and from this, land-use change, deforestation, fossil fuel use, fertilizer application, and livestock management rank as some of the key sources (Nsabiyeze et al. 2024; Lawrence et al. 2021). Specifically, enteric fermentation accounts for approximately 40% of anthropogenic CH_4, 70% of NH_3, and 80% of N_2O emissions (Xuan et al. 2023).

These emissions are aggravated by the fossil fuels from the agricultural sector for irrigation, machinery, processing, and storage, which makes the sector (also) energy price sensitive and leads to environmental degradation. Meanwhile, agriculture is sensitive to climate-induced pressures, including droughts, flooding, and erratic rainfall, which negatively impact output from crops alongside food production systems (Ul-Haq et al. 2022; Clarke et al. 2022; Meena et al. 2024). Because of the dual role, integrating the mitigation strategies is significantly important for sustainable systems. One fruitful approach is to use renewable energy technologies in agriculture, which provides opportunities for carbon credits with reduced emissions.

Alternative practical renewable energy sources such as solar, wind, biogas, and biomass are available for several agricultural activities such as water pumping, greenhouse heating, machinery operation, cold storage, and onsite electricity generation (Kabeyi and Olanrewaju 2022). The technologies will also create an opportunity for carbon-credit generation through the estimation of the avoided emissions and monetization through carbon trade in markets. In this manner, the interested farmers pursuing the clean energy base of agriculture may have the opportunity to obtain financial benefits and hence, reconcile the aspect of environmental sustainability with the benefit of economic reward.

In addition, human-based solutions to carbon sequestration include agroforestry, no-till farming, improved livestock feeding systems, manure management, and application of organic manure, which can be complemented with clean energies to increase the resilience to climate change (Albrecht and Kandji 2003; Abbas et al. 2017; Blanco-Canqui 2021; Jhariya et al. 2022; Meena et al. 2024a). These sustainable practices help in soil carbon storage and GHG reduction, thus facilitating the objectives of international frameworks such as the Kyoto Protocol and the Paris Agreement (UNFCCC 2015; Salma et al. 2024; Jangir et al. 2024). Such frameworks aim to reduce global emissions by 50% in 2030 and net-zero emissions by 2050, with agriculture occupying the central position to ensure delivery of these goals.

However, the full potential of renewable energy and carbon-credit generator farms in agriculture (specifically smallholder farmers in the Global South) is ongoing, with many challenges. These are a few access issues to financing, technical knowledge deficiency, poor infrastructure, and policy gaps (Rapsomanikis 2015). To overcome these hurdles, coordinated work through conducive regulations, market-based mechanisms such as carbon pricing, and aimed technological interventions are required (Giller et al. 2021; Rose et al. 2022; Meena et al. 2024b).

Within this context, this chapter examines the convergence of renewable energy use and carbon-credit production in agriculture. It covers the enabling technologies, opportunities, and challenges, and analyzes how scaling up of these strategies can promote sustainable growth, energy security, and climate action.

11.2 Overview of Renewable Energy in Agriculture

Renewable energy technologies have become breakthrough instruments for tackling emissions from agriculture in recent years. Technologies which use main energy sources that are not depletable are referred to as renewable energy (Chel and Kaushik 2011). Solar, wind, geothermal, and biomass energy are a few examples of renewable technology. According to this concept and the yearly hydrologic cycle, which replenishes water supplies, hydropower is regarded as a component of the renewable technology mix (Frey and Linke 2002).

Direct energy from the Sun is known as solar energy, and it is the most plentiful energy source in the environment, second only to nuclear energy. The photovoltaic (PV) cell, which transforms sunlight into electricity directly, is the alternative energy source with the quickest growth rate, expanding by 50% annually. Every year, the Sun provides more than 10,000 times the energy for people's consumption (Chel and Kaushik 2011). Uneven heating of the Earth's surface by the Sun and the warm core is the source of wind power. The majority of contemporary wind power is produced as electricity with an electrical generator to transform the rotation of turbine blades into electrical current. A considerably older technology, windmills use wind energy to power mechanical devices that perform physical tasks like pumping water or crushing grain (Woelfle 2013). Likewise, hydropower is energy that comes from the water flow in rivers and seas (or other energy differentials). It can also be employed mechanically to do useful tasks or to create electricity through the turbines. It is a widely used resource. The natural flow of heat from the Earth is immediately captured by geothermal power. Particularly during the day hours, the energy available from the natural decay of radioactive materials in the Earth's crust and mantle is roughly equivalent to that of incoming solar energy (Chel and Kaushik 2011). Another sustainable energy source is alcohol made from corn, sugar cane, switch grass, etc. In the same way, non-renewable diesel can be replaced with oils derived from plants and seeds (Reiter and Lindorfer 2015).

Reduced dependency on fossil fuels in agriculture makes renewable energy critical in rendering the sector more sustainable. Agriculture could instantaneously cut the rate of greenhouse gas emissions, maintain operating costs, and enhance energy security through the use of renewable sources of energy (Kabeyi and Olanrewaju 2022; Meena et al. 2023). Renewable energy is being used to drive several agricultural activities, such as solar irrigation, including wind-powered water pumping systems and biogas digesters for site energy generation and geothermal heating of greenhouses. For instance, solar photovoltaic (PV) systems are expanding in the operation of water pumps in irrigation systems to lessen the volume of diesel used. Wind turbines are used for farm lighting, milling grains, or refrigeration in off-grid rural areas. Biogas systems turn livestock and crop waste into clean energy that could fuel farm equipment or home cooking amenities, helping utilize waste and energize. Cold chain infrastructure presents renewable-powered cold storage systems that enable the preservation of perishable produce, leading to a reduction in post-harvest losses and enhanced food security. Even products that are value added through processes like drying or grinding can be processed profitably with renewable electricity.

These applications improve work efficiency and reduce the carbon footprint of the farming systems. Besides, renewable energy instils climate resilience that achieves the world's vision in terms of sustainability by allowing farmers to participate in carbon-credit schemes that pay for reduced emissions. This dual benefit serves environmental goals and provides economic incentives that help to promote sustainable farming systems across varied agricultural landscapes.

11.3 Concept of Carbon Credits in Agriculture

Global carbon markets have been introduced by the UNFCCC Kyoto Protocol (1997), comprising mechanisms such as the Clean Development Mechanism, joint implementation, and the international emission trading system (ETS). Because of scalability issues, the Paris Agreement (2015) initiated the international voluntary carbon market to overcome these challenges in line with the change in the climate action regime (Fig. 11.1).

Forecasts reveal that carbon markets can potentially generate US$5–50 billion in value by 2030, with the voluntary carbon market alone expanding at an annual rate of at least 35% to reach at least US$15 billion (Khurana et al. 2024a, b; Meena and Pradhan 2023). Carbon markets have compliance carbon markets, which are controlled by the government or any international organization, and voluntary carbon markets, where organizations buy credits for sustainability purposes. Tools, including carbon taxes and cap-and-trade systems that were first adopted in Finland in 1990 are very effective instruments in emissions reductions (Marchant et al. 2022).

Alongside carbon capture and storage (CCS) technology, the carbon market is an emerging tool for promoting and facilitating emission reductions in multiple sectors, including agriculture (Ashley and Johnson 2018). The concept that underlies the

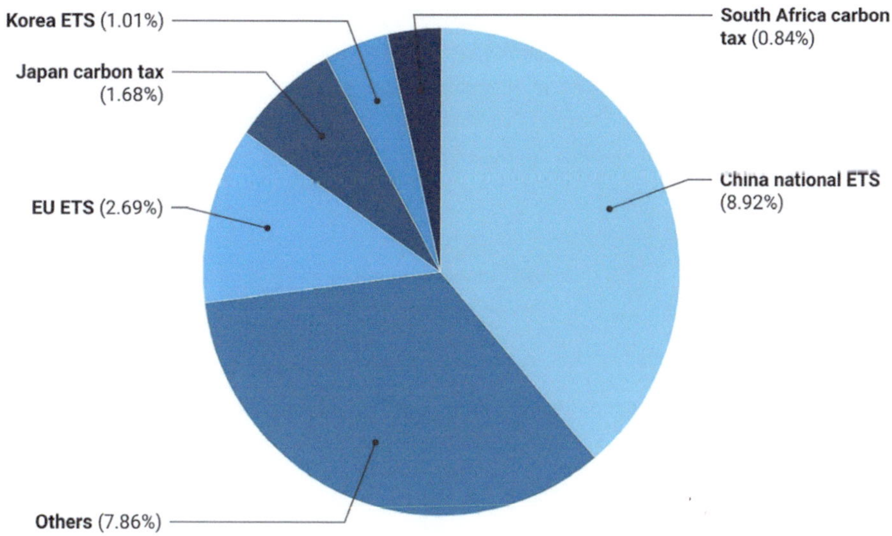

Fig. 11.1 GHG emissions covered by different carbon taxes and ETS. (*Data Source*: World Bank 2023)

carbon market (emission trading system) is to make CO_2 equivalent (CO_2eq) in the same way that other commodities are traded, and give a price to each ton of emissions. It creates possibilities to trade carbon emissions and reductions, also known as offsets (Reyes and Gilbertson 2010). An intermediary known as the marketplace exists between the buyer and seller to achieve this exchange. A carbon credit that can be bought in this marketplace is the right to emit one ton of CO_2. However, a carbon offset is attained for CCS when one ton of CO_2 is sequestered using any of the processes. The carbon market is developed through linking the credits and offsets, as pointed out by Calel (2013). The money made from this transaction is used to fund CCS initiatives and maintain marketplace management. As a representative answer to this problem, carbon credits have been paid attention to in many aspects (Brown et al. 2023). By merging economic stimulus with environmental concern, these credits offer an innovative financial approach to battle climate change (Tang 2023). Carbon credits encourage the general public, businesses, and governments to take measurable actions to reduce emissions by assigning a quantifiable value to each ton of CO_2eq emitted (Ghosh and Ghosh 2023; Pradhan and Meena 2023). Therefore, it reduces the environmental problems and creates new market opportunities (Fig. 11.2).

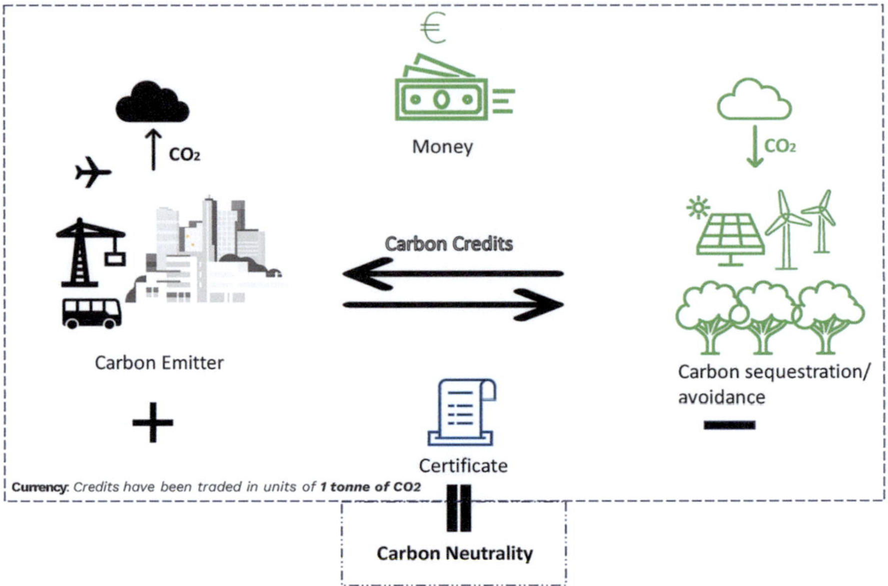

Fig. 11.2 Simplified mechanism of the carbon-credit system. (Salma et al. 2024)

11.4 Green Renewable Energy and Carbon Credits in Agriculture

The realization of sustainable agriculture development is increasingly dependent on green renewable energy sources and carbon-credit mechanisms to create a climate-resilient sector. Standard agricultural techniques (highly intensive tillage, overuse of fertilizers, and dependency on machinery driven by fossil fuels) are a major contributor to GHG emissions (Agbelusi et al. 2024; Pradhan and Meena 2023a; Leharwan et al. 2023). The renewable energy technologies (solar, wind, biogas, small hydroelectric, and geothermal) significantly diminish these and leave ecosystems intact.

Green renewable energy systems have natural replenishment and operational sustainability features. For agriculture, we have technologies, like solar-powered irrigation systems, wind-powered devices, and biogas digesters, that convert organic waste into a usable energy source. Fossil fuel consumption through these systems not only gets substituted but also facilitates energy independence, reduces production cost, and promotes farm readiness against energy market volatility (Rahman et al. 2022).

Importantly, the use of renewable energy makes it possible for farms to measure and prove GHG reductions and thus be eligible for carbon credits. For example, switching more diesel-powered irrigation pumps to solar systems can generate verifiable emission reductions, and those are credited to trading in the voluntary or compliance carbon markets (Lee 2017). This is another source of income for farmers and an incentive to take up clean technology. Moreover, under the Paris Agreement article six mechanisms, there are platforms created to enable carbon credits from renewable energy projects from agriculture, and it will encourage wider participation in the mitigation of climate change (Navarro Jiménez 2024).

In addition to downward trends in emissions, renewable energy promotes the general preservation of the environment. It supports soil and water conservation practices, preserves biodiversity, and adapts agricultural activities to emerging national and global emission reduction targets (Nemethy and Molnár 2014). As the need for energy increases in agriculture, strategic integration of renewable energy presents a way out of the situation and a source of profit for these ventures through carbon credits.

11.5 Scope and Future Prospects of Green Renewable Energy in Agriculture

Green renewable energy develops the agricultural sector by implementing cleaner, sustainable alternatives that replace conventional energy practices. The utilization of renewable energy surpasses basic power switching to establish opportunities that boost agriculture productivity, protect farms against climate change, and support long-term, environmentally friendly businesses (Chel and Kaushik 2011; Pradhan et al. 2023). Agricultural farmers reduce operational costs and environmental impact through solar-based irrigation systems, waste-to-energy technologies, and decentralized power distribution networks. This emerging innovative landscape encompasses energy storage system technologies with smart grids and precision agriculture systems, operating through renewable energy power (Rahman et al. 2022). These novel technologies provide businesses with the increasing energy usage in combination with operational enhancements in a wider range of farm activities. Green energy infrastructure investment with supportive government policies and the growing need for carbon-neutral food production will drive adoption rates of sustainable energy practices (Saeed and Siraj 2024). The upcoming agricultural model will unite power generation with food production into a single system to achieve symbiotic operation. It will be decisive to surpass barriers linked to initial expenses combined with technical skill deficits and uneven rural technology accessibility. Green renewable energy exhibits potential to build itself as an important support for future agricultural sustainability while driving growth toward worldwide climate and development goals through present research, industrial expansions, and industry alliances.

11.6 Potential of Earning Carbon Credits through Renewable Energy Applications

Green renewable energy in agricultural practices has a high potential for carbon credits, allowing farmers to reduce greenhouse gas (GHG) and receive a financial incentive. Furthermore, renewable energy technologies help to measure and certify reduced carbon emission levels in existing carbon markets and farms while deploying sustainable power solutions.

Solar Irrigation Systems
The alteration from diesel-powered to solar-powered irrigation systems significantly reduces emissions. For example, US irrigation pump energy consumption generated 12.6 million metric tons of CO_2eq. in 2018, with groundwater being the main contributor. Substituting diesel pumps with solar options reduces a considerable degree of these emissions, thus having great potential to affect carbon credits.

Biogas Digesters
Biogas systems in agriculture support GHG emissions reductions. According to a study, biogas power generation cut down on emissions by about 3316.8 tons of CO_2eq annually (more than the whole of Canada) relative to conventional fossil energies. In addition, biogas production can save as much as 240% of GHG emissions from EU fossil fuels, thus proving the effectiveness of biogas production for emission mitigation.

Wind Energy
Wind turbines are very effective at reducing their carbon footprint. A standard wind turbine pays back its carbon intensity within a few months and provides emission-free electricity during the rest of its (20–30 years) life cycle. The rapid offset period validates the viability of wind power to support carbon-credit generation in an agricultural environment.

Soil Carbon Sequestration
Other than energy production, practices such as better soil management can improve carbon sequestration. For instance, an Australian Queensland grazier received 85,262 Australian Carbon Credit Units (ACCUs) from better land management, earning one CO_2eq sequestrated per ACCU. This approach improves the health of the land and yields a huge financial gain in the form of carbon credits.

Strengthening verification requirements is important to develop targeted farmer education platforms and make certification platforms accessible for the benefit of carbon credits in renewable energies in agriculture. Such initiatives will equip farmers with the capacity to act on carbon markets, thus enhancing global climate mitigation strategies.

11.7 Strategies to Enhance Carbon-Credit Generation via Renewable Energy in Agriculture

The effort to maximize the production of carbon credits in agriculture with the help of green renewable energy calls for directed strategies along technology, finance, policy, and education. One of the major steps is the general outspread of decentralized renewable energy technologies, including solar-powered irrigation systems, wind turbines, and biogas digestors to address the reliance on fossil fuels and reduce directly the number of greenhouse gases emitted by farming activities (Mperejekumana et al. 2024). These systems replace carbon-intensive energy sources and are also eligible to benefit from the carbon markets' carbon offset schemes.

The other strategy is to utilize precision agriculture technologies like Internet of Things (IoT)-based smart sensors, artificial intelligence (AI)-powered analytics, and automated machinery to promote energy savings, optimize input utilization, lower emission levels, and increase eligibility for carbon credits (Jawad et al. 2017; Kumar et al. 2020). Moreover, a combination of real-time emission monitoring tools with renewables provides proper (MRV) monitoring, reporting, and verification, a key prerequisite for carbon credits.

When it comes to the financial front, the most crucial strategy will be to make green finance accessible, whether in the form of low-interest loans, grants, or carbon credit-based investment models for the smallholder and the mid-sized farmers (Taghizadeh-Hesary and Yoshino 2020). Financial mechanisms that connect the flow of carbon-credit income streams with renewable energy investments enhance the bankability of projects and make private sector participation more attractive (Meitern 2024). Bundled financing-based packages of initial setup, maintenance support, and credit registration services can yet again fast-track adoption.

From a policy perspective, governments can increase the production of carbon credits through well-defined incentive structures such as feed-in tariffs, popularization of renewable energy subsidies, and incentives in the form of tax relief for sustainable energy transitions in Agriculture (Samuel 2025). Besides, imposing standardized carbon accounting procedures and developing agriculture-specific MRV frameworks guarantee credibility and market value of carbon credits earned from farming exercises (Woo et al. 2021). Synchronizing renewable energy targets with the country's agricultural strategies will facilitate an integrated development route (Yucheng 2024).

Finally, a key strategy is in the education and capacity-building programs that aim to empower farmers, enabling them to adopt, maintain, and derive useful information from renewable technologies. These should be training on system operation, carbon-credit registration procedure, and market entry strategies. The digital platform and extension services are how continuous learning and dissemination of best practices can be achieved (Lee 2017). These approaches form a complete framework that utilizes green renewable energy to enhance carbon-credit capacities in agricultural sectors. The combination of technological advancements in financial backing from governments, with supportive policies and educational programs, will allow farmers to decrease carbon emissions and enjoy new profitable prospects from carbon trading.

11.8 Optimizing Carbon Credits Through Smart Utilization of Renewable Energy

Renewable energy also has a very significant contribution towards the reduction in the carbon footprint in agriculture due to the change from fossil fuel energy sources to clean energy sources. Irrigation, crop processing, and agricultural machinery all use diesel and electricity from non-renewable sources, which generate high GHG emissions. It is understood that technologies, like solar power generation, wind energy generation, and biomass-based systems in agriculture, can reduce the emission intensity as they improve the energy use efficiency (Rahman et al. 2022). This change lowers the carbon intensity and helps farmers achieve their input to climate change goals.

Renewable energy technologies, such as photovoltaics, solar heating panels, and biomass-fired Organic Rankine Cycles, have received a lot of attention in recent years (Montazerinejad and Eicker 2022; Bagherian and Mehranzamir 2020). All of these technologies, however, are still not cost-competitive with fossil fuel-fired units (such as boilers and internal combustion engines); for this reason, some tax on CO_2 emissions from conventional technologies and/or subsidies for renewable energy technologies are required. It incorporates several stakeholders, techno-economic concerns, and social consequences, determining the best policy for incentives and the carbon tax (Martelli et al. 2020).

Carbon credits are, therefore, an important approach in evaluating the emission reduction that will come and the level of renewable energy adoption in agriculture. These credits, standardized measurements, and documented reductions in emissions can be sold in compliance or voluntary carbon markets, giving revenues for renewable energy. For example, carbon accounting frameworks include the life carbon emissions associated with the reduced solar water pumps, biogas plants, and wind mills, which make it easier for project proponents to report their positive impact on the carbon footprint. Thus, well-documented renewable energy projects in the agricultural sector spell out real carbon savings, thereby mobilizing investment and facilitating participation in carbon markets.

11.9 Research and Development Priorities

The source for improving green renewable energy technology, accompanied by agricultural carbon-credit system optimization, lies in the research and development (R&D) process. Conducting research is a requirement to generate elastic, sustainable solutions that work well for specific agricultural settings and preserve low cost-effectiveness because the agriculture sector desires to diminish carbon footprints without compromising food production. Scientific research works to form modern renewable energy systems, which consist of solar-powered cold storage, low-emission biomass digesters, and hybrid renewable platforms planned precisely for agricultural uses. Modern technologies connect with digital monitoring systems that offer real-time energy use and carbon savings tracking, thus improving the legitimacy of carbon-credit information (Woo et al. 2021).

Research in soil science alongside agroecology and climate-smart practices simultaneously broadens our comprehension regarding carbon sequestrating techniques across different farming systems (Bayata and Mulatu 2024). The transparent verified carbon trading methodologies require research that calculates region-specific emission data, life cycle performance criteria, and carbon accounting standards (Cowie et al. 2012). Interdisciplinary research, which combines renewable energy science, economics, and environmental studies, permits the conception of solid revenue forecast models for carbon credits that support policy selections and investment distributions.

Current research and development efforts are integrating renewable power solutions with agricultural actions for refining carbon capture abilities while lessening greenhouse gas releases. Wang et al. (2023) studied CO_2 emission patterns in 38 sub-Saharan African countries from 2000 to 2019, and countries that consumed more renewable energy decreased their carbon footprint especially in regions with minimal income. The research maintained the significance of proper governance structures to achieve maximum environmental benefits through renewable energy applications in agricultural settings. The investigation of greenhouse gas emissions from climate-smart agriculture practices in South Africa worked with stochastic frontier analysis according to Haya and Hasegawa (2024). Amended soil nutrients and optimized animal feeding systems offer the most important paths to reduce emissions, which require substitutes to receive carbon credits (Kumar et al. 2022a, 2023). Tao et al. (2023) studied how environmental protocols with renewable energy R&D investments affect carbon emissions. The research recognized that merging higher spending on renewable energy R&D with firm environmental rules generates the main carbon emission, which proved policy sustainability as a crucial factor for encouraging environmentally sustainable agricultural systems. The study by Khurana et al. (2024a; b) tested the implementation potential of carbon-credit programs in farming operations. The researchers found that traditional farming methods combined with zero tillage technology, precision land leveling, and effective water resource management make agriculture a greenhouse gas-absorbing system. The authors endorsed the setup of complete MRV platforms, which enable farmers to join carbon market transactions. These findings demonstrate why focused research and development serve as a fundamental element for creating specific renewable energy technologies that ought to work alongside carbon trading systems serving the agricultural industry. The success of these innovations depends on supportive policies and governance structures that should exist to make scalability possible.

The rapid advancement of innovation happens through combined efforts between universities, research institutes, private enterprises, and international agencies, because they help spread knowledge and technological advancements to real-world use. R&D investments directed at underfunded agricultural regions will drive the complete adoption of green renewable energy and enhance carbon-credit accessibility for global farming communities.

11.10 Policy Frameworks and Institutional Roles

The implementation of green renewable energy in agriculture requires a well-developed policy structure for achieving optimal carbon credits. Multiple current policies and international frameworks enable the alignment of green renewable energy with carbon-credit systems in agricultural settings. In India, the *Pradhan Mantri Kisan Urja Suraksha evam Utthaan Mahabhiyan* (PM-KUSUM) scheme with its National Mission on Sustainable Agriculture (NMSA) promotes solar-based irrigation systems and independent solar power facilities through financial assistance (MNRE 2023). The Rural Energy for America Program of the United States Department of Agriculture provides funds, loans, and guarantees to farming operations and rural businesses who want to build renewable energy structures using solar, wind, and biomass systems (USDA 2024). The Paris Contract and UN Sustainable Development Goals (SDGs) support sustainable agriculture with climate resilience through clean energy access worldwide (UNFCCC 2015; United Nations 2015; Punia et al. 2020). The International Renewable Energy Agency (IRENA) issued rules for renewable energy policies in agri-food systems while representing the possibilities for reducing carbon emissions and developing rural areas (IRENA 2021).

The adoption of renewable energy in agriculture obtains organizational backing through financial support that consists of tax incentives and subsidies to get solar-powered irrigation and wind turbines, as well as the funding of biogas digesters. Implementing agriculture within emissions trading systems and voluntary carbon markets supports manufacturers in getting paid for employing practices that trap carbon and decrease carbon emissions. A combination of national carbon registries and verification systems ensures plain transparency and trustworthiness within carbon-credit frameworks (Santier 2023). Proper coordination between policies must exist to provide rural, renewably installed systems with grid connectivity and to establish decentralized energy systems that strengthen local farming communities. The policy framework needs to promote private–public cooperation while investing in rural energy technology development; it should also integrate agricultural programs into climate-related funding, such as the Green Climate Fund. Consolidated policies between agriculture farming and energy production and environmental stewardship lead to sustained effects (Pretty 2008). The correct policy implementation pathway allows agriculture to evolve into a low-carbon sector, which advances the country's energy targets and enables carbon-credit opportunities for financial returns.

11.11 Future Outlook

Renewable energy is currently used at a slow but steadily growing pace and its global consumption grew by 3.7% in 2019 (Rahman et al. 2022), thus indicating its role in energy systems; still, agriculture use of renewable energy is lower and will have to be considerably scaled up to meet the growing energy demand of the sector sustainably.

The most recent International Renewable Energy Agency (IRENA) study states that hydropower produced the most renewable electricity (63%), followed by onshore wind (18.1%) and solar (8.3%) (de Vries 2023). China was the largest consumer of renewable energy in 2019 (using roughly 6.63 EJ), followed by the United States (5.83 EJ) (Xue 2017). The agriculture sector accounts for a minute part of total energy use in many developed nations. In Australia, for example, it accounted for just 1.9% of the nation's overall energy consumption for the 2017–2018 economic year (Vadiee and Martin 2013). Similarly, the agriculture sector of developing countries such as Bangladesh consumed 2.42% of the total energy from 2019 to 2020, reported by Ravishankar et al. (2020). However, the energy demand of this industry increases every year; in Australia, that increase was by 1.1% (Vadiee and Martin 2013). For sustainable development to occur, it is obvious that the energy requirements in the industries have to be met through renewable sources.

In the future, employing green renewable energy to get the most from carbon credits in agriculture has the massive capability to change the prospects of the agriculture sector and advance global sustainable development objectives. Solar, wind, and biomass electricity technologies as investments and renewable energy systems will be cheaper and available due to increased climate awareness and technological advancement to enable farmers to adopt them widely. These systems' efficiency and scalability are achieved through energy storage advancement and precision agriculture. Increased farmer participation in carbon-credit markets results from blockchain incorporation and digital technology into carbon accounting, which enhances transparency and dependability in tracking emission reductions. The economic feasibility of these activities is also anticipated to be strengthened by developing legislation, such as increased carbon pricing systems and incentives for renewable energy.

Hopefully, international relationships supported by treaties such as the Paris Agreement will provide financial assistance and extension services to technologies in developing countries to embrace renewable energy. The sustainable use of renewable energy in agricultural work enhances the transformation of the consensus towards climate adaptation, sustainable farming, and resource recycling. Besides increasing the success rate of investment returns, carbon credits are also optimized.

11.12 Conclusion

The use of renewable energy in agricultural activities represents a practical approach to facilitate the enhancement of climate resilience, decreased emissions, and an increase in carbon-credit possibilities. Agriculture accounts for approximately 5.1–5.8 Gt CO_2 equivalent per annum, yet has great potential for carbon sequestration through improved land and energy use. Renewable energy applications (solar-powered irrigation, which could reduce diesel consumption by 30%, and biogas digesters, which have been shown to reduce methane by up to 50%) act as a mechanism to move the sector away from fossil fuels and measure these reductions that are therefore eligible for carbon-credit programs. Such a role on dual fronts boosts performance on the environment and energy efficiency. Breadth of adoption requires synchronization in the financing mechanisms, policy alignment, and targeted outreach to educate. For instance, countries using green energy subsidies in agriculture experience up to a 25% increased pace of implementation of clean technologies. The changes in carbon accounting and monitoring regimes, such as MRV systems, continue to enhance traceability and monetization of emission reductions. The Paris Agreement's alignment with Sustainable Development Goals (SDG 7 and SDG 13) underpins the structural integration of green energy in agriculture. The sustainable agricultural future hinged on collaborative frameworks across sectors: technology providers, policymakers, farmers, and research institutions to accelerate large-scale adoption of renewable energy solutions to maximize the generation of carbon credits.

References

Abbas F, Hammad HM, Fahad S, Cerdà A, Rizwan M, Farhad W, Ehsan S, Bakhat HF (2017) Agroforestry: a sustainable environmental practice for carbon sequestration under the climate change scenarios—a review. Environ Sci Pollut Res 24:11177–11191

Agbelusi J, Arowosegbe OB, Alomaja OA, Odunfa OA, Ballali C (2024) Strategies for minimizing carbon footprint in the agricultural supply chain: leveraging sustainable practices and emerging technologies. World J Adv Res Rev 23(3):2625–2646

Albrecht A, Kandji ST (2003) Carbon sequestration in tropical agroforestry systems. Agric Ecosyst Environ 99(1–3):15–27

Anderegg WR, Trugman AT, Badgley G, Anderson CM, Bartuska A, Ciais P, Cullenward D, Field CB, Freeman J, Goetz SJ, Hicke JA (2020) Climate-driven risks to the climate mitigation potential of forests. Science 368(6497):eaaz7005

Ashley MJ, Johnson MS (2018) Establishing a secure, transparent, and autonomous blockchain of custody for renewable energy credits and carbon credits. IEEE Eng Manag Rev 46(4):100–102
Bagherian MA, Mehranzamir K (2020) A comprehensive review on renewable energy integration for combined heat and power production. Energy Convers Manag 224:113454
Bayata A, Mulatu G (2024) Carbon farming, climate smart agriculture practice and current climate change mitigation strategy-in the case of Ethiopia. Int J 12(6):149–156
Bellassen V, Angers D, Kowalczewski T, Olesen A (2022) Soil carbon is the blind spot of European national GHG inventories. Nat Clim Chang 12(4):324–331
Blanco-Canqui H (2021) No-till technology has limited potential to store carbon: how can we enhance such potential? Agric Ecosyst Environ 313:107352
Brown M, Schoen JH, Gross J, Omary RA, Hanneman K (2023) Climate change and radiology: impetus for change and a toolkit for action. Radiology 307(4):e230229
Calel R (2013) Carbon markets: a historical overview. Wiley Interdiscip Rev Clim Chang 4(2): 107–119
Chel A, Kaushik G (2011) Renewable energy for sustainable agriculture. Agron Sustain Dev 31: 91–118
Clarke B, Otto F, Stuart-Smith R, Harrington L (2022) Extreme weather impacts of climate change: an attribution perspective. Environ Res Clim 1(1):012001
Cowie A, Eckard R, Eady S (2012) Greenhouse gas accounting for inventory, emissions trading and life cycle assessment in the land-based sector: a review. Crop Pasture Sci 63(3):284–296
de Vries DHX (2023) Investigating the potential role of pumped hydro storage in the Ethiopian energy system using OSeMOSYS. Doctoral dissertation, TU Delft
Frey GW, Linke DM (2002) Hydropower is a renewable and sustainable energy resource meeting global energy challenges in a reasonable way. Energy Policy 30(14):1261–1265
Ghosh SK, Ghosh SK (eds) (2023) Circular economy adoption: catalysing decarbonisation through policy instruments. Springer
Giller KE, Hijbeek R, Andersson JA, Sumberg J (2021) Regenerative agriculture: an agronomic perspective. Outlook Agric 50(1):13–25
Hayo L, Hasegawa H (2024) Enhancing emission reductions in South African agriculture: the crucial role of carbon credits in incentivizing climate-smart farming practices. Sustain Fut 8: 100260
International Renewable Energy Agency (IRENA) (2021) Renewable energy for agri-food systems: towards the sustainable development goals and the Paris Agreement. https://www.irena.org/publications/2021/Nov/Renewable-Energy-for-Agri-food-Systems
Jangir CK, Sangwan PS, Panghaal D, Kumar S, Meena RS, Bharti J, R.D. and Singh, N. (2024) Spatial variability and statistical analysis of soil properties in the rice wheat-based systems of North-West India. Commun Soil Sci Plant Anal 55(8):1205–1223
Jawad HM, Nordin R, Gharghan SK, Jawad AM, Ismail M (2017) Energy-efficient wireless sensor networks for precision agriculture: A review. Sensors 17(8):1781
Jhariya MK, Raj A, Banerjee A, Meena RS, Bargali SS, Kumar S, Nema S, Poonam, Oraon PR (2022) Plan and policies for soil organic carbon management under agroforestry system. In: Plans and policies for soil organic carbon management in agriculture. Springer, Singapore, pp 191–219
Kabeyi MJB, Olanrewaju OA (2022) Sustainable energy transition for renewable and low carbon grid electricity generation and supply. Front Energy Res 9:743114
Khurana A, Kajale D, Cariappa AA, Krishna VV (2024a) Shaping India's climate future: a perspective on harnessing carbon credits from agriculture. Outlook Agric:00307270241240778
Khurana S, Kumar D, Khurana A (2024b) Sustainable agriculture and carbon credits: integrating environmental and economic objectives. Indian J Agric Econ 53. https://doi.org/10.1177/00307270241240778
Kumar R, Punia H, Kumar S (2020) Nanomaterials for agriculture input use efficiency. Resources use efficiency in agriculture. Springer, Singapore. https://doi.org/10.1007/978-981-15-6953-1_5

Kumar S, Bamboriya SD, Rani K, Meena RS, Sheoran S, Loyal A, Kumawat A, Jhariya MK (2022) Grain legumes: a diversified diet for sustainable livelihood, food, and nutritional security. In: Advances in legumes for sustainable intensification. Academic Press, pp 157–178

Kumar S, Sharma SK, Thakral SK, Bhardwaj KK, Jhariya MK, Meena RS, Jangir CK, Bedwal S, Jat RD, Gaber A, Atta AA (2022a) Integrated nutrient management improves the productivity and nutrient use efficiency of *Lens culinaris* Medik. Sustainability 14(3):1284

Kumar S, Sharma SK, Dhaka AK, Bedwal S, Sheoran S, Meena RS, Jangir CK, Kumar D, Kumar R, Jat RD, Meena AK (2023) Efficient nutrient management for enhancing crop productivity, quality and nutrient dynamics in lentil (*Lens culinaris* Medik.) in the semi-arid region of northern India. PLoS One 18(2):e0280636

Kumar S, Meena RS, Kumar S, Pradhan G, Jangir CK, Ghosh S, Punia H, Sheoran P, Meena R, Ahmad MA, Goyal SK (2025) Designing a diversified Indian mustard production system for energy-carbon-cum-heat use efficiency and sowing dates assessment. GCB Bioenergy 17(6): e70044

Lawrence NC, Tenesaca CG, VanLoocke A, Hall SJ (2021) Nitrous oxide emissions from agricultural soils challenge climate sustainability in the US Corn Belt. Proc Natl Acad Sci 118(46): e2112108118

Lee J (2017) Farmer participation in a climate-smart future: evidence from the Kenya agricultural carbon project. Land Use Policy 68:72–79

Leharwan M, Kumar Y, Kumar R, Kumar Saraswat P, Kumar R, Veetil AKT, Bhattacharjee S, Kumar A, Kumar S (2023) Assessing the effects of conservation tillage and in-situ crop residue management on crop yield and soil properties in rice–wheat cropping system. Sustainability 15(17):12736

Marchant GE, Cooper Z, Gough-Stone PJ (2022) Bringing technological transparency to tenebrous markets: the case for using blockchain to validate carbon credit trading markets. Nat Resour J 62:159

Martelli E, Freschini M, Zatti M (2020) Optimization of renewable energy subsidy and carbon tax for multi energy systems using bilevel programming. Appl Energy 267:115089

Meena RS, Pradhan G (2023) Industrial garbage-derived biocompost enhances soil organic carbon fractions, CO_2 biosequestration, potential carbon credits, and sustainability index in a rice-wheat ecosystem. Environ Res:116525. https://doi.org/10.1016/j.envres.2023.116525

Meena RS, Pradhan G, Kumar S, Lal R (2023) Using industrial wastes for rice-wheat cropping and food-energy-carbon-water-economic nexus to the sustainable food system. Renew Sust Energ Rev 187:113756. https://doi.org/10.1016/j.rser.2023.113756

Meena RS, Pradhan G, Nalani, Singh K, Singh AK, Verma SK, Mina KK, Kumar S, Chaturvedi RK (2024) Diversified cropping modules designed for soil restoration, CO_2 sequestration, and generating carbon credits. Land Degrad Dev 35. https://doi.org/10.1002/ldr.5224

Meena RS, Pradhan G, Singh K, Kumar S, Singh AK, Shashidhar KS, Mina KK, Rao CS (2024a) Agriculture models for restoring degraded land to enhance CO_2 biosequestration and carbon credits in the Vindhyan region of India. Sci Total Environ 929:172661. https://doi.org/10.1016/j.scitotenv.2024.172661

Meena RS, Lal R, Kumar S, Pradhana G, Rao CS, Singh AK, Pathak H, Abhilash PC, Kumar A, Sharma SK, Jat ML, Singh S (2024b) Potential of Indian agriculture for capturing atmospheric CO_2 and monetizing carbon credits to the farmers: an overview and policy framework. Adv Agron 188:101–206. https://doi.org/10.1016/bs.agron.2024.06.001

Meena RS, Pradhan G, Kumar S (2025) Energy flow, eco-efficiency, and economic circulation with recycled industrial waste compost application in wheat and subsequent rice farming. Sci Total Environ 967:178779. https://doi.org/10.1016/j.scitotenv.2025.178779

Meitern M (2024) Unlocking carbon finance: empowering energy communities for mutual benefit. Renew Sust Energ Rev 199:114499

Ministry of New and Renewable Energy (MNRE) (2023) Pradhan Mantri Kisan Urja Suraksha evam Utthaan Mahabhiyan (PM-KUSUM). https://mnre.gov.in/en/pradhan-mantri-kisan-urja-suraksha-evam-utthaan-mahabhiyaan-pm-kusum

Montazerinejad H, Eicker U (2022) Recent development of heat and power generation using renewable fuels: A comprehensive review. Renew Sust Energ Rev 165:112578

Mperejekumana P, Shen L, Zhong S, Gaballah MS, Muhirwa F (2024) Exploring the potential of decentralized renewable energy conversion systems on water, energy, and food security in africa. Energy Convers Manag 315:118757

Navarro Jiménez A (2024) Leveraging Article 6 of the Paris agreement: enhancing Costa Rica's carbon market and decarbonization goals. Available at SSRN 4899823

Nemethy SA, Molnár G (2014) Sustainable management of lakes in connection with mitigation of adverse effects of climate change, agriculture and development of green micro regions based on renewable energy production. EQA Int J Environ Qual 13:21–35

Nsabiyeze A, Ma R, Li J, Luo H, Zhao Q, Tomka J, Zhang M (2024) Tackling climate change in agriculture: A global evaluation of the effectiveness of carbon emission reduction policies. J Clean Prod 468:142973

Pradhan G, Meena RS (2023) Interaction impacts of biocompost on nutrient dynamics and relations with soil biota, carbon fractions index, societal value of CO_2 equivalent, and ecosystem services in the wheat-rice farming. Chemosphere 139695. https://doi.org/10.1016/j.chemosphere.2023.139695

Pradhan G, Meena RS (2023a) Utilizing waste compost to improve the atmospheric CO_2 capturing in the rice-wheat cropping system and energy-cum-carbon credit audit for a circular economy. Sci Total Environ 164572. https://doi.org/10.1016/j.scitotenv.2023.164572

Pradhan G, Meena RS, Kumar S, Lal R (2023) Utilizing industrial wastes as compost in wheat-rice production to improve the above and below-ground ecosystem services. Agric Ecosyst Environ 358:108704. https://doi.org/10.1016/j.agee.2023.108704

Pretty J (2008) Agricultural sustainability: concepts, principles and evidence. Philos Trans R Soc B Biol Sci 363(1491):447–465

Punia H, Tokas J, Malik A, Satpal, Rani A, Gupta P, Kumari A, Mor VS, Bhuker A, Kumar S (2020) Solar radiation and nitrogen use efficiency for sustainable agriculture. Resources use efficiency in agriculture, pp 177–212

Rahman MM, Khan I, Field DL, Techato K, Alameh K (2022) Powering agriculture: present status, future potential, and challenges of renewable energy applications. Renew Energy 188:731–749

Rapsomanikis G (2015) Small farms big picture: smallholder agriculture and structural transformation. Development 58:242–255

Ravishankar E, Booth RE, Saravitz C, Sederoff H, Ade HW, O'Connor BT (2020) Achieving net zero energy greenhouses by integrating semitransparent organic solar cells. Joule 4(2):490–506

Reiter G, Lindorfer J (2015) Global warming potential of hydrogen and methane production from renewable electricity via power-to-gas technology. Int J Life Cycle Assess 20:477–489

Reyes O, Gilbertson T (2010) Carbon trading: how it works and why it fails. Soundings 45(45):89–100

Rose DC, Bhattacharya M, de Boon A, Dhulipala RK, Price C, Schillings J (2022) The fourth agricultural revolution: technological developments in primary food production. In: A research agenda for food systems. Edward Elgar Publishing, pp 151–174

Saeed S, Siraj T (2024) Global renewable energy infrastructure: pathways to carbon neutrality and sustainability. Solar Energy Sustain Dev J 13(2):183–203

Salma A, Fryda L, Djelal H (2024) Biochar: a key player in carbon credits and climate mitigation. Resources 13(2):31

Samuel SA (2025) Carbon pricing mechanisms for reducing greenhouse gas emissions and encouraging sustainable industrial practices. World J Adv Res Rev 25(02):001–024

Santier T (2023) The next priorities of the voluntary carbon market for mass adoption: the need for new technologies, carbon policy frameworks, and a meta registry. Doctoral dissertation, Université Paris-Saclay

Taghizadeh-Hesary F, Yoshino N (2020) Sustainable solutions for green financing and investment in renewable energy projects. Energies 13(4):788

Tang KHD (2023) A review of environmental, social and governance (ESG) regulatory frameworks: their implications on Malaysia. Trop Aquat Soil Poll 3(2):168–183

Tao R, Wang X, Ren L (2023) Exploring the impact of environmental regulation and renewable energy R&D on carbon emissions: evidence from panel data analysis. Sustainability 15(18): 13345. https://doi.org/10.3390/su151813345

Ul-Haq Z, Mehmood U, Tariq S, Qayyum F, Azhar A, Nawaz H (2022) Analyzing the role of meteorological parameters and CO2 emissions towards crop production: empirical evidence from South Asian countries. Environ Sci Pollut Res 29(29):44199–44206

United Nations (2015) Sustainable development goals. https://sdgs.un.org/goals

United Nations Framework Convention on Climate Change (UNFCCC) (2015) The Paris agreement. https://unfccc.int/process-and-meetings/the-paris-agreement/the-paris-agreement

United States Department of Agriculture (USDA) (2024) Rural energy for America Program Renewable Energy Systems & Energy Efficiency Improvement Guaranteed Loans & Grants. https://www.rd.usda.gov/programs-services/energy-programs/rural-energy-america-program-renewable-energy-systems-energy-efficiency-improvement-guaranteed-loans

Vadiee A, Martin V (2013) Energy analysis and thermoeconomic assessment of the closed greenhouse–the largest commercial solar building. Appl Energy 102:1256–1266

Wang Q, Li R, Mensah IA (2023) Renewable energy consumption and CO_2 emissions in sub-Saharan Africa: the role of governance and economic development. Sustain Environ Res 33:11. https://doi.org/10.1186/s42834-023-00193-8

Woelfle G (2013) The wind at work: an activity guide to windmills. Chicago Review Press

Woo J, Fatima R, Kibert CJ, Newman RE, Tian Y, Srinivasan RS (2021) Applying blockchain technology for building energy performance measurement, reporting, and verification (MRV) and the carbon credit market: a review of the literature. Build Environ 205:108199

World Bank (2023) State and Trends of Carbon Pricing 2023. Washington, DC: World Bank. https://doi.org/10.1596/978-1-4648-2006-9. License: Creative Commons Attribution CC BY 3.0 IGO

Xuan X, Zhang F, Deng X, Bai Y (2023) Measurement and spatio-temporal transfer of greenhouse gas emissions from agricultural sources in China: a food trade perspective. Resour Conserv Recycl 197:107100

Xue J (2017) Photovoltaic agriculture opportunity for photovoltaic applications in China. Renew Sust Energ Rev 73:1–9

Yucheng Z (2024) Application of carbon market mechanisms in agricultural economics and their roles in sustainable development. Acad J Bus Manag 6(9):92–96

Chapter 12
Wind-Powered Irrigation: Revolutionizing Agriculture and Carbon Credit in Arid and Semi-Arid Regions

Peter Osei Boamah, Raymond Gyaang, Jacqueline Onumah, Dennis Agyemanh Nana Gookyi, Fortunatus Aabangbio Wulnye, Samuel Dodobatia Wetajega, Newlove A. Afoakwah, Sufyan Yakubu, and Tayari Salifu

Abstract Agricultural irrigation in arid and semi-arid regions is often constrained by limited energy access, high operational cost, and environmental concerns tied to fossil fuel use. Wind-powered irrigation systems stand out not only as cost-efficient but also as renewable-sourced solutions to further improve water-use efficiency and significantly reduce greenhouse gas (GHG) emissions. Wind energy systems can replace diesel-powered pumps and save up to 6000 L and mitigate up to 16.08 tons of carbon dioxide (tCO_2) per unit annually, translating into measurable carbon credits valued at approximately US$ 320 per year. This chapter examines the intersection of wind energy, sustainable irrigation, and carbon credit generation. It

P. O. Boamah (✉) · J. Onumah
Department of Ecological Agriculture, Bolgatanga Technical University, Bolgatanga, Ghana
e-mail: pboamah@bolgatu.edu.gh; jonumah@bolgatu.edu.gh

R. Gyaang · S. D. Wetajega · S. Yakubu
Department of Electrical and Electronics Engineering, Bolgatanga Technical University, Bolgatanga, Ghana
e-mail: rgyaang@bolgatu.edu.gh; swetajega@bolgatu.edu.gh; syakubu@bolgatu.edu.gh

D. A. N. Gookyi
Institute for Scientific and Technological Information, Council for Scientific and Industrial Research, New Delhi, Delhi, India

F. A. Wulnye
Department of Telecommunication Engineering, Kwame Nkrumah University of Science and Technology, Kumasi, Ghana

N. A. Afoakwah
Department of Food Science and Technology, Faculty of Agriculture, Food and Consumer Sciences, Nyankpala Campus, University for Development Studies, Tamale, Ghana
e-mail: nafoakwah@uds.edu.gh

T. Salifu
Department of Agricultural Engineering, Bolgatanga Technical University, Bolgatanga, Ghana
e-mail: tsalifu@bolgatu.edu.gh

R. S. Meena et al. (eds.), *Carbon-Negative Agriculture*, Sustainability Sciences in Asia and Africa, https://doi.org/10.1007/978-981-95-1483-0_12

draws on established methodologies, including Intergovernmental Panel on Climate Change (IPCC) and European Union (EU) Innovation Fund frameworks, and presents a 10-year emission projection which quantifies emission reductions and supports potential revenue pathways. Co-benefits like agroforestry-based carbon sequestration, hybrid solar-wind options, and savings of up to 60%–80% compared to a fuel generator have also been discussed. Underutilization of wind-powered irrigation policy solutions, such as Clean Development Mechanism/Verified Carbon Standard (CDM/VCS) certification, subsidization, educational outreach programs, and economic adoption resilience strategies, was also explored. Wind-powered irrigation positions itself as a refreshing climate-smart innovation, connecting sustainable agriculture to carbon finance, thereby laying the groundwork for climate-resilient agricultural development.

Keywords Agricultural output · Carbon credit · Irrigation techniques · Renewable energy · Water-use efficiency

Abbreviations

CDM	Clean Development Mechanism
CERs	Certified energy reductions
CO_2	Carbon dioxide
CO_2e	Carbon dioxide equivalent
DOEs	Designated operational entities
FAO	Food and Agriculture Organization
GHG	Greenhouse gas
GPS	Global Positioning System
IoT	Internet of Things
IPCC	Intergovernmental Panel on Climate Change
IRENA	International Renewable Energy Agency
R&D	Research and development
UNFCCC	United Nations Framework Convention on Climate Change
VCS	Verified Carbon Standard
WHO	World Health Organization
WUE	Water-use efficiency

12.1 Introduction

Agriculture is among the major drivers of economic growth in countries worldwide. Farmers, however, still face difficulties in growing quality products, with the wrong or old farming practices causing low-quality yields. Pests, plant diseases particular to some crops, and natural hazards lead to huge losses in agriculture every year (Garg et al. 2021; Meena et al. 2025). The World Food Programme's 2024 Global Report on Food Crises states rising world hunger. As of 2023, it is estimated that there were 282 million people across 59 nations facing severe levels of food insecurity and in desperate need of help with food and livelihoods (James 2024). This represents an increase of 24 million people from 2022, therefore highlighting the urgency of a historic strategy tackling the underlying causes of food insecurity and malnutrition.

Climate change, conflict, increased food prices, and the coronavirus disease 2019 (COVID-19) pandemic have all combined to cause one of the world's worst food crises in decades, especially on the African Continent (WHO 2024). In the wake of the COVID-19 epidemic, the health, socioeconomic conditions, and food security of Africans have all suffered significantly (Mohamed et al. 2021) due to a series of government-led restrictions as measures to curb the virus from spreading throughout the continent consequently a reduction in agricultural activities as farmers were unable to reach their farms during the lockdown.

The globe accelerates its quest for sustainable agriculture as food insecurity, land degradation, and environmental degradation confront agricultural systems, particularly the semi-arid and arid tropics. Traditional irrigation systems, particularly surface and sprinkler irrigation that involve high energy levels, use of fossil fuels, and release of greenhouse gases and land degradation, exacerbate the issues (Jamali et al. 2021; Meena et al. 2024). Inefficient energy use in agriculture increases environmental degradation and undermines food security by increasing production costs and decreasing the availability of safe water resources. Renewable energy resources like wind, solar, and biomass offer a sustainable option using naturally replenished resources with negligible environmental impact (Ponta et al. 2007; Hossain 2019). Among them, one can utilize wind power to power irrigation pumps in water-scarce regions, reducing greenhouse gas emissions and facilitating sustainable agriculture. Such innovations must be adopted both by engineers and farmers at the same time, innovation moderated by pragmatism, to fight the food, land, and environmental crises combined.

Although the era of fossil fuels is not yet over, their dominance is waning. As global energy demand rises, the share of fossil fuels is gradually decreasing. Renewable energy technologies have accounted for half the new capacity installed to meet this growing demand in recent years. Wind and hydropower are at the forefront of the shift toward renewable energy (Kuriqi et al. 2021). The proportion of renewables in the global energy mix often depends on oil prices. When oil prices climb, renewables like wind and solar tend to make up a larger share of energy production. On the other hand, cheaper oil can lead to a dip in renewable energy use, even though high oil costs typically reduce reliance on other fossil fuels, such as coal and gas (Ackermann and Söder 2000). As a result, wind energy is becoming increasingly popular as a scalable solution, especially when oil prices rise.

Energy from renewable sources can save the environment from being permanently destroyed. A viable substitute is wind-powered irrigation, which uses sustainable wind energy to move water for farming. Wind-powered irrigation utilizes the wind's kinetic energy to power water pumps with mechanical or electrical energy.

With the help of air streams, wind energy is converted to electrical or mechanical energy for human use. It is a clean, renewable energy source found everywhere in abundance, free from global trade wars, and gentle on the environment, compared to oil and coal. With today's technology, that is, grid-connected wind farms, the intermittency of winds is reduced, and it is a reliable source of energy on a wide variety of timescales and an attractive option for green power.

With the global emphasis on climate change mitigation, advancing wind power technology is crucial for diversifying the energy mix and promoting industry growth, particularly in the agricultural sector. Wind-powered irrigation leverages wind energy to power water pumps for agricultural purposes. This new and green method is a pollution-free alternative to conventional irrigation, typically a sun-derived or fossil fuel-based system. Crop farmers can employ a consistent source of water supply for their farms by tapping into wind power, especially in agricultural lands that are not connected to the power grid.

Wind-powered irrigation makes use of wind turbines to pump water through wind energy as a renewable substitute for diesel-based systems for the irrigation of arid and semi-arid lands. Converting wind kinetic energy into mechanical energy, the systems pump water from wells or reservoirs and provide water to crops with reduced use of fossil fuel. The practice decreases fuel expense, greenhouse gas emissions by 0.54–1.35 tCO_2e every year per pump, and increases sustainable agriculture during periods of water shortage.

12.2 Wind-Powered Irrigation and Agriculture

Using wind turbines to generate electricity provides a current way of using wind energy in agriculture for equipment operation, irrigation, and other agricultural activities. Wind, as a renewable and green energy source, is flexible to various agricultural operations. Providing a cleaner alternative to fossil fuels encourages a decrease in greenhouse gas emissions and preserves environmental responsibility. Wind power is obtained from the movement of air, where its kinetic energy spins turbines to produce power. Wind power is the practical application of this energy. It is the conversion of wind power into mechanical or electric power. Windmills are seen everywhere in most agrarian settings, and wind power has been exploited for centuries. Earlier, the major work of windmills was grain grinding and water supply (Fig. 12.1).

Several issues, such as inadequate water supply and unreliable electricity supply, plague semi-dry and dry farming (Dos Santos Isaías et al. 2019). Expensive and environment-degrading diesel pumps dominate traditional irrigation (Usman et al. 2024). The alternative is more environmentally friendly wind-powered irrigation that utilizes the dominant winds in such regions to increase the productivity of crops at low environmental degradation.

About 10–20 inches (25–50 cm) of rainfall falls in semi-arid regions each year, and under 10 inches (25 cm) in arid regions. Arid regions may have similar geology

Fig. 12.1 Image of a 1 kW wind turbine in operation in Accra (Adapted from Bui Power Authority (2021))

to other regions, but have loose surficial materials and exposed ground upon which erosion is a governing process in their land. Water shortage is a catastrophic situation for arid and semi-arid areas, especially in areas like North Africa and Central West Asia, where it is simple to identify more than one issue (Dehghanisanij et al. 2006).

- Surface water is dwindling rapidly, pushing communities to rely more heavily on groundwater.
- Excessive use of groundwater is leading to water tables plummeting.
- Farming practices tend to prioritize high yields at the expense of water conservation, hence profligate use.
- Increasing world populations, particularly in the Third World, are generating demand for water and food and creating scarcity.
- Industry, domestic use, and the environment increasingly vie for water.

These pressing water challenges need to be solved by new technologies through sustainable agriculture, for example, cleaner energies to handle water. One such potential substitute is wind irrigation, where large turbines supply electricity to water pumps (Usman et al. 2024). Such an option saves running costs, reduces greenhouse gas emissions, and gives farmers greater energy autonomy.

The potential of wind power for small farmers in arid regions, particularly those suffering from power shortages, has been widely investigated (Usman et al. 2024).

The research presents the environmental significance of using generators and recommends wind power as a green source of irrigation, water storage, and water management in agriculture. A study conducted in Zaria Local Government, Kaduna State, Nigeria, examined a small-scale wind energy system and how the speed of wind influences electrical parameters like voltage, current, frequency, and power. Based on the outcome of the study, there was significant fluctuation of these parameters, but contrary to this, it proved that converter systems can level the output of energy to facilitate its incorporation into the grid. Employing wind energy to power irrigation in arid regions presents a sequence of benefits: it extends the harvest, enhances water efficiency, and ensures farmers' financial futures. Reducing electricity shortages with a clean substitute, wind power reduces reliance on costly, dirty generators and allows for abundant harvests during periods of shortage.

Past research has already proven the viability and efficiency of wind-powered irrigation in semi-arid and arid regions. For instance, the study of Zaria Local Government, Kaduna State, Nigeria, proves that small-scale wind energy conversion systems can be used to effectively supply farmers' irrigation requirements (Usman et al. 2024). It found that variations in wind speed led to variations in electrical parameters, but converter systems made the output stable to be fed to the grid.

Wind irrigation is a powerful ecological benefit by saving the use of fossil fuels and greenhouse gas emissions. Wind irrigation is an economic option for farmers as it is a cheaper alternative that will save them money by reducing their dependence on expensive diesel fuel and allowing them to weather unstable energy prices (Usman et al. 2024). There are also some drawbacks, including the initial cost, maintenance, and unstable winds. Future research is needed to develop the performance and power of wind turbines, considering hybrid technology such as wind combined with other clean technologies, and triggering policy and incentives for farmers. Wind irrigation can revolutionize agriculture for dry and semi-dry climatic conditions using a low-cost and lasting solution to the scarcity of water and energy (Meena et al. 2024a). Crop yields are enhanced, environmental contamination is minimized, and economic foundations are established through the use of wind power by farmers.

12.3 The Promise of Carbon Credits Generation Through Wind-Powered Irrigation

Crop production in arid and semi-arid regions is being threatened by water shortage and unstable energy, which in turn leads to the utilization of diesel irrigation pumps that emit a lot of greenhouse gases. Wind irrigation is a green technology that uses wind energy to pump water, thus reducing its impact on the environment. The system involves the transfer of wind energy into electricity for use by water pumps and presents an excellent alternative in many off-grid locations that are not connected to limited or zero-grid connections. Others include the possibility of lower running costs, lower emissions, and greater energy independence for farmers (Usman et al. 2024; Meena et al. 2024b). The Poldaw wind pump systems are an

Fig. 12.2 Image of a wind turbine from a wind energy project located in Ghana (Adapted from Namara et al. (2010))

Table 12.1 Generic features of the Poldaw wind pump

Feature	Parameters
Tower height (m)	9–10
Number of blades	12
Typical rotor speed (rpm)	40–80
Starting wind speed (m/s)	2.5
Survival wind speed (m/s)	50
Service life (years)	20

example of this technology, having been installed by the relevant bodies at various sites in the Upper East and Upper West regions of Ghana, as shown in Fig. 12.2, and further explained in Table 12.1. This section quantifies carbon credit potential on an empirical basis and through methodical means of the Clean Development Mechanism (CDM) and Verified Carbon Standard (VCS) with emission reductions, sequestration, and economic benefits.

12.3.1 Emission Reductions from Fossil Fuel Displacement

The carbon released may be reduced by a switch from diesel to wind pumps; it may be calculated and converted to carbon credits. The farmers also benefit from a clean and economic environment. Diesel-powered irrigation pumps for small-scale (5 kW) systems running 1000 h per year consume 0.2–0.5 L of diesel per hour, or 200–500 L per pump (Usman et al. 2024). This is 540–1350 kg CO_2e (0.54–1.35 tCO_2e) per pump per annum for an emission factor of 2.7 kg $CO2e$ per liter (IPCC 2006). The Poldaw wind pump replaces 810 kg of $CO2e$ (0.81 tCO_2e) per annum by replacing 300 L of diesel (Namara et al. 2010). According to a pilot case study in Nigeria's Kaduna State as part of the Zaria local government, small-scale wind systems can successfully supply irrigation requirements to improve agricultural yields without any negative impacts on the environment. Converter systems facilitated steady power for grid connection for a reliable irrigation supply regardless of the variations in electricity generation caused by different wind speeds (Usman et al. 2024). Scaling this up to 100 pumps, this equates to savings of 50–120 tCO_2e annually, which is the same as taking 10–25 gasoline-powered vehicles off the road (EPA 2023).

12.3.2 *Quantifying Carbon Credit Potential from Wind-Powered Irrigation*

Wind-powered irrigation systems contribute not only to climate mitigation but also suitably displace diesel-powered water pumps, thereby reducing greenhouse gas (GHG) emissions. This section creates a credible link between these systems and carbon credit generation by quantifying GHG emission avoidance, outlining the applicable methods and presenting a sample consistent calculation based on the CDM and European Union (EU) Innovation Fund Practices.

12.3.2.1 Emission Reduction Methodology

To estimate the emission reduction potential of wind-powered irrigation, we apply the absolute GHG emission avoidance equation, adopted from Annexe A of the EU Innovation Fund (Edwards et al. 2020). Methodology:

$$\Delta \mathrm{GHG_{abs}} = \sum_{y=1}^{10} \left(\mathrm{Ref}_y - \mathrm{Proj}_y\right) \tag{12.1}$$

where:

- $\mathrm{Ref_y}$: emissions from baseline (diesel-powered pumping) in year y.
- $\mathrm{Proj_y}$: emissions from wind-powered pumping in a year y (typically negligible).
- $\Delta\mathrm{GHG_{abs}}$: net avoided emissions over 10 years, in tCO_2e.

According to the Risø report (Ringius et al. 2002), the standard emission factor for fossil-based electricity or diesel substitution in developing countries ranges from

0.5496 to 0.6868 tCO_2/MWh. For diesel pumps, the Intergovernmental Panel on Climate Change (IPCC) defaults recommend 2.68 kg CO_2/L of diesel (Lapshin 2017).

12.3.2.2 Sample Calculation

Let us consider a hypothetical scenario of replacing a diesel-powered irrigation pump with a wind-powered system:

Assumptions:

- Diesel pump consumption: 5 L/h
- Daily usage: 6 h
- Operational days/year: 200
- Total fuel saved annually: 5*6*200 = 6000 L/year

GHG Emission Avoided:
6000 l/year * 2.68 kg CO_2/l = 16,080 kg CO_2/year = 16.08 TCO_2/year
Carbon Credit Revenue (assuming US$ 20/tCO_2):
16.08 * 20 = US$ 321.6/year
Over 10 years:

- Total CO_2 avoided: 160.8 tCO_2
- Total carbon revenue: US$ 3216

This conservative estimate demonstrates that even a modest wind-powered system can generate verifiable emission reductions eligible for Certified Emission Reductions (CERs) under the CDM or equivalent voluntary carbon markets.

12.3.2.3 Applicability and Verification

Wind-powered irrigation systems qualify as small-scale renewable energy CDM projects under the United Nations Framework Convention on Climate Change (UNFCCC). According to the Marrakesh Accords (Peake 2002) and methodologies discussed by Ringius et al. (2002), such projects must establish:

- Baseline: The default case is continued use of the diesel generator
- Additionality: The project would not have occurred without carbon finance.
- Monitoring: Diesel displacement is tracked through energy meters or pump operation logs.

Verification is performed by third-party Designated Operational Entities (DOEs) following CDM or Gold Standard protocols. A summary table of estimated annual emission reduction and carbon revenue from replacing a diesel pump with a wind-powered irrigation system is presented in Table 12.2.

Table 12.2 Emission reduction and carbon credit estimate for wind-powered irrigation

Parameter	Values
Diesel saved/year	6000 l
CO_2 emission factor (IPCC)	2.68 kg CO_2/l
Annual CO_2 avoided	16.08 tCO_2
Carbon price (assumed)	US$ 20/tCO_2
Annual carbon credit revenue	US$ 321.60
Total (10 years)	160.8 tCO_2/US$ 3216

12.3.3 Carbon Sequestration Synergies

Wind-powered irrigation subsidizes carbon sequestering activities like conservation tillage and agroforestry. Semi-arid agroforestry sequesters 1–3 tCO_2e/ha/year of tree biomass and soil carbon (Singh et al. 2024). A wind-powered irrigated 10-ha farm agroforested holding 10–30 tCO_2e/year audited under VCS methodology VM0017. Adding emission reduction (0.81 tCO2e/pump) and sequestration (10–30 tCO_2e/farm), a farm can earn up to 11–31 tCO_2e value of credit annually.

12.3.4 Economic and Environmental Benefits

Wind systems render operation more affordable by reducing diesel cost, with the farmer making a saving of between US$100 and US$300 per pump annually (at the cost of US$1/l diesel). Wind pumps use 0.1–0.2 kWh/m^3 water pumped as opposed to 0.5–1 kWh/m^3 for diesel pumps, reducing energy use by 60%–80% (IRENA 2019). With a case study in Nigeria, there were 30%–50% cost savings from wind-powered irrigation that created economic resilience (Usman et al. 2024). Carbon credits are another source of income that makes IPCC levels for climate change mitigation achievable (IPCC 2006).

12.3.5 Challenges and Solutions

Despite the potential, wind irrigation is accompanied by the expense of capital outlay, maintenance needs, and fluctuating speeds of wind. There is a need for better efficiency and dependability in turbines, the hybridization of wind with other alternative energy sources into the power supply, and extension and incentives to farmers in the years to come, so that farmers will adopt such systems. Better equipment and infrastructure can increase the use of wind-powered irrigation in terms of being more expensive in dry and arid regions (Smith et al. 2020). Subsidies and low-interest loans can reduce financial barriers, while hybrid wind-solar systems ensure consistent power (Smith et al. 2020). Community training programs enhance maintenance capacity (Ali et al. 2024). Wind irrigation is one of the strategies utilized to augment farm productivity and sustainability in the desert. The systems are emission-reducing, carbon credit-generating, and economic revenue-generating to farmers as they become one of the best tools to mitigate climate change and stimulate environmental conservation.

12.4 Enhancement of Water-Energy Efficiency in the Arid and Sub-Humid Regions

Low water and weak energy dominate dry and arid regions, presenting a significant hindrance to agricultural productivity (Frimpong et al. 2023). Sustainable agriculture and environmental protection require utmost efficiency in water and energy use. New technologies and intelligent water management need to be identified in this achievement.

Drip irrigation and precision agriculture are the optimum methods for enhancing the water productivity of plants in desert conditions (Singh et al. 2022; Frimpong et al. 2023). For instance, drip irrigation is utilized for effective water supply to crop roots, whereas precision agriculture is based on sensors and Internet of Things (IoT) technologies to reach higher optimality (Zeggaf et al. 2020). These are quite complemented by water-use-effective types of renewable sources like solar and wind, as indicated in Table 12.3.

12.4.1 Smart Farming

Using climate-responsive food cropping systems, water-saving techniques, equipment, and technology to produce food sustainably is known as water-smart farming (Meena et al. 2023). Chief strategies:

- Conservation tillage and organic matter incorporation to enhance the capacity of the soil to hold and retain water.
- Implementation of rainwater harvesting, storage, and use systems as a sustainable substitute source.

Table 12.3 Features of water-efficient irrigation techniques

Technique	WUE improvement	Energy source	Applicability in arid regions
Drip irrigation	High	Wind/solar	Highly applicable
Sprinkler systems	Moderate	Wind/solar/fossil fuels	Moderate
Surface irrigation	Low	Fossil fuels	Low
Precision agriculture	Very high	Wind/solar/fossil fuels	Highly applicable

- Cultivar development, research, and promotion of crop cultivars that survive and continue to produce under water-limiting conditions.
- Adoption of agricultural methods appropriate for drylands, such as the application of cover crops and intercropping for replenishing and preserving the soil.
- Adopting wet-dry cycling irrigation techniques to conserve water without reducing plant yields.

12.4.2 Superb Energy Irrigation Systems

- To reduce the consumption of energy in agriculture, the adoption of energy-efficient irrigation systems is inevitable. Some of the practices include:
- Solar pumps save fossil fuel because they are driven by solar power.
- Wind irrigation utilizes a clean and renewable source of water compared to diesel pumps.
- Efficient time of irrigation saves water and energy based on crop needs and the level of moisture in the ground.
- Precision agriculture uses sensors and the Global Positioning System (GPS) to monitor and control irrigation better.

12.4.3 Managing Water and Energy Together

It is important to think about water and energy as a package deal to get the most out of the limited resources available to us. As such, they must be handled side by side to avoid waste and boost efficiency (Zeggaf et al. 2020; Meena and Pradhan 2023). This can be done through:

- Digging into the water—energy connection by looking closely at how water and energy use overlap. Spotting where we can cut back or where one can help the other shine.
- Pushing smarter practices—get policies and incentives to nudge people toward methods that save water and energy.
- Spreading the word—teach farmers and others because using water and energy wisely pays off, think workshops, talks, or even just chatting over coffee.

Getting smart about water and energy is a must in dry or semi-arid regions. These places face harsh conditions, but by adopting approaches like combined water-energy management, water-saving farming tricks, or irrigation systems that do not guzzle power, they can grow more crops, lighten their environmental footprint, and better handle whatever climate change throws their way. It is a win-win for the planet and the people working the land.

12.5 Research on Smarter Use of Water and Carbon Credits for Dry Regions

Research and development (R&D) play an important role in sustaining arid and semi-arid region farming. Of utmost importance are enhanced water-use efficiency (WUE) and the generation of carbon credits via wind power-based irrigation for climate change mitigation. Through wind power complemented with the latest irrigation and land management methods, researchers anticipate enhanced farm resilience and generating financial returns via carbon markets. This chapter evaluates these innovations, quantifying GHG savings and carbon credit worth in empirical parameters and procedures like the CDM and VCS, with continued focus on WUE as a secondary measure.

12.5.1 More Crop with Less Water for Water-Use Efficiency

Water-use efficiency, or yield per unit of water applied, is a priority activity in arid environments where water limitation constrains farm output. Energy-efficient irrigation, augmented by wind power systems, reduces energy demands and GHG emissions, enabling carbon credit qualification. Such WUE advantages, augmented with wind power, enhance carbon credit value by saving fossil fuel use in irrigation. For example, a 5-ha farm with precision irrigation, most likely powered by wind energy, would be able to conserve 1–2 tCO_2e annually through energy and pumping water savings (IRENA 2019). The main R&D activities are:

- Researchers are raising more efficient crop types with fewer water requirements, thereby reducing irrigation. Usman et al. (2024) demonstrated that such crops on dry Nigerian lands irrigated by wind-powered pumps reduced 20%–30% with the same amount of yield. This reduces energy emissions by 0.3–0.5 tCO_2e/ha/year, which has economic prospects in generating carbon credits and CDM strategies (Usman et al. 2024; Pradhan and Meena 2023).
- GPS- and sensor-based systems provide water with precision, minimizing wastage. Precision irrigation in semi-arid regions on renewable energy like wind saves 20%–40% water and 50% energy (Zeggaf et al. 2020). This reduces emissions by 0.2–0.4 tCO_2e/ha/year, which could be validated against carbon credits (Zeggaf et al. 2020; IRENA 2019).

Rainwater harvesting systems reduce the groundwater reliance, complementing wind-powered irrigation. Zeggaf et al. (2020) also contributed that rainwater harvesting systems in drylands reduce pumping energy needs. This translates into 0.1–0.3 tCO_2e/ha/year reduced emissions that are CDM eligible (Zeggaf et al. 2020).

12.5.2 *Managing Carbon Credits for Advancing Carbon Sequestration and Lowering Emissions*

Carbon credits earned per metric ton of CO_2e sequestering or mitigating is a decent way to prevent climate change in drylands. R&D is being pursued on establishing emission reduction of wind-powered irrigation and augmented sequestration through sustainable land management using CDM and VCS guidelines still in force. Some of the major research lines are:

- Wind-powered alternatives to diesel pumps save GHGs, and savings are measured through studies. Usman et al. (2024) monitored a decrease of 0.5–1.2 tCO_2e/year in the production of small wind turbines installed in Nigeria's Zaria. These carbon credits with verified reductions by CDM methodology Approved Methodology for Small-Scale projects (AMS)-Type I: Renewable Energy Project Activities (I). The sub-category Grid connected renewable electricity generation (D) are equal to 50 pumps with a 25–60 tCO_2e/year output that can be sold at US$ 5–$20/$tCO_2e$ when sold at US$ 125– US$ 1200 of income (Usman et al. 2024; Kollmuss et al. 2008).
- Fruit trees with crops in them sequester carbon and contribute to the fertility of the soil. Boufous et al. (2022) stated that agroforestry in the semi-arid region sequesters 1–3 tCO_2e/ha/year, VCS methodology VM0017 certified. Ten hectares of farm supported by wind energy for irrigation and agroforestry would produce 10–30 tCO_2e as credits per annum, doubling the income when taken in conjunction with emission reduction (Boufous et al. 2022; Singh et al. 2024).
- Carbon sequestration in the soil through lower tillage returns 0.5–1 tCO_2e/ha/year. Boufous et al. (2022) clarified that wind-powered irrigation enables such practices by bringing the right amount of water that supports sequestration potential.
- The research examines the policies for the adoption of carbon credits to increase. Boufous et al. (2022) suggested a subsidy for installing wind turbines and harvesting the carbon market for rewarding more farmers' inclusion in VCS schemes. An Indian pilot demonstrated how policy incentives for wind-based irrigation transformed 1200 tCO_2e/year on 1000 ha, giving US$ 6000–US$ 24,000 (VCS 2022).

These R&D endeavors are at the core of climate change mitigation and climate-resilient agriculture in dry environments.

12.6 Plan for Wind-Powered Irrigation of Arid Lands

Utilization of wind-powered irrigation in arid and semi-arid regions provides a green path to enhance crop production and fight climate change by generating carbon credits. This chapter demonstrates an upscaling framework for wind-powered irrigation systems, connecting them with carbon markets and eliminating implementation barriers. Based on empirical evidence and standard processes like the CDM and VCS, the approach aims at attaining maximum GHG emission reduction and economic benefits to farmers.

12.6.1 Strategies for Greener Wind-Powered Irrigation

To create wind-powered irrigation, governments and stakeholders need to prioritize affordable technology and infrastructure.

- Implement 2–5 ha-sized farms with small windmills (1–5 kW), as implemented in Nigeria's Zaria province, where systems reduced emissions by 0.5–1.2 tCO_2e per pump annually (Usman et al. 2024). Implementing it in 1000 pumps would eliminate 500–1200 tCO_2e yearly, the same as taking 100–250 cars off the road (Environmental Protection Agency 2023).
- Integrate solar with wind to offer a constant power, reconciling the variability of the wind speed. Hybrid systems in semi-arid areas achieve 90% reliability, maintaining irrigation and entitlement to carbon credits (Smith et al. 2020).
- Farmer training: Maintenance and adoption are boosted by local programs, which decrease downtime by 30%–50% (Ali et al. 2024).

12.6.2 Policy Incentives for Carbon Credit Integration

Policy support will be essential in promoting wind-powered irrigation and entry into the carbon market. Induced policies are:

- Initial investment subsidization of 50%–70% (US$2000–US$5000/system) can initiate adoption because in India's wind irrigation project of 1200 tCO_2e/year over 1000 ha (VCS 2022).
- Farmers may get US$5–US$20/tCO_2e, or US$125–US$1200 annually for 50 pumps, with simplified VCS and CDM scheme certification (Kollmuss et al. 2008).
- Low-cost turbine and sequestration R & D method improves the viability of carbon credit, for example, agroforestry (1–3 tCO_2e/ha/year) (Boufous et al. 2022; Singh et al. 2024).

According to the Intergovernmental Panel on Climate Change (2006), these are the IPCC's sustainable agriculture goals and mitigation goals.

12.6.3 Overcoming Implementation Barriers

Drawbacks of wind irrigation are high upfront capital needs, maintenance needs, and variable wind speeds. Solutions are:

- Mechanisms for funding: Carbon credit returns and interest-free loans fund the cost, where 10-ha farming is sustained by 11–31 tCO_2e annually through agroforestry and carbon savings (Singh et al. 2024).
- Technological advances: 20-year-lasting high-tech turbines ensure reliability (Smith et al. 2020).
- Policy frameworks: Incentives and extension services increase adoption in dryland systems by 20%–30% (Boufous et al. 2022; Pradhan and Meena 2023a).

This policy leverages wind-driven irrigation to reduce emissions, produce carbon credits, and enhance economic robustness to render dry lands captains of sustainable farming.

12.7 The Future of Dry Land Wind-Powered Irrigation

Wind irrigation can be the only savior of semi-arid and dry land agriculture, and with new technologies and ongoing research on the technology, things could not possibly get better. Some hopeful ways of reaping the full benefits of this technology in challenging dry lands and enabling farmers to harvest more without compromising sustainability are outlined below. Integrating wind-powered irrigation with carbon credit generation, these innovations can reduce GHG emissions and provide economic benefits, certified under frameworks like the CDM and VCS.

12.7.1 Smart Technology to Save Water

Imagine smart irrigation. Smart irrigation can be made possible through the Internet of Things (IoT) and intelligent sensors, making windmills smart windmills. They monitor the moisture level of soil, weather, and plant health, and the farmers irrigate only as and when required. Singh et al. (2022) found in a study that such IoT improves water management and yield. These wind turbine-based systems conserve energy consumption by 40%–60%, and 0.2–0.4 tCO_2e/ha/year can be conserved, which can be verified in terms of carbon credits under CDM methodology AMS-I.D (Singh et al. 2022). If a 5-ha farm uses IoT-based wind irrigation, it can earn 1–2 tCO_2e in credits annually, which can be sold at US$5–US$20 per tCO_2e and bring in US$5–US$40 in revenue (Kollmuss et al. 2008).

12.7.2 Smart Wind Turbines for Difficult Sites

Not all wind turbines are created equal, especially when the winds are a bit more capricious in dry weather. Researchers are engineering special turbines to catch up with these anomalous winds' patterns, like variable-speed turbines or aerodynamic, streamlined turbines that catch every gasp of a breeze. The new turbine design can be adapted to more dependable power for water pumps under unfavorable conditions. Kumar et al. (2023) explain why these new-generation turbines can be a blockbuster in drylands. These turbines replace diesel pumps, reducing emissions by 0.5–1.2 tCO_2e per pump annually, as seen in Nigeria's Zaria region (Usman et al. 2024). Scaling to 100 turbines could avoid 50–120 tCO_2e annually, equivalent to removing 10–25 cars from the road (Environmental Protection Agency 2023).

12.7.3 Winning Over Communities

Technology is great, but it is the individuals who get things done. The key to making sure the locals and farmers are not left behind is making sure they learn how to operate and take care of the wind installations. Practice exercises, training, workshops, or even Internet sites where people can exchange tips can significantly enhance learning outcomes. When people feel that they belong and are empowered, they will be more likely to get things done. Trainers, workshops, or facilities where people can exchange tricks. Whenever people are motivated and dedicated, they will adapt. Ali et al. (2024) note that, however much the people's involvement is, it leads to making renewable energy plans sustainable. People-oriented schemes boost adoption by 20%–30% and reduce downtime to sustain by 30%–50%, making carbon credits more dependable (Ali et al. 2024). For instance, Indian villages with wind-based irrigation systems paid back 1200 tCO_2e annually in 1000 ha and received US$6000–US$24,000 (VCS 2022).

12.7.4 Teaming Up for Big Ideas

No one is taking this apart solo. Putting farmers, scientists, and industry representatives in a single room can start some serious creativity. Think about the networks wherein they exchange between themselves what will not work and what will work, and cross-pollinate success stories. These networks can propel the new ideas further and get wind-driven irrigation viral earlier through publication and knowledge sharing between local, small, and large-scale farmers for that improvement to occur. Collective networks can accelerate carbon credit programs, such as in Nigeria, where collaborations with stakeholders accelerated wind irrigation to avert 25–60

tCO_2e annually over 50 pumps (Usman et al. 2024). These activities align with IPCC mitigation objectives, raising world sustainability (Intergovernmental Panel on Climate Change 2006).

12.7.5 Overcoming the Challenges

The wind is not always a certainty; at times, it is too windy or there is none, and it has a tendency to disrupt the power grid (Dehghanisanij et al. 2006). The remedy is to mix wind with solar or other resources so that it becomes stabilized (Kumar et al. 2023). And then there is the price tag; installing the systems does not come cheap, and not all people are tech-hungry enough to follow the trend. The solution required is to request government subsidies to fund the investment cost, better turbine design to minimize maintenance cost, and awareness programs to educate and involve communities (Ali et al. 2024; Pradhan et al. 2023). This wind-irrigated project is as much a demonstration of flashy technology as of empowering desert farmers to thrive despite scarce water resources and an unforgiving climate. Through these practices, the future can now see a world where sustainable agriculture is not just feasible but also affordable. Hybrid systems have 90% reliability, reducing emissions by 0.5–1 tCO_2e/ha/year when combined with agroforestry, certified under VCS methodology VM0017 (Kumar et al. 2023; Singh et al. 2024). Subsidies on costs by 50%–70% (US\$2000–US\$5000 per system) can offset investments, with carbon credits offering an added 11–31 tCO_2e each year for a 10-ha farm (VCS 2022) (Table 12.4).

The future of wind-powered irrigation in dry regions is bright, driven by technology, community involvement, and cooperative efforts. Adopting these initiatives allows one to optimize wind energy, increase the resilience of agriculture, and improve a more sustainable and prosperous future for such regions. Generating carbon credits (e.g., 50–120 tCO_2e/100 turbines, 11–31 tCO_2e/10-ha farm), these systems achieve global climate goals, being economically and environmentally beneficial.

Table 12.4 Overcoming obstacles and exploring strategies in wind-powered irrigation

Obstacles	Effect	R&D strategy
Fluctuating wind speeds	Causes an inconsistent power supply	Develop hybrid systems to stabilize energy output
High upfront costs	Limits adoption by farmers	Provide subsidies and financial incentives
Frequent maintenance needs	Leads to more breakdowns and downtime	Design more durable and efficient turbines
Limited community awareness	Slows down the adoption of wind-powered irrigation	Launch community education and engagement programs

12.8 Conclusion and Future Outlook

Wind irrigation offers a critical solution to arid and semi-arid region irrigation, offering a sustainable method of enhancing water management and environmental sustainability. Using wind power, the regions will achieve maximum crop yields with minimal environmental degradation, without consuming the vast energy employed in traditional irrigation. These technologies substitute fossil fuel-powered pumps, with strong greenhouse gas advantages and enabling farmers to generate carbon credits under frameworks such as the Clean Development Mechanism and Verified Carbon Standard. An individual wind turbine is estimated to save can save approximately 6000 l of diesel annually, avoiding 16.08 tCO_2 yearly when it replaces a diesel generator. When scaled to 100 pumps, this creates 1608 tCO_2 reduced and US$32,160 in revenue yearly. The overall savings over a decade reach 16,080 tCO_2, and the potential revenue from carbon credits cumulates to US $321,600.

There is a second benefit: farmers receive payment in the form of carbon credits when they use green low-emission practices and sell the credits to earn extra income, and this increases economic viability and reduces world climate change. Wind-powered irrigation in Nigeria's Zaria replaced 50 pumps' emissions and raked in sufficient carbon credits, and India's programs proved staggering environmental and financial returns on 1000 ha. Researchers also keep researching to develop tools to enable the usage of water in an optimized manner with the aid of technology solutions such as IoT sensors tracking weather and soil to offer accurate irrigation. These smart devices, wind-powered, are energy conservers, conserving 0.2–0.4 tCO_2e/ha/year, tested under the CDM methodology AMS-I.D Experts also plan to harvest all gusts in windy and desert areas, generating power round-the-clock. High-quality turbines yield 0.5–1.2 tCO_2e savings per pump, giving value to carbon credits.

End-users and farmers still require training, empowerment, and maintenance of the system through workshops or training centers. Empowered users promote adoption and reduce downtime in order to gain carbon credits, with policy guarantees providing grants or incentives for renewable energy applications to incentivize farmers to adopt these systems, enhancing the diffusion of sustainable agriculture.

Policy incentive via subsidy and exposure to the carbon market makes wind irrigation cost-effective, high-carbon-producing farms through saving and agroforestry, as verified under the VCS methodology VM0017. Wind irrigation is an economical and cost-effective means of addressing desert water and energy problems. By leveraging advanced research, new technology, and collaboration among policymakers, researchers, and farmers, wind-powered irrigation can drive a new era of sustainable and climate-resilient agriculture. In support of global climate targets, wind-powered irrigation places dry lands at the center of sustainable agriculture with environmental and economic benefits through carbon credit mechanisms.

References

Ackermann T, Söder L (2000) Wind energy technology and current status: a review. Renew Sust Energ Rev 4(4):315–374

Ali M, Khan MA, Khan MA (2024) Community engagement for renewable energy adoption: a systematic review. Renew Sust Energ Rev 176:112759. https://doi.org/10.1016/j.rser.2023.112759

Boufous S, Hurd B, Ghimire R (2022) Carbon credit markets in agriculture. New Mexico State University

Bui Power Authority (2021) All posts—Page 6. https://buipower.com/all-posts/page/6/

Dehghanisanij H, Oweis T, Qureshi AS (2006) Agricultural water use and management in arid and semi-arid areas: current situation and improvement measures. Arid Zone Annal 45(3/4):355

Dos Santos Isaías DH, Cuamba BC, Leão AJ (2019) A review on renewable energy systems for irrigation in arid and semi-arid regions. J Pow Energy Eng 7(10):21

Edwards R, Rejtharova J, Padella M, O'Connell A, Pereira L, Kantamaneni R, Wachsmuth J, Lehmann S (2020) Methodology for calculation of GHG emission avoidance. Adv Sci 8(12): 18–24. https://climate.ec.europa.eu/system/files/2020-01/20200205_ghg_en.pdf

Environmental Protection Agency (2023) Greenhouse gas equivalencies calculator. United States Environmental Protection Agency https://www.epa.gov/energy/greenhouse-gas-equivalencies-calculator

Frimpong F, Asante MD, Peprah CO, Amankwaa-Yeboah P, Danquah EO, Ribeiro PF, Aidoo AK, Agyeman K, Osei Asante MO, Keteku A (2023) Water-smart farming: review strategies, technologies, and practices for sustainable agricultural water management in a changing climate in West Africa. Front Sustainable Food Syst 7:1110179. https://doi.org/10.3389/fsufs.2023.1110179

Garg G, Gupta S, Mishra P, Vidyarthi A, Singh A, Ali A (2021) CROPCARE: an intelligent real-time sustainable IoT system for crop disease detection using mobile vision. IEEE Internet Things J:2840–2851

Hossain S (2019) Comparative study on horizontal axis wind turbines and vertical axis wind turbines. Bachelor's Thesis, World University of Bangladesh, Dhaka

IPCC (2006) IPCC guidelines for national greenhouse gas inventories. Eggleston H S, Buendia L, Miwa K, Ngara T and Tanabe K (eds). IGES, Hayama, Japan

IRENA (2019) Global energy transformation: The REmap transition pathway (Background report to 2019 edition). International Renewable Energy Agency, Abu Dhabi

Jamali M, Soufizadeh S, Yeganeh B, Emam Y (2021) A comparative study of irrigation techniques for energy flow and greenhouse gas (GHG) emissions in wheat agroecosystems under contrasting environments in southern Iran. Renew Sust Energ Rev 139:110704

James B (2024, April 24) News releases. Retrieved from World Food Program: Program: Programme: https://www.wfp.org/news/global-report-food-crises-acute-hunger-remains-persistently-high-59-countries-1-5-people?form=MG0AV3

Kollmuss A, Zink H, Polycarp C (2008) Making sense of the voluntary carbon market: a comparison of carbon offset standards. World Wildlife Fund

Kumar A, Singh A, Singh S (2023) Advanced wind turbine technologies for arid regions: A review. Renew Energy 199:1139–1152. https://doi.org/10.1016/j.renene.2023.05.123

Kuriqi A, Pinheiro AN, Sordo-Ward A, Bejarano MD, Garrote L (2021) Ecological impacts of run-of-river hydropower plants—current status and prospects on the brink of energy transition. Renew Sust Energ Rev 142:110833

Lapshin OV (2017) Nonstationary combustion wave. Concise encyclopedia of self-propagating high-temperature synthesis: history, theory, technology, and products, pp 224–226. https://doi.org/10.1016/B978-0-12-804173-4.00102-2

Meena RS, Pradhan G (2023) Industrial garbage-derived biocompost enhances soil organic carbon fractions, CO_2 biosequestration, potential carbon credits, and sustainability index in a rice-wheat ecosystem. Environ Res 116525. https://doi.org/10.1016/j.envres.2023.116525

Meena RS, Pradhan G, Kumar S, Lal R (2023) Using industrial wastes for rice-wheat cropping and food-energy-carbon-water-economic nexus to the sustainable food system. Renew Sust Energ Rev 187:113756. https://doi.org/10.1016/j.rser.2023.113756

Meena RS, Pradhan G, Nalani, Singh K, Singh AK, Verma SK, Mina KK, Kumar S, Chaturvedi RK (2024) Diversified cropping modules designed for soil restoration, CO_2 sequestration, and generating carbon credits. Land Degrad Dev 35. https://doi.org/10.1002/ldr.5224

Meena RS, Pradhan G, Singh K, Kumar S, Singh AK, Shashidhar KS, Mina KK, Rao CS (2024a) Agriculture models for restoring degraded land to enhance CO_2 biosequestration and carbon credits in the Vindhyan region of India. Sci Total Environ 929:172661. https://doi.org/10.1016/j.scitotenv.2024.172661

Meena RS, Lal R, Kumar S, Pradhana G, Rao CS, Singh AK, Pathak H, Abhilash PC, Kumar A, Sharma SK, Jat ML, Singh S (2024b) Potential of Indian agriculture for capturing atmospheric CO_2 and monetizing carbon credits to the farmers: an overview and policy framework. Adv Agron 188:101–206. https://doi.org/10.1016/bs.agron.2024.06.001

Meena RS, Pradhan G, Kumar S (2025) Energy flow, eco-efficiency, and economic circulation with recycled industrial waste compost application in wheat and subsequent rice farming. Sci Total Environ 967:178779. https://doi.org/10.1016/j.scitotenv.2025.178779

Mohamed EMA, Abdallah SMA, Ahmadi A, Lucero-Prisno DE III (2021) Food security and COVID-19 in Africa: implications and recommendations. Am J Trop Med Hygiene 104(5):1613

Namara R, Horowitz L, Kolavalli S, Kranjac-Berisavljevic G, Busia N, Dawuni B, Barry M, Giordano M (2010) Typology of irrigation systems in Ghana. International Water Management Institute. https://www.researchgate.net/publication/259487757_Typology_of_Irrigation_Systems_in_Ghana

Peake S (2002) The Marrakesh Accords: the value of carbon after COP7. Refocus 3(1):20–23. https://doi.org/10.1016/S1471-0846(02)80006-3

Ponta FL, Seminara JJ, Otero AD (2007) On the aerodynamics of variable-geometry oval-trajectory Darrieus wind turbines. Renew Energy 32(1):35–56

Pradhan G, Meena RS (2023) Interaction impacts of biocompost on nutrient dynamics and relations with soil biota, carbon fractions index, societal value of CO_2 equivalent, and ecosystem services in the wheat-rice farming. Chemosphere 139695. https://doi.org/10.1016/j.chemosphere.2023.139695

Pradhan G, Meena RS (2023a) Utilizing waste compost to improve the atmospheric CO_2 capturing in the rice-wheat cropping system and energy-cum-carbon credit audit for a circular economy. Sci Total Environ 164572. https://doi.org/10.1016/j.scitotenv.2023.164572

Pradhan G, Meena RS, Kumar S, Lal R (2023) Utilizing industrial wastes as compost in wheat-rice production to improve the above and below-ground ecosystem services. Agric Ecosyst Environ 358:108704. https://doi.org/10.1016/j.agee.2023.108704

Ringius L, Grohnheit PE, Henrik L, Olivier A, Painuly J (2002) Wind power projects in the CDM: methodologies and tools for baselines, carbon financing and sustainability analysis. Manag Serv 1380(December)

Singh R, Singh A, Singh S (2022) IoT-based irrigation systems for sustainable agriculture: a review. Comput Electron Agric 196:106763. https://doi.org/10.1016/j.compag.2022.106763

Singh MK, Yadav SK, Rajput BS, Sharma P (2024) Carbon storage and economic efficiency of fruit-based systems in semi-arid regions: a symbiotic approach for sustainable agriculture and climate resilience. Carbon Res 3:33

Smith P, Adams J, Gregory P (2020) Carbon credits and agricultural sustainability. Environ Res Lett 15(4):045016

Usman HM, Mahmud M, Yahaya MS, Saminu S (2024) Wind-powered agriculture: enhancing crop production and economic prosperity in arid regions. Elektrika 16(1):10–19

Verified Carbon Standard (2022) VCS project registry: wind-powered irrigation project in India. Verra. https://registry.verra.org/

WHO (2024) Drought and food insecurity in the greater Horn of Africa. Retrieved from World Health Organization: https://www.who.int/emergencies/situations/drought-food-insecurity-greater-horn-of-africa

Zeggaf A, Carmi G, Ünlü M (2020) Promising water management strategies for arid and semi-arid environments. In: Landscape architecture – processes and practices towards sustainable development. IntechOpen, p 72574. https://doi.org/10.5772/intechopen.87103

Chapter 13
Digital Agriculture for Carbon-Negative Emissions

Anamika Kandel and Ritika Lamichhane

Abstract The agriculture sector acts as both a source and sink of greenhouse gases (GHG), as various agricultural activities emit these gases into the atmosphere but also capture them through soil carbon sequestration in the form of soil organic carbon. Amidst the climate challenges that the world is facing, digitalization in agriculture has appeared as a pathway to a low-emission food system. Digital technologies greatly help in mitigating GHG emissions, improving soil carbon sequestration, and thus playing a vital role in achieving carbon-neutral goals through the carbon credit mechanism. Digital agriculture technologies enhance resource use efficiency as they allow site-specific management of water, nutrients, and pesticides. Tools like remote sensing, geographic information systems (GIS), satellite imaging, and Internet of Things (IoT) offer real-time monitoring of crops and soil conditions, hence enhancing productivity and reducing emissions. Machine guidance (MG) and controlled traffic farming optimize the inputs and preserve the soil structure, which contributes to a reduced carbon footprint. Beyond the contribution to low-carbon agricultural production, digital tools also serve as a way forward in making the food supply chain more efficient. They support transparency and traceability across the supply chain, facilitating market access. Utilizing digital tools such as remote sensing, IoT sensors, artificial intelligence (AI), and machine learning supports accurate monitoring, quantification, and verification of emission reductions and soil carbon enhancements. When the reduced carbon emissions can be quantified, it can be monetized in the form of carbon credits, which can generate carbon revenue for farmers and entrepreneurs. This earned revenue holds strong potential to further reinforce this mechanism and make agriculture more climate-resilient and sustainable. However, a supportive policy and strategic approach from the government and policymakers are crucial for fully leveraging the economic potential of digital agriculture and achieving carbon neutrality.

A. Kandel (✉)
Department of Agricultural and Environmental Biology, The University of Tokyo, Tokyo, Japan

R. Lamichhane
Nebraska Extension, University of Nebraska-Lincoln, Lincoln, NE, USA

R. S. Meena et al. (eds.), *Carbon-Negative Agriculture*, Sustainability Sciences in Asia and Africa, https://doi.org/10.1007/978-981-95-1483-0_13

Keywords Digital agriculture · Carbon credits · Carbon sequestration · Climate neutrality · Precision farming

13.1 Introduction

Global warming is one of the biggest challenges that humankind is facing at present. Greenhouse gases such as carbon dioxide (CO_2), methane (CH_4), nitrous oxide (N_2O), and sulfur dioxide (SO_2) play a major role in retaining high temperatures in the lower atmosphere of our planet Earth, thus preventing the heat from escaping back to space and contributing to global warming (Yoro and Daramola 2020; Meena et al. 2025). Agriculture is both a direct and indirect contributor to global warming. The direct contribution of agriculture to climate change includes the emission of CH_4 and N_2O in crop and livestock production, while the indirect effect includes its role in net CO_2 release through its impact on soil, forests, and other land uses (Blandford and Hassapoyannes 2018).

Over the past few decades, digital agriculture (DA) has emerged as a transformative approach that aims to boost agricultural productivity and holds potential towards sustainable agriculture by mitigating greenhouse gas emissions. Key digital technologies in agriculture are precision farming, remote sensing, Internet of Things (IoT), sensors, drones, robotics, cloud computing, artificial intelligence (AI), and machine learning (ML). These technologies support sustainable agriculture through effective monitoring, decision-making, and management of agricultural practices, which leads to more efficient use of resources and reduced environmental impact. In addition, digital tools also support practices that enhance soil carbon sequestration by monitoring the soil health, weather, crop requirements, and patterns (Balasundram et al. 2023; Balafoutis et al. 2017; Meena et al. 2024c).

The environmental benefits offered by digital agriculture can possess a high economic value under the carbon market mechanism. Carbon credits, which are the tradable certificates, can be generated by any project or activity that can reduce, avoid, destroy, or capture GHG emissions. The digital agricultural practices that are leveraged on the farm, to achieve the capture of carbon in soil through minimalized tillage, or to support the low emission through enhanced resource use efficiency, etc., are directly and indirectly preventing the GHG emission, which can be recognized as the carbon credit. This will provide farmers with the incentive to adopt low-emission, sustainable agricultural practices and practices to sequester more carbon in soil. In this way, the carbon credit generated from reduced emissions and enhanced sequestration can balance the amount of CO_2 emitted from agricultural operations, which contributes to achieving carbon neutrality in agriculture. In this chapter, we will discuss various components of digital agriculture and explore the link between digitalization and reduced carbon emissions that contribute to the carbon credit generation.

13.2 Digital Agriculture: A Global Overview

Agriculture, like many other sectors, is undergoing rapid transformation. The method we used to manage our farming system in the past has evolved significantly, marking considerable progress over the years. In the last few decades alone, innovations in agriculture have streamlined fieldwork and increased production. In the past 30 years, the agriculture sector has seen a lot of changes with the use of automation control systems, data processing software, web-based applications, mobile tools, global positioning system (GPS), ground-based sensing platforms, satellite maps, and local sensing devices such as data loggers. The recent emergence of tools like unmanned aerial vehicles (UAVs), IoT gadgets, and robotics has further shifted the concept of digitalization in agriculture (Balafoutis et al. 2017) (Fig. 13.1).

Shepherd et al. (2020) defined digital agriculture as an approach that uses digital data to inform decisions throughout the agricultural value chain. They describe digital agriculture as a technology that extends beyond farm or livestock operations and encompasses the entire production chain, from the moment a farmer plants a crop until it reaches the consumer's table. Shen et al. (2010) described digital agriculture as an expansion of the concept of "digital earth" from the 90 s and the idea of "precision farming," which aims to enhance agricultural productivity. It serves as a comprehensive database that contains information on various components of agriculture, such as soil conditions, crop yield, weather data, and market analysis, optimizing our decision-making.

To better understand the concept of digital agriculture, we can look into the framework constructed by Tang et al. (2002) in their study. The framework starts with a basic information database that provides data on farmland, crops, climate, and socioeconomic factors. This is followed by a real-time information-collecting system

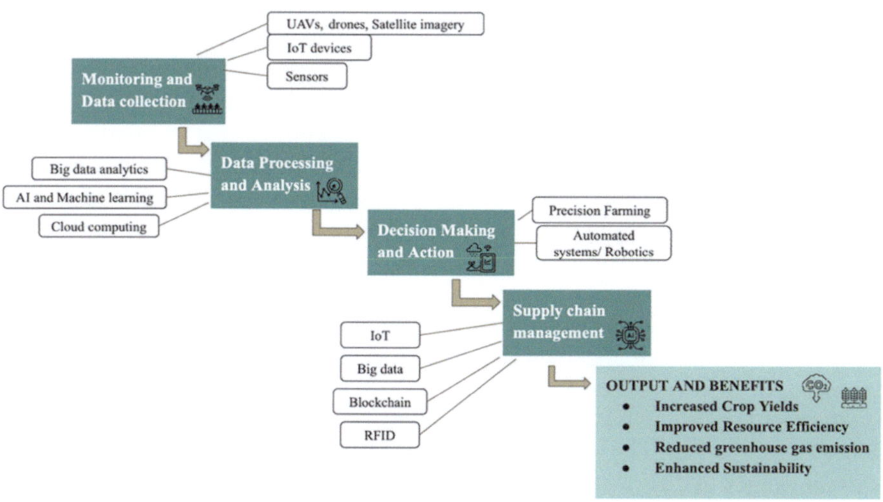

Fig. 13.1 Flowchart showing recently advanced digital tools and their role in agriculture

that monitors agricultural activities through sensors and keeps the database updated. The information is shared through a digital network transmission system. A central processing system also analyzes the data using geographic information systems (GIS). Finally, digital agriculture machines use this information to improve farming processes and make them more efficient (Tang et al. 2002).

Digital agriculture uses technologies to monitor crop, soil, and resource conditions. It contributes to sustainable food production and economic development through increased water and land use efficiency. Beyond enhanced productivity and efficient resource use, we will explore digital agriculture's role in reducing GHG emissions and generating carbon credits, contributing to carbon-neutral agriculture.

13.3 Concept of Digital Agriculture in the Context of Climate Change

Climate change is one of the biggest challenges facing humankind today. Climate change refers to the shift in average atmospheric temperatures and weather patterns, such as precipitation, over an extended period. Agriculture is one of the sectors most vulnerable to the impacts of climate change. Climate change affects agriculture in various ways. One way is that high temperatures accelerate plant development and drive them to maturity early, but grain production is reduced. High temperatures also diminish a plant's ability to absorb, retain, and utilize soil water due to increased evapotranspiration. Other effects include higher disease and pest infestations, a shift in weed flora, shortened crop duration, and decreased microbial activity in the soil (Malhi et al. 2021; Meena et al. 2024a). Changes in climatic events, such as temperature and rainfall, significantly impact crop yields. Rising temperatures are found to reduce crop yields, while increased rainfall may offset the effects of higher temperatures (Adams et al. 1998). Conversely, waterlogging or water scarcity caused by excessive rainfall or drought can also be detrimental to plant growth and yield. When soil and atmospheric temperatures rise, insects and pests become more prevalent, affecting crop health and leading to a sharp decline in yield (Rosenzweig and Liverman 1992). Simultaneously, crops' vulnerability to various pests, diseases, and weeds is also expected to change with shifts in climatic patterns. It has been predicted that insect pest infestations will result in a 10–25% increase in losses with a temperature rise of one degree (Shrestha 2019).

Agriculture is not only a sector highly susceptible to climate change impacts, but it is also a direct and indirect contributor to climate change. The direct contribution of agriculture to climate change includes the emissions of CH_4 and N_2O in crop and livestock production, while the indirect effects include its role in net CO_2 release through its impact on soil, forests, and other land uses (Blandford and Hassapoyannes 2018). Despite agriculture's contribution to global anthropogenic emissions, it plays a more pivotal role in reducing global emissions than its current and projected future contributions. As long as humans exist, the demand for food persists, and the process of food production cannot occur without emitting GHGs. However, emissions need to be minimized, and the contribution of Agriculture, Forestry, and Other Land Uses (AFOLU) to mitigation needs to be enhanced by optimizing carbon sinks and uptakes. AFOLU plays a role in mitigation not only through agricultural production but also by leveraging agricultural soil's ability to sequester carbon and taking advantage of the mitigation potential of forests and other land uses (Blandford and Hassapoyannes 2018; Meena et al. 2024b). Digitalization in agriculture serves as a crucial tool not only to aid the mitigation of climate change's impact on agriculture but also to lower the environmental impact generated by agricultural activity, thereby steering agriculture toward sustainability. In addition to improved agricultural productivity, agricultural digitalization can benefit the environment by reducing the pressure on scarce resources (Balasundram et al. 2023). This can be achieved by improving food safety through increased traceability (Rejeb et al. 2019) and, more importantly, by combating climate change (Balafoutis et al. 2017).

Digitalization is a rapidly growing trend within agriculture. Key technologies include remote sensing, IoT, sensors, drones, robotics, cloud computing, artificial intelligence (AI), decision support software (DSS), and blockchain to optimize agricultural production processes and value chains (Haji et al. 2020; Rejeb et al. 2019; Siikavirta et al. 2002; Cariappa et al. 2024). These technologies support sustainable agriculture through effective monitoring, decision-making, and communication (Mouratiadou et al. 2021). Remote sensing technologies, such as satellite imaging and UAVs, together with AI and machine learning, can be used to assess changes in land use over large geographic areas. This will help to monitor the compliance and efficacy of the policy and predict future trends to encourage positive actions in the future (Angelopoulou et al. 2019). Similarly, IoT technology, together with remote sensing, allows the real-time monitoring of soil and atmospheric conditions. When these atmospheric factors that affect carbon cycling and storage are effectively monitored, it promotes sustainable agriculture and carbon management practices (Saini and Saini 2020). Technologies like variable-rate technologies (VRT), yield monitoring, GPS tractor navigation, controlled traffic farming (CTF), and cloud computing enhance the decision support system, contribute to climate-smart practice by reducing emissions through resource efficiency, and protect the soil from nutrient imbalance and soil compaction (Lieder and Schröter-Schlaack 2021). Communication-enhancing technologies like Radio Frequency Identification (RFID), IoT, blockchain, and QR codes, and digital marketplaces (e-commerce) enhance the traceability of products and transparency on production conditions and optimize supply chain logistics (Haji et al. 2020; Anshari et al. 2019) (Fig. 13.2).

Sustainable agriculture practices like conservation agriculture and regenerative agriculture practices, which are recognized as viable and practical climate-smart solutions, contribute to sustainability by retaining and restoring the soil health that had been degraded by conventional tillage, reducing GHGs emissions by minimizing fertilizer use, promoting carbon sequestration, and emphasizing long-term organic agriculture. Growing evidence shows that the utilization of digital

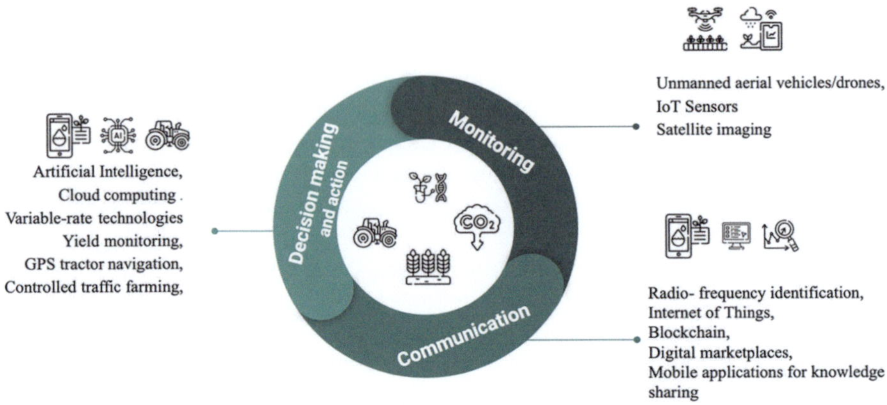

Fig. 13.2 Various digital technologies that contribute to the sustainable agriculture outcome through effective monitoring, decision making, actions, and communication

technologies would have a greater impact on such sustainable agricultural practices (Sharma et al. 2024). A study in Ghana by Quarshie et al. (2023) showed us that digital agriculture tools and services facilitated the climate-smart agricultural practices among farmers by promoting enhanced adaptation, GHGs emissions mitigation, and increased productivity or smallholding agriculture. Another study in Nigeria attributed adopting climate-smart agriculture practices to digitization efforts, which helped small-scale farmers through an enhanced value chain, enhanced information sharing, and improved soil management (Awoniyi et al. 2023).

13.4 Digital Agriculture in Developed and Developing Nations

Digitalization is transforming agriculture in both developed and developing countries, but on different scales with different impacts and challenges. In Europe, there is a relatively high level of adoption and use of digital agriculture (DA) technologies, particularly in France, Germany, and the Netherlands. The European Union has also been promoting the adoption of PA practices through its Common Agricultural Policy (Bolfe et al. 2020). In North and South America, the adoption and use of DA technologies are relatively high, and the United States, Brazil, and Canada are leading countries in this. In Asia, the adoption and use of DA technologies are growing rapidly, especially in China with the Chinese government's promotion to the use of PA technologies (Benyam et al. 2021). In contrast, the adoption and use of DA technologies in African countries are still relatively low, due to various factors, but there are initiatives aimed at promoting the use of DA technologies in Africa, such as the African Agricultural Technology Foundation's Digital Agriculture Services project (Yatribi 2020).

Overall, the adoption of digital agriculture technologies in developed and developing countries varies, and that is influenced by factors such as access to technology and infrastructure, government policies and initiatives, and investment in research and development. In developed nations, digital agricultural tools are often advanced, like autonomous farm equipment, AI-powered systems, machine learning, and data analytics for precision farming. Developing countries, on the other hand, tend to focus on simpler digital tools, such as mobile applications, online networking, basic data analysis, online market access, and crop information.

Most of the agriculture in developing economies is characterized by small-scale agriculture. Lowder et al. (2021) reported that there are more than 608 million farms in the world. Out of which more than 90% are family farms, and they occupy around 70–80% of farmland, producing roughly 80% of the world's food in value terms. Small farms, on the other hand, account for 84% of all farms worldwide but operate on only about 12% of all agricultural land, which is estimated to produce roughly 35% of the world's food. Large farms are leading the way in digital agriculture. A study by McFadden et al. (2023) in the USA explored the adoption of digital tools in larger farms (that are in the top 60% of acreage) and smaller farms (that are in the bottom 20% of acreage) and revealed that advanced tools like yield maps, soil maps, VRT, and guidance systems are commonly used in larges farms, whereas less than 25% of smaller farms use them. This is because the adoption of these advanced technologies is much easier at a large scale and is not designed for the scale at which smallholder farmers can adopt. This creates unequal adoption between small-scale and large-scale farmers, with a corresponding disparity between developed and developing countries (Trendov et al. 2019). Reviewing the trends of digital agriculture in Europe and Central Asia, Sinitsa et al. (2021) summarized that the success of a digital agricultural system depends on key factors like good broadband coverage, information security and data protection, digital platforms, and financial support.

In developing countries with smaller and fragmented farmlands, accessing the digital tools in agriculture is more challenging due to limited capital, insufficient digital infrastructure, and high cost of technology. A review article by Gumbi et al. (2023) revealed that digital agriculture is still very underdeveloped for smallholder farmers, mostly due to the challenges concerning infrastructure, affordability, and digital literacy or skills. However, there are a few early adopters who utilize the online information related to agriculture, selling, and marketing. With the access to information becoming more affordable, the gap is expected to narrow. Smallholder farmers in developing countries are finding creative ways to utilize information and communication technologies (ICTs) like smartphones, the Internet, and information-providing applications to boost their return despite limited financial and human resources (Deichmann et al. 2016). A study showed that the use of smartphone technologies providing agricultural information in sub-Saharan Africa and India has led to a 4% increase in yields and a 22% higher likelihood of adopting recommended agrochemical inputs (Fabregas et al. 2019; Meena et al. 2023). A case study was done in Chongzhou County in China on how smallholder farmers are involved in digital agriculture. The findings emphasized the importance of forming cooperatives and outsourcing production to service providers that offer integrated digital agricultural platforms. The focus for small-scale farmers should shift from the scale of farmland to effectively utilizing the limited resources that are already out there (Xie et al. 2021). Such findings demonstrate the potential of digital tools to empower farmers despite limited resources.

13.5 Role of Digital Agriculture in Achieving Low-Carbon and Sustainable Food Systems

The rise in atmospheric temperature is attributed to the GHG effect resulting from gases such as CH_4, CO_2, and N_2O. that drive ozone depletion in the atmosphere. Among these, CO_2 emissions account for the maximum proportion of GHGs (Malhi et al. 2021). However, CH_4 and N_2O have 28–36 and 265–298 times more global warming potential, respectively, than that of CO_2 in a 100-year time horizon, making them the major anthropogenic GHGs from agriculture (Vallero 2019) (Table 13.1). Hence, reducing the emission of these GHGs becomes the utmost concern.

Using digital technology also influences the farmers' low-carbon production behaviors directly and indirectly. Directly, digital tools like mobile apps and online resources provide vital information on crop management, disease prevention, and resource management enabling farmers to reduce pesticide use and adopt practices that lower carbon emissions. On the other hand, digitalization indirectly affects the farmers' low-carbon production behavior by affecting their understanding of environmental and climate risks. Studies show that digital technology notably improves farmers' awareness of environmental, health, and safety risks associated with agricultural practices, underscoring its critical role in promoting sustainable agriculture (Huang et al. 2022; Meena and Pradhan 2023).

Precision agriculture technologies have proved to be a promising and relevant agronomic practice for climate change mitigation over the years. This technology enables the variable rate application of water, fertilizers, pesticides, etc., which ensures the application of these inputs only where it is necessary and prevents the unnecessary waste of resources. Another aspect of precision technology is guidance technology, which includes automatic steering and guidance systems for tractors and self-propelled agricultural machinery. Technologies like driver assistance, machine guidance (MG), and controlled traffic farming (CTF) reduce unnecessary in-field

Table 13.1 Global warming potential (GWP) of four major agricultural greenhouse gases

GHG	CO_2	CO	N_2O	CH_4
Global warming potential (GWP)	1	1.57	300	25

Source: Forster et al. (2007)
The GWP100 characterization factors, where the amount of emissions is converted into a common unit; CO_2 equivalent

operations that lower fuel consumption, and prevent soil compaction caused by random traffic (Soto et al. 2019; Kroulik 2012). In this way precision farming technologies help lower GHG emissions in several ways. First, they enhance the soil's capacity to act as a carbon stock reservoir by promoting practices like reduced tillage and decreased nitrogen application. Second, they reduce fuel use by limiting unnecessary in-field operations. Thirdly, precision technologies indirectly cut emissions through more efficient use of inputs (seeds, fertilizers, water, and pesticides) (Šarauskis et al. 2022; Pradhan and Meena 2023b). In the next subtopic of the chapter, we will be discussing in detail how precision agriculture technologies are playing a pivotal role in reducing the GHG emissions in agri-food production and the supply chain.

13.6 Leveraging Digital Technologies for Low-Emission Agriculture

13.6.1 Precision Agriculture for Resource-Efficient, Low-Carbon Farming

Precision agriculture, which uses various advanced digital tools, technologies and management practices, is known for its ability to achieve higher agricultural productivity with reduced input requirements. Not only does the reduced input need but the judicious use of water and nutrients make precision farming an efficient farming practice. Given that improper irrigation and nutrient management practices are significant contributors to CH_4 and N_2O emissions, enhancing resource use efficiency through precision management of water, nutrients, and cultural practices is pivotal in mitigating these emissions, hence contributing to generating carbon credits.

13.6.1.1 Precision Water Management

Precision water management is the accurate assessment and quantification of soil moisture state and the crop water requirements, followed by the site-specific and time-specific application of the optimum amount of water (Smith et al. 2010; Pradhan and Meena 2023a). For the successful implementation and output of the precision irrigation system, there should be a balance on key factors, including real-time monitoring of crop and soil conditions, effective irrigation scheduling, and precise control of irrigation application equipment (Adeyemi et al. 2017). Precision irrigation is based on the hypothesis that crop water needs vary spatially within a field (Adeyemi et al. 2017). So, a key tool component of precision water management (PWM) is variable-rate irrigation (VRI), which involves applying water at different rates along the center pivot instead of uniformly across the entire system.

This approach relies on inputs from GPS, GIS technologies, and drone monitoring to ensure that the correct amount of water is distributed at varying rates throughout the field (Brahmanand and Singh 2022). In other words, variable-rate irrigation uses data and sensor feedback to deliver water more accurately through techniques such as drip, trickle, or micro-spraying (Takacs-Gyorgy and Takács 2022). Real-time monitoring of soil, crops, and weather parameters using advanced models and sensor-based automated irrigation is an important tool in accessing real-time data for ensuring timely and better decision support systems of precision irrigation water management systems (Brahmanand and Singh 2022).

Besides assisting farmers in making data-driven decisions regarding when and what quantity of water to apply to crops through the analysis of this data (Bhakta et al. 2019), precision irrigation also allows farmers to better cope with changing weather conditions and extreme events like droughts and floods by improving irrigation efficiency and reducing unnecessary water use (Balasundram et al. 2023). Precision water management can help protect aquifers, rivers, and other water resources by reducing water waste, removing unnecessary burdens on these resources, and enhancing irrigation methods, which ensures that these vital resources are available during periods of drought (Raihan 2024). Adjusting water applications to match the soil properties can help in avoiding water wastage due to surface runoff and can prevent soil erosion, which ultimately prevents not only the nutrient loss from the soil through nitrogen leaching but also the groundwater pollution (Sadler et al. 2005; Qasim et al. 2021). In this way, by adjusting irrigation according to the soil's infiltration rate and water storage capacity, water use efficiency can be enhanced while minimizing the negative environmental impacts linked to traditional irrigation methods (Sadler et al. 2005).

Furthermore, the contribution of VRI to the reduction in GHG emissions is primarily attributed to its ability to decrease water usage, which in turn reduces the energy required for pumping. Additionally, VRI facilitates precise irrigation scheduling, which prevents excessive soil moisture levels that could otherwise lead to increased N_2O emissions (Balafoutis et al. 2017). Being specific to a crop, rice, which is one of the major crops on earth, is cultivated under flooded soil conditions, and that results in higher water usage and GHG emissions, contributing approximately 30% of global agricultural CH_4 and 11% of N_2O emissions (Gupta et al. 2021). It has even been indicated that the majority of emissions in rice farming can be attributed to the irrigation system, surpassing those from farm machinery, fuel usage, agrochemicals, and animal labor (Maraseni et al. 2009). This highlights that efficient water management plays a crucial role in managing GHGs emissions from the rice cultivation system.

Another study suggested that the traditional method of cultivating rice under continuous flooding throughout most of the growing season is not ideal. Innovative irrigation practices such as Alternate Wetting and Drying (AWD), drip irrigation, and intermittent irrigation using advanced technologies like the use of automated irrigation, GIS, GPS, remote sensing, and laser land levelling can be employed to reduce water usage in rice fields (Nayak 2022). A study by Linquist et al. (2015) reported that AWD has the potential to reduce the global warming impact by cutting

CH_4 and N_2O emissions by 45–90% (Linquist et al. 2015). Also, the drip irrigation technology has the potential to reduce the CH_4 and N_2O emissions by 68.6% and 34.4% compared to flooded rice farming systems (Ramesh and Rathika 2020). The reduced emissions by implementing precise water management can be documented through standardized methodologies, allowing them to generate carbon credits.

13.6.1.2 Precision Nutrient Management

Precision nutrient management, a major aspect of precision agriculture, is a practice that involves the application of fertilizers and other nutrients in variable amounts across a field based on the requirements instead of applying the fertilizers uniformly without considering site-specific needs. Such site-specific nutrient application optimizes the nutrient use as it prevents the overuse and underuse of the fertilizers. This practice leads to the enhancement of crop yield because the nutrient input aligns with the specific needs of the crop and the soil's nutrient supply capacity (Balasundram et al. 2023; Pradhan et al. 2023). For the precision management of nutrients, advanced and innovative monitoring tools and site-specific automation technologies are used that address the spatial as well as temporal variations in soil fertility status and facilitate the needy application, overall contributing to improving productivity, efficiency, and profitability in agricultural production systems (Brar and Kaur 2016). Key tools that are used in Precision nutrient management (PNM) are remote sensing, GIS, GPS, and proximal sensors to accurately detect soil and crop nutrient statuses. Proximal crop sensors, such as chlorophyll meters, leaf color charts, and GreenSeekers, help to identify the crop's needs. For example, nitrogen deficiency can be identified through vegetation indices like the Normalized Difference Vegetation Index (NDVI). This accurate identification can be synchronized with fertilizer application, which reduces the excess nutrient application and related emissions (Brar and Kaur 2016; Gorai et al. 2021). Additionally, tools like GIS and GPS facilitate the development of spatial variability maps and zones, which makes it easier to optimize nutrient applications, hence enhancing nitrogen use efficiency (NUE). The real-time assessments of soil properties using soil sensing technologies support more precise and efficient fertilization (Gorai et al. 2021). Chemical fertilizer is a major source of agricultural GHG emission; there are several studies around it, mainly focusing on the N fertilizer over other fertilizers because nitrogen fertilization is the leading contributor to GHG emissions in agriculture, as the production of synthetic nitrogen fertilizers releases CO_2 and their application in soils leads to significant N_2O emissions.

Several studies have shown that the GHG-reducing capability of PNM technology is mostly associated with efficient N fertilization. Variable-rate fertilizer and manure application enable the precise application of nitrogen based on crop requirements, thereby reducing the overall amount of fertilizer or manure used and lowering associated GHGs (CO_2 from fuel, N_2O from N-fertilizer, and CH_4 from manure) emissions. Furthermore, GHG emissions can be further minimized by combining nitrogen fertilization with rain forecasting or optimized irrigation scheduling

(Balafoutis et al. 2017). Integrating both soil and crop data for VR nutrient management significantly enhances nitrogen use efficiency (NUE), maintains soil health, minimizes nutrient wastage, and reduces nitrogen pollution in the environment while sustaining crop productivity (Longchamps and Khosla 2015; Gorai et al. 2021). The precision practice also leads to a decrease in the off-site movement of agricultural chemicals through surface runoffs, subsurface drainage, and leaching (Baker et al. 2005).

13.6.1.3 Precision Land Use Optimization

With the increasing global population, the burden on the fertile land is increasing, and the need for sustainable and efficient use of land and natural resources is becoming a pressing issue. Precision land use management leverages advanced tools such as GIS to evaluate land quality and optimize use, and offers a potential solution to the issue of land scarcity and inefficiency. Digital agriculture can help in achieving a balance between the demand to maximize agricultural productivity in a limited available land and the need to preserve the ecosystem for future generations. Through detailed analysis of factors like soil type, texture, fertility, underground water levels, hydraulic conductivity, climate, and topography, GIS technology helps land managers gain a comprehensive understanding of how these variables interact (Ghosh and Kumpatla 2022). This information allows land managers to better understand the ability of particular land and make informed decisions about the most suitable land uses, ensuring that resources are used effectively while maintaining environmental sustainability (Balasundram et al. 2023). For precision land management, tools like GIS, GPS gather and analyze the data and provide the data-based insights, and variable-rate technology (VRT) uses those insights about soil conditions, yield, and pest infestations to determine the precise amounts of fertilizers, pesticides, and herbicides needed in specific areas. Such site-specific intervention ensures the application of resources at the right time and right place, which saves the input cost, improves input efficiency, and minimizes the potential for runoff and soil degradation (Ghosh and Kumpatla 2022). Other digital tools like satellite imagery and machine learning algorithms can assist in monitoring soil health, crop growth patterns, and weed growth patterns. A study by Taghizadeh-Mehrjardi et al. (2020) found that land suitability maps generated using machine learning were more accurate compared to the maps obtained by the traditional approach.

Also, the areas in the field that are prone to issues like soil erosion or nutrient depletion can be easily identified with these technologies, which assist in implementing climate-resilient strategies (Lipper et al. 2017). One major way in which digital land use management contributes to GHG mitigation is through the use of precision irrigation systems. Site-specific irrigation systems lead to reduced water waste and help cope with water shortages and droughts (Balasundram et al. 2023). Precision land use management serves as a crucial tool for both boosting agricultural output and reducing the environmental impact of farming. The reduced impact on the environment has the potential to serve as a carbon credit in a carbon market.

13.6.1.4 Variable Rate Pesticide Application

Pesticide application by conventional spray techniques uses a sprayer that can uniformly discharge the chemical, hence leading to consistent application throughout the field, where each plant receives the same dosage of the pesticide. However, each plant or tree has different canopy characteristics, height, width, and shape that demand varying amounts of pesticide for an effective result. When pesticides are applied uniformly, irrespective of the varying needs of each plant, they get wasted in inter-tree spaces, above smaller plants, and around tree trunks beneath canopies. On the other hand, a significant portion of insecticide gets lost due to off-target factors like airborne drift, runoff, and evaporation (Nackley et al. 2021). Variable rate pesticide application becomes a solution that can address this challenge. This technology applies precise amounts of pesticides effectively targeting diseases, pests, and weeds. It not only ensures the effective control of pests and weeds but also maintains the yield. When comparing this variable-rate pesticide application technology with the constant-rate application, research showed that the variable-rate sprayer mode used 67–74% less volume of pesticides in the orchard and 61–80% less volume in the vineyard (Nackley et al. 2021). Another study recorded that real-time sensor-based precision spraying reduced pesticides (including herbicides, insecticides, and fungicides) by 12–96% in soybean and maize crops, with greater reductions occurring in areas with less vegetation cover (Zanin et al. 2022).

Pesticides can be applied either by map-based VR technology that uses GPS and a prescription map to adjust input rates as the applicator moves through the field or a sensor-based VR technology that measures crop characteristics in real-time and adjusts inputs accordingly, without needing a map (Grisso et al. 2011). Sensor-based technology is mainly used in an innovative real-time robotic spraying system that is designed for precise and targeted pesticide applications. This system consists of a robotic unmanned vehicle, an ultrasonic wave-based sensor, a microcontroller board, an electronic pump, a nozzle for dispensing pesticides, and a solenoid valve unit (Pramod et al. 2024).

Besides, chemical waste, land and water pollution due to pesticides are other persistent problems in modern agriculture. Pesticides contaminate the atmosphere through various pathways, including GHG emissions. So, the reduced use of pesticides in the field is significantly associated with lowered GHG emissions and better environmental safety. The GHG reduction potential of variable-rate pesticide application (VRPA) also comes from its indirect role in reducing industrial production of pesticides resulting from the lowered pesticide use in agriculture (Soto et al. 2019).

13.6.1.5 Variable Rate Seeding and Planting

Variable rate planting (VRP) and seeding (VRS) are precise agricultural technologies in which planting and seeding rates are adjusted according to the variability in local soil potential, meteorological conditions, and other factors, and non-uniform application of seed and plants is applied across the field. In highly heterogeneous fields, where soil characteristics like organic matter and water retention are not consistent across the field, VRP/VRS optimizes seeding by matching plant density to each area's specific conditions. Also, the VRS innovations enable us to combine two different plant varieties into the same field. For this, a multilayer seeding method can be used where seeds or two different varieties are distributed differently in the same field at the same time (Fulton 2018). This allows for a high-yield variety to be seeded in productive zones, while a stress-tolerant one is planted in less fertile sections, maximizing growth potential across varied conditions.

The appropriate seeding rate for each field zone is decided based on site-specific data layers of soil texture, electrical conductivity, pH, and yield maps. Various proximal and remote sensor systems, contact and contactless equipment, mapping, and VRS modeling technologies are used to determine soil and crop variability (Šarauskis et al. 2022). There are soil organic matter (SOM) sensors that detect different levels of organic matter and adjust the plant population rate accordingly. The site-specific sowing method (with a variable sowing rate for each field area) allows the optimization of crop density to obtain the best agronomic and economic results (Šarauskis et al. 2022). When the seeding density is optimized using VRS, it results in efficient use of resources due to fewer seeds and lower fuel usage. But VRS doesn't only reduce GHG emission directly through lower seed rates and fuels, but it also offers GHG reduction potential through yield gains and cost savings as a lower amount of fuel is required to produce the same yield (Soto et al. 2019). Variable-rate planting/seeding (VRP/VRS) shows lower GHG mitigation potential compared to other precision technologies (Table 13.2), but it is not insignificant (Soto et al. 2019).

Table 13.2 Greenhouse reduction potential of various precision agriculture technologies

Technologies	Scale
Variable-rate nitrogen application (VRNT)	5
Variable-rate irrigation (VRI)	3
Controlled traffic farming (CTF)	2
Machine guidance (MG)	2
Variable-rate pesticide application (VRPA)	2
Variable-rate planting/seeding (VRP/VRS)	1
Precision physical weeding (PPW)	1

Source: Soto et al. (2019)

Scale of importance of GHG reduction potential (Likert-type scale identified by the authors): *5* very high potential, *4* high potential, *3* moderate potential, *2* slight potential, *1* low potential

13.6.1.6 Precision Weed Management

Precision weed management, which is also known as site-specific weed management, is a weed management practice that is designed according to weed location, density, and population, and depending on those factors, weed management practice is applied unevenly. Similar to variable rate pesticide application, weedicides can also be applied either by map-based or sensor-based VR technology (Grisso et al. 2011). These practices are accomplished using VRTs that spray herbicide in real-time as decided by the central brain of the machine, the vision-based image processing algorithm (Rai et al. 2023). There have been several research projects on novel techniques, such as deep learning, which are revolutionizing weed detection for site-specific weed management and taking site-specific weed management a step further. The proximal-based weed detection or remote sensing-based image acquisition is followed using DL techniques applied to process the images for weed detection and mapping (Rai et al. 2023).

Another precision weed management method is precision physical weeding, which avoids chemical usage, instead of controlling the weed using techniques like mechanical weeding and flame-based treatments, ensuring minimal crop damage (Balafoutis et al. 2017). Precision technologies enable changes to the configuration of mechanical weeders (e.g., in the position of or the resistance exerted by the times of a harrow) during weeding, to match weed presence and/or density in the field (Soto et al. 2019).

This VR technology contributes to GHG mitigation just like other pesticides through reduced herbicides, as the site-specific management lowers the levels of herbicide being applied to the fields (Grisso et al. 2011). Precision physical weeding, on the other hand, could reduce GHG emissions by avoiding the use of chemical pesticides (Balafoutis et al. 2017). Precision thermal weed control reduces fuel usage by only targeting weed patches, in contrast to conventional burning that covers the entire field. However, if thermal weed control replaces mechanical tillage, the GHG emissions from burning may offset the benefit obtained from pesticide reductions (Soto et al. 2019). Mechanical weeding results in lower fuel consumption, as less aggressive tilling is required due to reduced draught forces when using precision harrows (Peteinatos et al. 2015). Overall, the impact of precision physical weeding on GHG emissions is expected to be minimal, similar to the limited effects seen with variable-rate pesticide application.

13.6.1.7 Machine Guidance and Automation

Machine guidance, a significant advancement in modern agriculture, utilizes Global Navigation Satellite Systems (GNSS) to improve how farming equipment is operated. Two main types of systems exist: driver assistance and fully automated machine guidance. Driver assistance systems help the machine operator stay on the correct path using simple tools like light-emitting diodes (LEDs), while automated systems handle steering completely without manual intervention (Balafoutis et al. 2017).

Auto-guidance systems, or auto-steering, fully automate the steering of agricultural equipment using GPS data (Shockley et al. 2011). Equipped with mechanical or electro-hydraulic controls, the tractor can follow predetermined paths across the field with incredible accuracy (Bora et al. 2012). Advanced auto-guidance systems are structured into four main units. The detection unit measures the vehicle's position and orientation in the field using a GPS receiver and antenna, while the control unit consists of a controller that plans the vehicle's path and ensures accurate tracking. The execution unit handles steering by adjusting the wheels based on the control unit's instructions. Finally, the monitoring unit, or field computer, serves as the operator interface, allowing for monitoring and adjustments during fieldwork. Some systems also feature data loggers or audio alerts to help operators stay informed during operation (Hu et al. 2015; Bora et al. 2012).

The use of machine guidance systems in agriculture offers numerous benefits that contribute to both efficiency and sustainability. By utilizing GPS-guided technology, farmers can significantly reduce the overlap and omission in field operations, which leads to a decrease in the area that is unnecessarily run over by machinery. This reduction in random traffic across the field minimizes soil compaction and the environmental impact often associated with conventional farming practices. Additionally, machine guidance helps identify areas of frequent machine passes, enabling farmers to optimize their routes and further minimize repeated passes (Kroulik 2012).

Another significant advantage of these systems is that they help in reducing the input waste by ensuring precise placement of seeds, fertilizers, pesticides, etc., plus this also reduces the fuel consumption and machinery costs (Shockley et al. 2011; Bora et al. 2012). In this way, machine guidance technology offers the potential to substantially reduce GHG emissions. For instance, one study on corn and soybean farming in Kentucky demonstrated savings of 2.2% in fertilizer, and 10.4% in tractor fuel by machine guidance during planting and fertilization, which was translated to mitigation of GHG emission through savings (Shockley et al. 2011). Similarly, farms that used GPS guidance systems reported a 6.04% reduction in machine time and a 6.32% decrease in fuel consumption, while farms using auto-steering systems saw further reductions of 5.75% in machine time and 5.33% in fuel use (Bora et al. 2012).

Farmers adopting digital tools like auto-guidance systems to ensure fewer passes over the field, less fuel consumption, and optimized input application can be eligible to generate carbon credits for reducing CO_2 emissions from diesel use and optimizing nitrogen-based fertilizer application, which lowers N_2O emissions.

13.6.1.8 Controlled Traffic Farming

Agricultural mechanization has revolutionized agricultural commercialization over the past few decades. The use of heavy machinery and vehicles on the farm has been an integral part of the production system. However, there have been several studies showing how traffic-induced soil compaction, directly and indirectly, impacts the soil's physical properties like decreasing air and water transport to the subsoil, increasing soil bulk density, and reducing the hydraulic conductivity and soil gas diffusivity (Obour and Ugarte 2021). Compaction restricts crop root functions and growth, consequently reducing crop yield (Obour and Ugarte 2021; Gasso et al. 2013). Moreover, compaction contributes to increased emissions of GHGs with higher losses of nutrients and agrochemicals via runoff (Soane and Van Ouwerkerk 1995). Soil compaction not only raises the CO_2 emissions from the soil, but it also boosts CH_4 production due to the creation of anaerobic conditions, with CH_4 having 28 times the GHGs potential of CO_2 (Shubha et al. 2020; Sitaula et al. 2000). The reduced soil gas diffusivity caused by compaction prolongs the retention of N_2O in the soil, which accelerates the denitrification process and thereby increases N_2O production (Shubha et al. 2020).

The CTF is a system that confines machinery movement to the smallest possible area in the field by designating permanent traffic lanes, aiming to reduce soil compaction caused by heavy vehicle loads. Compared to random traffic farming (RTF), CTF significantly lowered environmental impacts, including reductions in 21–45% N_2O emissions, 372–2100% CH_4 emissions, 27–42% water runoff, and 23% direct emissions from in-field operations (Gasso et al. 2013; Shubha et al. 2020). In addition to reducing GHG emissions, research has demonstrated a substantial improvement in crop yield across various production systems globally when compared to random traffic farming (Shubha et al. 2020). Gasso et al. (2014) noticed that the CTF system reduced environmental burden, which was attributed to increased grain yields, reduced soil compaction, minimized fertilizer and pesticide overlaps, causing a low phosphorus runoff and in-field emissions of N_2O and ammonia (NH_3). In the study by Tullberg (2014), it was estimated that CTF reduces 40% of fuel requirements with uncontrolled traffic zeros tillage and up to 70% with controlled traffic zero tillage. Additionally, the fertilizer costs were reduced by 10–15% as the application on permanent wheel tracks was minimized. Moreover, the nitrogen use efficiency was increased by 40–80% due to lower soil compaction and enhanced soil biological activity. In addition, the problem of denitrification could also be addressed with the strategic application of nitrogen. Application of nitrogen fertilizer to compacted and moist seed zones often leads to high denitrification and subsequent N_2O emissions. However, CTF allows split fertilizer placement by reducing the seed zone compaction and waterlogging, which results in low denitrification and N_2O emissions. This contributes not only to improved crop productivity but also to environmental sustainability by reducing GHG emissions (Tullberg 2014). The avoided GHG emissions can be traded or sold in the form of carbon credits, which offers an additional income stream to farmers while contributing to national and global carbon neutrality targets.

13.6.2 Digital Supply Chain Optimization for Emission Reduction

The effectiveness of digitalization in farming is widely studied with main focus on production system to solve the issue of low productivity and economic return but the key challenge in supply chain persists leading to substantial food loss due to improper post-harvest handling, lack of careful monitoring, and lack of traceability (Ranganathan et al. 2022). Weak logistical management and various supply chain issues like road blockages exacerbate food waste, and this waste is a significant contributor to food insecurity (Chandrasekaran and Ranganathan 2017). The direct food waste, along with the indirect waste of energy and resources that were used to produce the food, translates into economic losses and GHG emissions.

The agricultural supply chain is complex as it encompasses a wide range of stakeholders, including farmers, processors, distributors, and consumers. As it is energy-intensive, the agri-food supply chain must be closely monitored for its contribution to GHG emissions (El Hathat et al. 2024). Digitalization in supply chain management has been a critical step in food system optimization. Technologies like real-time tracking, online procurement, and the delivery of essential market-related data and information to farmers allow stakeholders to make better decisions and improve risk management abilities. Farmers in many parts of the world can now access real-time climate data, pricing information, early warnings, etc., which are helping them to mitigate risks and reduce unnecessary wastage (Chandrasekaran and Ranganathan 2017).

Furthermore, advanced tools like IoT and blockchain have emerged in recent years as powerful tech to enhance logistical efficiency, transparency, and sustainability in the food supply chain. IoT devices can track and monitor products throughout the supply chain, while blockchain records every data point and transaction, promoting trust and reducing fraud. Digital marketplaces further improve accessibility and easily connect producers and consumers, removing the middleman's activities. Research has shown that these digital innovations in supply chain optimization collectively contribute to a more sustainable, low-emission food supply chain by lowering the overall carbon footprint of agricultural activities. These environmentally friendly supply chains ultimately contribute to global climate-neutral goals when environmental benefits obtained from the digitalization can be translated into a tradable carbon certificate.

13.6.2.1 Internet of Things for Supply Chain Monitoring and Logistical Efficiency

IoT is a technology that connects any physical objects or things with the help of embedded sensors and software. These are termed as "smart" objects, and these smart objects connect with other systems over the Internet and exchange data that enables better monitoring at every stage of the agricultural supply chain from farm to fork. Tools like sensors, RFID tags, and real-time data analytics comprise key IoT tools that enable farmers and businesses to track the products, which help cut down on losses. Sensors can detect the real-time information on storage temperature, location, etc., which allows a business to manage its inventory efficiently. RFID tags help in tracking items, which improves the transparency. Machine learning and real-time data analytics can help in avoiding possible problems and making appropriate decisions. This combination of IoT technologies leads to greater visibility, responsiveness, which ultimately, leads to a more competitive and sustainable supply chain (Haji et al. 2020; Udeh et al. 2024).

IoT can play a significant role in logistics management in agri-food systems. When perishable items such as fruits vegetables and dairy products are monitored with IoT sensors, tracking the temperature and humidity during storage and transportation, it makes sure that food is stored under optimal conditions which ultimately minimizes spoilage and waste (Haji et al. 2020). In this way, IoT ensures sustainability by reducing the loss and optimizing resource use as the technology helps manage consumption of resources such as energy, water, and raw materials in real-time. When inefficiencies are identified, companies can make targeted improvements and reduce the waste in the supply chain.

IoT contributes to lowering the environmental footprint of agriculture. But beyond just contributing to low GHG emissions, IoT also directly contributes to monitoring carbon credits. In modern agricultural practices, IoT-equipped sensors and devices can be used in documenting a reduction in water usage and monitoring the carbon emissions involved in livestock and crop management systems. Connectivity devices like wireless system-on-modules (SOMs) and radio modules, IoT gateways, and industrial cellular routers enable this data to be shared and documented properly for credit acquisition.

Beyond monitoring, IoT devices also help automate tasks in the field that include checking the soil moisture and tank levels to control the automated irrigation, or guiding the self-driving tractors. This makes the data collection easier and reduces the need to rely on fuel vehicles just to collect data. Some farming practices also utilize IoT for self-driving tractors, harvesting robots, and monitoring weather metrics and forecasts, etc., that optimize planting, fertilizing, and irrigation schedules. Altogether, IoT does not just make farming smarter, but it also contributes to making agriculture carbon neutral by contributing to the carbon offsetting procedure.

13.6.2.2 Blockchain Technology for Transparent and Efficient Supply Chain

Blockchain is an advanced database mechanism that securely tracks and records the data of all kinds of trade and transactions and allows transparent information sharing within a business network that cannot be altered. In the agricultural supply chain, blockchains have contributed by offering a transparent and traceable management of data. The immutable nature of blockchain ensures that once the data is recorded, it cannot be altered, making it transparent and trustworthy, which minimizes the risk of fraud (Rejeb et al. 2019).

Blockchain in a food supply chain allows real-time tracking of food items during their movement from farm to table. This traceability helps monitor the origin, handling of food, ensuring safety and reducing waste. Blockchain, when integrated with IoT devices such as sensors and trackers, captures the real-time data on product conditions and more. This data can then be circulated in a blockchain where it is securely kept and monitored. When the supply chain partners have access to original and real-time data visibility, it ensures the efficient management of the supply chain and an instant solution if any problem arises (Rejeb et al. 2019; Udeh et al. 2024).

Another key contribution of blockchain to a sustainable supply chain is its ability to track the carbon footprint of products during different stages of food production and transportation (Chandan et al. 2023). For example, private blockchain frameworks like Hyperledger Fabric can be used to create a transparent and verifiable record of GHG emissions associated with the movement of goods. This allows various supply chain stakeholders to record data on emissions such as CO_2, CH_4, and N_2O as products move in the supply chain. By monitoring these emissions, enterprises can identify opportunities to reduce their carbon footprint and contribute to a more sustainable supply chain (El Hathat et al. 2024). A specific example of this is a cluster-based carbon emission record-keeping system proposed by Shakhbulatov et al. (2019) that uses a blockchain to track the environmental impact of transporting food. This system records important information such as mileage, carbon footprint, and cluster history, helping businesses make more informed decisions to minimize their environmental impact (Shakhbulatov et al. 2019). But not only limited to that, such effective monitoring and recording of carbon footprint data is a key step towards the issuance of carbon credits for the carbon trading to achieve carbon-neutral goals.

13.6.2.3 Digital Marketplaces and Platforms for Sustainable Trade

Buying and selling of goods and services through e-commerce are reshaping the way consumers access the market. Various online platforms are winning the hearts of customers for the convenience they provide. e-Commerce covers a wide range of business activities, including retail and wholesale, and has now expanded to all kinds of industries, including agriculture.

The rise of digital marketplaces in agriculture is now essential for businesses looking to stay competitive. e-Commerce provides an avenue for direct-to-consumer (D2C) sales, which reduces the layers of intermediaries and directly helps connect investors, landowners, farmers, and consumers through a single mobile marketplace. By using smartphones, consumers can interact directly with producers, such as farmers and primary processors, which foster transparency and reduce middlemen in the supply chain (Anshari et al. 2019). These online platforms also improve access to essential agricultural inputs such as seeds, fertilizers, and equipment. In addition, digital financial services such as mobile banking and online credit boost financial inclusion and liquidity for farmers. Online training and webinars also help farmers upgrade their skills in modern practices and market trends. Additionally, data analytics provide insights into market trends and consumer preferences, allowing for more informed decision-making and enhanced productivity (Vavekanand and Kumar 2024).

e-Commerce can play a role in reducing GHG emissions through more efficient delivery models and logistics. Also e-Commerce supports better communication and coordination among companies and between supply chain stakeholders, which facilitates the efficient handling of agricultural products. One way in which e-commerce contributes to sustainability is that when there is good coordination in a market chain, transportation becomes efficient; for instance, trucks and other vehicles carry full loads. This means fewer trips are needed, which help reduce transportation-related emissions (Siikavirta et al. 2002). Another significant benefit of e-commerce in agriculture is its ability to avoid overproduction and waste. When demand and supply are aligned, energy used in storage gets minimized, which lowers the missions from waste disposal. A case study by Siikavirta et al. (2002) on electronic grocery shopping (e-grocery) showed that home delivery services can cut GHG emissions by 18–87% compared to traditional in-store shopping.

Digital marketplaces also play a role in achieving climate-neutral agriculture by providing smallholder farmers with a platform to trade their carbon credits that are generated from reduced GHG emissions. When farmers adopt climate-smart agricultural practices, they not only contribute to climate mitigation goals but they can also create income generation, which can further reinforce their commitment to sustainable farming practices (Tumushabe et al. 2023). In this way, e-commerce and digital platforms are contributing to a more sustainable and environmentally responsible food system.

13.6.2.4 Other Emerging Digital Technologies in the Supply Chain Management

There are various other digital technologies across the agricultural supply chain which are making the pursuit of carbon-neutral agriculture easier. Besides improving the efficiency, reducing the emissions, these tools also support carbon credit farming by enabling accurate tracking of emissions and reduction and also the reporting and verification of sustainable practices.

RFID is one of the commonly adopted technologies used in the supply chain which ensures traceability and transparency of food products (Haji et al. 2020) reducing the chance of delay and unnecessary waste which is one of key factors that contribute to carbon emissions. When emissions and waste can be tracked accurately, it helps carbon accounting be more precise and improves the carbon documentation to earn carbon credits.

Alongside RFID, innovations like UAVs and autonomous vehicles contribute to speeding up the delivery process, reducing labor and transportation costs, which, ultimately, lowers the carbon footprint by making logistics more efficient. (Haji et al.

2020). When these data are integrated with carbon tracking platforms, these emission reductions can be quantified and submitted for carbon credit certification.

Three-dimensional printing (3DP) is another emerging innovation that makes sure of demand-based food production and also facilitates localized production. This technology prevents the overproduction and long supply chains which lowers the wastage and emission. These reduced emissions can be documented using IoT sensors or digital ledgers, and it can help food producers building evidence for carbon credit schemes. Big data and Cloud Computing can monitor and analyze where and how energy and resources are consumed across the supply chain. There are some cloud-based platforms which can monitor real-time energy so that they can optimize the management. Big data, on the other hand, helps forecast the demand and prevents overproduction. These technologies altogether lower the environmental footprint (Wang et al. 2023) and these environmental improvements have potential to be turned into carbon credit if can be verified through digital monitoring.

Similarly, artificial intelligence (AI) supports carbon-neutral goals by automating decisions that improve energy use and reduce material waste (Wang et al. 2023). This intelligent AI-driven system manages resources and results in a more sustainable and efficient food production system. These efficiencies can contribute to measurable carbon savings and credit eligibility.

Overall, the integration of digital tools like IoT, RFID, UAV, AI, machine learning, and big data analytics makes the agricultural supply chain smarter and greener and also lays a foundation for tradable carbon credits. As there is a shift in the farming system, farmers and other stakeholders across the food supply chain are better equipped to contribute to carbon marketing and zero-emission agriculture (Zhou 2024).

13.6.3 Digitalization for Enhanced Carbon Sequestration

Carbon sequestration is the process of capturing and storing atmospheric CO_2. Agriculture is one of the key sectors that has the potential to store carbon, especially in soil, which is known as soil carbon sequestration. Soil carbon sequestration has become a crucial tool to reduce GHGs and mitigate climate change, as studied by several researchers. To maximize the benefit of soil carbon sequestration, assessment of soil organic carbon (SOC) is very important. This quantification of soil organic carbon (SOC) will contribute to agricultural management activities, monitoring land degradation and erosion, predicting future sequestration capacities, and making carbon sequestration and GHG emission trading plans (Pouladi et al. 2023). Emerging digital technologies like remote sensing, IoT, AI, and machine learning a have been appearing as transformative tools to facilitate the monitoring, modeling, tracking, and management of soil carbon. And also, these technologies can assist with the prediction of future sequestration capacities, hence providing the scientific basis for the carbon credit market and emission trading schemes.

Sustainable agricultural concepts like regenerative agriculture, when combined with digital technologies, are found to not only have improved environmental outcomes but also unlock the economic opportunities for farmers through participation in carbon market schemes (Sharma et al. 2024).

13.6.3.1 Remote Sensing and IoT for Soil Carbon Monitoring

Remote sensing technology allows for the frequent and large-scale observation of variables related to vegetation, topography, climate, and geological indices that directly or indirectly influence soil carbon sequestration. High-resolution satellite imagery and aerial drones equipped with multispectral and hyperspectral sensors gather detailed data over vast areas that help to evaluate the spatial variability of SOC stocks. Evaluating the changes in the observed parameters over time, carbon sequestration trends of agriculture and forest landscapes can be tracked and predicted for future patterns (Angelopoulou et al. 2019).

Remote sensing is being widely used in the carbon market commercially. A study on smallholder carbon projects in Kenya showed that remote sensing technology offers low-cost carbon monitoring, while it lacks transparency and community engagement (Okoli and Birkenberg 2024).

The impact of remote sensing can be enhanced through its cooperation with IoT technology, as it provides real-time, ground-level data on soil and atmospheric conditions. IoT devices such as soil moisture sensors, temperature probes, and climate stations can be strategically used across agricultural fields to collect data that affects carbon cycling. This data helps us understand the environmental factors that influence carbon storage, such as soil moisture, temperature, and nutrient availability (Saini and Saini 2020). Mateev and Marinova (2021) developed a distributed IoT system for CO_2 monitoring, which utilizes a network of ground sensors to accurately pinpoint gas emission sources and enable real-time monitoring of emissions data (Mateev and Marinova 2021). Polymeni et al. (2024) identified how 6g-enabled IoT offers a transformative solution for reducing GHG emissions in agriculture by enabling better emission source identification and long-term environmental monitoring.

13.6.3.2 AI and Machine Learning for Predictive Sequestration Models

A key marker in the carbon credit generation from agriculture is the soil organic carbon (SOC). Quantification and management of SOC using the emerging AI technology in integration with satellite imaging and hyperspectral sensors allow real-time monitoring and data analysis. This data-driven decision-making is essential for the accurate assessment of carbon sequestration.

Big and complex data collected through remote sensing and IoT systems are handled and analyzed using machine learning models, which can detect and decode the complicated data that could not be done with traditional statistical approaches.

Machine learning models use past data and predict the future carbon sequestration potential under varying environmental conditions and farming systems based on the previous patterns. AI techniques like deep learning can filter and model the data to identify patterns. This makes informed decisions backed up with facts and data (Natcvetova 2021).

A study by Abdoli et al. (2023) explored the use of AI and machine learning using the remotely sensed data obtained from various satellites (Landsat-8, Sentinel-2, Sentinel-3, and MODIS). Machine learning tools such as artificial neural networks (ANN), support vector regression (SVR), and genetic expression programming (GEP) analyzed the data to predict SOC. Lamichhane et al. (2019) also reported that there has been a shift from linear models to ML models for predictive mapping of SOC recently. Such predictive models developed by AI and ML have practical use for decision-making in agriculture and land management. Besides farmers, policymakers can also use these models to evaluate the potential carbon sequestration benefits of different agricultural and land management practices, which will allow them to make informed decisions. When a model can predict the long-term impact of using certain cropping systems on the SOC level, it subsequently results in actionable recommendations for whether any adjustments are needed to enhance soil carbon storage while maintaining crop productivity. Grunwald (2022) also states that the AI-driven models in predicting SOC dynamics and soil respiration are much more reliable and superior to traditional techniques. These AI-enhanced tools are essential not only for improving agricultural sustainability, but they can also be utilized in ensuring transparency and credibility in carbon credit certification processes. The utilization of AI and ML in the accurate estimation and verification of carbon sequestration can contribute meaningfully to the development of carbon-neutral farming systems.

13.7 Roadmap to Carbon Generation Through Digital Agriculture

Carbon credits can be issued by any project that can reduce, avoid, destroy, or capture emissions. Carbon can be captured and stored in various ways, including carbon sequestration in agricultural soil. Plant photosynthesis facilitates carbon sequestration in the ground in the form of organic carbon (SOC) or carbonates. The agricultural practices that are leveraged to achieve the capture of carbon in the soil include crop rotation, no-tillage, or conservation tillage practices that enhance water use efficiency, application of biochar, mulching, promoting native vegetation, cover cropping to improve soil structure, adding manure and organic matters, retaining crop stubble after harvest, planting perennial pastures with deep root system, etc. that enhance soil sequestration (Kyriakarakos et al. 2024).

As we have discussed earlier in the chapter, the use of digital tools at various stages of production, supply chain and also in soil carbon sequestration results in low emissions from agriculture and also digital agriculture facilitates monitoring, prediction and reporting of soil organic carbon stocks in soil and enhances the carbon capture. Research has already shown that the incorporation of digitalization in regenerative agriculture practices results in better sequestration outcomes (Sharma et al. 2024). Also, the use of tools like remote sensing and sensors lowers the cost of measuring soil carbon sequestration in small-scale farms out of other monitoring tools (Okoli and Birkenberg 2024).

These tools serve as scalable, data-driven solutions that enhance transparency and traceability and support the monitoring and verification, which are key requirements for participating in carbon offset programs. By ensuring credible and standardized reporting of the environmental footprint of agricultural and marketing activities, digital tools support the development of measurement, reporting, and verification (MRV) systems, which are key components for issuing tradable carbon credits.

13.8 Economic Benefits of Digitalization in Agriculture

Environmental benefits of digitalization in agriculture are very well documented, and its adoption is also widely expanding, depending upon the economic viability of the farm. When making decisions about whether or not to embrace digital transformation, farmers and agri-entrepreneurs have to critically think about financial sustainability.

One of the major ways in which digitalization ensures economic benefit is through the efficient use of resources such as water, fertilizers, and pesticides. Precision technologies, which use GPS, drones, IoT sensors, etc., facilitate the site-specific management of farms. Precision agriculture technologies allowed by GPS, drones, and IoT sensors facilitate site-specific management of farm activities; Inputs are applied precisely where needed, minimizing waste and maximizing yield. Several studies have shown that variable rate technologies (VRTs) reduce input use without compromising productivity. According to the use of GreenSeeker technology in Maryland, USA, led to an 8% increase in wheat yields when use of Nitrogen fertilizer was lowered by 10%, and a 5% increase in corn yield when use of nitrogen fertilizer was lowered by 21% (Biggar et al. 2013). Likewise, variable rate irrigation (VRI) saved around €30/ha in water and energy costs in lettuce production in an area that was usually over-irrigated (Daccache et al. 2015). Also in weed management, although site-specific technologies are costly at first, they ultimately reduce herbicide expenses and improve cost efficiency, as shown by Gerhards and Sökefeld (2003).

Input saving provides a huge economic benefit to the farmers. But, beyond input savings, digital agriculture supports economic resilience by mitigating production-related risks. AI and machine learning tools can forecast weather, predict soil fertility and moisture conditions, and optimize cropping decisions, which reduce the likelihood of crop failure caused by unexpected climate.

More importantly, digital tools are contributing to the economic strengthening of farmers and entrepreneurs through the carbon economy. As discussed earlier in the chapter, digital technologies facilitate soil carbon sequestration by supporting conservation agriculture, soil monitoring, and precision nutrient and water management. Also, digital tools are leveraged to monitor soil organic carbon stocks. Such sequestrated carbon, when verified using digital MRV systems, can generate certifiable carbon credits. A study conducted in Bihar and Punjab, India, revealed that conservation agriculture could earn farmers between US$18 and US$30 per hectare through carbon credit mechanisms (Cariappa et al. 2024). When conservation agriculture is incorporated with digitalization, it holds even greater potential to generate economic sustainability. Such income diversification is particularly valuable for smallholder farmers.

However, as easy as it seems to discuss it, there are several challenges too. The cost of implementing digital tools can be a major barrier, especially for small-scale farmers. Site-specific input management may improve yields and have environmental and economic benefits, but the concern will persist: "The economic gains may not always justify the high capital costs-Particularly for precision irrigation and AI-powered systems" (Lu et al. 2005). For small-scale farmers, the expenses associated with infrastructure, sensors, and decision support systems can be limitations, preventing widespread adoption. Additional research is needed to better align the economic value of precision irrigation with its costs (Adeyemi et al. 2017; Raihan 2024).

One study showed that the profitability of variable rate seeding depends on field conditions. The economic benefits can be significantly noticed in fields where there is variability in soil conditions such as moisture, fertility, and productivity. However, in more uniform fields, the economic return is minimal (Šarauskis et al. 2022).

In conclusion, digitalization in agriculture has appeared as an innovation with many economic opportunities for farmers through resource use efficiency and its effect on productivity and yield. But besides just the efficiency and productivity gains, digitalization serves as a vital tool in carbon farming and resulting in a tradable carbon certification which presents farmers with opportunities to generate additional income. Despite many environmental and economic benefits of digital agriculture, key issues like affordability and access should be addressed, especially for the benefit of smallholder farmers and for them to fully realize the economic potential of digital intervention in agriculture.

13.9 Policy Perspective for Low and Net-Zero Carbon Emissions

The adoption of digital tools in agriculture in developing countries is often limited because of limited digital literacy, limited infrastructure, and high initial cost (Quarshie et al. 2023). To fully utilize the potential of digital agriculture in contributing to carbon management and carbon credit generation, clear and supportive policy frameworks are essential (Huang et al. 2022). Governments should start planning and implementing farmer-centric initiatives, such as educating small-scale farmers through training and demonstration on the potential role of digitalization in carbon sequestration and emissions reduction by minimizing the overuse of inputs (Huang et al. 2022).

The carbon market ecosystem has been facing difficulties in tracking emissions, and emission-reduction advanced digital monitoring systems are addressing these issues by supporting accurate monitoring, reporting, and verification of greenhouse gas emissions and carbon sequestration. Technologies like data analytical methods and blockchain verification platforms can help in achieving more accurate MRV processes and higher precision in their assessment and verification methods for participation in carbon markets. (Mungroo et al. 2024).

The government now must develop policy mechanisms to not only promote the adoption of digital agriculture but also to establish an institutional framework to support and guide farmers in accessing voluntary and compliance carbon markets. The government should encourage farmers to certify reduced emissions and participate in carbon credit schemes.

The role of digital agriculture has received significant policy attention in terms of its role in enhancing crop productivity and farming efficiency. However, its contribution to sustainable goals such as environmental and soil benefits, human health benefits, and the economic gain through environmental benefits needs more attention in policy (MacPherson et al. 2022). The ability of digital technology in agriculture to provide sustainability benefits depends on the policy and legal settings that drive its implementation. If the policy is not favorable, digitalization can also become a commercial element to promote economic efficiency rather than environmental sustainability.

A progressive example of a policy supporting digitalization in agriculture and climate-neutral goals is the European Union's Green Deal. It targets to reduce net GHG emissions by a minimum of 55% by 2030 and achieve climate neutrality by 2050, and also promotes digitalization in agriculture to reduce emissions and support environmental sustainability (European Commission 2019). Similar strategies and policies that support digitalization, sustainability, and carbon credit facilitation programs should be implemented.

13.10 Conclusion

Digital agriculture can be a tool to drive the next big transformation in agriculture because of its potential in achieving resource-efficient, low-emission, sustainable, and carbon-neutral agriculture goals in this era of pressing climate challenges. Over the past few decades, digital innovations such as precision agriculture, automated systems, UAVs, IoT, and AI, have totally revolutionized agriculture by enhancing productivity and climate resilience. Precision agriculture, involving precise management of water, nutrients, pesticides, as well as land use systems, is a key enabler of reduced GHGs because it supports the site-specific management of farm activities and hence the efficient use of resources. Beyond contribution in agriculture production, digital tools also play a vital role in efficient management of the food supply chain by contributing to better decision making, reduced waste, and reduced carbon footprints. Environmental benefits presented by digital technologies can be transcribed into an economic value in the system of voluntary and compliance carbon markets. While the indirect environmental benefits of digitalization can be monetized, there is a direct way in which digital tools support carbon sequestration and the generation of carbon credits. Technologies such as remote sensing, IoT sensors, AI, and machine learning support accurate monitoring, quantification, and verification of soil carbon stocks. Remote sensing data paired with AI and machine learning helps to make predictions about future carbon sequestration rates under varying environmental conditions, providing actionable recommendations to maximize the sequestration benefit. By encouraging and facilitating farmers to participate in the carbon market, digitalization contributes to climate-neutral goals and, side by side, it opens up income streams for farmers. In this way, mitigation of climate change can turn into a financially viable enterprise. However, supportive policies and strategic approaches will be necessary to fully unlock this potential. This might include digital literacy, financial incentives for adopting the technologies, and regulatory frameworks for carbon trading. In conclusion, digitalization in agriculture holds immense environmental and economic potential if the digital innovations are aligned with robust policy frameworks and well-defined carbon market mechanisms. Integrating digital agriculture and carbon farming creates a roadmap for achieving global net-zero targets and carbon-neutral food systems.

References

Abdoli P, Khanmirzaei A, Hamzeh S, Rezaei S, Moghimi S (2023) Use of remote sensing data to predict soil organic carbon in some agricultural soils of Iran. Remote Sens Appl Soc Environ 30: 100969. https://doi.org/10.1016/j.rsase.2023.100969

Adams RM, Hurd BH, Lenhart S, Leary N (1998) Effects of global climate change on agriculture: an interpretative review. Clim Res 11(1):19–30. https://doi.org/10.3354/cr011019

Adeyemi O, Grove I, Peets S, Norton T (2017) Advanced monitoring and management systems for improving sustainability in precision irrigation. Sustainability 9(3):353. https://doi.org/10.3390/su9030353

Angelopoulou T, Tziolas N, Balafoutis A, Zalidis G, Bochtis D (2019) Remote sensing techniques for soil organic carbon estimation: a review. Remote Sens 11(6):676. https://doi.org/10.3390/rs11060676

Anshari M, Almunawar MN, Masri M, Hamdan M (2019) Digital marketplace and fin tech to support agriculture sustainability. Energy Procedia 156:234–238. https://doi.org/10.1016/j.egypro.2018.11.134

Awoniyi S, Osunmakinde M, Abdulkadir A, Olatundun S (2023) Digitalization of agriculture: what relevance and challenges in enhancing climate smart agriculture in Nigeria. FARA Res Rep 7(7):48–58. https://doi.org/10.59101/frr072307

Baker JL, David MB, Lemke DW (2005) Understanding nutrient fate and transport, including the importance of hydrology in determining losses, and potential implications on management systems to reduce those losses. In: Proceedings of the Gulf hypoxia and local water quality concerns workshop. Upper Mississippi River Sub-basin Hypoxia Nutrient Committee, pp 11–26

Balafoutis A, Beck B, Fountas S, Vangeyte J, Van der Wal T, Soto I, Gómez-Barbero M, Barnes A, Eory V (2017) Precision agriculture technologies positively contributing to GHG emissions mitigation, farm productivity and economics. Sustainability 9(8):1339. https://doi.org/10.3390/su9081339

Balasundram SK, Shamshiri RR, Sridhara S, Rizan N (2023) The role of digital agriculture in mitigating climate change and ensuring food security: an overview. Sustainability 15(6):5325. https://doi.org/10.3390/su15065325

Benyam A, Soma T, Fraser E (2021) Digital agricultural technologies for food loss and waste prevention and reduction: Global trends, adoption opportunities and barriers. Journal of Cleaner Production 323: 129099. https://doi.org/10.1016/j.jclepro.2021.129099

Bhakta I, Phadikar S, Majumder K (2019) State-of-the-art technologies in precision agriculture: a systematic review. https://doi.org/10.1002/jsfa.9693

Biggar S, Man D, Moffroid K, Pape D, Riley-Gilbert M, Steele R, Thompson V (2013) Greenhouse gas mitigation options and costs for agricultural land and animal production within the United States. ICF International, Department of Agriculture Climate Change Program Office, Washington, DC. https://www.usda.gov/sites/default/files/documents/GHG_Mitigation_Options.pdf

Blandford D, Hassapoyannes K (2018) The role of agriculture in global GHG mitigation. https://doi.org/10.1787/da017ae2-en

Bolfe EL, Jorge LA, Sanches ID, Luchiari Júnior A, Costa CC, Victoria DC, Inamasu RY, Grego CR, Ferreira VR, Ramirez AR (2020) Precision and Digital Agriculture: Adoption of Technologies and Perception of Brazilian Farmers. Agriculture 10(12): 653. https://doi.org/10.3390/agriculture10120653

Bora GC, Nowatzki JF, Roberts DC (2012) Energy savings by adopting precision agriculture in rural USA. Energy Sustain Soc 2:1–5. https://doi.org/10.1186/2192-0567-2-22

Brahmanand PS, Singh AK (2022) Precision irrigation water management-current status, scope and challenges. Indian J Fertil 18:372–380. https://www.researchgate.net/publication/360119350_Precision_Irrigation_Water_Management_-_Current_Status_Scope_and_Challenges

Brar BS, Kaur A (2016) Precision nutrient management: a review. Indian J Fertil 1:15. https://www.researchgate.net/profile/Bijay-Singh-3/publication/310605305_Precision_nutrient_management_A_review/links/583326d208ae138f1c0a8ae9/Precision-nutrient-management-A-review.pdf

Cariappa AA, Konath NC, Sapkota TB, Krishna VV (2024) Evaluating the potential and eligibility of conservation agriculture practices for carbon credits. Sci Rep 14(1):9193. https://doi.org/10.1038/s41598-024-59262-6

Chandan A, John M, Potdar V (2023) Achieving UN SDGs in food supply chain using blockchain technology. Sustainability 15(3):2109. https://doi.org/10.3390/su15032109

Chandrasekaran M, Ranganathan R (2017) Modelling and optimisation of Indian traditional agriculture supply chain to reduce post-harvest loss and CO_2 emission. Ind Manag Data Syst 117(9):1817–1841. https://doi.org/10.1108/IMDS-09-2016-0383

Daccache A, Knox JW, Weatherhead EK, Daneshkhah A, Hess TM (2015) Implementing precision irrigation in a humid climate–recent experiences and on-going challenges. Agric Water Manag 147:135–143. https://doi.org/10.1016/j.agwat.2014.05.018

Deichmann U, Goyal A, Mishra D (2016) Will digital technologies transform agriculture in developing countries? Agric Econ 47(S1):21–33. https://doi.org/10.1111/agec.12300

El Hathat Z, Venkatesh VG, Sreedharan VR, Zouadi T, Manimuthu A, Shi Y, Srinivas SS (2024) Leveraging greenhouse gas emissions traceability in the groundnut supply chain: Blockchain-enabled off-chain machine learning as a driver of sustainability. Inf Syst Front:1–18. https://doi.org/10.1007/s10796-024-10514-w

European Commission (2019) A European green deal: striving to be the first climate-neutral continent. https://commission.europa.eu/strategy-and-policy/priorities-2019-2024/story-von-der-leyen-commission/european-green-deal_en#:~:text=The%20European%20Green%20Deal%20was,course%20of%20action%20is%20critical

Fabregas R, Kremer M, Schilbach F (2019) Realizing the potential of digital development: the case of agricultural advice. Science 366(6471). https://doi.org/10.1126/science.aay3038

Forster P, Ramaswamy V, Artaxo P, Berntsen T, Betts R, Fahey D, Haywood J, Lean J, Lowe D, Myhre G, Nganga J, Prinn R, Raga G, Schulz M, Dorland R, Bodeker G, Boucher O, Collins W, Whorf T (2007) Changes in atmospheric constituents and in radiative forcing. https://elib.dlr.de/51416/

Fulton J (2018) Variable-rate seeding systems for precision agriculture, pp 285–297 https://www.taylorfrancis.com/chapters/edit/10.1201/9781351114592-16/variable-rate-seeding-systems-precision-agriculture-john-fulton-ohio-state-university-usa

Gasso V, Sørensen CA, Oudshoorn FW, Green O (2013) Controlled traffic farming: a review of the environmental impacts. Eur J Agron 48:66–73. https://doi.org/10.1016/j.eja.2013.02.002

Gasso V, Oudshoorn FW, Sørensen CA, Pedersen HH (2014) An environmental life cycle assessment of controlled traffic farming. J Clean Prod 73:175–182. https://doi.org/10.1016/j.jclepro.2013.10.044

Gerhards R, Sökefeld M (2003) Precision farming in weed control-system components and economic benefits. In: Precision agriculture. Wageningen Academic, pp 229–234. https://doi.org/10.3920/9789086865147_033

Ghosh P, Kumpatla SP (2022) GIS applications in agriculture. In: Geographic information systems and applications in coastal studies. IntechOpen. https://www.intechopen.com/chapters/81685

Gorai T, Yadav PK, Choudhary GL, Kumar A (2021) Site-specific crop nutrient management for precision agriculture – a review. Curr J Appl Sci Technol 40(10):37–52. https://doi.org/10.9734/cjast/2021/v40i1031357

Grisso RD, Alley MM, Thomason WE, Holshouser DL, Roberson GT (2011) Precision farming tools: variable-rate application. https://vtechworks.lib.vt.edu/server/api/core/bitstreams/9df8cff9-74f8-4826-8d5e-700e302ee2ee/content

Grunwald S (2022) Artificial intelligence and soil carbon modeling demystified: power, potentials, and perils. Carbon Footpr 1(1):6. https://doi.org/10.20517/cf.2022.03

Gumbi N, Gumbi L, Twinomurinzi H (2023) Towards sustainable digital agriculture for smallholder farmers: a systematic literature review. Sustainability 15(16):12530. https://doi.org/10.3390/su151612530

Gupta K, Kumar R, Baruah KK, Hazarika S, Karmakar S, Bordoloi N (2021) Greenhouse gas emission from rice fields: a review from Indian context. Environ Sci Pollut Res 28(24):30551–30572. https://doi.org/10.1007/s11356-021-13935-1

Haji M, Kerbache L, Muhammad M, Al-Ansari T (2020) Roles of technology in improving perishable food supply chains. Logistics 4(4):33. https://doi.org/10.3390/logistics4040033

Hu J, Gao L, Bai X, Li T, Liu X (2015) Review of research on automatic guidance of agricultural vehicles. Trans Chin Soc Agric Eng 31(10):1–10. https://doi.org/10.11975/j.issn.1002-6819.2015.10.001

Huang X, Yang F, Fahad S (2022) The impact of digital technology use on farmers' low-carbon production behavior under the background of carbon emission peak and carbon neutrality goals. Front Environ Sci 10:1002181. https://doi.org/10.3389/fenvs.2022.1002181

Kroulik M (2012) Benefits of GPS agricultural guidance for sustainable agriculture. Agrociencia Uruguay 16(3):107–116. https://doi.org/10.31285/AGRO.16.663

Kyriakarakos G, Petropoulos T, Marinoudi V, Berruto R, Bochtis D (2024) Carbon farming: bridging technology development with policy goals. Sustainability 16(5):1903. https://doi.org/10.3390/su16051903

Lamichhane S, Kumar L, Wilson B (2019) Digital soil mapping algorithms and covariates for soil organic carbon mapping and their implications: a review. Geoderma 352:395–413. https://doi.org/10.1016/j.geoderma.2019.05.031

Lieder S, Schröter-Schlaack C (2021) Smart farming technologies in arable farming: towards a holistic assessment of opportunities and risks. Sustainability 13(12):6783. https://doi.org/10.3390/su13126783

Linquist BA, Anders MM, Adviento-Borbe MAA, Chaney RL, Nalley LL, Da Rosa EF, Van Kessel C (2015) Reducing greenhouse gas emissions, water use, and grain arsenic levels in rice systems. Glob Change Biol 21(1):407–417. https://doi.org/10.1111/gcb.12701

Lipper L, McCarthy N, Zilberman D, Asfaw S, Branca G (2017) Climate smart agriculture: building resilience to climate change. Springer Nature 630. https://doi.org/10.1007/978-3-319-61194-5

Longchamps L, Khosla R (2015) Improving N use efficiency by integrating soil and crop properties for variable rate N management. In: Precision agriculture'15. Wageningen Academic, pp 249–256. https://doi.org/10.3920/978-90-8686-814-8_30

Lowder SK, Sánchez MV, Bertini R (2021) Which farms feed the world and has farmland become more concentrated? World Dev 142:105455. https://doi.org/10.1016/j.worlddev.2021.105455

Lu YC, Sadler EJ, Camp CR (2005) Economic feasibility study of variable irrigation of corn production in southeast coastal plain. J Sustain Agric 26(3):69–81. https://doi.org/10.1300/J064v26n03_08

MacPherson J, Voglhuber-Slavinsky A, Olbrisch M, Schöbel P, Dönitz E, Mouratiadou I, Helming K (2022) Future agricultural systems and the role of digitalization for achieving sustainability goals: a review. Agron Sustain Dev 42(4):70. https://doi.org/10.1007/s13593-022-00792-6

Malhi GS, Kaur M, Kaushik P (2021) Impact of climate change on agriculture and its mitigation strategies: a review. Sustainability 13(3):1318. https://doi.org/10.3390/su13031318

Maraseni TN, Mushtaq S, Maroulis J (2009) Greenhouse gas emissions from rice farming inputs: a cross-country assessment. J Agric Sci 147(2):117–126. https://doi.org/10.1017/S0021859608008411

Mateev V, Marinova I (2021) Distributed Internet of Things system for CO_2 monitoring with LoRaWAN. In: 2021 12th national conference with international participation (electronica). IEEE, pp 1–5. https://doi.org/10.1109/ELECTRONICA52725.2021.9513682

McFadden J, Njuki E, Griffin T (2023) Precision agriculture in the digital era: recent adoption on US farms. https://doi.org/10.22004/ag.econ.333550

Meena RS, Pradhan G (2023) Industrial garbage-derived biocompost enhances soil organic carbon fractions, CO_2 biosequestration, potential carbon credits, and sustainability index in a rice-wheat ecosystem. Environ Res 116525. https://doi.org/10.1016/j.envres.2023.116525

Meena RS, Pradhan G, Kumar S, Lal R (2023) Using industrial wastes for rice-wheat cropping and food-energy-carbon-water-economic nexus to the sustainable food system. Renew Sust Energ Rev 187:113756. https://doi.org/10.1016/j.rser.2023.113756

Meena RS, Pradhan G, Singh K, Kumar S, Singh AK, Shashidhar KS, Mina KK, Rao CS (2024a) Agriculture models for restoring degraded land to enhance CO_2 biosequestration and carbon credits in the Vindhyan region of India. Sci Total Environ 929:172661. https://doi.org/10.1016/j.scitotenv.2024.172661

Meena RS, Lal R, Kumar S, Pradhan G, Rao CS, Singh AK, Pathak H, Abhilash PC, Kumar A, Sharma SK, Jat ML, Singh S (2024b) Potential of Indian agriculture for capturing atmospheric CO_2 and monetizing carbon credits to the farmers: an overview and policy framework. Adv Agron 188:101–206. https://doi.org/10.1016/bs.agron.2024.06.001

Meena RS, Pradhan G, Nalani Singh K, Singh AK, Verma SK, Mina KK, Kumar S, Chaturvedi RK (2024c) Diversified cropping modules designed for soil restoration, CO_2 sequestration, and generating carbon credits. Land Degrad Dev 35. https://doi.org/10.1002/ldr.5224

Meena RS, Pradhan G, Kumar S (2025) Energy flow, eco-efficiency, and economic circulation with recycled industrial waste compost application in wheat and subsequent rice farming. Sci Total Environ 967:178779. https://doi.org/10.1016/j.scitotenv.2025.178779

Mouratiadou I, Lemke N, Zander P, Shaaban M, Macpherson J, Gaiser T, Bellingrath-Kimura SD (2021) Digital agricultural knowledge and information system. The DAKIS decision support platform for management design and ecosystem services provision. Landscape 16:20–23. https://doi.org/10.1016/j.ese.2023.100274

Mungroo ZBA, Tupsee RS, Pierre JD (2024) Addressing leakage in mechanism methodologies (v. 01.0): case study of technology adoption in carbon market MRV systems [Internet]. UNFCCC, Bonn. https://unfccc.int/sites/default/files/resource/MEP004_CfI_Addressing_Leakage_Mungroo.pdf

Nackley LL, Warneke B, Fessler L, Pscheidt JW, Lockwood D, Wright WC, Fulcher A (2021) Variable-rate spray technology optimizes pesticide application by adjusting for seasonal shifts in deciduous perennial crops. HortTechnology 31(4):479–489. https://doi.org/10.21273/HORTTECH04794-21

Natcvetova A (2021) The role of artificial intelligence in measuring and modelling soil organic carbon in agricultural lands. https://urn.fi/URN:NBN:fi-fe2021061838806

Nayak AK (2022) Advanced techniques for precision farming in rice. In: Climate resilient technologies for rice based production systems in eastern India. ICAR, Odisha

Obour PB, Ugarte CM (2021) A meta-analysis of the impact of traffic-induced compaction on soil physical properties and grain yield. Soil Tillage Res 211:105019. https://doi.org/10.1016/j.still.2021.105019

Okoli AO, Birkenberg A (2024) Monitoring soil carbon in smallholder carbon projects: insights from Kenya. Clim Chang 177(9):143. https://doi.org/10.1007/s10584-024-03796-1

Peteinatos GG, Rueda-Ayala V, Gerhards R, Andujar D (2015) Precision harrowing with a flexible tine harrow and an ultrasonic sensor. In: Precision agriculture, vol 15. Wageningen Academic, pp 579–586. https://doi.org/10.3920/978-90-8686-814-8_72

Polymeni S, Skoutas DN, Sarigiannidis P, Kormentzas G, Skianis C (2024) Smart agriculture and greenhouse gas emission mitigation: a 6G-IoT perspective. Electronics 13(8):1480. https://doi.org/10.3390/electronics13081480

Pouladi N, Gholizadeh A, Khosravi V, Borůvka L (2023) Digital mapping of soil organic carbon using remote sensing data: a systematic review. Catena 232:107409. https://doi.org/10.1016/j.catena.2023.107409

Pradhan G, Meena RS (2023a) Utilizing waste compost to improve the atmospheric CO_2 capturing in the rice-wheat cropping system and energy-cum-carbon credit audit for a circular economy. Sci Total Environ:164572. https://doi.org/10.1016/j.scitotenv.2023.164572

Pradhan G, Meena RS (2023b) Interaction impacts of biocompost on nutrient dynamics and relations with soil biota, carbon fractions index, societal value of CO_2 equivalent, and ecosystem services in the wheat-rice farming. Chemosphere:139695. https://doi.org/10.1016/j.chemosphere.2023.139695

Pradhan G, Meena RS, Kumar S, Lal R (2023) Utilizing industrial wastes as compost in wheat-rice production to improve the above and below-ground ecosystem services. Agric Ecosyst Environ 358:108704. https://doi.org/10.1016/j.agee.2023.108704

Pramod PP, Kashinath KJ, Raosaheb GN, Madhukar NS (2024) Development of automated pesticide application system. https://doi.org/10.51470/PLANTARCHIVES.2024.v24.no.1.057

Qasim W, Xia L, Lin S, Wan L, Zhao Y, Butterbach-Bahl K (2021) Global greenhouse vegetable production systems are hotspots of soil N_2O emissions and nitrogen leaching: a meta-analysis. Environ Pollut 272:116372. https://doi.org/10.1016/j.envpol.2020.116372

Quarshie PT, Abdulai AR, Duncan E, Krishna Bahadur KC, Roth R, Sneyd A, Fraser ED (2023) Myth or reality? The digitalization of climate-smart agriculture (DCSA) practices in smallholding agriculture in the bono east region of Ghana. Clim Risk Manag 42:100553. https://doi.org/10.1016/j.crm.2023.100553

Rai N, Zhang Y, Ram BG, Schumacher L, Yellavajjala RK, Bajwa S, Sun X (2023) Applications of deep learning in precision weed management: a review. Comput Electron Agric 206:107698. https://doi.org/10.1016/j.compag.2023.107698

Raihan A (2024) A review of digital agriculture toward food security and environmental sustainability. In Proceedings of the international conference on digital agriculture, food security, and environmental sustainability

Ramesh T, Rathika S (2020) Evaluation of rice cultivation systems for greenhouse gases emission and productivity. Int J Ecol Environ Sci 2:49–54

Ranganathan V, Kumar P, Kaur U, Li SH, Chakraborty T, Chandra R (2022) Re-inventing the food supply chain with IoT: a data-driven solution to reduce food loss. IEEE Internet Things Mag 5(1):41–47. https://doi.org/10.1109/IOTM.003.2200025

Rejeb A, Keogh JG, Treiblmaier H (2019) Leveraging the internet of things and blockchain technology in supply chain management. Future Internet 11(7):161. https://doi.org/10.3390/fi11070161

Rosenzweig C, Liverman D (1992) Predicted effects of climate change on agriculture: a comparison of temperate and tropical regions. In: Global climate change: implications, challenges, and mitigation measures. Pennsylvania Academy of Science, Easton, p 342

Sadler EJ, Evans RG, Stone KC, Camp CR (2005) Opportunities for conservation with precision irrigation. J Soil Water Conserv 60(6):371–378

Saini MK, Saini RK (2020) Agriculture monitoring and prediction using Internet of Things (IoT). In: 2020 sixth international conference on parallel, distributed and grid computing (PDGC). IEEE, pp 53–56. https://doi.org/10.1109/PDGC50313.2020.9315836

Šarauskis E, Kazlauskas M, Naujokienė V, Bručienė I, Steponavičius D, Romaneckas K, Jasinskas A (2022) Variable rate seeding in precision agriculture: recent advances and future perspectives. Agriculture 12(2):305. https://doi.org/10.3390/agriculture12020305

Shakhbulatov D, Arora A, Dong Z, Rojas-Cessa R (2019) Blockchain implementation for analysis of carbon footprint across food supply chain. In: 2019 IEEE international conference on blockchain (blockchain), vol 49. IEEE, pp 546–551. https://doi.org/10.1109/Blockchain.2019.00079

Sharma C, Pathak P, Kumar A, Gautam S (2024) Sustainable regenerative agriculture allied with digital agri-technologies and future perspectives for transforming Indian agriculture. Environ Dev Sustain:1–36. https://doi.org/10.1007/s10668-024-05231-y

Shen S, Basist A, Howard A (2010) Structure of a digital agriculture system and agricultural risks due to climate changes. Agric Agric Sci Procedia 1:42–51. https://doi.org/10.1016/j.aaspro.2010.09.006

Shepherd M, Turner JA, Small B, Wheeler D (2020) Priorities for science to overcome hurdles thwarting the full promise of the 'digital agriculture' revolution. J Sci Food Agric 100(14):5083–5092. https://doi.org/10.1002/jsfa.9346

Shockley JM, Dillon CR, Stombaugh TS (2011) A whole farm analysis of the influence of auto-steer navigation on net returns, risk, and production practices. J Agric Appl Econ 43(1):57–75. https://doi.org/10.22004/ag.econ.100640

Shrestha S (2019) Effects of climate change in agricultural insect pest. Acta Sci Agric 3(12):74–80. https://doi.org/10.31080/ASAG.2019.03.0727

Shubha K, Singh NR, Mukherjee A, Maity A, Dubey R (2020) Controlled traffic farming. Curr Sci 119(11):1760–1766

Siikavirta H, Punakivi M, Kärkkäinen M, Linnanen L (2002) Effects of e-commerce on greenhouse gas emissions: a case study of grocery home delivery in Finland. J Ind Ecol 6(2):83–97. https://doi.org/10.1162/108819802763471807

Sinitsa Y, Borodina O, Gvozdeva O, Kolbneva E (2021) Trends in the development of digital agriculture: a review of international practices. In: BIO web of conferences, vol 37. EDP Sciences, p 00172. https://doi.org/10.1051/bioconf/20213700172

Sitaula BK, Hansen S, Sitaula JIB, Bakken LR (2000) Methane oxidation potentials and fluxes in agricultural soil: effects of fertilisation and soil compaction. Biogeochemistry 48:323–339. https://doi.org/10.1023/A:1006262404600

Smith RJ, Baillie JN, McCarthy AC, Raine SR, Baillie CP (2010) Review of precision irrigation technologies and their application. National Centre for Engineering in Agriculture Publication, 1003017(1)

Soane BD, Van Ouwerkerk C (1995) Implications of soil compaction in crop production for the quality of the environment. Soil Tillage Res 35(1–2):5–22. https://doi.org/10.1016/0167-1987(95)00475-8

Soto I, Barnes A, Balafoutis A, Beck B, Sánchez B, Vangeyte J, Gómez-Barbero M (2019) The contribution of precision agriculture technologies to farm productivity and the mitigation of greenhouse gas emissions in the EU. Publications Office of the European Union, Luxembourg. https://doi.org/10.2760/016263

Taghizadeh-Mehrjardi R, Nabiollahi K, Rasoli L, Kerry R, Scholten T (2020) Land suitability assessment and agricultural production sustainability using machine learning models. Agronomy 10(4):573. https://doi.org/10.3390/agronomy10040573

Takacs-Gyorgy K, Takács I (2022) Towards climate smart agriculture: how does innovation meet sustainability? Ecocycles 8(1):61–72. https://doi.org/10.19040/ecocycles.v8i1.220

Tang S, Zhu Q, Zhou X, Liu S, Wu M (2002) A conception of digital agriculture. In: IEEE international geoscience and remote sensing symposium, vol 5, pp 3026–3028. https://doi.org/10.1109/IGARSS.2002.1026858

Trendov NM, Varas S, Zeng M (2019) Digital technologies in agriculture and rural areas. https://openknowledge.fao.org/handle/20.500.14283/CA4887EN

Tullberg JN (2014) CTF and global warming. In 5th Australian controlled traffic and precision agriculture conference, p 214

Tumushabe JT, Turyasingura B, Chavula P (2023) The sustainability of carbon markets for climate-smart agriculture among smallholder farmers in Uganda. Asian J Res Agric For 9(4):337–345. https://doi.org/10.9734/ajraf/2023/v9i4263

Udeh EO, Amajuoyi P, Adeusi KB, Scott AO (2024) The role of IoT in boosting supply chain transparency and efficiency. https://doi.org/10.30574/msarr.2024.11.1.0081

Vallero D (2019) Air pollution biogeochemistry, air pollution calculations. https://doi.org/10.1016/B978-0-12-814934-8.00008-9

Vavekanand R, Kumar S (2024) Rural agricultural development through e-commerce platforms. Authorea Preprints. https://doi.org/10.36227/techrxiv.171439779.97867951/v1

Wang Y, Yang Y, Qin Z, Yang Y, Li J (2023) A literature review on the application of digital technology in achieving green supply chain management. Sustainability 15(11):8564. https://doi.org/10.3390/su15118564

Xie L, Luo B, Zhong W (2021) How are smallholder farmers involved in digital agriculture in developing countries: a case study from China. Land 10(3):245. https://doi.org/10.3390/land10030245

Yatribi T (2020) Factors Affecting Precision Agriculture Adoption: A Systematic Litterature Review. ECONOMICSINNOVATIVE AND ECONOMICS RESEARCH JOURNAL 8(2): 103–121. https://doi.org/10.2478/eoik-2020-0013

Yoro KO, Daramola MO (2020) CO2 emission sources, greenhouse gases, and the global warming effect. In: Advances in carbon capture, pp 3–28. https://doi.org/10.1016/B978-0-12-819657-1.00001-3

Zanin ARA, Neves DC, Teodoro LPR, da Silva Júnior CA, da Silva SP, Teodoro PE, Baio FHR (2022) Reduction of pesticide application via real-time precision spraying. Sci Rep 12(1):5638. https://doi.org/10.1038/s41598-022-09607-w

Zhou F (2024) The impact of digital economy development on agricultural carbon emissions——empirical research based on Chinese data. Appl Econ Lett:1–8. https://doi.org/10.1080/13504851.2024.2310060

Chapter 14
Using Biophysical Modeling to Predict Carbon Sequestration Dynamics in Cropping Systems

Anannya Dhar, Saikat Dey, Sukamal Sarkar, and Ram Swaroop Meena

Abstract The sustainability of agricultural lands is at risk owing to intensive farming practices coupled with undesirable activities like deforestation, soil erosion, and overgrazing, which serve as the drivers of the reduction in soil organic carbon (SOC). To support the ever-growing global population with food and nutrition, the indiscriminate application of inorganic chemical fertilizers not only exacerbates the ecological balance but also increases the discharge of greenhouse gases like carbon dioxide (CO_2), nitrous oxide (N_2O), and methane (CH_4) from the soil to the open atmosphere. This results in the reduction in the potential of the soils to produce healthy and nutritious crops for mankind. Carbon sequestration proves to be a climate-resilient strategy for capturing and storing soil carbon stocks, thereby enhancing the physicochemical and biological properties of soil and emerging as a carbon-neutral country in the near future. Hence, proper estimation of carbon sequestration potential and its prediction in the future in various cropping systems is vital to maintaining the sustainability of the soil ecosystem. Biophysical models like APSIM, DNDC, DSSAT, ROTH-C, EPIC, CENTURY, and SWAT-C are some of the often-used models in organic carbon simulation in long-term field experiments. By considering drivers like climate, soil properties, agronomic practices, and plant physiology, they can accurately simulate carbon fluxes, the dynamics of soil organic matter (SOM), and the impacts of land use changes on carbon sequestration. Moreover, these models enable scenario analysis, facilitating the assessment of various management strategies and their potential impacts on carbon sequestration, thus aiding in the development of sustainable agricultural practices aimed at mitigating climate change. Thus, a clear understanding of these soil carbon simulation models is crucial in modern-day agriculture for upscaling the degraded lands and preserving the degrading ones.

A. Dhar · S. Dey · S. Sarkar (✉)
Division of Agronomy, School of Agriculture and Rural Development, Ramakrishna Mission Vivekananda Educational and Research Institute, Narendrapur, West Bengal, India
e-mail: sukamal.sarkar@gm.rkmvu.ac.in

R. S. Meena
Department of Agronomy, Institute of Agricultural Sciences, Banaras Hindu University, Varanasi, Uttar Pradesh, India

R. S. Meena et al. (eds.), *Carbon-Negative Agriculture*, Sustainability Sciences in Asia and Africa, https://doi.org/10.1007/978-981-95-1483-0_14

Keywords APSIM · Carbon flux · Carbon neutrality · Soil organic carbon · Sustainable agriculture

14.1 Introduction

Soil organic carbon (SOC) is imperative for refining soil fertility and cultivating its physicochemical and biological characteristics (Mohanty et al. 2020; Meena et al. 2025). Soil acts both as a sink and a source of carbon, necessitating precise management to maintain its organic content. Historically, activities like rigorous tillage, deforestation, excessive use of chemicals, and inadequate application of quality manure have significantly pressured soil systems, resulting in a gradual reduction in soil carbon content and challenging soil fertility (Manna et al. 2013; Choudhary et al. 2016; Meena et al. 2024c). Fertile soil is essential for supporting the growing population with quality foodstuffs. Poor management practices have depleted soil organic content by 50–70% (Sanderman et al. 2017). Consequently, the primary focus has shifted to increasing SOC through soil carbon sequestration (CS). Carbon sequestration (CS) encompasses the long-term capture and storage of carbon dioxide (CO_2) or other carbon forms in various reservoirs, including oceans, soils, vegetation (notably forests), and geological formations, to mitigate or delay global warming (Nair et al. 2015). Soil carbon sequestration specifically refers to capturing atmospheric CO_2 and incorporating it into the soil within a specific land unit. This process leverages crops, crop residues, and other organic manures, contributing to the creation and preservation of soil organic matter (SOM) (Chenu et al. 2019). Soil carbon sequestration can occur in two ways: direct and indirect methods. Direct soil CS involves inorganic chemical reactions that transform CO_2 into soil inorganic carbon compounds, such as magnesium and calcium carbonates. In contrast, indirect CS occurs when plants photosynthesize atmospheric CO_2 into their dry matter, incorporating carbon into their structure and contributing to SOC through residue and root exudates (Nair et al. 2015; Meena et al. 2024a). Implementing various carbon-neutral technologies is crucial to improving soil carbon storage. Balanced application of chemical fertilizers in conjunction with quality organic manure, as part of an integrated nutrient management approach, has proven highly effective in improving soil carbon stocks (Mohanty et al. 2020; Louhar et al. 2020; Kumar et al. 2022a, 2023). Additionally, conservation agriculture strategies, including reduced tillage and retention of crop residues, contribute significantly to building SOC stocks. These strategies not only support soil fertility and health but also play a substantial role in modifying the influence of climate change by enhancing soil carbon sequestration in agricultural soils. Estimating the organic carbon level in the soil and the rate of carbon sequestration is essential for farmers to evaluate their management practices. Soil organic carbon stock is typically calculated by multiplying the soil's carbon content by its bulk density and soil depth (Pearson et al. 2007; Meena et al. 2024b). However, understanding the various carbon pools, such as labile and non-labile carbon, enhances our comprehension of soil carbon dynamics. Carbon simulation models, once calibrated and validated with field experimental data spanning two or more years, are widely used to estimate and quantify long-term carbon accumulation in soil. Agricultural simulation models are widely used by groups of researchers and scientists in predicting the carbon stock, carbon sequestration dynamics, and carbon flux of the soil. Though manual soil sampling and organic carbon estimation methods present more accurate data, they seem to be more laborious and time-consuming, which prevents the monitoring of long-term data (Liu et al. 2023). After calibration and parameterization, these carbon models can simulate the dynamics of carbon stock in the soil and cropping systems (He et al. 2021). The biophysical models that simulate SOC integrate various scenarios like crop, soil, and prevailing management and macro and micro-climate (Aditi et al. 2023). After parameterization, the models can detect the actual interaction between plant growth, soil, moisture, nutrients, the management system of the farmer, and edaphic-climatic conditions (Gaydon et al. 2017). In the recent era of agricultural research, numerous process-based site-calibrated system simulation models like APSIM, CENTURY, DNDC, DSSAT, C-QUEST, ROTH-C, and EPIC are used in the dynamics of the SOC pool in response to diverse agronomic practices, climatic, and edaphic conditions (Mohanty et al. 2020). APSIM (Agricultural Production Systems Simulator) is a biophysical, robust, component-based software model that predicts long-term results of several farmland practices, crop yield, soil nutrients, SOM, rotation in cropping systems, soil moisture, etc. (Sarkar et al. 2024; Aditi et al. 2023). The model calculates the daily dynamics based on regular climatic and management data (Godde et al. 2016). They are used in numerous environmental conditions to detect the SOC pool dynamics (Luo et al. 2014). APSIM Manager Modules like APSIM-Surface OM and APSIM-Soil N are used in forecasting the dynamics of carbon sequestration of the cropping system (Godde et al. 2016). Organic matter is distinguished as fresh organic matter, BIOM, the more labile microbial pool, and humified organic matter (HUM), the remaining SOM (Ranatunga et al. 2001). The DNDC (Denitrification and Decomposition) model is a robust process-based model that is employed to estimate the nitrogen and carbon cycles, emissions of greenhouse gas (GHG), crop growth and production, soil carbon sequestration, nutrient, and water use efficiency, global warming potential, etc., of a cropping system (Shaukat et al. 2023; Meena et al. 2023). This simulation model can operate at regional, national, and site-scale levels (Sleutel et al. 2006). One major benefit enjoyed by this model is that it does not require any external crop-based inputs, as these features are built into the software. DSSAT (Decision Support System for Agro-Technology Transfer) is a web-based interface that is extensively used for the estimation of soil carbon and nitrogen balance, precision nutrient management, climate change, GHG emissions, etc., at different spatiotemporal scales (He et al. 2021; Punia et al. 2020). This mechanistic robust model integrates two soil carbon and nitrogen models, that is, CERES and CENTURY-based soil models for quantification of carbon flux in soils (Hoogenboom et al. 2019). ROTH-C (Rothamsted Soil Carbon Turnover Model) is a software model that is applied to measure the turnover of SOC. All types of carbon transformations that occur within a single soil layer are considered in this model (Ranatunga et al. 2001). Most frequently used carbon-simulating models (Falloon and Smith 2002), though the

ROTH-C model is easy to implement, it cannot estimate the SOC level in waterlogged soils accurately (Geremew et al. 2024). The CENTURY model is considered to forecast the complex interactions and dynamics of carbon along with other macronutrients within various plant-soil ecosystems. It can model a varied range of environments, including grasslands, agricultural crop fields, forests, and savannas (Parton 1996). By accounting for the intricate biochemical processes and nutrient cycling within these systems, the CENTURY model provides valuable insights into how different ecosystems respond to various environmental changes and management practices. The textural class of the soil is an external input for the model. It operates on the three organic matter pools in the form of active, passive, and slow under the SOM sub-model (Ranatunga et al. 2001; Meena and Pradhan 2023). The EPIC (Erosion Policy Integrated Climate) is a robust software model that can be employed to predict the carbon flux in soil (Izaurralde et al. 2006). Initially, the EPIC model was used for agroecosystem study; later it was merged with the CENTURY model and became an easier web interface in the quantification of soil carbon, soil moisture, etc. in response to climate (Longo et al. 2023). Implementing a carbon simulation model will greatly enhance our understanding of carbon dynamics and fluxes, enabling farmers to adopt precise soil and crop management practices. This will enhance soil carbon stock, thereby refining soil quality and fertility, and combating the adverse influences of climate change and global warming by lowering atmospheric GHGs.

14.2 Activities Causing Emissions of Carbon from Agricultural Fields

The escalating levels of carbon emissions pose a negative threat to the global environment, with the agricultural sector emerging as a substantial contributor to this issue. Agriculture's activities, including deforestation, soil cultivation, and livestock production, release considerable amounts of CO_2 and other GHGs directly into the atmosphere. This heightened concentration of CO_2 contributes to the greenhouse effect, leading to a notable increase in global temperatures. The resultant global warming precipitates a cascade of climate changes, encompassing more frequent and severe weather events, shifting precipitation patterns and adverse impacts on ecosystems and biodiversity. Several factors influence carbon emissions from soil, with temperature being a critical one. The CO_2 emissions exhibit a logarithmic increase between 20 °C and 40 °C, peaking in mid-July due to heightened root activity and accelerated decomposition of organic matter. However, carbon emissions decline sharply above 50 °C, primarily because microbial respiration is inhibited and biological oxidation systems become deactivated at such high temperatures (Rastogi et al. 2013). Soil texture also performs a significant role in CO_2 formation by regulating air and moisture supply, thereby affecting water infiltration and gas diffusion rates (Hashimoto and Komatsu 2006). The pH level of the soil is another crucial factor, with low pH levels inhibiting microbial respiration and subsequently reducing carbon emissions. Excessive soil salinity disrupts essential soil processes, leading to a lessening of the carbon discharges as well. Soil moisture levels have an enormous influence on carbon emissions: an initial increase in soil moisture enhances CO_2 emissions up to an optimal level, beyond which they begin to decrease. Periodic drying and rewetting events further influence carbon emissions, as rewetting stimulates microbial activity and leads to more CO_2 release.

Intensive agricultural practices, characterized by high levels of soil manipulation through conventional and deep tillage, significantly contribute to the weakening of soil structure. This intensive soil disturbance disrupts soil aggregates and increases the susceptibility of fertile, carbon-rich topsoil to erosion by water and wind (Deng et al. 2016). Tilled soil produces more CO_2 because the disturbance creates conditions that favor microbial decomposition of organic matter. Reducing tillage intensity is therefore recommended to minimize carbon loss from the soil and help maintain soil carbon levels.

As a result, SOCs are lost along with the eroded soil particles (Meena et al. 2020; Pradhan and Meena 2023b). Moreover, deforestation, a critical issue in global carbon emissions, exacerbates soil degradation. The removal of trees leaves the soil fallow and devoid of the anchorage provided by deep tree roots, leading to further destabilization. Trees contribute more stable carbon to the soil compared to crop plants, making their removal particularly detrimental (Benbi 2013). Transforming forested regions into agricultural land or other uses releases large amounts of carbon that were previously sequestered in trees and vegetation, significantly impacting soil carbon dynamics.

The rapid decomposition of carbon added by crop residues, compared to the more stable carbon inputs from agroforestry systems, results in greater carbon loss from agricultural soils (Helfrich et al. 2006; Jhariya et al. 2022). In response to the growing global food demand, the overuse of chemical fertilizers has become widespread. This practice not only depletes soil health but also increases GHG emissions, including methane (CH_4) and CO_2, from agricultural fields (Sah and Devakumar 2018; Kumar et al. 2022b; Pradhan and Meena 2023a). Additionally, the lack of residue retention and the common practice of burning crop stubble to clear fields exacerbate the problem. Open burning of crop residues releases substantial amounts of CO_2 back into the air, causing global warming. Furthermore, the minimization of residue retention exposes the soil surface to the impact of falling raindrops, which leads to the washing away of soil carbon and other essential plant nutrients (Katterer and Bolinder 2023).

Forest fires are another major source of carbon emissions from soils. During combustion, these fires release extensive amounts of CO_2 into the air, further resulting in climatic variability (Cramer et al. 2004). Additionally, the activation of soil microbes due to intensive agricultural practices accelerates the degradation of SOM, leading to substantial CO_2 emissions from the soil. This microbial activity,

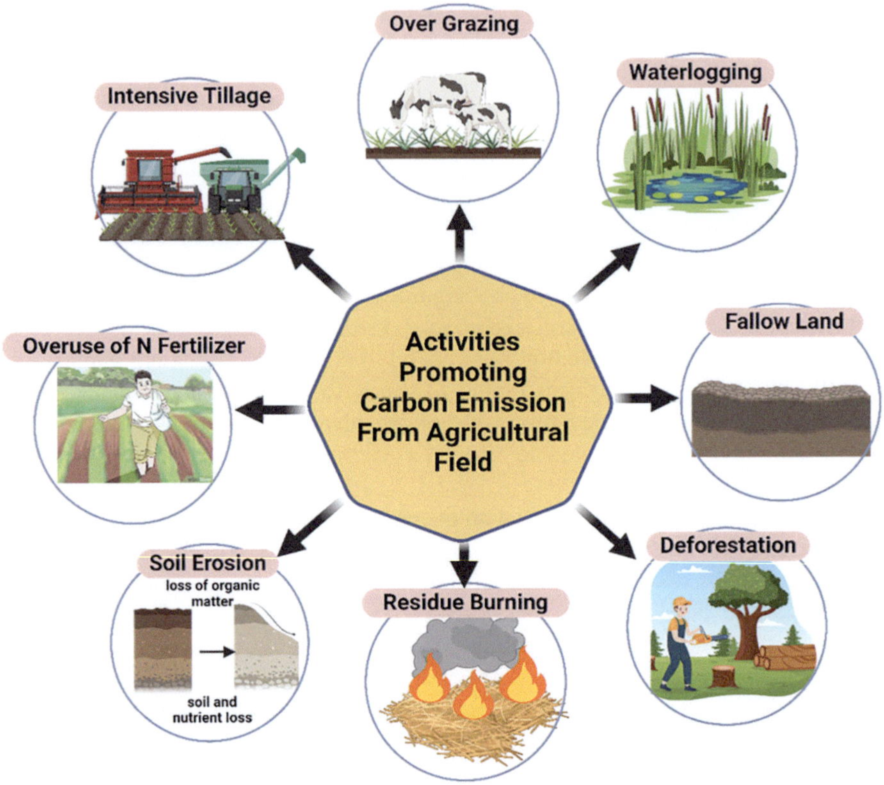

Fig. 14.1 Agricultural activities promoting carbon emissions. (Image created with BioRender.com; license number: ID273KA5IG)

while essential for nutrient cycling, can bring about the rapid depletion of the OC of soil when soil management practices are not sustainable.

The multitude of factors including intensive tillage, deforestation, non-judicious use of chemical fertilizers, crop residue burning, forest fires, temperature, soil texture, pH, salinity, organic manure and fertilizer application, soil moisture, and the presence of crops, significantly impact soil carbon dynamics and contribute to greenhouse gas emissions (Fig. 14.1). Sustainable soil management strategies, like reduce tillage, agroforestry, residue retention, judicious use of fertilizers, optimizing temperature conditions, maintaining appropriate soil texture and pH, controlling salinity, managing soil moisture, and reducing tillage intensity, are essential to mitigate these adverse effects and enhance soil carbon sequestration. Addressing these challenges requires a holistic approach that integrates soil conservation, sustainable agriculture, and forest management to protect soil health and mitigate climate change impacts. Consequently, addressing carbon emissions from agriculture is crucial in mitigating climate change and ensuring environmental sustainability. Further research and policy interventions are imperative to develop and implement sustainable agricultural practices that reduce carbon emissions while maintaining food security and economic viability.

14.3 Soil Carbon Sequestration: Concept and Enhancement Strategies

Greenhouse gases like CO_2 are potential agents in increasing the global air temperature of the Earth. To keep the average atmospheric temperature increase of the globe below 2 °C, and within 1.5 °C, 196 parties signed the Paris Agreement at the gathering of COP 21 (Conference of Parties) in December 2015 (Katterer and Bolinder 2023). Hence, some negative carbon emission strategies were considered, including direct capture and storage of CO_2 from the atmosphere, enhancement of weathering, planting more trees, fertilization of oceans, carbon capture and storage with the help of bioenergy, and sequestration of organic carbon (Minx et al. 2018; Pradhan et al. , b; Kumar et al. 2025). After the oceans, soil holds the second position as the largest carbon sink in the world (Dadhich and Meena 2014). The carbon equilibrium indicates that the concentration of organic carbon (OC) in the soil is nearly twice that found in the atmosphere. Therefore, by enhancing the soil carbon concentration, atmospheric carbon levels can be reduced, contributing to the mitigation of climate variability and the warming of the globe (Meena et al. 2020).

Soil organic carbon sequestration refers to the method of capturing atmospheric CO_2 and incorporating it into the soil within a specific land unit. This process utilizes crops, crop residues, and other organic manures, which contribute to the development and retention of organic matter present in the soil (Chenu et al. 2019). Carbon sequestration can be classified into organic carbon sequestration and inorganic carbon sequestration. Organic carbon sequestration relies on biological processes and stores carbon in organic forms within ecosystems. At the same time, inorganic carbon sequestration involves chemical (calcium carbonate) and pedogenic processes that store carbon in stable, non-living forms for much longer periods (Meena et al. 2020). In organic carbon sequestration, organic carbon enters the soil in three ways: (i) as aboveground crop biomass or residues, (ii) underground crop biomass, and (iii) through the process of rhizodeposition (Kogel-Knabner et al. 2023). Cellulose, lignin, hemicellulose, cutin, and suberin present in the foliage, stems, twigs, and roots of crops are large polymers of organic matter that mainly constitute the crop biomass grown above and below the surface of the soil (Kögel-Knabner 2017). They are generally added to the surface or different rooting depths and contribute to soil carbon flux mainly through bioturbation or leaching of their dissolved constituents. In the case of rhizodeposition (deposition by roots), compounds of low molecular weight are produced by the roots into the soil. The distribution of root biomass is 75% at 40 cm depth, 50% at 20 cm depth, and only 30% biomass at 10 cm depth; hence, with the increase in root depth, the root leftover, and rhizodeposition increase (Kogel-Knabner et al. 2023). The carbon derived from the roots of the crops rather than the aboveground parts has a longer turnover time (Katterer et al. 2011). Hence, plant breeding for developing deep and robust rhizospheric systems is of primary importance to soil carbon sequestration (Eckersten et al. 2017). Fast decomposition occurs with the added organic matter in the soil through crop residues, rhizodeposition, and recycled organic wastes like vermicompost, farmyard manure, municipal solid waste compost, poultry manure, turkey, and goat manure, though only a fraction of this organic matter is stabilized by various physicochemical and biological processes depending upon the traits of the matter and prevailing edaphic and climatological factors (Basile-Doelsch et al. 2020).

SOC exists in various pools, including dissolved organic carbon (DOC), particulate organic carbon (POC), microbial biomass carbon (MBC), light fraction organic carbon (LFOC), and mineral-bound organic carbon (MBOC). MBOC represents the stable fraction of organic carbon, encompassing humic acid, fulvic acid, and humin. Among these pools, DOC and MBC are the most labile, meaning they are easily altered in the soil and respond quickly to daily fluctuations in temperature, moisture, and soil management practices. POC and LFOC exhibit changes on an annual or seasonal basis. The relative stability of these SOC pools follows the order: DOC $<$ MBC $<$ POC $<$ LFOC $<$ MBOC (Abdelrahman et al. 2023). In field and laboratory experiments conducted over short durations, it is generally feasible to estimate the labile pools of carbon, such as microbial biomass carbon (MBC) and dissolved organic carbon (DOC). These labile pools are responsive to immediate changes in environmental conditions and agronomic practices. However, to quantify the long-standing dynamics of all soil carbon pools, including more stable forms such as light fraction organic carbon (LFOC), particulate organic carbon (POC), and mineral-bound organic carbon (MBOC), organic carbon simulation models are required. These models enable researchers to project changes over extended periods, providing a comprehensive understanding of the complex interactions and stability of different SOC pools under varying environmental and management scenarios.

The principles of soil carbon sequestration have four pillars: (a) less disturbance of soil through its mechanical manipulation leading to enhanced stability of carbon in the aggregates of soil, (b) increasing the quantity and quality of applied organic matter, such as manures, to facilitate greater carbon sequestration, (c) enhancing the quantity, variety, and functionality of soil microbial biota to improve carbon storage, and (d) avoiding fallow periods to prevent carbon emissions from the topsoil (Meena et al. 2020).

Many activities promote organic carbon sequestration in soil. Among them, the prominent ones are residue retention, the addition of recycled organic waste (organic manure), reduced/minimum tillage, crop rotations, agroforestry systems, irrigation management, integrated nutrient management, etc. (Fig. 14.2).

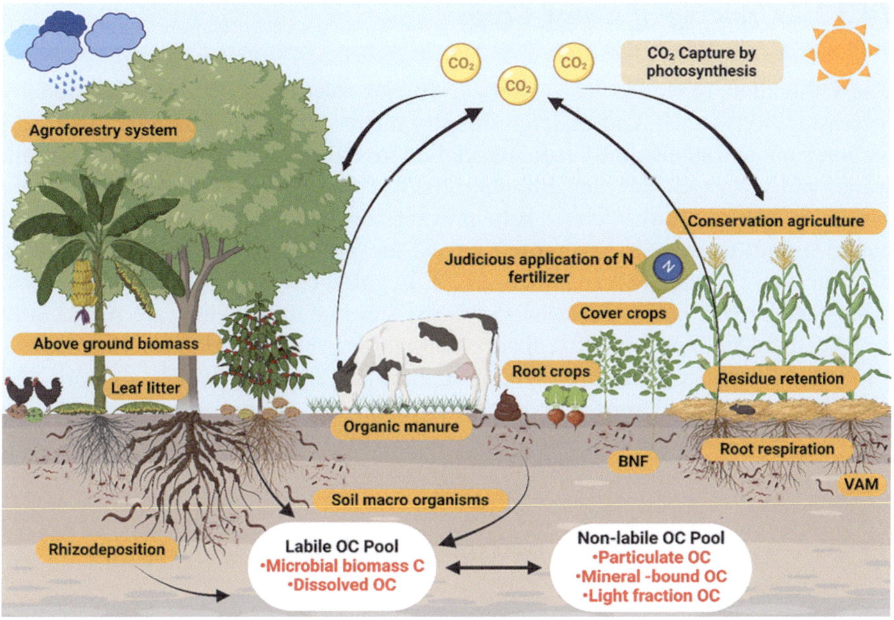

Fig. 14.2 Activities enhancing sequestration of organic carbon in soil

14.3.1 Reduced Tillage Practices

Conservation tillage practices in the form of reduced, minimum, or zero tillage are acknowledged to sink carbon in agricultural soils worldwide (Paustian et al. 1997; Leharwan et al. 2023). Intensive manipulation of soil disturbs soil more leading to disintegration of soil structure and erosion of carbon-rich topsoil (Lal and Kimble 1997). Implementation of reduced and less intensive tillage prevents damage from falling rain and maintains the integrity of soil structural colloids (Meena et al. 2020). Less disturbance to soil and deep placement of carbon in deeper horizons lead to an increment of aggregate formation by the microorganism (Lal and Kimble 1997). Incorporation of crop residue results in the enhancement of SOC. Owing to reduced manipulation, the fast microbial degradation of SOC is lowered, and simultaneously carbon emission is lowered, and sink is more (Meena et al. 2020). However, some meta-analyses have depicted that conservation tillage enhancement of organic carbon is up to the upper layers only and it gradually decreases in the deeper soil horizons (Ogle et al. 2019; Meurer et al. 2018).

14.3.2 Growing of Cover Crops

The cultivation of cover crops plays a crucial role in carbon sequestration by increasing both above- and underground dry matter production, thereby increasing carbon input and significantly reducing carbon loss from soil through erosion by soil, water, and wind (Katterer and Bolinder 2023). In addition to contributing organic carbon to the soil, cover crops help prevent soil carbon loss by anchoring soil particles with their roots and protecting the topsoil from the impact of raindrops on their dense foliage (Meena et al. 2020). Legume cover crops, such as cowpea, moth bean, green gram, dhaincha (Sesbania spp.), and sunhemp, are particularly beneficial due to their ability to enhance biological diversity, enrich soil flux, and fix atmospheric nitrogen through symbiotic nitrogen fixation, which supports the growth of subsequent crops (Lal 2004). By mitigating mineralization and reducing erosion, cover crops create a carbon-rich zone within the soil (Lal 2016).

14.3.3 Addition of Organic Manure and Recycled Organic Matter

Organic manure, specifically the bulky organic manure (farmyard manure, compost, vermicompost, green manure, etc.), serves as a brilliant source of SOC. Apart from supplying essential plant nutrients, organic manure enhances soil fertility and productivity by improving the physicochemical and biological properties of the soil (Noorzai et al. 2017). When the optimum dose of nutrients for the crops is met through only organic manures, the content of SOC, the activity of the soil-dwelling microbial population, and the concentration of SOC in the soil are increased compared to the sole application of inorganic fertilizer (Meena et al. 2020). Additionally, the application of recycled organic matter, such as municipal solid waste compost, sewage, and sludge, can sequester a significant amount of carbon in the soil (Aguilera et al. 2013). Generally, SOC stock is increased by manure application, but it depends on several factors, including site characteristics such as the initial organic carbon content of the soil, its texture, and prevailing climatic conditions; manure properties like nature, quantity, and quality; and agronomic decisions such as tillage, dose, and duration of manure application (Gross and Glaser 2021). However, in long-term experiments, it was observed that the continuous application of sole organic manure (in the form of FYM) increased the SOC content, but the carbon sequestration efficiency was recorded as low compared to inorganic chemical fertilizers (Kumar et al. 2022b; Huang et al. 2022). Therefore, integrated nutrient management strategies that replace specific doses of chemical inorganic fertilizers with bulky organic manure, along with other carbon-resilient strategies like conservation tillage and retention of crop residues, are advantageous (Meena et al. 2020).

14.3.4 Irrigation Management

Organic matter deposition in soil and carbon sequestration are influenced by irrigation management, which has both desirable and undesirable effects on SOM (Anton et al. 2022). Providing irrigation to crops at the right stages to maintain desirable soil moisture levels promotes better aboveground and underground biomass production. This, in turn, increases SOC stocks (Meena et al. 2020). Sensible irrigation management supports crop production, leading to increased carbon addition in the soil through aboveground biomass and rhizodeposition (Kochsiek et al. 2009). However, maintaining favorable moisture levels for microorganisms through irrigation can also enhance microbial activity. This increased activity results in the breakdown of organic soil matter and higher emissions of greenhouse gases, ultimately reducing SOC content (Trost et al. 2013; Anton et al. 2022). Therefore, while irrigation can boost SOC by supporting biomass production, it can also have the counterproductive effect of accelerating organic matter decomposition and greenhouse gas emissions, highlighting the need for balanced and strategic irrigation practices.

14.3.5 Residue Retention

Residue retention, which means returning back the harvested biomass back to the field, is a climate-smart strategy recommended all over the globe to improve the sequestration of SOC (Wang et al. 2020). Retention of aboveground residues is a brilliant strategy of conservation agriculture that helps to develop the soil's organic carbon stock besides offering other benefits like nutrient management and conservation of soil and water. They also lessen the loss of carbon from the field, employing erosion of soil by water, wind, etc. (Katterer and Bolinder 2023). In sugarcane, residue retention treatment was reported to record the highest carbon sequestration and carbon retention efficiency, besides enhancing the yield by 38% as compared to no residue retention treatment (Pradhan et al. 2023a).

14.3.6 Agroforestry System

Agroforestry significantly contributes to the mitigation and sequestration of atmospheric carbon by integrating trees or other woody perennials into agricultural landscapes (Meena et al. 2020). This integration can occur through the retention of existing vegetation, the introduction of new trees, or their intermixing with crops, pastures, or livestock. Agroforestry sequesters carbon through a combination of mechanisms, including increased biomass production, heightened soil fertility, and reduced soil erosion (Kumar et al. 2018). By incorporating trees into agricultural systems, agroforestry promotes the growth of woody biomass, which acts as a long-term carbon sink by storing atmospheric CO_2 through photosynthesis (Nair et al. 2010). Moreover, agroforestry systems often comprise diverse plant species, contributing to improved soil fertility and the accumulation of organic matter (OM). The leaves and other organic residues from these trees add to the soil's organic carbon content, further augmenting its capacity for carbon storage. Thus, agroforestry plays a pivotal role in enhancing carbon sequestration and promoting sustainable farming practices.

Soil is a chief carbon sink, and its potential to sequester organic carbon depends on several factors. These include the soil's ability to prevent organic matter degradation, the chemical composition of SOM, and the accumulation of both labile and non-labile organic matter fractions (Swift 2001; Meena et al. 2020). Agricultural soils can sequester between 0.4 and 1.2 gigatons of carbon annually (Lal 2004). Several factors influence this potential, including soil depth, the type of vegetation growing over it, nutrient content, and prevailing macro and microclimate conditions. Additionally, the relative proportions of soil moisture and soil air are crucial in determining the carbon sequestration rate in global agricultural soils (Swift 2001). Managing these factors effectively is essential for enhancing soil carbon sequestration and mitigating climate change through sustainable farming practices.

14.4 Basics of Biophysical Models and Their Advantages in Predicting the Carbon Sequestration Dynamics

Biophysical models simulate a wide range of trajectories of organic carbon fluxes and sequestration within predictable forthcoming climate change circumstances (Smith et al. 2005; Wan et al. 2011). Another advantage of the biophysical model is its ability to evaluate not only the present status but also to predict future SOC sequestration in a location (Hashimoto et al. 2012). Various types of research have been directed to determine the SOC pools locally and globally based on future climatic circumstances (Gupta and Kumar 2017). A biophysical model in predicting SOC dynamics integrates biological, physical, and chemical processes to simulate the behavior and changes of SOC over time. These models are crucial for understanding how different factors, such as climate change, land use, and agronomic practices, affect the sink of soil carbon and emissions. Several simulation models are performed in varied ways based on their characteristics, sub-modules present in the framework, and trajectories toward future climatic projections. The SOC predicting models are typically divided into several pools based on their decomposition rates, such as fast, slow, and passive (Liu et al. 2023; Gupta and Kumar 2017). Each pool represents organic matter at various stages of decomposition, from fresh crop residues to stable humus. Multiple factors affect the decomposition rates of crop inputs, such as the C:N ratio, lignin content, and other compositions, and are influenced by several ecological aspects such as moisture, temperature, and microbial activity (Izaurralde et al. 2006). These models incorporate the consequences of soil temperature, moisture, pH, and texture on the dynamics of SOC. Environmental factors can either accelerate or decelerate the decomposition process. The models evaluate the influence of agricultural practices such as mechanical soil manipulation, crop rotation, fertilization, and cover cropping on SOC dynamics (Zhao et al. 2013; Robertson et al. 2015). These models involve carbon inputs like above- and below-ground plant litter, root exudates, manure, and fertilizer, while outputs may include erosion, leaching loss, and CO2 emissions through microbial respiration (Tang et al. 2012; Ntamwira et al. 2023). There is multiple simulation models used to predict soil organic dynamics, including APSIM, DSSAT, CENTURY, ROTH-C, DNDC, EPIC, and SWAT-C. By simulating the influence of various land management practices, biophysical models help in developing strategies to enhance SOC storage and mitigate greenhouse gas emissions. They provide insights into the complex interactions between soil, plants, and microorganisms, improving our understanding of soil carbon cycling processes. These models can be applied at different spatial scales (from field to global) and temporal scales (seasonal to decadal), making them versatile tools for researchers and policymakers. Each of these models has exceptional features and applications, making them appropriate for various research and management objectives related to SOC dynamics.

14.5 Mechanism of Biophysical Models for Carbon Sequestration

Soil organic carbon is a key factor in augmenting soil quality, health, and yield of crops, which has been regulated through several soil and agronomic practices, including irrigation management, tillage, crop covers, and applications of organic amendments in the soil. The deposition of SOM is a long-term process, while manually predicting soil carbon pools over time becomes troublesome; therefore, the integration of several biophysical models in estimating soil carbon dynamics provides insight to the scientific community.

14.5.1 Agricultural Production Systems sIMulator (APSIM)

The Agricultural Production Systems sIMulator or APSIM model (www.apsim.info) has been instrumental in advancing our understanding of SOC sequestration dynamics across various agroecosystems and regions. Numerous studies have demonstrated its utility in modeling and predicting the outcome of diverse agricultural interventions on pools of soil C. The “SoilN” module within the APSIM environment addresses the nitrogen and carbon dynamics in the soil (Luo et al. 2011). Unlike the CERES model, APSIM divides SOM into two pools: BIOM, which signifies the more labile soil microbial biomass products, and HUM, which

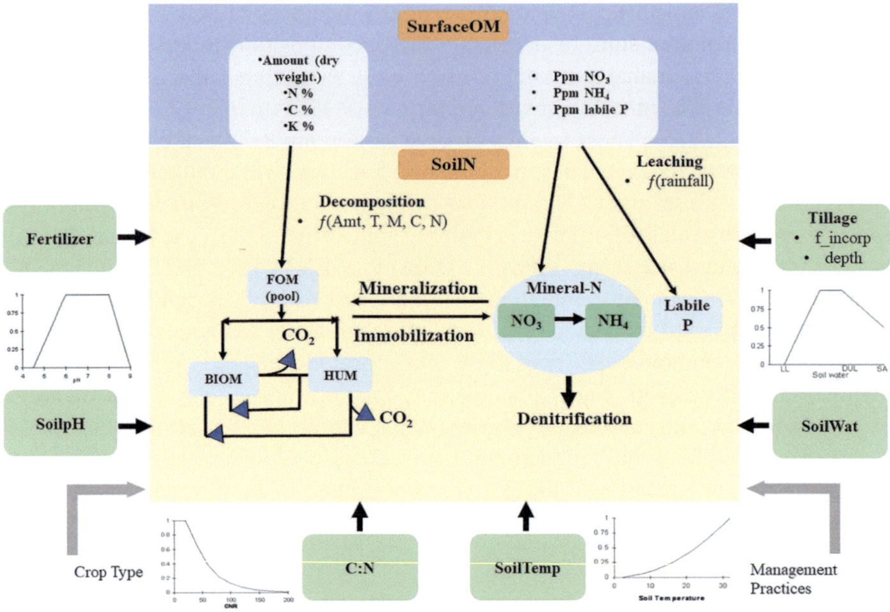

Fig. 14.3 Schematic mechanism of APSIM to simulate SOC sequestration in cropping system. (Modified and redrawn from www.apsim.info)

comprises the rest of the SOM (Saddique et al. 2019; Feleke et al. 2021). The flows among these various carbon fluxes are measured in terms of C, with the corresponding nitrogen flows depending on the C: N ratio of the receiving pool. Carbon dynamics within APSIM are affected by multiple environmental features (such as edaphic temperature, moisture, and pH) and agronomic factors (crop type, cultivation practices, and tillage management). The decomposition mechanism and carbon dynamics within APSIM are also significantly affected by the "SurfaceOM" and "SurfaceOM" modules, which regulate the processing, decomposition, and mineralization patterns of aboveground organic material (Mohanty et al. 2012). The detailed mechanisms and interrelationships within the modules related to carbon dynamics in APSIM are presented in Fig. 14.3.

The APSIM was predominantly used to simulate and predict the carbon dynamics for various cropping systems. Luo et al. (2013) designed a meta-model using APSIM to assess the potential of SOC sequestration in major cereal-growing areas of Australia. Their findings revealed that soil pH significantly influences SOC levels, with lower pH leading to higher SOC. They also identified that under optimal management, including the optimal fertilizers application and complete retention of crop residues, these regions could sequester approximately 1.2 Pg of SOC, with 0.12 Pg specifically in wheat croplands. Similarly, Mohanty et al. (2020) utilized APSIM to estimate long-term SOC fluctuation in vertisols of central India under varied nitrogen and FYM applications. Their study highlighted that integrating N with FYM had the maximum potential to increase SOC, with an optimum N rate for wheat estimated at 155 kg ha^{-1}, leading to a 28.6% rise in SOC over the initial concentration. Another study by Bana et al. (2023) employed APSIM to explore the consequences of sustainable agriculture on crop yield, water use efficiency, and sequestration of OC in a rice–wheat system in the (Indo-Gangetic Plains) IGPs (Jangir et al. 2024). They found that higher residue levels in conservation agriculture significantly improved crop productivity and SOC levels rather to intensive agricultural practices. Furthermore, Wang et al. (2010) demonstrated through APSIM modeling that optimal nitrogen supply along with 100% crop residue retention in an annual wheat system could significantly enhance SOC sequestration in the Murray-Darling Basin of Australia (Patial et al. 2017). Similarly, Kumari et al. (2023) assessed the potential for carbon sequestration under various residue and tillage practices in semi-arid tropical drylands of India, using APSIM. They concluded that adding crop residues significantly enhanced the content of OC under conventional as well as reduced tillage systems. These studies underscore the versatility and robustness of APSIM in assessing the effects of several agronomic management practices on the sequestration of OC in soil. They collectively indicate that strategic management of residues, nitrogen, and conservation tillage can substantially augment soil carbon stock, combating climate change and sustainable agricultural productivity.

14.5.2 Decision Support System for Agrotechnology Transfer

The Decision Support System for Agrotechnology Transfer (DSSAT) model is widely established for estimating crop growth and SOC dynamics under various agricultural management practices. Several studies have employed the DSSAT model to determine the impact of different strategies on carbon sequestration. Nicoloso et al. (2020) used the DSSAT-CENTURY model to assess no-till and organic fertilization strategies, discovering that no-till soils can sequester more than 0.5 Mg C ha^{-1} $year^{-1}$ under corn crops with medium-low production. However, the SOC accrual is a small amount under crop rotations of soybean/wheat. Organic inputs and crop rotation improved SOC in both Mollisol and Oxisol soils. Similarly, Soler et al. (2011) used DSSAT to predict SOC dynamics in West Africa, noting slight SOC increases for sorghum-fallow rotation with manure, though other rotations generally resulted in SOC declines. Attia et al. (2021) demonstrated the capability of this model to simulate corn production and dynamics of SOC in arid Mediterranean climates, emphasizing the sensitivity of SOC and crop productivity to soil genotype and agronomic inputs. Musinguzi et al. (2014) emphasized the effectiveness of this model in showing high SOC decline rates with continuous tillage and no fertilization in low-input maize systems, advocating for conservation agriculture to maintain soil fertility. The DSSAT model has also been modified to better simulate low-input agricultural systems, as described by Gijsman et al. (2002), who incorporated a crop input module from the CENTURY interface, improving the working results of the model in measuring the dynamics of OC in soil.

Shrivastava (2006) used DSSAT to simulate carbon sequestration at the micro-watershed scale, finding that practices like no-tillage and double organic manure addition significantly increased SOC over 30 years. Overall, the DSSAT model is a valuable framework for simulating long-term OC stocks and assessing the efficacy of various agricultural management practices in enhancing carbon sequestration, contributing to more sustainable and productive agroecosystems.

14.5.3 The Rothamsted Carbon Model

The Rothamsted Carbon Model (ROTH-C) is a popular, competent, framework-based model adopted throughout the world for several agroecosystems such as grazing, forests, and arable lands (http://www.rothamsted.ac.uk/sustainable-soils-and-grassland-systems/rothamsted-carbon-model-rothc) (Jordon et al. 2022; Geremew et al. 2024). This model was generally designed to predict soil organic turnover under non-waterlogged soil at Rothamsted, London. Subsequently, it has been adapted to various management and climatic scenarios (Cerri et al. 2007; Jiang et al. 2014). The ROTH-C model categorizes SOC into various pools based on the varying degrees of resistance to the degradation of plant materials, namely resistant plant material (RPM), decomposable plant material (DPM), humified organic matter (HUM), and microbial biomass (BIO). Additionally, there is a minor fraction called inert organic matter (IOM), which is extremely resilient to breakdown (Li et al. 2016; Geremew et al. 2024). However, the rate of decomposition of the plant materials is subject to soil temperature, moisture content, soil types, and crop covers (Yao et al. 2017). The model lacks an important factor: carbon inputs from plant biomass in the soil. This model is not designed for the simulation of crop growth, so it lacks a sub-model for the development and harvest prediction of crops (Coleman and Jenkinson 2014). Therefore, carbon input data from crop biomass is required from other models. The ROTH-C model calculates SOC dynamics monthly. The major dataset needed to calibrate ROTH-C includes the physical and chemical properties of soil (bulk density, content of organic carbon and clay, IOM), several weather parameters (total evaporation, average air temperature, and total precipitation on a monthly time step), and agronomic management practices (organic matter, crop covers, plant residue inputs) (Li et al. 2016; Geremew et al. 2024).

14.5.4 CENTURY

The CENTURY model is a sophisticated process-based simulation model that has gained significant attention from researchers due to its ability to estimate the continuing conversion and cycling of SOC and crucial plant nutrients, including nitrogen, phosphorus, and potassium (Liu et al. 2023). This model works on a monthly time step and can simulate these dynamics over extended periods, including several centuries. The CENTURY model (http://www.nrel.colostate.edu/projects/century/) was designed to estimate an inclusive array of complex agricultural management practices, including nutrient applications, irrigation management, crop diversification, tillage operations, biomass production, and crop harvesting (Primo et al. 2023). By accounting for these various factors, CENTURY provides a comprehensive tool for analyzing and optimizing agricultural systems. Initially, the CENTURY model was developed to perform by emphasizing previous land use patterns for establishing primary amounts of the SOC pool, and the basic climatic data, including the quantity of rainfall and minimum and maximum temperatures, are required to run the model at the initial stage (Lacoste et al. 2015). The CENTURY model comprises several sub-models, including a SOM model, a water budget model, and a plant production model (Gupta and Kumar 2017). The SOM sub-model is further differentiated into three distinct pools—active pool, passive pool, and slow pool—on the basis of their respective rates of decomposition over time. Additionally, SOM is categorized into two litter pools: aboveground structural litter pools and underground structural litter pools, as well as microbial pools (Liu et al. 2023; Gupta and Kumar 2017; Lacoste et al. 2015). Active SOC primarily comprises microbial load and exhibits rapid monthly to yearly decomposition. In contrast, a slow carbon pool possesses medium steadiness and time taken for turnover, persisting for decades. The passive carbon pool constitutes the most recalcitrant fraction, with decomposition occurring over centuries. The SOM sub-model simulates the total pool of organic carbon, which is dynamically affected by the rate of decomposition of organic matter and the soil's physical properties. The CENTURY model works based on the disk operating system model from file formation to the accomplishment of the model. CENTURY was initially parameterized using several important datasets such as soil features, climate, and agronomic practices, including crop and land management, along with the temporal and quantitative additions of organic materials (Gupta and Kumar 2017). Then the model must be calibrated and validated based on the observed datasets and ultimately is used to estimate the future scenario of the organic carbon stocks and organic matter in soil.

14.5.5 Denitrification-Decomposition (DNDC)

The Denitrification-Decomposition or DNDC model is a computer-based simulation tool designed to mimic agricultural ecosystems' nitrogen and carbon cycles (http://www.dndc.sr.unh.edu) (Levy et al. 2007; Gilhespy et al. 2014). It is composed of six sub-models: soil climate, plant growth, organic matter decomposition, fermentation, nitrification, and denitrification. The plant growth sub-model predicts how vegetation changes over time by considering plant types, their needs (physiology), soil conditions, weather, and farming practices (Ding et al. 2023). The soil climate sub-model utilizes edaphic and weather data as inputs to predict variations in the soil hydrothermal regime (moisture and temperature), acid–base chemistry (pH), and redox potential (oxygen availability) within the soil profile. The organic matter decomposition sub-module involves measuring the disintegration of SOM. Moreover, SOM consists of four main pools, including microbial biomass, plant residue (litter), active biomass, and passive biomass, which are further sub-sectioned into two to three sub-pools based on the rate of breakdown (Ding et al. 2023; Khalil et al. 2020; Li et al. 2017). The nitrification sub-module outlines the process of oxidation of NH_4 N in the soil, predominantly facilitated by bacteria causing nitrification, resulting in nitrate (NO_{3-}) and nitrite (NO_{2-}) as primary ions, with nitrogen monoxide (NO) and nitrous oxide (N_2O) forming along the way. This sub-model focuses on denitrification, a process where specific bacteria consume nitrate (NO_{3-}). As they break down the nitrate, they release a sequence of transitional compounds, including nitric oxide (NO), nitrite (NO_{2-}), and N_2O, before finally converting it all into nitrogen gas (N_2) (Gilhespy et al. 2014; Li et al. 2017). The fermentation sub-model predicts how much CH_4 is produced, used up, and released by the soil, considering the amount of active SOC and environmental circumstances. It has advantages over ROTH-C in that it does not require any plant-based inputs externally to feed the model. However, if the crop-derived inputs are improper, then the whole soil organic estimation may give incorrect information (Li et al. 2016).

14.5.6 Erosion Policy Integrated Climate (EPIC)

The Erosion Policy Integrated Climate or EPIC model has been into existence since the 1980s with the help of the United State Department of Agriculture (USDA), and it was initially built to evaluate the outcome of soil erosion on soil productivity (http://epicapex.tamu.edu/epic/) (Williams et al. 1984; Longo et al. 2023). Later, several modules were added to the EPIC model in response to simulating an extensive array of agroecosystems, including plant growth, harvest, nitrogen (N), phosphorus (P), and potassium (K) cycling, and other agronomic practices (Gassman et al. 2004). EPIC has focused on the response of plants and soil to various climatic events, creating a more comprehensive simulation interface among biophysical models used in agroecosystems (Farina et al. 2011). The model is categorized into two litter sub-models, which are structural and metabolic litter. Additionally, soil carbon is subdivided into three modules: slow humus, passive humus, and microbial biomass. These models differ among themselves in function, size, and time taken in turnover (Izaurralde et al. 2006). Therefore, the immobilization and mineralization rates vary with variations in the C:N ratio of the entire microbial load. Generally, four basic differences exist between the EPIC and CENTURY models in terms of the conversion of organic matter. First, the EPIC model involves leaching equations that use soil water content and a linear partition coefficient to determine the movement and transfer of organic components from aboveground litter to subsurface layers. Second, EPIC calculates the effects of temperature and water on transformation rates using specific equations. Third, EPIC includes a slow compartment in the surface litter fraction, along with the structural and metabolic litter components found in

CENTURY (Balkovič et al. 2013). Finally, EPIC models lignin concentration as a sigmoidal function of the age of the plant, whereas CENTURY (Izaurralde et al. 2006; Paustian et al. 2019) bases it on annual precipitation. The model must be parameterized with several parameters, including daily weather data for measuring the potential evapotranspiration by assessing the Penman-Monteith equation (Longo et al. 2023). Soil profiling is done up to a 1.5 m soil depth, wherein the upper 0–50 cm of the soil is used to measure the size distribution of particles, soil pH, bulk density, and concentration of organic carbon. A soil profile below 50 cm is required to estimate the model's other soil parameters. After the parameterization of data, calibration and validation programs were accomplished by comparing the investigational values with simulated values of climate, soil, and crop yields. Recently, the model has been modified with features taken from the CENTURY model, making it easier for users (Gerik et al. 2015).

14.5.7 Soil and Water Assessment Tools (SWAT-C)

The Soil and Water Assessment Tools, or SWAT model, is a globally recognized hydrological tool developed for the analysis of water dynamics and the transport of agricultural chemicals in extensive water bodies and sediments (https://swat.tamu.edu) (Liang et al. 2022). It is extensively utilized worldwide to measure pollution from non-point sources and to observe the influences of continuously changing climate, land use patterns, and farming practices on water quality and resources (Fu et al. 2019; Risal and Parajuli 2019; Wang et al. 2021). The development of the carbon sub-module in the SWAT interface began with the integration of a multi-layer, single-pool soil organic model (Kemanian et al. 2011). Subsequently, an updated version of the CENTURY soil carbon sequestration model (Parton et al. 1994) was incorporated into SWAT, resulting in the development of the SWAT-Carbon (SWAT-C) model (Zhang 2018; Zhang et al. 2013). Researchers improved the model (Du et al. 2019; Qi et al. 2020a, b) to include how carbon cycles within rivers, allowing it to simulate carbon storage and movement across entire watersheds. While previous tests showed the model's accuracy in aspects like CO_2 flux, soil erosion, and river carbon transport, a comprehensive evaluation of its ability to replicate long-term changes in SOC is still needed. In this model, SOC turnover rates rely on non-linear partitioning, which is regulated by management and environmental factors. It is subdivided into five major pools, such as microbial biomass, structural and metabolic litter, along with slow and passive humus. The rate of decomposition and conversion of carbon and nitrogen in these sub-modules depends on moisture content, soil temperature, texture, and other management practices (Liang et al. 2022).

14.6 Practical Examples of Biophysical Models in Sequestering Carbon

Biophysical models were used in different sections of the globe to determine and predict the sequestration of organic carbon and carbon flux in the soil. Irrespective of the situation, crop, and management practices, different biophysical models have certain advantages as well as limitations. Most of the work related to the simulation and prediction of SOC sequestration has focused on different field crops like wheat, maize, cotton, soybean, and pigeon pea (Sheoran et al. 2022). A detailed description related to the utilization of biophysical models in estimating SOC sequestration has been provided in Table 14.1. From these examples, it is evident that researchers primarily focus on tillage management practices (conventional or conservation), crop rotation adjustments (including the introduction of leguminous or non-leguminous crops in cereal-based systems), and nutrient management options. Various models have demonstrated satisfactory performance in addressing complex research questions; however, among these models, APSIM and DSSAT (including DSSAT-CENTURY and DSSAT-CERES) have shown more robust performance compared to other biophysical models.

14.7 Conclusion

Maintaining soil health is paramount for agricultural sustainability and mitigating climatic variability. The sustainability of agricultural lands is increasingly threatened by intensive farming practices, deforestation, soil erosion, and overgrazing, which collectively drive the reduction in the carbon level in soil. Fortunately, SOC sequestration emerges as a powerful strategy to combat this decline. Biophysical models such as APSIM, DNDC, DSSAT, ROTH-C, EPIC, CENTURY, and SWAT-C play a critical role by accurately simulating carbon fluxes and SOC dynamics under various cropping systems and management practices. These models offer a significant advantage over traditional methods by incorporating crucial factors like climate, soil properties, and crop management. This holistic approach enables the simulation of carbon fluxes and soil organic matter dynamics with greater accuracy. Furthermore, their predictive power facilitates scenario analysis, allowing us to evaluate the influence of various management strategies on organic carbon sequestration before real-world implementation. Ultimately, these models emphasize the development of sustainable agricultural practices that enhance soil health, mitigate climate change, and contribute to achieving carbon neutrality. Embracing and continuously refining these models are essential for restoring degraded lands, preserving soil health, and securing food security for future generations.

Table 14.1 Examples of biophysical models in estimating soil organic carbon sequestration

Model used	Location	Crops/ cropping system	Question addressed	Major findings	References
APSIM	Australia	Wheat	How the organic carbon stock of soil is affected by climate, soil, and management?	APSIM identified temperature, rainfall, and soil water-retention capacity as primary factors influencing SOC dynamics. The initial quantity of OC, including both active and inert organic carbon, introduces significant uncertainty in the sequestration of SOC	Luo et al. (2013)
	India	Soybean–wheat	What is the influence of nitrogen and FYM on the organic carbon sequestration of the cropping system?	APSIM forecasted that improved nitrogen and FYM management strategies significantly increased SOC sequestration, with the optimum nitrogen rate for maximum organic carbon sequestration estimated at 155 kg/ha for wheat	Mohanty et al. (2020)
	India	Rice–wheat	What is the ideal level of residue retention in conservation agriculture to reduce the carbon footprint?	APSIM predicted a 30–95% increment in organic carbon concentration with a carbon sequestration rate of 0.1–0.37 tons/ha/year	Bana et al. (2023)
	Australia	Wheat	How will the dynamics of soil carbon affect the application of nitrogen and residue retention?	APSIM predicted that the 100% residue retention, along with the application of N at an optimal dose, will increase the SOC to 8 Mg/ha in 100 years	Wang et al. (2010)
	India				

(continued)

Table 14.1 (continued)

Model used	Location	Crops/ cropping system	Question addressed	Major findings	References
		Maize–chickpea and maize–pigeon pea	What is the consequence of residue management and tillage on OC sequestration in the soil of both systems?	APSIM simulation predicted that maize–chickpea sequential cropping performs well with the incorporation of previous residue and minimum tillage	Aditi et al. (2023)
	China	Maize	What is the effect of nitrogen application and residue retention on the emission of N_2O and SOC change?	APSIM effectively simulated crop yield, N_2O emissions, and SOC changes, and the findings underscore the potential to further optimize nitrogen and residue management strategies, to reduce GHG emissions while maintaining crop yield	Li et al. (2019)
	Australia	Wheat	What are the primary management strategies and environmental factors affecting the dynamics of organic carbon in a constant wheat-based cropping system in cereal-growing regions of Australia?	APSIM predicted that changes in organic carbon during the simulation were affected by agronomic strategies such as residue retention and nitrogen fertilization, as well as ecological factors including initial OC content and temperature	Zhao et al. (2013)
DSSAT-CENTURY	USA and southern Brazil	Multiple crops	What will be the long SOC content affected by tillage, nitrogen, and rotation of crops?	Long-term simulations (100 years) showed that while accumulation of organic carbon in no-tilled soils (0–15 cm depth) increases, it is moderately	Nicoloso et al. (2020)

(continued)

Table 14.1 (continued)

Model used	Location	Crops/ cropping system	Question addressed	Major findings	References
				counterbalanced by losses in the soil depth of 15–30 cm under little carbon input conditions	
	China	Wheat–maize	What will be the SOC stock under the cropping system based on the N application?	DSSAT failed to provide reliable predictions for the N_{150} treatment, but it did suggest that agronomic strategies with low nitrogen application (N_0 and N_{150}) would be unsustainable	Liu et al. (2017)
	India	Pigeon pea	Which system among BBF and hats is better in the sequestration of SOC?	The BBF system was found to be more effective in sequestering organic carbon in soil	Shrivastava (2006)
DSSAT	West Africa	Cotton, sorghum, maize, groundnut	What factors influence the dynamics of organic carbon in soil?	The simulated organic carbon increased slightly over the years in the sorghum–fallow rotation with manure application, while it decreased in all other rotations, except for prolonged fallow, where it remained stable	Soler et al. (2011)
DSSAT (CERES-Maize)	Egypt	Maize	What is the result of compost application and residue retention on the SOC dynamics?	Retaining maize residue along with compost application under a zero-till system is a viable agronomic strategy for improving agronomic performance and ensuring	Attia et al. (2011)

(continued)

Table 14.1 (continued)

Model used	Location	Crops/ cropping system	Question addressed	Major findings	References
				environmentally sustainable maize production	
ROTH-C		Barley	Which cropping system among mono-cropped barley and Norflok crop rotation, sequesters more SOC?	The projection indicated that carbon stocks were reduced under monoculture, whereas incorporation of straw and intercrops demonstrated sustainability and an increase in carbon stocks across all modeled climate scenarios	Prudil et al. (2023)
	Ethiopia	–	What factors are projected to contribute to the increase in SOC under different future scenarios, according to the analysis?	The model projected that the increase in organic carbon across various future scenarios can be attributed to changes in climate and land use, long-term conservation efforts of soil and water, and improved land and soil management within the watershed	Geremew et al. (2024)
	Japan	–	How did the SOC content change over a 25-year simulation in Japanese arable soils under different management scenarios using the Rothamsted carbon (ROTH-C) model?	The total SOC content declined in all management scenarios because the carbon input was insufficient to maintain the baseline SOC level from 1990	Yokozawa et al. (2010)
SWAT-C	USA	Maize	What are the parameters that influence the	The SOC dynamics depend upon soil depth, site	Liang et al. (2023)

(continued)

Table 14.1 (continued)

Model used	Location	Crops/ cropping system	Question addressed	Major findings	References
			carbon sequestration dynamics?	characteristics, and tillage intensities	
DNDC	Pakistan	Rice–wheat	How do organic and chemical fertilizers influence the carbon sequestration dynamics in cropping systems?	The DNDC model suggested that integrating inorganic and organic fertilization can enhance carbon content in biomass and improve organic carbon sequestration in cropping systems	Shaukat et al. (2023)
	China	Rice	How does reduced tillage combined with crop residue retention affect the sequestration of organic carbon compared to conventional tillage without crop residue retention?	Zero-tillage with straw retention was the most effective, while conventional tillage without straw retention was significantly less effective than the other methods	Huang et al. (2012)
EPIC	USA	Maize, soybean	What are the impacts of soil and crop management on the SOC stock?	Sequestration rates of organic carbon at 0–20 cm depth were evaluated to be 0.50–0.63 mg ha/ha/year, depending on climate and edaphic factors	Causarano et al. (2008)
	Cambodia	Rice, soybean, cassava	How is the sequestration of SOC influenced by tillage practice and conservation agriculture?	The 20-year future simulations by EPIC indicated a decreasing tendency in organic carbon content for all CT strategies, while most CA treatments showed an increasing trend	Le et al. (2018)

References

Abdelrahman H, Hofmann D, Sleighter RL, Olk DC, Berns AE, Miano T, Cocozza C (2023) Molecular composition and possible transformations of labile soil organic matter fractions in Mediterranean arable soils: relevance and implications. Environ Res 232:116315

Aditi K, Abbhishek K, Chander G, Singh A, Falk T, Mequanint MB, Nagaraji S (2023) Assessing residue and tillage management options for carbon sequestration in future climate change scenarios. Curr Res Environ Sustain 5:100210

Aguilera E, Lassaletta L, Gattinger A, Gimeno BS (2013) Managing soil carbon for climate change mitigation and adaptation in Mediterranean cropping systems: A meta-analysis. Agric Ecosyst Environ 168:25–36

Anton R, Derrien D, Urmeneta H, van der Heijden G, Enrique A, Virto I (2022) Organic carbon storage and dynamics as affected by the adoption of irrigation in a cultivated calcareous mediterranean soil. Front Soil Sci 2:831775

Attia A, El-Hendawy S, Al-Suhaibani N, Alotaibi M, Tahir MU, Kamal KY (2021) Evaluating deficit irrigation scheduling strategies to improve yield and water productivity of maize in arid environment using simulation. Agric Water Manag 249:106812

Balkovič J, van der Velde M, Schmid E, Skalský R, Khabarov N, Obersteiner M, Stürmer B, Xiong W (2013) Pan-European crop modelling with EPIC: implementation, up-scaling and regional crop yield validation. Agric Syst 120:61–75

Bana RS, Bamboriya SD, Kumar V, Godara S, Gaydon DS, Laing AM, Dhakar R, Shivay YS, Meena VS, Singh D (2023) Identifying optimum residue levels for stable crop and water productivity and carbon sequestration under a conservation agriculture-based rice-wheat system. Soil Tillage Res 232:105745

Basile-Doelsch I, Balesdent J, Pellerin S (2020) Reviews and syntheses: the mechanisms underlying carbon storage in soil. Biogeosciences 17(21):5223–5242

Benbi DK (2013) Greenhouse gas emissions from agricultural soils: sources and mitigation potential. J Crop Improv 27(6):752–772

Causarano HJ, Doraiswamy PC, McCarty GW, Hatfield JL, Milak S, Stern AJ (2008) EPIC modeling of soil organic carbon sequestration in croplands of Iowa. J Environ Qual 37(4): 1345–1353

Cerri CEP, Easter M, Paustian K, Killian K, Coleman K, Bernoux M, Cerri CC (2007) Predicted soil organic carbon stocks and changes in the Brazilian Amazon between 2000 and 2030. Agric Ecosyst Environ 122(1):58–72

Chenu C, Angers DA, Barré P, Derrien D, Arrouays D, Balesdent J (2019) Increasing organic stocks in agricultural soils: knowledge gaps and potential innovations. Soil Till Res 188:41–52

Choudhary M, Rana KS, Rana DS, Bana RS (2016) Tillage and crop residue effects in rainfed pearl millet (*Pennisetum glaucum*) in conjunction with sulphur fertilization under pearl millet–Indian mustard (*Brassica juncea*) cropping system. Indian J Agron 61(1):15–19

Coleman K, Jenkinson DS (2014) RothC-A model for the turnover of carbon in soil Model description and users guide (Windows version)(updated June 2014)

Cramer W, Bondeau A, Schaphoff S, Lucht W, Smith B, Sitch S (2004) Tropical forests and the global carbon cycle: impacts of atmospheric carbon dioxide, climate change and rate of deforestation. Philos Trans R Soc Lond Ser B Biol Sci 359(1443):331–343

Dadhich RK, Meena RS (2014) Performance of Indian mustard (*Brassica juncea* L.) in response to foliar spray of thiourea and thioglycolic acid under different irrigation levels. Indian J Ecol 41(2):376–378

Deng L, Zhu GY, Tang ZS, Shangguan ZP (2016) Global patterns of the effects of land-use changes on soil carbon stocks. Glob Ecol Conserv 5:127–138

Ding W, Chang N, Zhang G, Kang J, Yi X, Zhang J, Li H (2023) Soil organic carbon changes in China's croplands: A newly estimation based on DNDC model. Sci Total Environ 905:167107

Du X, Zhang X, Mukundan R, Hoang L, Owens EM (2019) Integrating terrestrial and aquatic processes toward watershed scale modeling of dissolved organic carbon fluxes. Environ Pollut 249:125–135

Eckersten H, Bolinder M, Johnsson H, Kätterer T, Mårtensson K, Collentine D, Tidåker P, Marstorp H (2017) Nitrogen leaching and soil organic carbon sequestration of a barley crop with improved N use efficiency—a regional case study. Acta Agric Scand B Soil Plant Sci 67(7): 615–627

Falloon P, Smith P (2002) Simulating SOC changes in long-term experiments with RothC and CENTURY: model evaluation for a regional scale application. Soil Use Manag 18(2):101–111

Farina R, Seddaiu G, Orsini R, Steglich E, Roggero PP, Francaviglia R (2011) Soil carbon dynamics and crop productivity as influenced by climate change in a rainfed cereal system under contrasting tillage using EPIC. Soil Tillage Res 112(1):36–46

Feleke HG, Savage MJ, Tesfaye K (2021) Calibration and validation of APSIM–Maize, DSSAT CERES–Maize and AquaCrop models for Ethiopian tropical environments. S Afr J Plant Soil 38(1):36–51

Fu B, Merritt WS, Croke BF, Weber TR, Jakeman AJ (2019) A review of catchment-scale water quality and erosion models and a synthesis of future prospects. Environ Model Softw 114:75–97

Gassman PW, Williams JR, Benson VW, Izaurralde RC, Hauck LM, Jones CA et al (2004) Historical development and applications of the EPIC and APEX models. In: 2004 ASAE annual meeting. American Society of Agricultural and Biological Engineers, p 1

Gaydon DS, Balwinder-Singh, Wang E, Poulton PL, Ahmad B, Ahmed F, Akhter S, Ali I, Amarasingha R, Chaki AK et al (2017) Evaluation of the APSIM model in cropping systems of Asia. Field Crop Res 204:52–75

Gerik T, Williams JR, Dagitz S, Magre M, Meinardus A, Steglich E, Taylor R (2015) Environmental Policy Integrated Climate Model, User Manual Version 0810, September 1015

Geremew B, Tadesse T, Bedadi B, Gollany HT, Tesfaye K, Aschalew A, Abera W (2024) Evaluation of RothC model for predicting soil organic carbon stock in north-west Ethiopia. Environ Challenges 15:100909

Gijsman AJ, Hoogenboom G, Parton WJ, Kerridge PC (2002) Modifying DSSAT crop models for low-input agricultural systems using a soil organic matter–residue module from CENTURY. Agron J 94(3):462–474

Gilhespy SL, Anthony S, Cardenas L, Chadwick D, del Prado A, Li C, Yeluripati JB (2014) First 20 years of DNDC (DeNitrification DeComposition): model evolution. Ecol Model 292:51–62

Godde CM, Thorburn PJ, Biggs JS, Meier EA (2016) Understanding the impacts of soil, climate, and farming practices on soil organic carbon sequestration: a simulation study in Australia. Front Plant Sci 7:661

Gross A, Glaser B (2021) Meta-analysis on how manure application changes soil organic carbon storage. Sci Rep 11(1):5516

Gupta S, Kumar S (2017) Simulating climate change impact on soil carbon sequestration in agro-ecosystem of mid-Himalayan landscape using CENTURY model. Environ Earth Sci 76:1–15

Hashimoto S, Komatsu H (2006) Relationships between soil CO2 concentration and CO2 production, temperature, water content, and gas diffusivity: implications for field studies through sensitivity analyses. J For Res 11:41–50

Hashimoto S, Ugawa S, Morisada K, Wattenbach M, Smith P, Matsuura Y (2012) Potential carbon stock in Japanese forest soils–simulated impact of forest management and climate change using the CENTURY model. Soil Use Manag 28(1):45–53

He W, Grant BB, Jing Q, Lemke R, Luce MS, Jiang R, Smith WN (2021) Measuring and modeling soil carbon sequestration under diverse cropping systems in the semiarid prairies of western Canada. J Clean Prod 328:129614

Helfrich M, Ludwig B, Buurman P, Flessa H (2006) Effect of land use on the composition of soil organic matter in density and aggregate fractions as revealed by solid-state 13C NMR spectroscopy. Geoderma 136(1–2):331–341

Hoogenboom G, Porter CH, Shelia V, Boote KJ, Singh U, White JW, Hunt LA, Ogoshi R, Lizaso JI, Koo J, Asseng S, Singels A, Moreno LP, Jones JW (2019) Decision support system for agrotechnology transfer (DSSAT), vol 5. DSSAT Foundation, Gainesville. Version 4.7. https://DSSAT.net

Huang X, Tang B, Xia X, Zhang H (2012) Evaluation of soil carbon sequestration potential under conservation tillage in double rice cropping region with DNDC model. In: 2012 2nd international conference on remote sensing, environment and transportation engineering. IEEE, Piscataway, pp 1–4

Huang X, Jia Z, Jiao X, Wang J, Huang X (2022) Long-term manure applications to increase carbon sequestration and macroaggregate-stabilized carbon. Soil Biol Biochem 174:108827

Izaurralde RC, Williams JR, McGill WB, Rosenberg NJ, Jakas MQ (2006) Simulating soil C dynamics with EPIC: model description and testing against long-term data. Ecol Model 192(3–4):362–384

Jangir CK, Sangwan PS, Panghaal D, Kumar S, Meena RS, Bharti, Jat RD, Singh N (2024) Spatial variability and statistical analysis of soil properties in the rice wheat-based systems of northwest India. Commun Soil Sci Plant Anal 55(8):1205–1223

Jhariya MK, Raj A, Banerjee A, Meena RS, Bargali SS, Kumar S, Nema S, Poonam, Oraon PR (2022) Plan and policies for soil organic carbon management under agroforestry system. In: Plans and policies for soil organic carbon management in agriculture. Singapore, Springer, pp 191–219

Jiang G, Xu M, He X, Zhang W, Huang S, Yang X et al (2014) Soil organic carbon sequestration in upland soils of northern China under variable fertilizer management and climate change scenarios. Glob Biogeochem Cycles 28(3):319–333

Jordon MW, Smith P, Long PR, Bürkner PC, Petrokofsky G, Willis KJ (2022) Can regenerative agriculture increase national soil carbon stocks? Simulated country-scale adoption of reduced tillage, cover cropping, and ley-arable integration using RothC. Sci Total Environ 825:153955

Katterer T, Bolinder MA (2023) Agriculture practices to improve soil carbon storage in upland soil. In: Rumpel C (ed) Agriculture practices to improve soil carbon storage in upland soil. Burleigh Dodds Science Publishing, Cambridge

Katterer T, Bolinder MA, Andrén O, Kirchmann H, Menichetti L (2011) Roots contribute more to refractory soil organic matter than above-ground crop residues, as revealed by a long-term field experiment. Agric Ecosyst Environ 141:184–192

Kemanian AR, Julich S, Manoranjan VS, Arnold JR (2011) Integrating soil carbon cycling with that of nitrogen and phosphorus in the watershed model SWAT: theory and model testing. Ecol Model 222(12):1913–1921

Khalil MI, Formara DA, Osborne B (2020) Simulation and validation of long-term changes in soil organic carbon under permanent grassland using the DNDC model. Geoderma 361:114014

Kochsiek AE, Knops JMH, Walters DT, Arkebauer TJ (2009) Impacts of management on decomposition balance in irrigated and rainfed no till agricultural systems. Agric For Meteorol 149: 1983–1993

Kögel-Knabner I (2017) The macromolecular organic composition of plant and microbial residues as inputs to soil organic matter: Fourteen years on. Soil Biol Biochem 105:A3–A8

Kogel-Knabner I, Wiesmeier M, Mayer S (2023) Mechanisms of soil organic carbon sequestration and implications for management. In: Rumpel C (ed) Agriculture practices to improve soil carbon storage in upland soil. Burleigh Dodds Science Publishing, Cambridge

Kumar S, Meena RS, Lal R, Yadav GS, Mitran T, Meena BL, Dotaniya ML, EL-Sabagh A (2018) Role of legumes in soil carbon sequestration. In: Meena RS et al (eds) Legumes for soil health and sustainable management. Springer, Singapore. https://doi.org/10.1007/978-981-13-0253-4_4

Kumar S, Sharma SK, Thakral SK, Bhardwaj KK, Jhariya MK, Meena RS, Jangir CK, Bedwal S, Jat RD, Gaber A, Atta AA (2022a) Integrated nutrient management improves the productivity and nutrient use efficiency of *Lens culinaris* Medik. Sustainability 14(3):1284

Kumar S, Bamboriya SD, Rani K, Meena RS, Sheoran S, Loyal A, Kumawat A, Jhariya MK (2022b) Grain legumes: a diversified diet for sustainable livelihood, food, and nutritional security. In: Advances in legumes for sustainable intensification. Academic Press, London, pp 157–178

Kumar S, Sharma SK, Dhaka AK, Bedwal S, Sheoran S, Meena RS, Jangir CK, Kumar D, Kumar R, Jat RD, Meena AK (2023) Efficient nutrient management for enhancing crop productivity, quality and nutrient dynamics in lentil (*Lens culinaris* Medik.) in the semi-arid region of northern India. PLoS One 18(2):e0280636

Kumar S, Meena RS, Kumar S, Pradhan G, Jangir CK, Ghosh S, Punia H, Sheoran P, Meena R, Ahmad MA, Goyal SK (2025) Designing a diversified Indian mustard production system for energy-carbon-cum-heat use efficiency and sowing dates assessment. GCB Bioenergy 17(6): e70044

Kumari A, Kumar A, Chander G, Singh A, Falk T, Mequanint MB et al (2023) Assessing residue and tillage management options for carbon sequestration in future climate change scenarios. Curr Res Environ Sustainability 5:100210

Lacoste M, Viaud V, Michot D, Walter C (2015) Landscape-scale modelling of erosion processes and soil carbon dynamics under land-use and climate change in agroecosystems. Eur J Soil Sci 66(4):780–791

Lal R (2004) Soil C sequestration impacts on global climate change and food security. Science 304: 1623–1627

Lal R (2016) Soil C sequestration and aggregation by cover cropping. J Soil Water Conserv 70(6): 329–339. https://doi.org/10.2489/jswc.70.6.329

Lal R, Kimble JM (1997) Conservation tillage for C sequestration. Nutr Cycl Agroecosyst 49:243–253

Le KN, Jha MK, Reyes MR, Jeong J, Doro L, Gassman PW, Boulakia S (2018) Evaluating carbon sequestration for conservation agriculture and tillage systems in Cambodia using the EPIC model. Agric Ecosyst Environ 251:37–47

Leharwan M, Kumar Y, Kumar R, Kumar Saraswat P, Kumar R, Kumar Thaliyil Veetil A, Bhattacharjee S, Kumar A, Kumar S (2023) Assessing the effects of conservation tillage and in-situ crop residue management on crop yield and soil properties in rice–wheat cropping system. Sustainability 15(17):12736

Levy PE, Mobbs DC, Jones SK, Milne R, Campbell C, Sutton MA (2007) Simulation of fluxes of greenhouse gases from European grasslands using the DNDC model. Agric Ecosyst Environ 121(1–2):186–192

Li S, Li J, Li C, Huang S, Li X, Li S, Ma Y (2016) Testing the RothC and DNDC models against long-term dynamics of soil organic carbon stock observed at cropping field soils in North China. Soil Tillage Res 163:290–297

Li H, Wang L, Li J, Gao M, Zhang J, Zhang J, Frolking S (2017) The development of China-DNDC and review of its applications for sustaining Chinese agriculture. Ecol Model 348:1–13

Li J, Luo Z, Wang Y, Li H, Xing H, Wang L et al (2019) Optimizing nitrogen and residue management to reduce GHG emissions while maintaining crop yield: A case study in a mono-cropping system of northeast China. Sustainability 11(18):5015

Liang K, Qi J, Zhang X, Deng J (2022) Replicating measured site-scale soil organic carbon dynamics in the US Corn Belt using the SWAT-C model. Environ Model Softw 158:105553

Liang F, Li S, Jie F, Ge Y, Liu N, Jia G (2023) The development of a coupled soil water assessment tool-MODFLOW model for studying the impact of irrigation on a regional water cycle. Water 15(20):3542

Liu HL, Liu HB, Lei QL, Zhai LM, Wang HY, Zhang JZ et al (2017) Using the DSSAT model to simulate wheat yield and soil organic carbon under a wheat-maize cropping system in the North China Plain. J Integr Agric 16(10):2300–2307

Liu X, Chen Y, Liu Y, Wang S, Jin J, Zhao Y, Yu D (2023) A framework combining CENTURY modeling and chronosequences sampling to estimate soil organic carbon stock in an agricultural region with large land use change. Agronomy 13(4):1055

Longo M, Dal Ferro N, Izaurralde RC, Furlan L, Chiarini F, Morari F (2023) Deep SOC stock dynamics under contrasting management systems: is the EPIC model ready for carbon farming implementation? Eur J Agron 145:126771

Louhar G, Bana RS, Kumar V, Kumar H (2020) Nutrient management technologies of millets for higher productivity and nutritional security. Indian J Agric Sci 90(12):2243–2250

Luo Z, Wang E, Sun OJ, Smith CJ, Probert ME (2011) Modeling long-term soil carbon dynamics and sequestration potential in semi-arid agro-ecosystems. Agric For Meteorol 151(12): 1529–1544

Luo Z, Wang E, Bryan BA, King D, Zhao G, Pan X, Bende-Michl U (2013) Meta–modeling soil organic carbon sequestration potential and its application at regional scale. Ecol Appl 23(2): 408–420

Luo Z, Wang E, Fillery IRP, Macdonald LM, Huth N, Baldock J (2014) Modelling soil carbon and nitrogen dynamics using measurable and conceptual soil organic matter pools in APSIM. Agric Ecosyst Environ 186:94–104

Manna MC, Bhattacharyya P, Adhya TK, Singh M, Wanjari RH, Ramana S et al (2013) Carbon fractions and productivity under changed climate scenario in soybean–wheat system. Field Crop Res 145:10–20

Meena RS, Pradhan G (2023) Industrial garbage-derived biocompost enhances soil organic carbon fractions, CO_2 biosequestration, potential carbon credits, and sustainability index in a rice-wheat ecosystem. Environ Res:116525. https://doi.org/10.1016/j.envres.2023.116525

Meena RS, Kumar S, Yadav GS (2020) Soil carbon sequestration in crop production. In: Nutrient dynamics for sustainable crop production. Springer, Singapore, pp 1–39

Meena RS, Pradhan G, Kumar S, Lal R (2023) Using industrial wastes for rice-wheat cropping and food-energy-carbon-water-economic nexus to the sustainable food system. Renew Sust Energ Rev 187:113756. https://doi.org/10.1016/j.rser.2023.113756

Meena RS, Pradhan G, Singh K, Kumar S, Singh AK, Shashidhar KS, Mina KK, Rao CS (2024a) Agriculture models for restoring degraded land to enhance CO_2 biosequestration and carbon credits in the Vindhyan region of India. Sci Total Environ 929:172661. https://doi.org/10.1016/j.scitotenv.2024.172661

Meena RS, Lal R, Kumar S, Pradhana G, Rao CS, Singh AK, Pathak H, Abhilash PC, Kumar A, Sharma SK, Jat ML, Singh S (2024b) Potential of Indian agriculture for capturing atmospheric CO_2 and monetizing carbon credits to the farmers: an overview and policy framework. Adv Agron 188:101–206. https://doi.org/10.1016/bs.agron.2024.06.001

Meena RS, Pradhan G, Nalani SK, Singh AK, Verma SK, Mina KK, Kumar S, Chaturvedi RK (2024c) Diversified cropping modules designed for soil restoration, CO_2 sequestration, and generating carbon credits. Land Degrad Dev 35. https://doi.org/10.1002/ldr.5224

Meena RS, Pradhan G, Kumar S (2025) Energy flow, eco-efficiency, and economic circulation with recycled industrial waste compost application in wheat and subsequent rice farming. Sci Total Environ 967:178779. https://doi.org/10.1016/j.scitotenv.2025.178779

Meurer KHE, Haddaway NR, Bolinder MA, Kätterer T (2018) Tillage intensity affects total SOC stocks in boreo-temperate regions only in the topsoil—a systematic review using an ESM approach. Earth Sci Rev 177:613–622

Minx JC, Lamb WF, Callaghan MW, Fuss S, Hilaire J, Creutzig F, Amann T, Beringer T, de Oliveira Garcia W, Hartmann J, Khanna T, Lenzi D, Luderer G, Nemet GF, Rogelj J, Smith P, Vicente Vicente JL, Wilcox J, del Mar Zamora Dominguez, M. (2018) Negative emissions—part 1: research landscape and synthesis. Environ Res Lett 13(6):063001

Mohanty M, Probert ME, Reddy KS, Dalal RC, Mishra AK, Rao AS, Menzies NW (2012) Simulating soybean–wheat cropping system: APSIM model parameterization and validation. Agric Ecosyst Environ 152:68–78

Mohanty M, Sinha NK, Somasundaram J, McDermid SS, Patra AK, Singh M, Dwivedi AK, Reddy KS, Rao CS, Prabhakar M, Hati KM (2020) Soil carbon sequestration potential in a vertisol in Central India – results from a 43-year long-term experiment and APSIM modeling. Agric Syst 184:102906

Musinguzi P, Ebanyat P, Tenywa JS, Mwanjalolo M, Basamba TA, Tenywa MM, Porter C (2014) Using DSSAT-CENTURY model to simulate soil organic carbon dynamics under a low-input maize cropping system. https://doi.org/10.5539/jas.v6n5p120

Nair PR, Nair VD, Kumar BM, Showalter JM (2010) Carbon sequestration in agroforestry systems. Adv Agron 108:237–307

Nair R, Mehta CR, Sharma S (2015) Carbon sequestration in soils – a review. Agric Rev 36(2): 81–99

Nicoloso RS, Amado TJ, Rice CW (2020) Assessing strategies to enhance soil carbon sequestration with the DSSAT-CENTURY model. Eur J Soil Sci 71(6):1034–1049

Noorzai AU, Choudhary AK, Bana RS, Prasad R (2017) Growth behaviour, productivity and profitability of promising mungbean varieties in semi-arid region of Afghanistan. Ann Agric Res 38(1):78–86

Ntamwira J, Ocimati W, Blomme G, Lubobo AK, Mwarabu Lolonga Pyame D, Dhed'a Djailo B (2023) Innovative agroecological practices can restore degraded farmlands and revive crop yields. Front Sustain Food Syst 7:1017341

Ogle SM, Alsaker C, Baldock J, Bernoux M, Breidt FJ, McConkey B, Regina K, Vazquez-Amabile GG (2019) Climate and soil characteristics determine where no-till management can store carbon in soils and mitigate greenhouse gas emissions. Sci Rep 9(1):11665

Parton WJ (1996) The CENTURY model. In: Powlson DS, Smith P, Smith JU (eds) Evaluation of soil organic matter models. NATO ASI series, vol 38. Springer, Berlin/Heidelberg. https://doi.org/10.1007/978-3-642-61094-3_23

Parton WJ, Ojima DS, Cole CV, Schimel DS (1994) A general model for soil organic matter dynamics: sensitivity to litter chemistry, texture and management. Quant Model Soil For Processes 39:147–167

Patial M, Dharam Pal, Thakur A, Bana RS, Patial S (2017) Double haploidy technique in wheat (*Triticum aestivum* L.): an overview. Proc Natl Acad Sci India Sect B Biol Sci 89. https://doi.org/10.1007/s40011-017-0870-z

Paustian K, Andrén O, Janzen HH, Lal R, Smith P, Tian G, Tiessen H, Noordwijk M, Woomer PL (1997) Agricultural soils as a sink to mitigate CO2 emissions. Soil Use Manag 13(s4):230–244

Paustian K, Collier S, Baldock J, Burgess R, Creque J, DeLonge M et al (2019) Quantifying carbon for agricultural soil management: from the current status toward a global soil information system. Carbon Manag 10(6):567–587

Pearson TRH, Brown SL, Birdsey RA (2007) Measurement guidelines for the sequestration of forest carbon. In: Gen. Tech. Rep. NRS-18. U. S. Department of Agriculture, Forest Service, Northern Research Station, Newtown Square, p 42

Pradhan G, Meena RS (2023a) Utilizing waste compost to improve the atmospheric CO_2 capturing in the rice-wheat cropping system and energy-cum-carbon credit audit for a circular economy. Sci Total Environ 164572. https://doi.org/10.1016/j.scitotenv.2023.164572

Pradhan G, Meena RS (2023b) Interaction impacts of biocompost on nutrient dynamics and relations with soil biota, carbon fractions index, societal value of CO_2 equivalent, and ecosystem services in the wheat-rice farming. Chemosphere:139695. https://doi.org/10.1016/j.chemosphere.2023.139695

Pradhan A, Wakchaure GC, Shid D, Minhas PS, Biswas AK, Reddy KS (2023a) Impact of residue retention and nutrient management on carbon sequestration, soil biological properties, and yield in multi-ratoon sugarcane. Front Sustain Food Syst 7:1288569

Pradhan G, Meena RS, Kumar S, Lal R (2023b) Utilizing industrial wastes as compost in wheat-rice production to improve the above and below-ground ecosystem services. Agric Ecosyst Environ 358:108704. https://doi.org/10.1016/j.agee.2023.108704

Primo AA, de Araújo Neto RA, Zeferino LB, Fernandes FÉP, de Araújo Filho JA, Cerri CEP, de Oliveira TS (2023) Slash and burn management and permanent or rotation agroforestry systems: a comparative study for C sequestration by century model simulation. J Environ Manag 336: 117594

Prudil J, Pospíšilová L, Dryšlová T, Barančíková G, Smutný V, Sedlák L, Rábek M (2023) Assessment of carbon sequestration as affected by different management practices using the RothC model. Plant Soil Environ 69(11):532

Punia H, Tokas J, Malik A, Satpal, Rani A, Gupta P, Kumari A, Mor VS, Bhuker A, Kumar S (2020) Solar radiation and nitrogen use efficiency for sustainable agriculture. In: Resources use efficiency in agriculture. Springer, Singapore, pp 177–212

Qi J, Du X, Zhang X, Lee S, Wu Y, Deng J, McCarty GW (2020a) Modeling riverine dissolved and particulate organic carbon fluxes from two small watersheds in the northeastern United States. Environ Model Softw 124:104601

Qi J, Zhang X, Lee S, Wu Y, Moglen GE, McCarty GW (2020b) Modeling sediment diagenesis processes on riverbed to better quantify aquatic carbon fluxes and stocks in a small watershed of the Mid-Atlantic region. Carbon Balance Manag 15:1–14

Ranatunga K, Hill MJ, Probert ME, Dalal RC (2001) Comparative application of APSIM, RothC and century to predict soil carbon dynamics. In Proceedings of the international congress on modelling and simulation (MODSIM'01), pp 733–738

Rastogi M, Singh S, Pathak H (2013) Emission of carbon dioxide from soil. Curr Sci 105(12): 1622–16292

Risal A, Parajuli PB (2019) Quantification and simulation of nutrient sources at watershed scale in Mississippi. Sci Total Environ 670:633–643

Robertson F, Armstrong R, Partington D, Perris R, Oliver I, Aumann C (2015) Effect of cropping practices on soil organic carbon: evidence from long-term field experiments, vol 53. CSIRO Publishing, Victoria, pp 636–646

Saddique Q, Ji J, Ajaz A, Jiatun X, Yufeng Z, He J, Cai H (2019) Performance comparison of the APSIM and CERES-wheat models in Guanzhong plain, China. In: 2019 ASABE annual international meeting. American Society of Agricultural and Biological Engineers, St. Joseph, p 1

Sah D, Devakumar AS (2018) The carbon footprint of agricultural crop cultivation in India. Carbon Manag 9(3):213–225

Sanderman J, Creamer C, Baisden WT, Farrell M, Fallon S (2017) Greater soil carbon stocks and faster turnover rates with increasing agricultural productivity. Soil 3(1):1–16

Sarkar S, Gaydon DS, Dey S, Chaki AK, Brahmachari K, Dhar A et al (2024) Evaluation of the APSIM model in rice-lentil cropping system in a complex coastal saline environment. J Indian Soc Coast Agric Res 42(1):53–72

Shaukat M, Hoshide AK, Muhammad S, Arshad IA, Mushtaq M, de Abreu DC (2023) Predicting soil carbon sequestration and harvestable C-biomass of rice and wheat by DNDC model. Crops 3(3):220–240

Sheoran S, Gupta M, Kumari S, Kumar S, Rakshit S (2022) Meta-QTL analysis and candidate genes identification for various abiotic stresses in maize (*Zea mays* L.) and their implications in breeding programs. Mol Breed 42(5):26

Shrivastava M (2006) Simulating carbon sequestration at micro-watershed scale with changes in cropping pattern and management systems. Doctoral dissertation. TERI School of Advanced Studies, New Delhi

Sleutel S, De Neve S, Beheydt D, Li C, Hofman G (2006) Regional simulation of long-term organic carbon stock changes in cropland soils using the DNDC model: 1. Large-scale model validation against a spatially explicit data set. Soil Use Manag 22(4):342–351

Smith JO, Smith P, Wattenbach M, Zaehle S, Hiederer R, Jones RJ, Ewert F (2005) Projected changes in mineral soil carbon of European croplands and grasslands, 1990–2080. Glob Chang Biol 11(12):2141–2152

Soler CT, Bado VB, Traore K, Bostick WM, Jones JW, Hoogenboom G (2011) Soil organic carbon dynamics and crop yield for different crop rotations in a degraded ferruginous tropical soil in a semi-arid region: a simulation approach. J Agric Sci 149(5):579–593

Swift RS (2001) Sequestration of C by soil. Soil Sci 166(11):858–871

Tang X, Ellert BH, Hao X, Ma Y, Nakonechny E, Li J (2012) Temporal changes in soil organic carbon contents and δ13C values under long-term maize–wheat rotation systems with various soil and climate conditions. Geoderma 183:67–73

Trost B, Prochnow A, Drastig K, Meyer-Aurich A, Ellmer F, Baumecker M (2013) Irrigation, soil organic C and N2O emissions. A review. Agron Sustain Dev 33:733–749. https://doi.org/10.1007/s13593-013-0134-0

Wan Y, Lin E, Xiong W, Guo L (2011) Modeling the impact of climate change on soil organic carbon stock in upland soils in the 21st century in China. Agric Ecosyst Environ 141(1–2):23–31

Wang E, Luo Z, Smith CJ (2010) Potential change of soil carbon in Australian agro-ecosystems as affected by conservation management: data synthesis and modelling. In: Proceedings of the 19th world congress of soil science, pp 57–60

Wang X, He C, Liu B, Zhao X, Liu Y, Wang Q, Zhang H (2020) Effects of residue returning on soil organic carbon storage and sequestration rate in China's croplands: a meta-analysis. Agronomy 10(5):691

Wang Q, Qi J, Qiu H, Li J, Cole J, Waldhoff S, Zhang X (2021) Pronounced increases in future soil erosion and sediment deposition as influenced by Freeze–Thaw Cycles in the Upper Mississippi River Basin. Environ Sci Technol 55(14):9905–9915

Williams JR, Jones CA, Dyke PT (1984) The EPIC model and its application. In: Proceedings of the ICRISAT-IBSNATSYSS symposium on minimum data sets for agrotechnology transfer, pp 111–121

Yao C, Qian S, Mao Y, Li Z (2017) Decomposition of impacting factors of animal husbandry carbon emissions change and its spatial differences in China. Trans Chinese Soc Agric Eng 33(12):10–19

Yokozawa M, Shirato Y, Sakamoto T, Yonemura S, Nakai M, Ohkura T (2010) Use of the RothC model to estimate the carbon sequestration potential of organic matter application in Japanese arable soils. Soil Sci Plant Nutr 56(1):168–176

Zhang X (2018) Simulating eroded soil organic carbon with the SWAT-C model. Environ Model Softw 102:39–48

Zhang X, Izaurralde RC, Arnold JG, Williams JR, Srinivasan R (2013) Modifying the soil and water assessment tool to simulate cropland carbon flux: model development and initial evaluation. Sci Total Environ 463:810–822

Zhao G, Bryan BA, King D, Luo Z, Wang E, Song X, Yu Q (2013) Impact of agricultural management practices on soil organic carbon: simulation of Australian wheat systems. Glob Chang Biol 19(5):1585–1597

Chapter 15
Integrating Government Schemes for Carbon Incentives and Farmers' Benefits

Gongalla Sreeja Reddy, Suddamalla Manoj Kumar Reddy, Satya Priya, K. Srikanth Reddy, and K. Sreenivasa Reddy

Abstract Climate change necessitates all nations to make efforts to reduce greenhouse gas (GHG) emissions to limit the rise in global average temperatures to 1.5 °C above pre-industrial levels. Governments globally adopted various policy measures such as carbon tax, emission trading mechanisms, voluntary carbon markets, and compliance mechanisms to regulate emissions in their respective countries. Carbon markets that allow the trading of carbon credits became prominent as an opportunity for all the stakeholders to sequester and reduce GHG altogether, earning additional income. While carbon markets are open to all sectors, the participation of the agricultural sector is still lacking, given the magnitude of emissions arising from the sector. Farmers can also participate in such markets by adopting various sustainable agricultural practices that reduce emissions. Currently, farmers in India organized through private entities participate in voluntary carbon markets. To make the markets more inclusive to all, the Indian government has notified the Carbon Credit Trading Scheme to establish a national marketplace. While the implementation takes time, there is a need to understand the global experience in such markets and mold the policy to be more efficient. This chapter addresses this by explaining various carbon pricing mechanisms followed globally, agricultural practices that reduce GHGs, policies of carbon markets in India, learnings from global policies, and prospects of integrating government schemes with carbon incentives.

Keywords Carbon credits · Sustainable agricultural practices · Carbon tax · Climate change · Carbon Credit Trading Scheme

G. S. Reddy (✉) · S. M. K. Reddy · S. Priya
Division of Agricultural Extension, ICAR-Indian Agricultural Research Institute, New Delhi, India

K. S. Reddy · K. S. Reddy
Division of Agronomy, ICAR-Indian Agricultural Research Institute, New Delhi, India

R. S. Meena et al. (eds.), *Carbon-Negative Agriculture*, Sustainability Sciences in Asia and Africa, https://doi.org/10.1007/978-981-95-1483-0_15

Abbreviations

BEE	Bureau of Energy Efficiency
CA	Conservation agriculture
CCTS	Carbon Credit Trading Scheme
CO_2-e	Carbon dioxide equivalent
COP	Conference of Parties
ETS	Emission trading system
FPO	Farmer Producer Organization
GCF	Green Climate Fund
GHG	Greenhouse gases
Gt	Gigaton
NAPCC	National Action Plan on Climate Change
NDC	Nationally determined contributions
NGO	Non-governmental organization
NMSA	National Mission for Sustainable Agriculture
PKVY	Paramaparagat Krishi Vikas Yojana
SAP	Sustainable agricultural practices
SHG	Self-help groups
SOC	Soil organic carbon
USDA	United States Department of Agriculture
VCM	Voluntary carbon markets
VCS	Verified carbon standard

15.1 Introduction

Sustainable agriculture is an integrated farming system that balances social equity, economic viability, environmental conservation, and agricultural productivity. It emphasizes practices that sustain soil fertility, protect natural resources, and reduce environmental harm, ensuring long-term food security (Jhariya et al. 2022; Jangir et al. 2024). While sustainability is necessary to mitigate the adverse effects of climate change, prevailing agriculture systems are far from being sustainable. Indian agriculture is responsible for around 12.2% of global agricultural greenhouse gas (GHG) emissions, contributing around 658 teragrams (Tg) of carbon dioxide equivalent (CO_2e), ranking as the world's second-largest source (Maraseni and Qu 2016). Despite housing over 17% of the global population, India accounts for just 6.7% of worldwide emissions, with a per capita footprint of 1.7 tCO_2 annually, considerably lower than many other countries. Nearly 90% of India's total GHG emissions come from the energy and agriculture sectors, while industry and waste management make up the remaining 10% (Fig. 15.1) (Patra 2014). Within this, the energy sector is dominant, generating 75% of the emissions, largely from electricity production (68.3%) and transportation (17.8%) (Mohan et al. 2019). The industrial sector,

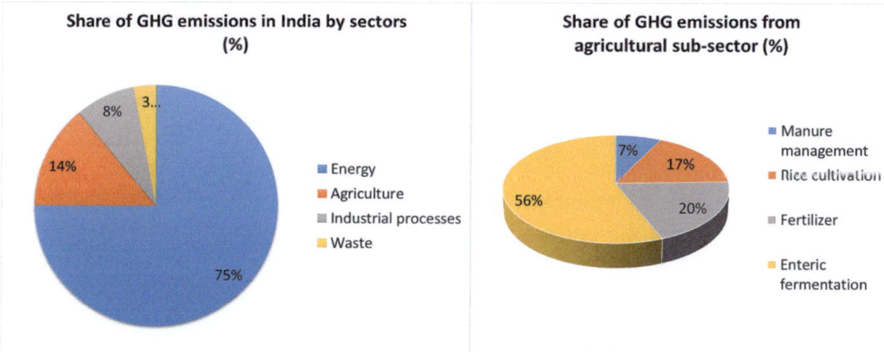

Fig. 15.1 Share of GHG emissions in India by sectors (%). (*Source*: MoEFCC 2021). Share of GHG emissions from the agricultural sub-sector (%). (*Source*: MoEFCC 2021)

particularly energy used for manufacturing chemicals, metals, and materials, contributes about 8% (MoEFCC 2021), while the waste sector, though only 3%, still requires careful management. In 2016, agriculture contributed 14% to India's overall GHG emissions—a 2.25% decline compared to 2014 (MoEFCC 2021). Within agriculture (Fig. 15.1), the main emission sources are enteric fermentation in livestock (54.6%), fertilizer application (19.1%), rice cultivation (17.5%), manure management (6.7%), and field burning of agricultural residues (FBAR) (2.2%) (MoEFCC 2021).

Carbon credits are generated through the adoption of eco-friendly technologies and sustainable business practices, which help reduce GHG emissions while fostering a profitable and dynamic industry. To curb the rise in atmospheric carbon dioxide (CO_2) concentrations, the United Nations (UN) established a carbon credit trading system under the framework of the Intergovernmental Panel on Climate Change (IPCC). This system, often termed a "market-driven approach" (Choudhary et al. 2017), allows companies to offset their emissions by acquiring carbon credits through investments in projects such as reforestation, conservation agriculture (CA), and renewable energy initiatives. Each carbon credit represents a verifiable reduction in CO_2 emissions and plays a role in advancing global sustainability objectives (Meena et al. 2023, 2025). Functioning as a type of "climate currency," carbon credits are governed by market forces of supply and demand. Since 1995, the United Nations has convened annual climate change negotiations known as the Conference of the Parties (COP). To meet the 1.5 °C target, global GHG emissions must be reduced by 50% by 2030 and achieve net-zero levels by 2050 (IPCC 2022). The Paris Agreement, finalized at COP21 on December 12, 2015, seeks to cap global warming below 2 °C, ideally aiming for 1.5 °C, as a critical step in climate change mitigation. This historic agreement, signed by 197 parties, requires regular five-year reviews to track and strengthen progress. As the detrimental effects of climate change on agriculture become more widely recognized, sustainability has emerged as a global imperative. At COP27, held in Egypt in 2022, developing nations—

including India—strongly advocated for the New Collective Quantified Goal on Climate Finance to address the pressing needs of global climate financing. The Global Soil Partnership collaborates with regional public–private initiatives to integrate carbon credit systems into long-term sustainability efforts. In India, both state and central governments are aligning policies on natural, regenerative, agro-forestry, and organic farming to encourage farmers' participation in carbon credit programs. Awareness campaigns are designed to educate farmers on the advantages of carbon farming and the potential financial rewards it offers. As of 2023, carbon credits are valued between USD 40 and 80 per ton, with demand anticipated to increase 15-fold by 2030 (Boyce and He 2023). This surge in demand is expected to drive up market prices considerably, creating lucrative opportunities for farmers and related stakeholders to gain income through carbon credit trading. Farmers in India can leverage such prospects in the carbon markets to move towards environmental and economic sustainability. Carbon markets offer farmers an opportunity to earn additional income by practicing various agricultural practices that reduce carbon emissions. However, given the socioeconomic landscape of farmers in India, wherein more than 80% of the farmers are small and marginal farmers, policies dealing with carbon farming should address the needs and hindrances these farmers may face in utilizing the markets. With that view, this chapter aims to address the factors that should be given adequate importance to make carbon policies more inclusive to farmers.

15.2 Status of Carbon Incentives at Global and Indian Levels

15.2.1 Global Status of Carbon Incentives

Carbon incentives have emerged as a key instrument in the global fight against climate change, aiming to cut GHG emissions across various sectors. These efforts operate through approaches like carbon pricing, rewards for carbon sequestration, and international collaborations, bolstered significantly by corporate initiatives and non-governmental organizations (NGOs) led programs. This integrated strategy underscores the crucial roles played by both public and private actors in driving the transition toward a low-carbon economy.

15.2.1.1 Carbon Pricing Mechanisms

Carbon pricing mechanisms have become a cornerstone strategy in tackling climate change, motivating industries to shift toward more sustainable practices. As seen in Fig. 15.2, globally, various nations have already put in place certain carbon credit mechanisms, while a few others are contemplating implementing them shortly.

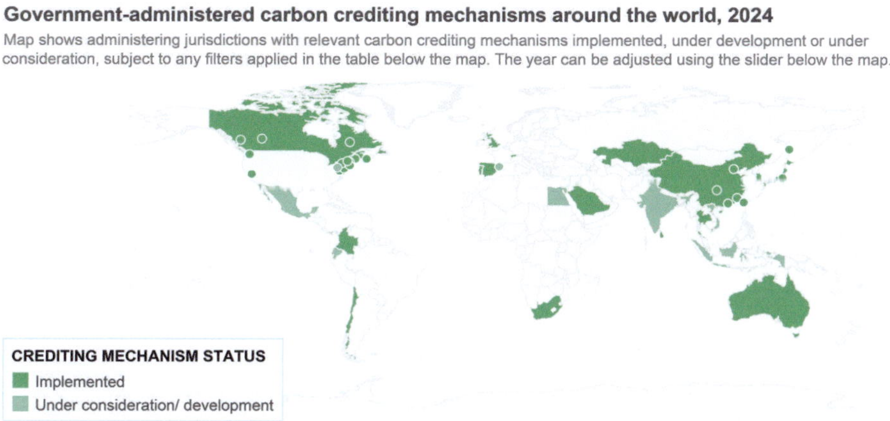

Fig. 15.2 Map showing the status of carbon crediting mechanisms in various countries. (*Source*: World Bank 2025)

Table 15.1 Minimum national carbon pollution price schedule (2023–2030)

Year	2023	2024	2025	2026	2027	2028	2029	2030
Minimum carbon pollution price ($ CAD/ton CO_2e)	$65	$80	$95	$110	$125	$140	$155	$170

Government of Canada (2023)

Carbon taxes, for instance, are increasingly adopted by countries seeking to reduce their carbon emissions. Sweden, a pioneer in this field, introduced a carbon tax back in 1991 at an initial rate of €26 per ton of CO_2. Over time, this rate has steadily climbed, reaching €125 per ton by 2024 (Government Offices of Sweden 2024), achieving notable emission reductions while sustaining economic growth. Similarly, Canada has put in place a strong carbon pricing system, including a federal carbon tax slated to increase gradually to CAD 170 per ton by 2030. As of 2023, Canada's national minimum carbon price under explicit pricing systems stood at CAD 65 per ton of CO_2e, with planned annual increases of CAD 15 until it reaches CAD 170 by 2030 (Government of Canada 2023; see Table 15.1). Alongside carbon taxes, emissions trading systems (ETS) play a crucial role in carbon pricing. The European Union (EU) ETS is a leading example, covering approximately 40% of the EU's GHG emissions. Despite the global energy crisis in 2022, carbon pricing instruments remained resilient: prices rose for half of these mechanisms and stayed stable for about a third, with only 15% experiencing declines. Notably, the EU ETS, linked with the Swiss ETS, saw a sharp price surge, crossing €100 (USD 109) per ton for the first time in March 2023 (World Bank 2023). ETSs allow industries to trade carbon allowances, spurring innovation in low-carbon technologies and encouraging sustainable agricultural methods. In addition, voluntary carbon markets (VCMs) complement these formal systems by enabling private sector actors to purchase carbon credits from projects in developing countries. Standards such as the verified carbon standard (VCS) and the Gold Standard have effectively backed agricultural carbon initiatives like soil carbon sequestration and agroforestry, generating both environmental gains and socioeconomic benefits for rural communities.

15.2.1.2 Carbon Sequestration Incentives

Incentives for carbon sequestration are vital for promoting sustainable agricultural practices (SAPs) that enhance soil health and draw down atmospheric carbon. The 4 per 1000 Initiative, launched at COP21 in Paris, has gained international acclaim for advocating soil-focused strategies that boost soil organic carbon (SOC) stocks. By promoting methods such as reduced tillage, cover cropping, and organic farming, the initiative aims to build long-term soil fertility and increase resilience to climate change, with a vision of achieving carbon-rich, healthy soils globally by 2050 (Soussana et al. 2019). The initiative emphasizes natural organic carbon sequestration, supports scientific advancement in soil carbon research, and encourages land managers to adopt diversified, productive, and resource-efficient agricultural and forestry systems (Kumar et al. 2025). These collective efforts empower farmers and foresters, strengthen the agricultural sector, and contribute to global economic development. Participation involves implementing sustainable practices that align with the initiative's objectives, thereby enhancing carbon storage while promoting diversified and resilient land use (Li et al. 2023; Meena et al. 2024c).

In the United States, the Department of Agriculture (USDA) has launched extensive programs under its climate-smart agriculture initiative, providing substantial funding for practices that boost soil carbon, such as rotational grazing, agroforestry, and biochar application (Meena et al. 2023, 2024a). At the heart of this effort is the USDA's Partnerships for Climate-Smart Commodities program, which has awarded over US$3.1 billion in grants to support 141 projects that integrate carbon sequestration with economic incentives for farmers. The initiative seeks to expand markets for climate-smart commodities, harness the GHG reduction potential of sustainable farming, and deliver meaningful support to a diverse array of farmers, ranchers, and private forest owners—with particular attention to small and historically underserved producers. As of January 2025, more than 21,000 farms are enrolled, gaining access to climate-smart markets, premium opportunities, and incentives aimed at strengthening both environmental resilience and economic sustainability. Over 5.2 million acres of working land are now part of the program, benefiting from financial and technical assistance to adopt GHG-reducing and carbon-sequestering practices. These efforts have already sequestered more than 550,000 metric tons of carbon, with projections indicating a total of over 60 million metric tons captured by the program's end in roughly 5 years, underscoring its transformative potential in advancing climate action and sustainable agriculture (USDA 2025).

15.2.1.3 International Agreements

A landmark international agreement, the Paris Agreement acknowledges the vital role that agriculture plays in lowering GHG emissions and improving carbon sequestration. Countries committed themselves to submitting nationally determined contributions (NDCs) outlining their plans for reducing GHG emissions under this agreement. Many of these NDC_S emphasize agriculture as a key focus area, prioritizing improvements in soil health, agroforestry, and methane reduction in livestock systems (IPCC 2022). International financial mechanisms like the Green Climate Fund (GCF) play a vital role in supporting the implementation of these commitments. The goal of the GCF, the largest climate fund in the world, is to encourage developing countries to move towards low emission, climate-resilient development. Serving the 2015 Paris Agreement and functioning under the auspices of the United Nations Framework Convention on Climate Change (UNFCCC), the GCF seeks to keep the increase in global temperatures well below 2 °C. With 286 initiatives totaling US$ 16 billion (US$ 61.5 billion including co-finance) in its portfolio, it is promoting revolutionary climate action in 133 developing nations. Through its Readiness Support Programme, the GCF has also allotted more than US$ 630 million to strengthen capacity-building initiatives and help nations develop long-term plans to tackle climate change (UNFCCC 2023).

15.2.1.4 Corporate and NGO Initiatives

Corporate entities and NGOs are playing a pivotal role in advancing the global drive toward carbon neutrality. Industry leaders like Microsoft and Google have set ambitious targets to achieve carbon neutrality or even become carbon-negative, supported by substantial investments in carbon offset initiatives. For example, to integrate environmental care with business advancement, Microsoft's US$1 billion Climate Innovation Fund supports initiatives in soil carbon capture, reforestation, and sustainable agriculture (Jat et al. 2022). In a similar vein, international groups and NGOs are making a significant contribution by integrating farmers into global carbon markets and enhancing capacity. Through the provision of technical training and the facilitation of access to carbon markets, initiatives like the World Resources Institute's Land Accelerator have enabled farmers and entrepreneurs throughout Africa, Asia, and Latin America (Jian et al. 2020). In the meantime, groups like Conservation International and the United Nations Food and Agriculture Organization (FAO) concentrate on teaching farmers about climate-smart agriculture and putting them in touch with financing sources for carbon sequestration projects (Lal 2004).

The alignment of policy frameworks, financial support, and grassroots initiatives underscores the global commitment to utilizing carbon incentives as a strategy to combat climate change. These collective efforts highlight the importance of continued collaboration among governments, corporations, NGOs, and local communities to address climate challenges and achieve sustainable development.

15.2.2 Status of Carbon Incentives in India

In order to encourage farmers to participate in carbon credit programs, the Indian government, both at the state and central levels, has taken steps to integrate current policies and schemes in natural, regenerative, agroforestry, and organic farming. In order to ensure greater involvement and access to these initiatives, efforts are being made to educate farmers about the advantages of carbon farming (Altieri et al. 2012). Demand for carbon credits is expected to increase by about 15 times by 2030, with prices ranging from US$ 40 to US$ 80 per ton in 2023 (Boyce and He 2023). Given the increasing demand for carbon credits and the anticipated rise in market prices, farmers can gain greatly from the carbon credit markets. The Indian government has started programs at the farm and community levels to improve carbon stocks and lower carbon emissions from a variety of cropping systems in order to promote climate-resilient agricultural practices (MoEFCC 2022). These practices contribute to ecological stability and global sustainability by improving soil, food, water, and climate security, among other ecosystem advantages (Aryal et al. 2015; Kumar et al. 2022a, b; Meena et al. 2024b). Food grain production could surpass 400 million tons by 2050, and carbon sequestration could rise by 70–130 Tg CO_2 equivalent (CO_2e) annually if best management practices are encouraged among India's small and marginal farmers (Nath et al. 2021). Despite many advancements, the Indian government has voiced concerns about international programs like the "4 per 1000" program because of the difficulties caused by the country's high temperatures and protracted fallow seasons. However, a way forward might be provided by targeted strategies to increase the stability of SOC (Meena et al. 2023).

India needs to concentrate on expanding carbon capture and storage research, development, and extension initiatives in order to successfully tackle climate change. Enhancing agronomic productivity and increasing landowners' carbon credits both depend on improved agricultural practices (MoEFCC 2022). The creation of carbon markets, soil health-focused programs, active private sector participation in voluntary markets, and governmental policies such as the governmental Action Plan on Climate Change (NAPCC) and the National Mission for Sustainable Agriculture (NMSA) are important efforts and tactics. Together, these actions seek to promote a resilient and ecologically viable agricultural ecosystem while reducing the effects of climate change.

15.2.2.1 National Policies and Programs

The NAPCC is the cornerstone of India's climate action framework and the basis for its strategic initiatives. Promoting climate-resilient farming practices is one of the

NMSA's eight primary missions. This mission prioritizes practices such as organic farming, agroforestry, and efficient water management, aiming to reduce GHG emissions while enhancing carbon sequestration. Complementing the NAPCC, State Action Plans on Climate Change enable states to design region-specific climate strategies. In order to promote sustainable agricultural practices, these plans emphasize activities such as implementing climate-resilient crop varieties, encouraging carbon-sequestering methods like applying biochar, and running awareness programs (Nath et al. 2021; Meena and Pradhan 2023).

15.2.2.2 Carbon Markets in India

The Indian government also proactively implemented certain carbon credit mechanisms with the potential to transform the agricultural sector. The Perform, Achieve, and Trade (PAT) scheme, primarily designed to enhance energy efficiency in industries, offers a framework that could be adapted to agriculture. It may incentivize sustainable practices such as energy-efficient irrigation and solar-powered storage solutions. Furthermore, India is in the process of developing a robust domestic carbon market in collaboration with the Bureau of Energy Efficiency (BEE) and the Ministry of Environment, Forest, and Climate Change (MoEFCC). This initiative seeks to integrate practices like no-till farming, cover cropping, and biochar application into the carbon market, enabling farmers to earn carbon credits while promoting sustainability (MoEFCC 2022).

15.2.2.3 Private Sector and Voluntary Markets

In the voluntary carbon markets, the participation of private entities is essential. For instance, the goal of Mahindra & Mahindra's 2007-launched Project Hariyali is to produce carbon offsets through extensive afforestation. The project's objectives are to empower underprivileged farmers, rectify ecological imbalances, and expand green cover. Mahindra & Mahindra contributed 1.59 million of the 2.21 million trees that the Mahindra Group planted during the fiscal year 2022–2023. Three thousand two hundred seventy-five tribal farmer families were supported financially by the substantial amount (1.31 million) that was planted in the Araku Valley. As a result of these efforts, the carbon footprint has decreased by more than 153 k tons of CO_2e (Jain et al. 2024). The planting of fast-growing trees on marginal lands is also encouraged by ITC's Social and Farm Forestry Programs, which significantly reduce carbon emissions (Chavan et al. 2023).

Internationally, India's carbon-positive farming methods are also supported by the VCS. VCS, the most popular GHG crediting program in the world, allocates money to initiatives that lower emissions, enhance livelihoods, and save the environment. More than one billion tons of carbon and other GHG emissions have been eliminated or decreased as a result of VCS projects (Indrajaya et al. 2024; Pradhan and Meena 2023b). With a focus on topics like blue carbon, carbon capture and

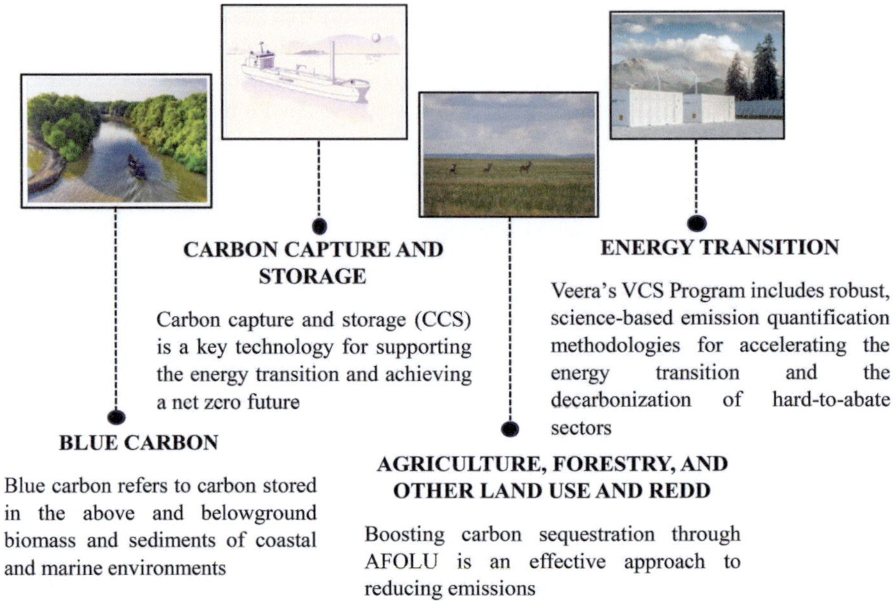

Fig. 15.3 Area of focus of the Verified Carbon Standard (VCS) Program. (Modified from Indrajaya et al. 2024)

storage, agriculture, forestry, and energy transition, the VCS program helps smallholder farmers in particular by providing technical assistance on sustainable farming methods like precision and no-till farming and by making carbon credits more accessible (Fig. 15.3).

15.2.2.4 Carbon Sequestration Initiatives

The Soil Health Card Scheme, introduced in 2015, aims to promote balanced fertilizer application and advance soil conservation by curbing the overuse of nitrogen fertilizers, thereby improving soil carbon retention. This initiative plays a key role in preserving soil health and reducing nitrous oxide emissions, a highly potent GHG. Complementing these efforts, the Sub-Mission on Agroforestry offers both financial and technical assistance to farmers, encouraging the integration of trees into cropping systems. This approach not only enhances carbon sequestration but also improves soil quality, boosts biodiversity, and fosters a more sustainable, carbon-positive agricultural landscape (Pretty et al. 2018; Pradhan and Meena 2023a).

15.3 Innovative Practices in Carbon Farming and Their Benefits

Carbon farming offers a sustainable method for producing food and related goods in an eco-friendly manner. Its goal is to integrate regenerative farming practices with commercially viable products (Blanco-Canqui and Ruis 2020; Sun et al. 2018). Techniques such as crop rotation, consistent soil coverage, cover crops, minimal soil disturbance, zero tillage, and CA are some of the promising approaches to carbon farming in the Indian context (Fig. 15.4). The following section provides an in-depth overview of various carbon farming practices and how they specifically benefit farmers.

15.3.1 Agroforestry

By adding trees and shrubs to agricultural areas, agroforestry increases soil and biomass carbon storage. There is a lot of promise for carbon farming with alley cropping systems, which blend tree cultivation and grassland farming (Gaujour et al. 2012). Higher levels of CO_2 can be sequestered by teak (Tectona grandis)-based land use systems, according to studies. Aboveground sequestration can reach 111 Mg ha^{-1} year^{-1}, belowground sequestration can reach 29 Mg ha^{-1} year^{-1}, and total sequestration can reach 140.0 Mg ha^{-1} year^{-1} (Roy et al. 2022). By combining agroforestry with conservation techniques that integrate agriculture, livestock, and forestry, soil health is improved, and long-term carbon sequestration is enhanced (Pacheco et al. 2013; Pradhan et al. 2023). This practice not only helps sequester substantial amounts of carbon but also provides farmers with multiple income sources. The carbon that trees store can be converted into carbon credits that farmers can sell in voluntary or compliance carbon markets. Additional revenue streams are provided by the production of timber, fruits, nuts, and other non-timber items by agroforestry systems. By adding more organic matter to the soil, these systems help improve soil fertility, which eventually results in higher agricultural yields. To further entice farmers to embrace agroforestry, numerous countries, such as India's Sub-Mission on Agroforestry, provide financial incentives and technical assistance.

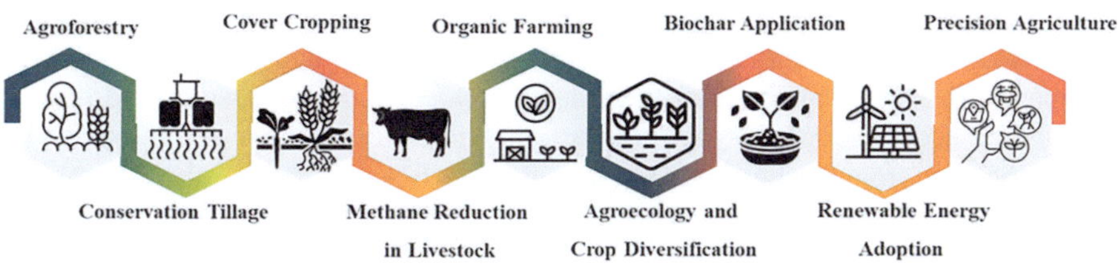

Fig. 15.4 Innovative practices for improving carbon farming

15.3.2 Conservation Tillage

Conservation tillage reduces soil disturbance, helping to preserve soil organic carbon and lower emissions compared to traditional tillage practices. Farmers who adopt this method can earn carbon credits for reducing emissions and enhancing soil carbon sequestration. This practice also cuts operational costs by decreasing fuel usage and labor demands. In addition to the economic benefits, conservation tillage enhances soil structure and water retention, strengthening crops' resistance and possibly increasing yields over time (Smith et al. 2008). Conservation tillage is an economically viable option since it is supported by financial incentives offered by several nations to farmers who use it (Leharwan et al. 2023).

15.3.3 Cover Cropping

Cover cropping involves planting species such as clover, rye, or vetch during fallow periods to protect and enrich soils. This practice increases soil organic matter, enhances carbon sequestration, and helps reduce GHG emissions. Farmers can earn carbon credits by boosting soil carbon levels, which can then be traded on carbon markets. Beyond climate benefits, cover crops naturally enhance soil fertility, reducing dependence on chemical fertilizers and lowering production costs. They also play a critical role in preventing soil erosion and suppressing weeds, contributing to overall farm sustainability. In temperate regions, winter cover crops like vetch and rye have been shown to significantly raise soil organic carbon, adding between 6.8 and 22.8 Mg ha^{-1} of carbon (Sainju et al. 2006). Similarly, in tropical and subtropical zones, summer cover crops such as sun hemp offer comparable advantages, contributing about 8 Mg ha^{-1} of organic carbon to the soil (Yan et al. 2011) (Table 15.2). These findings underscore the essential role of cover crops in promoting soil health and advancing climate-smart agriculture, particularly in regions facing soil fertility challenges and emission reduction pressures.

15.3.4 Organic Farming

Organic farming reduces or eliminates the use of synthetic inputs like chemical fertilizers and pesticides, focusing on natural nutrient cycles and biological processes (Kumar et al. 2023a, b). This approach leads to lower GHG emissions and greater soil carbon sequestration. Organic farming is believed to enhance soil fertility by increasing soil organic matter, which significantly benefits carbon dioxide sequestration in agriculture and related systems. Studies have shown that adopting organic farming leads to an annual increase of 2.2% in SOC levels (Leifeld et al. 2013). Compared to conventional systems, organic agriculture significantly increases

Table 15.2 Organic carbon yield from different cover crops across climatic regions

Region	Type of cover crops	Examples of cover crops	Organic carbon yield (Mg ha^{-1})	References
Temperate region	Winter cover crops	Rye (*Secale cereale*) Ryegrass (*Lolium perenne*) Oats (*Avena sativa*) Pea (*Pisum sativum*) Vetch (*Vicia sativa*) Clover (*Trifolium pratense*)	6.8–22.8 mg ha^{-1} (vetch and rye)	Sainju et al. (2006)
Tropical and sub-tropical regions	Summer cover crops	Sun hemp (*Crotalaria juncea*) Sorghum (*Sorghum bicolor*) Velvet bean (*Mucuna pruriens*) Sudan grass (sorghum × *drummondii*)	8 mg ha^{-1} (sun hemp)	Yan et al. (2011)

Modified from Sainju et al. (2006)

carbon accumulation in soils (Gattinger et al. 2012). Organic systems typically hold higher soil carbon stocks due to larger external carbon inputs such as manure, slurry, and compost, delivering about 1.20 Mg C ha^{-1} annually, compared to just 0.29 Mg C ha^{-1} in conventional farming (Leifeld et al. 2013). This consistent addition of organic materials boosts soil carbon concentrations in organic systems. Beyond soil benefits, numerous studies emphasize the broader environmental advantages of organic farming, including enhanced soil health, reduced nutrient runoff, and lower energy demands. Moreover, organic farmers often benefit from premium market prices for their products, increasing profitability. Practices like composting and using natural soil amendments also qualify for carbon credits, providing further financial incentives.

15.3.5 Agroecology and Crop Diversification

Agroecology promotes farming system diversification through approaches such as intercropping, crop rotation, and integrating multiple species. Notably, intercropping within woody cropping systems is emerging as a valuable strategy in India to combat soil degradation, reduce the effects of climate change, boost carbon sequestration, and strengthen food security (Kumawat et al. 2022). In the short term, incorporating perennial crops into low-input, rainfed woody systems has been shown to substantially enhance the annual soil carbon balance (Akca et al. 2022). Several studies have demonstrated that crop diversification can significantly improve the global carbon and nutrient cycles (Kumar et al. 2019; Quinton et al. 2010). These diversified systems sequester more carbon and reduce emissions, offering farmers the opportunity to earn carbon credits. Moreover, diversified farms are more resilient to climate-related shocks, ensuring more stable incomes even in challenging conditions. By growing a variety of crops on the same land, farmers can generate additional income from different harvests. Governments and NGOs frequently provide grants or technical support to assist farmers in transitioning to agroecological systems, reducing financial challenges and promoting long-term sustainability.

15.3.6 Methane Reduction in Livestock

Livestock farming contributes significantly to methane emissions, but several strategies can help reduce this impact. Improving feed efficiency is one such approach, as it lowers methane production in ruminants (Kumar et al. 2020; Punia et al. 2020; Sheoran et al. 2022). Additionally, manure management practices, such as installing biodigesters, capture methane emissions and convert them into biogas. Optimizing feeding practices and feed conversion efficiency is essential to minimizing methane output. By enhancing the energy content and digestibility of feed, less methane is produced, leading to higher productivity. Specific feed additives, including tannins, essential oils, and enzyme supplements, have been shown to effectively reduce enteric methane emissions in livestock (Hegarty et al. 2021). Additionally, research has produced anti-methanogenic feed additives, such as "Harit Dhara," that are safe to include in animal diets. Harit Dhara can cut methane emissions by up to 17–20%, according to feeding trials carried out at several Indian Council of Agricultural Research (ICAR) research institutes (Kumar et al. 2023a, b). By using these methods, farmers can prevent emissions and earn carbon credits that can be sold in carbon markets. In addition to lowering methane, biodigesters also create biogas, which may be sold as renewable energy or utilized on the farm to make extra money. In order to make methane-reducing devices more affordable to farmers, governments frequently offer subsidies. Long-term advantages result from these methods, which also encourage healthier animals and increased productivity.

15.3.7 Biochar Application

Biochar, a carbon-rich material created by heating biomass under low-oxygen conditions, is increasingly valued for its dual role in enhancing soil health and combating climate change. When incorporated into soils, biochar improves nutrient retention, boosts water-holding capacity, and supports healthier, more productive crops (Purakayastha et al. 2019). Importantly, biochar sequesters carbon in soils, where it can remain stable for centuries, effectively lowering atmospheric CO_2 levels. Farmers applying biochar can generate carbon credits for the carbon stored, offering a valuable financial incentive. Moreover, producing biochar from agricultural residues creates an additional income stream, as surplus biochar can be sold to other agricultural stakeholders. Research has highlighted biochar's potential to enhance CO_2 sequestration while improving crop yields, making it a promising strategy for mitigating global warming (Smith et al. 2008).

15.3.8 Renewable Energy Adoption

Incorporating renewable energy solutions like solar-powered irrigation pumps and wind turbines into farming delivers both environmental and economic benefits. By decreasing dependence on fossil fuels, these technologies help reduce GHG emissions, enabling farmers to actively support climate change mitigation (Majeed et al. 2023). Farmers adopting renewable energy systems can also generate carbon credits for the emissions they avoid, creating an additional income source through carbon markets. Beyond environmental gains, solar and wind energy lower operational expenses by supplying low-cost or even free electricity, thereby improving profit margins. In many regions, government subsidies and financial incentives ease the upfront investment burden, encouraging broader uptake (Babatunde et al. 2019). Moreover, any surplus energy produced can often be sold back to the grid, providing farmers with further revenue. This integrated approach not only enhances farm sustainability but also bolsters local economies and advances wider environmental goals.

15.3.9 Precision Agriculture

Precision agriculture utilizes advanced technologies like Global Positioning System (GPS), drones, sensors, and data analytics to optimize resource management on farms. By carefully managing inputs such as water, fertilizers, and pesticides, farmers can minimize waste and inefficiencies, leading to reduced GHG emissions. By using more effective techniques, precision agriculture allows farmers to receive carbon credits for the emissions they save. Furthermore, precision agriculture improves crop yields by establishing ideal growing conditions and reducing input costs by applying resources in the appropriate amounts at the right times. The use of these technologies can enhance farm profitability while supporting sustainable farming practices (Nowak 2021). Government programs and private organizations often provide subsidies or financial support to make precision farming tools more accessible to small and medium-sized farms. Beyond financial savings, precision agriculture strengthens the sustainability of farming systems, improves food security, and promotes long-term environmental health (Yost et al. 2019).

15.4 Policy Recommendations for Successful Agricultural Carbon Trading in India

The Indian government has been mooting to creation of a national carbon market for Indian players to trade their credits. As the prevailing VCMs are distant from being approached and utilized by most farmers, national-level markets would provide an easier place for farmers to trade. Heeding the need for such a market, the Indian government in June 2023 notified the Carbon Credit Trading Scheme (CCTS) under the Energy Conservation Act of 2001. The CCTS framework provides for a national market for carbon credit trading to incentivize GHG emission reductions. While the scheme is framed as a compliance mechanism, it is inclusive of all the sectors of the economy, including agriculture, which hitherto lacked a significant presence in the carbon markets in India. Non-obligated entities, such as farmers in the agricultural sector, can register their projects that reduce GHGs and earn carbon credits under the scheme. The onus for the formulation of rules and regulations for the implementation of the scheme was placed upon the BEE. Setting of emission reduction targets, issuing of certificates, onboarding agencies for verification, etc., will be done by the BEE. For the verification of the projects, agencies, namely, Accredited Carbon Verifiers, will be set up by BEE. A National Steering Committee will be established to monitor and oversee the market's entire functioning. While the notification was out in 2023, its implementation remains to be seen. While CCTS is under development, capitalizing on the VCMs will encourage our farmers to adopt sustainable agricultural practices (SAPs) that sequester carbon effectively. Recently, the government, in line with promoting carbon trading, has launched the "Framework for Voluntary Carbon Market in Agriculture Sector." This framework was developed as the demand for carbon credits was increasing exponentially, and prevailing voluntary markets were not able to meet such demand. While the CCTS and VCMs are yet to be implemented and scaled completely, understanding global experience in carbon markets and their bottlenecks facilitates the stakeholders to make the policies effective and efficient. Various such aspects that should be considered for an effective carbon policy are described in the following subheads:

15.4.1 Motivational Factors

Monetary benefits are generally considered the primary reasons for adopting carbon farming; however, this is true. Many researchers reveal that farmers adopt SAPs mainly due to the importance of maintaining soil health with a view to long-term farm sustainability. Farmers believe that following SAPs in itself will lead to farm profitability in the long run, and carbon credits and other incentives are added stacks on that (Barbato and Strong 2023). Also, in the majority of the cases, farmers who have already adopted various SAPs in their farm activities are the ones who are being enrolled for carbon credit programs. Therefore, any programs dealing with carbon credits should primarily focus on increasing the awareness and interest of the farmers in adopting SAPs. The focus should be on demonstrating various benefits to the farm economically, environmentally, and socially to the communities, rather than merely showcasing the potential income generation from carbon credit sales. However, monetary incentives should not be overlooked as they can indeed act as motivators for farmers who have never practiced SAPs on their farms. Additional income from carbon credits that adopt SAPs can act as compensation for any probable loss of returns.

15.4.2 Inclusion of Small and Marginal Farmers

Prevailing voluntary carbon markets often cater to the needs of large farmers who have the necessary financial and technical resources to spend and use, respectively. Many current carbon credit programs require farmers to maintain precise records of their land and the activities they are performing. Maintenance of such records for a small farmer may seem difficult and not worth pursuing, as he/she need to devote time and even other trained personnel to do for them, whereas for large farmers, the economies of scale will positively influence doing the same. Therefore, any carbon credit policy should address the bottlenecks that a resource-poor farmer may face while adopting the policies. Establishing easier mechanisms for less paperwork and utilizing the already existing land records and agri-stack database are required. Also, small and marginal farmers must be oriented towards such policies and their benefits at the local level by organizing training and meetings with the extension personnel.

15.4.3 Complexities for Leased Land Farmers

Leased land farmers have a substantial contribution to India's agricultural sector, with around 30% of the farming being cultivated on lands taken for lease. Around 8% of the total farmers in India are lease land farmers who lack their land to cultivate (GOI 2021). The absence of their land already hinders such farmers from availing various agricultural schemes such as crop insurance and formal credit. The interest of such landless farmers who continue cultivation despite many bottlenecks should be protected. Thereby, ensuring that these farmers can also avail the incentives from carbon farming by providing contracts with the land owners is required. A clear mechanism to ensure that the incentives are reaching the actual grower rather than the passive land owner should be implemented. In such a way, double counting of the same carbon credits by various parties can be avoided, which otherwise can result in overestimation of the emission reductions.

15.4.4 Addressing the Additionality Principle

The additionality principle requires that emission reductions arising from adopting SAPs are exclusively due to the incentives provided by the carbon markets. It mandates that emission reductions should be additional to what would have occurred in status quo scenarios. While prevailing policies globally feature various requirements regarding additionality principles, consideration should also be given to early adopters of SAPs who may not be able to avail of carbon credits (Barbato and Strong 2023). Early adopters of SAPs may have been practicing them for years back but the requirement of the additionality principle may disregard their contributions. Policies should account for such early adopters and ensure that they are also provided with an opportunity to enter the carbon marketplace.

15.4.5 Uncertainty of Payments

Any farmer adopting carbon farming with an outlook towards achieving carbon credits must be able to receive the relevant payments, which, if not done, would lead to discontinuation of practices by the farmers. Operating an effective monetary mechanism that provides farmers with a system to receive payments by selling their earned credits is crucial for the carbon credit markets. While all the policymakers understand its importance, various studies point out the prevalence of uncertainty in payments. While currently, few voluntary carbon markets in India are operational, most of the farmers under these markets are not receiving any monetary benefits. This lack of monetary benefits also led to the non-adoption among the farmers. Market uncertainty must also be addressed as the supply of credits should match the demand, which, if not, would leave the process astray. Also, the price of each carbon credit varies in different markets, and regulating and monitoring the pricing mechanisms are required to prevent the players from exploiting the farmers' hardships.

15.4.6 Maintaining Permanency

A key aspect behind the promotion of carbon farming is to mitigate the climate change consequences through GHG reductions. While farmers may adopt SAPs for monetary incentives, the policies should address that GHG reductions are permanent. Farmers indulging in SAPs will be able to sequester the carbon dioxide in the soil, while in the short term, this may remain in the soil, reverting to conventional farm practices can cause the sequestered carbon to enter the atmosphere again. Thereby, policies should include features to make farmers follow the practices for the long term, to not revert to old practices, deeming the work done to be futile.

Current voluntary markets address it by withholding part of the incentives conditional on the re-emissions in the coming years (Paul et al. 2023).

15.4.7 Preventing Greenwashing

Greenwashing has become a word of mouth nowadays as various organizations worldwide are obscurely following by obtaining carbon credits. Hatalis (2024) reveals that organizations purchasing carbon credits or other such offsetting certificates from markets still have higher GHG emissions arising from their firm's operations. In such ways, organizations are misleading people about working towards sustainability while they do not make efforts to reduce the emissions from their operations. As the key principle in the carbon credits market is to reduce emissions, greenwashing threatens the whole objective. There should be accountability held on the entities participating in trading credits to see that other real and significant emission reduction practices are in place and offsetting through carbon credits just complements it. Policies should emphasize the reduction in absolute emissions rather than relative emissions to combat climate change.

15.4.8 Leveraging FPOs for Carbon Credit Policies

While the notified CCTS scheme, upon successful implementation, will provide a national marketplace for the farmers to sell and earn through carbon credits, their successful adoption remains to be seen. Utilizing the CCTS scheme by the farmers at the individual level may face certain hindrances that arise due to a lack of digital, financial, and technical literacy among farming communities. The Government of India (GOI) should proactively leverage the existing farmer collectives, such as Farmer Producer Organizations (FPOs), and other farmer groups, wherein the organization can assist the farmers in accessing the markets and selling on behalf of the farmers as a group. FPOs, respectively, should create awareness about the prospects of carbon credits and their related financial outcomes that accrue upon adopting sustainable agricultural practices. Currently, various private organizations such as Veera are providing carbon credit certification programs to farmers. Upon certification, farmers can use those credits to sell in private markets. However, such markets are difficult for small individual farmers to reach because of the low scale of operations. Using such opportunities, certain organizations are organizing farmers into groups and help them adopt various SAPs and thereby avail the benefits of carbon credits. A private entity, namely, CarbonCoreX, has various such projects wherein farmers are supported to adopt SAPs that reduce GHGs. CarbonCoreX avails carbon credit certificates from Veera upon successful verification of the farmers' fields, which are later sold in the markets. Income generated from the selling of carbon credits is taken by CarbonCoreX, while the farmers benefit indirectly from the increase in soil quality, yields, productivity, mobilization of resources, etc. FPOs can also play a key role in utilizing the growing carbon markets by following similar processes as CoreCarbonX while also distributing the financial benefits among the registered farmers.

15.4.9 Globally Acceptable Standards and Methodologies

For the CCTS and voluntary carbon market in agriculture in India to be successful, the farmers must be able to trade their credits globally. To achieve this, the standards followed for project assessments, their registrations, and certification of credits must be in line with the acceptable standards of the other countries. Certain registries, such as Verra and Gold Standard, are accepted by most countries for purchasing their certified credits. Establishment of a national registry for the agricultural sector would be better as the compliance standards, costs of registration, etc., of prevailing registries may seem higher for the farmers (GOI 2024). Studying the practices and methodologies followed by globally reputed registries in tandem with adequate consideration of the socio-economic conditions of Indian farmers must be done while devising the policies.

15.4.10 Capacity Building

A primary challenge confronting India in the space of carbon markets is the lack of institutional and technological infrastructure and capacities. Establishing separate departments in the governments, training of personnel, technological developments to measure and verify the credits, creation of a platform for the trading activity, digital storage of the trading activities, etc., are paramount. It calls for strong political will to allocate the necessary funds to pursue the development of infrastructure and the capacities of personnel. To help developing countries develop their national carbon markets, the World Bank provides the necessary initial support. The World Bank, through its initiative Partnership for Market Readiness, assists developing countries with financial and technical readiness (World Bank 2023). India should utilize the assistance offered to pilot its policies before implementing them fully.

15.5 Scope for Convergence Between Government Initiatives and Carbon Incentives

The Indian government, from time to time, initiates new schemes to foster the growth and development of the agricultural sector and farmers. Since the initiation of NMSA in 2008, the government has implemented various schemes that address sustainability in the agricultural sector. Although these schemes are not intended to reduce GHGs, they indirectly reduce emissions. Hereunder, we discuss a few such initiatives of the Indian government that have the scope of providing carbon incentives to the farmers who are implementing them.

15.5.1 Paramaparagat Krishi Vikas Yojana

The Paramaparagat Krishi Vikas Yojana (PKVY) is a component launched under the Soil Health Management scheme of the National Mission on Sustainable Agriculture. This scheme involves farmers organized into voluntary groups with a minimum of 50 farmers, with a minimum farm cluster of 50 acres. Farmers are required to follow all the necessary practices that are required to receive Participatory Guarantee Scheme certification, ensuring chemical-free farming. GOI also provides an incentive of Rs. 50,000 per hectare per 3 years to the farmer to assist in the purchase of organic inputs and other farm-related practices. Along with organic farming practices, the scheme also included natural farming practices. Though the scheme mandate is primarily to organize farmers into clusters to increase the area covered under SAPs, it also reduces the emissions which otherwise arise. Relevant mechanisms can be established to monitor and certify the emission reductions from the clusters under PKVY and provide them with an opportunity to earn carbon credits. As the incentives given to the farmers for adopting PKVY come to a halt after the third year, farmers often revert to conventional practices for various reasons. Reports reveal that around 44% of the farmers under PKVY tend to discontinue the usage of SAPs due to a lack of incentives, poor markets for organic produce, etc. (Centre for Management in Agriculture 2023). Under such circumstances, integrating the PKVY scheme with a scope to earn carbon credits for farmers will facilitate the farmers to continue adopting the scheme.

15.5.2 Pradhan Mantri Krishi Sinchai Yojana

The Pradhan Mantri Krishi Sinchai Yojana (PMKSY) is a national micro-irrigation scheme launched in the year 2015 with the aim of water conservation and efficient management. A subcomponent under the scheme, that is, Per Drop More Crop, also has opportunities to be integrated with carbon credits. PDMC operates to replace conventional flood irrigation practices with more water-efficient practices such as drip and sprinkler irrigation. The scheme thereby also reduces the dependency on diesel-powered engines to pump groundwater. Till 2023–2024, the scheme covered an area of 83 lakh ha under micro irrigation. Various researchers have reported the potential GHG reductions due to adopting micro-irrigation practices rather than flooded irrigation practices. The scheme can be essentially integrated into the CCTS scheme to provide more benefits to the farmers.

15.5.3 Pradhan Mantri Kisan Urja Suraksha Evam Utthan Mahabhiyan

The Pradhan Mantri Kisan Urja Suraksha evam Utthan Mahabhiyan (PM-KUSUM) scheme was launched in the year 2019. The PM-KUSUM aims to promote the usage of solar-powered water pumps in place of diesel-powered pumps for irrigation. The scheme was proposed to result in the reduction of 32 million tons of CO_2 emissions per annum. The scheme with three components under it offers an opportunity for farmers to earn carbon credits. By replacing the existing diesel pumps with solar pumps and also by installing stand-alone solar pumps in new areas, the scheme results in the reduction in GHGs while also adding to renewable energy capacities.

Likewise, various such schemes implemented by the GOI for farmers result in the reduction and offsetting of GHGs. However, most such schemes involve financial assistance from the government to implement the schemes at the farmers' level. Establishing a central mechanism to monitor and calculate the emission reductions arising from such schemes per unit area can help farmers avail themselves of carbon credits.

15.6 Conclusion

Carbon farming is gaining prominence in the era of the sustainable development goals (SDGs) that mandate governments to meet them by 2030. As climate change is threatening all sectors around the globe, there is an imperative to limit it by reducing GHG emissions. Carbon marketplaces that provide carbon credits to farmers for following various practices that reduce or sequester GHG emissions have a key role to play in the context of climate change. Nations globally follow certain policies for carbon transition, such as carbon tax, emission trading systems, compliance mechanisms, and voluntary carbon markets. The Paris Agreement, too, binds the nations to submit individual NDCs wherein they are also required to detail the strategies that will be followed to reduce emissions. The Indian government, for years, has been initiating various schemes to address the sustainability concept in agriculture. However, a national policy to deal with carbon credits is still lacking in India, while a few other nations already have such policies in place. Also, the agriculture sector lacked the necessary attention in the carbon markets, though it is a major contributor to GHG emissions. Various agricultural practices such as agroforestry, conservation tillage, cover cropping, and biochar application have huge potential to reduce and sequester GHG emissions. However, for the farmers to adopt such practices, there is a need for incentives to be provided to them in the form of carbon incentives. Carbon credits offer the farmers such incentives when the carbon market is fully established. In 2023, the Indian government notified the CCTS scheme, intending to establish a national marketplace for Indian players to trade carbon credits. The agricultural sector is also allowed to trade under the market, which hitherto lacked such wide participation. While we wait for its implementation, policymakers need to consider various factors such as motivational factors for farmers to adopt SAPs, inclusion of marginal and landless farmers, the additionality principle, payment uncertainty, permanency, etc. By addressing such factors, policies can be made more efficient and effective. Also, various previous initiatives of the government, such as PKVY, PMKSY, and PM-KUSUM, reduce GHG emissions when adopted by the farmers. By integrating such schemes with carbon incentives, more opportunities can be provided to the farmers, as various farmers already implement SAPs under such schemes. In such ways, India has positive prospects to utilize the steeply growing carbon markets to benefit the farmers while also nurturing sustainability in the agricultural sector.

References

Akca E, Buyuk G, Inan M, Kırpık M (2022) Sustainable management of land degradation through legume-based cropping system. In: Advances in legumes for sustainable intensification. Elsevier, London, pp 267–280

Altieri MA, Funes-Monzote FR, Petersen P (2012) Agroecologically efficient agricultural systems for smallholder farmers: contributions to food sovereignty. Agron Sustain Dev 32:1–13

Aryal JP, Sapkota TB, Jat ML, Bishnoi DK (2015) On-farm economic and environmental impact of zero-tillage wheat: a case of north-west India. Exp Agric 51:1–16

Babatunde OM, Munda JL, Hamam Y (2019) Decarbonisation of electricity generation: Efforts and challenges. In Carbon Footprints: Case Studies from the Energy and Transport Sectors (pp. 47–77). Singapore: Springer Singapore

Barbato CT, Strong AL (2023) Farmer perspectives on carbon markets incentivizing agricultural soil carbon sequestration. Climate Action 2(1):26

Blanco-Canqui H, Ruis SJ (2020) Cover crop impacts on soil physical properties: a review. Soil Sci Soc Am J 84:1527–1576

Boyce S, He F (2023) Effects of government policy, socioeconomics, and weather on residential GHG emissions across subnational jurisdictions: the case of Canada. Energy Policy 182:113765

Center for Management in Agriculture (2023) Progress and uptake of sustainable agricultural practices. Indian Institute of Management, Ahmedabad, p 89

Chavan SB, Dhillon RS, Sirohi C, Uthappa AR, Jinger D, Jatav HS, Rajput VD (2023) Carbon sequestration potential of commercial agroforestry systems in Indo-Gangetic Plains of India: poplar and eucalyptus-based agroforestry systems. Forests 14(3):559

Choudhary M, Rana KS, Bana RS, Ghasal PC, Choudhary GL, Jakhar P, Verma RK (2017) Energy budgeting and carbon footprint of pearl millet–mustard cropping system under conventional and conservation agriculture in rainfed semi-arid agro-ecosystem. Energy 141:1052–1058

Gattinger A, Muller A, Haeni M, Skinner C, Fliessbach A, Buchmann N, Mader P, Stolze M, Smith P, Scilabba H (2012) Enhanced top soil carbon stocks under organic farming. Proc Natl Acad Sci 109:18226–18231

Gaujour E, Amiaud B, Mignolet C, Plantureux S (2012) Factors and processes affecting plant biodiversity in permanent grasslands. A review. Agron Sustain Dev 32:133–160

GOI (2021) NSS REPORT NO. 587: situation assessment of agricultural households and land and livestock holdings of households in rural India, 2019, January–December 2019 [online]. Available: https://www.pib.gov.in/PressReleseDetailm.aspx?PRID=1753856

GOI (2024) Framework for voluntary carbon market in agriculture sector. Department of Agriculture and Farmers Welfare, New Delhi, p 42

Government of Canada (2023) Update to the Pan-Canadian approach to carbon pollution pricing 2023–2030. Retrieved from: https://www.canada.ca. Accessed on 22 nd January 2024

Government Offices of Sweden (2024) Sweden's carbon tax. Retrieved from: https://www.government.se. Accessed on 22ndJanuary 2024

Hatalis M (2024) Greenwashing or going green? An empirical analysis of the drivers and the effects of carbon offsets and renewable energy certificates on firm performance. Am J Manag 24(2). https://doi.org/10.33423/ajm.v24i2.7108

Hegarty RS, Passetti RA, Dittmer KM, Wang Y, Shelton SW, Emmet-Booth J, Gurwick N (2021) An evaluation of emerging feed additives to reduce methane emissions from livestock. Edition 1. A report coordinated by Climate Change, Agriculture and Food Security (CCAFS) and the New Zealand Agricultural Greenhouse Gas Research Centre (NZAGRC) initiative of the Global Research Alliance (GRA)

Indrajaya Y, Weikard HP, Mohren F, van der Werf E (2024) Paying for forest carbon: cost-effectiveness of the verified carbon standard (VCS) remuneration scheme. Nat Resour Model 37(1):e12387

Intergovernmental Panel on Climate Change (IPCC) (2022) Climate change 2022: mitigation of climate change. Contribution of Working Group III to the Sixth Assessment Report of the Intergovernmental Panel on Climate Change. Retrieved from https://www.ipcc.ch/report/ar6/wg3/. Assessed on 17thJanuary 2024

Jain V, Jain K, Vedant V (2024) Sustainable practices of emerging market multinational enterprises: a study of select Indian automotive firms. FOCUS J Int Bus 11(1):66–83

Jangir CK, Sangwan PS, Panghaal D, Kumar S, Meena RS, Bharti, Jat RD, Singh N (2024) Spatial variability and statistical analysis of soil properties in the rice wheat-based systems of north-west India. Commun Soil Sci Plant Anal 55(8):1205–1223

Jat ML, Chakraborty D, Ladha JK, Parihar CM, Datta A, Mandal B, Nayak HS, Maity P, Rana DS, Chaudhari SK (2022) Carbon sequestration potential, challenges, and strategies towards climate action in smallholder agricultural systems of South Asia. Crop Environ 1:86–101

Jhariya MK, Raj A, Banerjee A, Meena RS, Bargali SS, Kumar S, Nema S, Poonam, Oraon PR (2022) Plan and policies for soil organic carbon management under agroforestry system. In: Plans and policies for soil organic carbon management in agriculture. Singapore, Springer, pp 191–219

Jian J, Du X, Reiter MS, Stewart RD (2020) A meta-analysis of global cropland soil carbon changes due to cover cropping. Soil Biol Biochem 143:107735

Kumar N, Nath CP, Hazra KK, Das K, Venkatesh MS, Singh MK, Singh SS, Praharaj CS, Singh NP (2019) Impact of zero-till residue management and crop diversification with legumes on soil aggregation and carbon sequestration. Soil Tillage Res 189:158–167

Kumar R, Punia H, Kumar S (2020) Nanomaterials for agriculture input use efficiency. In: Resources use efficiency in agriculture. Springer, Singapore. https://doi.org/10.1007/978-981-15-6953-1_5

Kumar S, Bamboriya SD, Rani K, Meena RS, Sheoran S, Loyal A, Kumawat A, Jhariya MK (2022a) Grain legumes: a diversified diet for sustainable livelihood, food, and nutritional security. In: Advances in legumes for sustainable intensification. Academic Press, London, pp 157–178

Kumar S, Sharma SK, Thakral SK, Bhardwaj KK, Jhariya MK, Meena RS, Jangir CK, Bedwal S, Jat RD, Gaber A, Atta AA (2022b) Integrated nutrient management improves the productivity and nutrient use efficiency of *Lens culinaris* Medik. Sustainability 14(3):1284

Kumar S, Sharma SK, Dhaka AK, Bedwal S, Sheoran S, Meena RS, Jangir CK, Kumar D, Kumar R, Jat RD (2023a) Efficient nutrient management for enhancing crop productivity, quality and nutrient dynamics in lentil (*Lens culinaris* Medik.) in the semi-arid region of northern India. PLoS One 18:e0280636

Kumar S, Sharma SK, Dhaka AK, Bedwal S, Sheoran S, Meena RS, Jangir CK, Kumar D, Kumar R, Jat RD, Meena AK (2023b) Efficient nutrient management for enhancing crop productivity, quality and nutrient dynamics in lentil (*Lens culinaris* Medik.) in the semi-arid region of northern India. PLoS One 18(2):e0280636

Kumar S, Meena RS, Kumar S, Pradhan G, Jangir CK, Ghosh S, Punia H, Sheoran P, Meena R, Ahmad MA, Goyal SK (2025) Designing a diversified Indian mustard production system for energy-carbon-cum-heat use efficiency and sowing dates assessment. GCB Bioenergy 17(6): e70044

Kumawat A, Vishwakarma AK, Wanjari RH, Sharma NK, Yadav D, Kumar D, Biswas AK (2022) Impact of levels of residue retention on soil properties under conservation agriculture in Vertisols of central India. Archives of Agronomy and Soil Science 68(3):368–382

Lal R (2004) Soil carbon sequestration impacts on global climate change and food security. Curr Sci 304:1623–1627

Leharwan M, Kumar Y, Kumar R, Kumar Saraswat P, Kumar R, Kumar Thaliyil Veetil A, Bhattacharjee S, Kumar A, Kumar S (2023) Assessing the effects of conservation tillage and in-situ crop residue management on crop yield and soil properties in rice–wheat cropping system. Sustainability 15(17):12736

Leifeld J, Angers DA, Chenu C, Fuhrer J, Katterer T, Powlson DS (2013) Organic farming gives no climate change benefit through soil carbon sequestration. Proc Natl Acad Sci 110:E984–E984

Li X, Zhang Y, Yang Z, Zhu Y, Li C, Li W (2023) Modeling choice behaviors for ride splitting under a carbon credit scheme. Sustainability 15:12241

Majeed Y, Khan MU, Waseem M, Zahid U, Mahmood F, Majeed F, Raza A (2023) Renewable energy as an alternative source for energy management in agriculture. Energy Rep 10:344–359

Maraseni TN, Qu J (2016) An international comparison of agricultural nitrous oxide emissions. J Clean Prod 135:1256–1266. https://doi.org/10.1016/j.jclepro.2016.07.035

Meena RS, Pradhan G (2023) Industrial garbage-derived biocompost enhances soil organic carbon fractions, CO_2 biosequestration, potential carbon credits, and sustainability index in a rice-wheat ecosystem. Environ Res:116525. https://doi.org/10.1016/j.envres.2023.116525

Meena RS, Pradhan G, Kumar S, Lal R (2023) Using industrial wastes for rice-wheat cropping and food-energy-carbon-water-economic nexus to the sustainable food system. Renew Sust Energ Rev 187:113756

Meena RS, Pradhan G, Singh K, Kumar S, Singh AK, Shashidhar KS, Mina KK, Rao CS (2024a) Agriculture models for restoring degraded land to enhance CO_2 biosequestration and carbon credits in the Vindhyan region of India. Sci Total Environ 929:172661. https://doi.org/10.1016/j.scitotenv.2024.172661

Meena RS, Lal R, Kumar S, Pradhana G, Rao CS, Singh AK, Pathak H, Abhilash PC, Kumar A, Sharma SK, Jat ML, Singh S (2024b) Potential of Indian agriculture for capturing atmospheric CO_2 and monetizing carbon credits to the farmers: an overview and policy framework. Adv Agron 188:101–206. https://doi.org/10.1016/bs.agron.2024.06.001

Meena RS, Pradhan G, Nalani SK, Singh AK, Verma SK, Mina KK, Kumar S, Chaturvedi RK (2024c) Diversified cropping modules designed for soil restoration, CO_2 sequestration, and generating carbon credits. Land Degrad Dev 35. https://doi.org/10.1002/ldr.5224

Meena RS, Pradhan G, Kumar S (2025) Energy flow, eco-efficiency, and economic circulation with recycled industrial waste compost application in wheat and subsequent rice farming. Sci Total Environ 967:178779. https://doi.org/10.1016/j.scitotenv.2025.178779

MoEFCC (2021) India: third biennial update report to the United Nations framework convention on climate change. Ministry of Environment, Forest and Climate Change, Government of India, New Delhi

MoEFCC (2022) India's long-term low-carbon development strategy. Ministry of Environment, Forest and Climate Change, Government of India, New Delhi

Mohan RR, Dharmala N, Ananthakumar MR, Kumar P, Bose A (2019) Greenhouse gas emission estimates from the energy sector in India at the sub-national level (version/edition 2.0). GHG Platform India Report-CSTEP, New Delhi

Nath AJ, Sileshi GW, Laskar SY, Pathak K, Reang D, Nath A, Das AK (2021) Quantifying carbon stocks and sequestration potential in agroforestry systems under divergent management scenarios relevant to India's nationally determined contribution. J Clean Prod 281:124831

Nowak B (2021) Precision agriculture: where do we stand? A review of the adoption of precision agriculture technologies on field crops farms in developed countries. Agric Res 10(4):515–522

Pacheco AR, Chaves RD, Nicoli CML (2013) Integration of crops, livestock, and forestry: a system of production for the Brazilian Cerrados. In: Eco-efficiency from vis. to real, pp 51–60

Patra AK (2014) Trends and projected estimates of GHG emissions from indian livestock in comparisons with GHG emissions from world and developing countries. Asian Australas J Anim Sci 27(4):592–599. https://doi.org/10.5713/ajas.2013.13342

Paul C, Bartkowski B, Dönmez C, Don A, Mayer S, Steffens M, Helming K (2023) Carbon farming: are soil carbon certificates a suitable tool for climate change mitigation? J Environ Manag 330:117142

Pradhan G, Meena RS (2023a) Utilizing waste compost to improve the atmospheric CO_2 capturing in the rice-wheat cropping system and energy-cum-carbon credit audit for a circular economy. Sci Total Environ:164572. https://doi.org/10.1016/j.scitotenv.2023.164572

Pradhan G, Meena RS (2023b) Interaction impacts of biocompost on nutrient dynamics and relations with soil biota, carbon fractions index, societal value of CO_2 equivalent, and ecosystem services in the wheat-rice farming. Chemosphere 139695. https://doi.org/10.1016/j.chemosphere.2023.139695

Pradhan G, Meena RS, Kumar S, Lal R (2023) Utilizing industrial wastes as compost in wheat-rice production to improve the above and below-ground ecosystem services. Agric Ecosyst Environ 358:108704. https://doi.org/10.1016/j.agee.2023.108704

Pretty J, Benton TG, Bharucha ZP, Dicks LV, Flora CB, Godfray HCJ, Goulson D, Hartley S, Lampkin N, Morris C (2018) Global assessment of agricultural system redesign for sustainable intensification. Nat Sustain 1:441–446

Punia H, Tokas J, Malik A, Satpal, Rani A, Gupta P, Kumari A, Mor VS, Bhuker A, Kumar S (2020) Solar radiation and nitrogen use efficiency for sustainable agriculture. In: Resources use efficiency in agriculture. Springer, Singapore, pp 177–212

Purakayastha TJ, Bera T, Bhaduri D, Sarkar B, Mandal S, Wade P, Kumari S, Biswas S, Menon M, Pathak H (2019) A review on biochar modulated soil condition improvements and nutrient dynamics concerning crop yields: pathways to climate change mitigation and global food security. Chemosphere 227:345–365

Quinton JN, Govers G, Van Oost K, Bardgett RD (2010) The impact of agricultural soil erosion on biogeochemical cycling. Nat Geosci 3:311–314

Roy O, Meena RS, Kumar S, Jhariya MK, Pradhan G (2022) Assessment of land use systems for CO_2 sequestration, carbon credit potential, and income security in Vindhyan region, India. Land Degrad Dev 33:670–682

Sainju UM, Singh BP, Whitehead WF, Wang S (2006) Carbon supply and storage in tilled and nontilled soils as influenced by cover crops and nitrogen fertilization. J Environ Qual 35:1507–1517

Sheoran S, Gupta M, Kumari S, Kumar S, Rakshit S (2022) Meta-QTL analysis and candidate genes identification for various abiotic stresses in maize (*Zea mays* L.) and their implications in breeding programs. Mol Breed 42(5):26

Smith P, Martino D, Cai Z, Gwary D, Janzen H, Kumar P, McCarl B, Ogle S, O'Mara F, Rice C (2008) Greenhouse gas mitigation in agriculture. Philos Trans R Soc Lond B Biol Sci 363:789–813

Soussana JF, Lutfalla S, Ehrhardt F, Rosenstock T, Lamanna C, Havlik P, Lal R (2019) Matching policy and science: rationale for the '4 per 1000-soils for food security and climate initiative. Soil Tillage Res 188:3–15

Sun L, Wang S, Zhang Y, Li J, Wang X, Wang R, Lyu W, Chen N, Wang Q (2018) Conservation agriculture based on crop rotation and tillage in the semi-arid loess plateau, China: effects on crop yield and soil water use. Agric Ecosyst Environ 251:67–77

U.S. Department of Agriculture (2025) Partnerships for climate-smart commodities. Retrieved from: https://www.usda.gov. Accessed on 14thJanuary 2024

UNFCCC (2023) The Paris Agreement. Retrieved from https://unfccc.int/process-and-meetings/the-paris-agreement/the-paris-agreement. Accessed on 14th January 2024

World Bank (2023) State and trends of carbon pricing 2023. World Bank, Washington, DC. https://doi.org/10.1596/978-1-4648-2006-9

World Bank (2025) Carbon pricing dashboard. World Bank. Retrieved from: carbonpricingdashboard.worldbank.org. Accessed on 21stMarch 2025

Yan X, Cai Z, Wang S, Smith P (2011) Direct measurement of soil organic carbon content change in the croplands of China. Glob Chang Biol 17:1487–1496

Yost MA, Sudduth KA, Walthall CL, Kitchen NR (2019) Public–private collaboration toward research, education and innovation opportunities in precision agriculture. Precis Agric 20:4–18

Chapter 16
Role of the Private and Industrial Sectors in Carbon Credit Initiatives

Abhay Kumar, M. S. Malik, P. R. Oraon, Swati Sabnam, Kerobim Lakra, Ram Gopal, Muneshwar Prasad, Dinesh Mahto, and Sunita Kumari

Abstract Currently, 22% (13 Gt CO_2-eq) of the total global greenhouse gas (GHG) emissions are contributed through agriculture, forestry and other land use. India is the world's fourth biggest emitter of carbon dioxide after China, the United States and the EU27 and in the year 2021 the total carbon emissions were 2648 million tonnes in India. According to the report 'State of Global Climate 2020' published by the World Meteorological Organization, in 2019, greenhouse gas concentrations reached new highs, with globally averaged mole fractions of carbon dioxide (CO_2) at 410.5 ppm, methane (CH_4) at 1877 ppb, and nitrous oxide (N_2O) at 332 ppb, respectively, which are increase of 148%, 260% and 123%, respectively, from pre-industrial (before 1750) levels. The Prime Minister of India announced at COP-26, held in Glasgow, United Kingdom, in 2021, the commitment to achieve net-zero carbon emissions by 2070. The Government of India plans to develop the Indian Carbon Market (ICM), where a national framework will be established to decarbonise the Indian economy by pricing the greenhouse gas (GHG) emission through trading of the Carbon Credit Certificates, and is developing the Carbon

A. Kumar (✉) · M. S. Malik · P. R. Oraon
Faculty of Forestry, Department of Silviculture and Agroforestry, Birsa Agricultural University, Ranchi, Jharkhand, India

S. Sabnam
School of Agriculture, Rural and Tribal Development, Ramakrishna Mission Ashrama, Ranchi, Jharkhand, India

K. Lakra
Department of Agricultural Economics, Ranchi Agriculture College, Birsa Agricultural University, Ranchi, Jharkhand, India

R. Gopal
North Eastern Hill University, Tura, Meghalaya, India

M. Prasad · D. Mahto
KVK, Jehanabad, Bihar Agricultural University, Bhagalpur, Bihar, India

S. Kumari
Faculty of Forestry, Department of FPU, Birsa Agricultural University, Ranchi, Jharkhand, India

R. S. Meena et al. (eds.), *Carbon-Negative Agriculture*, Sustainability Sciences in Asia and Africa, https://doi.org/10.1007/978-981-95-1483-0_16

Credit Trading Scheme. These agricultural practices like agroforestry, conservation agriculture, integrated nutrient management, agroecology, livestock management and land restoration help in improved soil health, increased biodiversity, reduced agrochemicals, reduced GHG, increased carbon sequestration, carbon stock, carbon credit etc. The carbon credit or certified emission reduction (CER) is the unit related to the reduction of one tonne of CO_2 emission. One tonne of mitigated carbon dioxide in the form of plant biomass is equal to one carbon credit. Currently, global carbon credit prices range from US$1 to US$139, with typical prices around US$20 and EU prices averaging between $7.78 and $25.19, with a global average of approximately US$16.21. The World Bank recommended that the carbon credit price should be US$40–80 in 2020. In India, different companies like ACC, Ambuja Cements, Hindustan Zinc, Infosys Limited, Dalmia Bharat Cement Group, Mahindra & Mahindra Ltd., Tata Steel, and Tata Chemicals set rate of carbon credit price as US $ 47.33, US$ 30.74, US$ 16.33, US$14.25, US$11, US$10, US$15 and US$20, respectively, which vary from US$ 10 to 47.33.

Keywords Carbon credit market · Carbon-credit-based farming · Carbon sequestration · Climate change · Policy framework

Abbreviations

ACC	Associated Cement Companies
AFOLU	Agriculture, forestry and other land use
AI	Artificial intelligence
Bt	Billion tonne
C	Carbon
CC	Climate change
CCS	Carbon capture and storage
CCT	Clean coal technologies
CCTS	Carbon credit trading scheme
CDM	Clean Development Mechanism
CER	Certified emission reduction
CH_4	Methane
CO_2	Carbon dioxide
CO_2e	Carbon dioxide equivalent
COP	Conference of the Parties
CSR	Corporate social responsibility
ETS	EU trading system
EU	European Union
EU ETS	European Union's Emissions Trading System
FAO	Food and Agriculture Organization
FYM	Farmyard manure
GEF	Global Environment Facility

GHGs	Greenhouse gases
ICM	Indian carbon market
IFPRI	International Food Policy Research Institute
IFS	Integrated farming system
INDC	Intended Nationally Determined Contributions
IPCC	Intergovernmental Panel on Climate Change
JI	Joint Implementation
MCX	Multi Commodity Exchange
MRV	Monitoring, reporting, and verification
N_2O	Nitrous oxide
NBS	Nature-based solutions
PPPs	Public–private partnerships
R&D	Research and development
REDD+	Reducing emissions from deforestation and degradation
SMEs	Small- and medium-sized firms
TCS	Tata Consultancy Services
UNFCCC	United Nations Framework Convention on Climate Change
VCM	Voluntary carbon market

16.1 Introduction

With the increase in international efforts to prevent climate change, the participation of business and industrial sectors has become more relevant in carbon credit projects. In a drive to reduce their carbon footprints, industry is purchasing from projects that offset emissions by either sequestering or reducing greenhouse gases (GHGs) (Meena et al. 2025). The regulations and corporate social responsibility (CSR) programmes mean that companies are increasingly watching to carbon credit markets as part of the solution to achieving climate change, mitigation, and adaptation and their sustainability goals. The farmers are one of the main people behind these markets, due to their huge potential for carbon sequestering in the way they use land. The convergence of agriculture and industry opens the door to new ways of addressing climate change as well as benefiting farmers financially.

A carbon credit is a tradable certificate or permit representing the right to emit one tonne of carbon or carbon dioxide equivalent (tCO_2e). One carbon credit is one tonne of carbon dioxide, or in some markets, carbon dioxide equivalent gases, after which it can be traded. An organisation which produces one tonne less of carbon dioxide equivalent than the standard level of carbon emission allowed for its outfit or activity earns a carbon credit. Countries which are signatories to the Kyoto Protocol under the United Nations Framework Convention on Climate Change (UNFCCC) have laid down gas emission norms for their companies to be met by 2012. In such cases, a company has two ways to reduce emissions.

(i) It can reduce the GHG (greenhouse gases) by adopting new technology or improving upon the existing technology to attain the new norms for emission of gases.
(ii) It can tie up with developing nations and help them set up new technology that is eco-friendly, thereby helping developing countries or their companies earn credits. This credit becomes a permit for the company to emit GHGs in its own country. However, only a portion of carbon credits of the company in a developing country can be transferred to the company in a developed country.

Developing countries like India and China are likely to emerge as the biggest sellers, and Europe is going to be the biggest buyer of carbon credits. Presently, almost global carbon credit trading was estimated at US$ 5 billion, with India's contribution at around US$ 1 billion. China is currently the largest seller of carbon credits, controlling about 70% of the market share. Multi Commodity Exchange (MCX) has become the first exchange in Asia to trade carbon credits.

Two types of carbon markets exist: the regulatory compliance and the voluntary markets. The compliance market is used by companies and governments that, by law, have to account for their GHG emissions. It is regulated by mandatory national, regional or international carbon reduction regimes. On the voluntary market, the trade of carbon credits is voluntary. The size of the two markets differs considerably. In 2008, on the regulated market, US$119 billion were traded, and on the voluntary market, US$704 million were traded (Hamilton et al. 2009). The three Kyoto Protocol mechanisms are very important for the regulatory market: the Clean Development Mechanism (CDM), Joint Implementation (JI) and the European Union (EU) Emissions Trading System (ETS). Some countries have not legally accepted the Kyoto Protocol, but have other legally binding state and regional GHG reduction schemes. Developing countries can only participate in the CDM. In general, for small-scale agriculture, forestry and other land use (AFOLU) projects in developing countries, the voluntary market is more interesting than the regulatory market because the CDM market has quite complex procedures and methodologies for project registration, and the majority of agriculture and forestry and Reducing Emissions from Deforestation and Degradation (REDD+) projects are excluded. However, a brief introduction is given to the CDM, because some possibilities for small-scale projects (e.g. renewable energy) exist.

Carbon trading is a mechanism through which countries have to achieve their emission reduction targets. Carbon trading means trading of units of carbon dioxide reduced in the environment. Countries which reduce carbon emissions earn Carbon Emission Certificates (CER), which are traded and are called carbon trading. One carbon credit is equal to 1 tonne of carbon dioxide reduced in the environment. There are various exchanges in which carbon emissions are traded, like Chicago Climate Exchange, National Association of Securities Dealers Automated Quotations (NASDAQ) OMX, POWER NEXT, and European Energy Exchange. In India, MCX and National Commodity & Derivatives Exchange Limited (NCDEX) are two exchanges which deal with the trading of carbon emissions in the form of

Carbon Emission Reduction Certificates, which countries earn through the completion of CDM projects or in the form of Emission Reduction Units (ERUs).

Farmers are being recognised as key force multipliers in the carbon sequestration landscape. Agricultural practices like regenerative agriculture, agroforestry, cover cropping and conservation tillage were also found to significantly increase the capacity of soil for storing carbon. The farmers generate carbon credits and sell them to companies looking to offset the emissions that are directly associated with their business operations. This method not only helps curb global climate change mitigation measures but also provides an additional source of income for farmers, smallholders and rural farms. Mainly, farmers through agroforestry have become a high-profile practice for carbon sequestration. Planting trees on agricultural land could help sequester carbon while improving the soil, retaining water and adding biodiversity (Meena et al. 2024c). These approaches can generate carbon credits, which are in demand by private sector organisations that wish to reduce their net CO_2 emissions. The private sector, that is, companies willing to pay for carbon sequestration projects, is increasingly seeking initiatives such as forestry-based carbon sequestration. Corporations are seeking carbon sequestered projects like forestry, soils and agroforestry to provide solutions for industries that have minute other option. This creates an entirely new market force for the farmer and large-scale land owner that benefits sustainable practices. Many companies also support farms beyond just buying credits by providing technical assistance, finance and infrastructure. Examples can include investing in technology and training programmes to help farmers monitor, verify and certify their carbon sequestration activities, ensuring these efforts are eligible for a sale of verified carbon credits. Farmers benefit from these relationships, and the prospective stability of decent-quality carbon credits provides another feather in the cap of business sustainability goals.

India is the world's fourth biggest emitter of carbon dioxide after China, United States and EU27 and in the year 2021 the total carbon emissions in India were 2648 million tonnes (Crippa et al. 2022). According to the report 'State of Global Climate 2022' published by the World Meteorological Organization, in 2019, greenhouse gas concentrations reached new highs, with globally averaged mole fractions of carbon dioxide (CO_2) at 410.5 ppm, methane (CH_4) at 1877 ppb and nitrous oxide (N_2O) at 332 ppb, respectively, which are increase of 148%, 260% and 123%, respectively, from pre-industrial (before 1750) levels. The Prime Minister of India announced at COP-26 held in Glasgow, United Kingdom, in 2021 and committed to a net-zero carbon emissions target by 2070. The Government of India plans to develop the Indian Carbon Market (ICM), where a national framework will be established to decarbonise the Indian economy by pricing the greenhouse gas (GHG) emission through trading of the Carbon Credit Certificates, and is developing the Carbon Credit Trading Scheme. These agricultural practices like agroforestry, conservation agriculture, integrated nutrient management, agroecology, livestock management and land restoration help in improved soil health, increased biodiversity, reduced agrochemicals, reduced GHG, increased carbon sequestration, carbon stock, carbon credit etc.

16.2 Financial Needs of the Agriculture Sector for Enhancing Carbon Credit-Based Farming

Globally, the agricultural sector is vigorous in reducing climate change through carbon sequestration, but the financial crisis is hindering its full potential in carbon credit-based farming (Meena et al. 2024a). Integrated and long-term farming system that increases carbon sequestration needs considerable financial investments in sustainable goals, investigation, new technology and other systems. The worldwide initiatives have recognised these issues, but the financial structures required to aid farmers in carbon credit-based farming remain underutilised. Currently, farmers are accepting carbon credit-based farming systems like climate resilient agriculture, agroforestry, plantation forestry, climate-smart agriculture, cover cropping and conservation tillage. These methods enhance carbon sequestration and improve soil, air and water quality. For example, agroforestry systems involve long-term tree planting, planning and maintenance costs; conservation agriculture, artificial agroforestry and cover crops require new technology, labour shifts and organic inputs. The Food and Agriculture Organisation (FAO) reported that in developing nations, the overall small and marginal holder farmers have limited access to wealth, and find major difficulty upfront in converting to sustainable techniques.

For agricultural research and development (R&D), financial resources are required to research more effective crops, cropping systems, fertilisers, pesticides, weedicides and soil management systems that maximise carbon sequestration strategies (Meena et al. 2024b). The International Food Policy Research Institute (IFPRI) emphasises the requirement of public-private sector partnership in funding R&D, mainly in areas where agriculture is significantly impacted by climate change. Without proper financial support, possible development of this area is not possible.

Additionally, soil carbon detection and monitoring technologies are challenging to implement for accurately determining the quantity of carbon sequestered and for ensuring the validity of carbon credits. Marginal-sized farmers frequently find such technology too expensive. According to the World Bank report, there is a need to allocate financial funds towards the development of low-cost monitoring, reporting and verification (MRV) systems that are available to all farmers, irrespective of their area or economic level.

Farmers require access to training, capacity building and technical help to adopt sustainable practices rapidly. To help farmers understand how to adapt their practices for carbon sequestration, extension services and demonstration projects require funding. The Global Environment Facility (GEF) sponsors capacity-building projects in various locations to train farmers with the information and need-based tools. However, much more financing is needed for these activities worldwide.

Besides, farmers want financial security to balance the risks involved with shifting to a carbon credit-based farming system, during which farmers require help to loss of revenue or reduced output. The potential to issue carbon credits depends on MRV technology that measures and quantifies the carbon sequestered by farming operations. MRV technology, including ground sampling, remote sensing, soil testing, satellite imaging and other methods, is essential to ensure the authenticity and transparency of carbon credits. According to the Carbon Trust, the prices associated with MRV might contribute to carbon markets impossible for small and marginal holders unless supported or subsidised by external financial sources. International organisations and governments must work to provide financial incentives that reduce the burden of MRV expenditures, allowing more farmers to be involved in carbon credit schemes.

Improving farming systems using carbon credits on a global scale necessitates significant financial resources spread across several sectors. To fully realise the agriculture sector's potential as a carbon sink, sustainable farming practices, capacity building, R&D investments, MRV systems and market access are essential. The corporate sector, international organisations and governments must work together to offer the financial resources needed to help farmers through this transformation.

16.3 Status of Private and Industrial Sectors in the Carbon Credit Market at the World Level

The global carbon credit market size was valued at US$ 480.11 billion in 2023 and is expected to reach around US$ 13,322.68 billion by 2033, rising at a Compounded Annual Growth Rate (CAGR) of 39.42% from 2024 to 2033. Consciousness about the reduction in greenhouse gas (GHG) emissions by major players, along with governments' severe policies, rules and regulations to trade the carbon credit, are the major driving factors of the market on a wider scale. The carbon credit or certified emission reduction (CER) is the unit related to reducing one tonne of CO_2 emissions. One tonne of mitigated carbon dioxide in the form of plant biomass is equal to one carbon credit. Currently, global carbon credit price varies from US$ 1–139, US$20, EU$ 7.78–25.19 average of US$16.21. The World Bank recommended that the carbon credit price should be US$40–80 in 2020. In India, different companies like ACC, Ambuja Cements, Hindustan Zinc, Infosys Limited, Dalmia Bharat Cement Group, Mahindra & Mahindra Ltd., Tata Steel and Tata Chemicals set rate of carbon credit price as US$47.33, US$30.74, US$16.33, US$14.25, US$11, US$10, US$15 and US$20, respectively, which vary from US$10 to 47.33.

In response to rising legal requirements and voluntary business pledges to combat desertification and climate change, the global carbon credit market has undergone a marvellous transformation. In this market, the commercial as well as industrial sectors are vital since they are the main buyers of carbon credits and the investors in projects that reduce emissions and sequester carbon (Meena et al. 2023). To accelerate international efforts to achieve net-zero carbon, these trades must participate. The carbon credit market is divided into two types: the compliance market and the voluntary carbon market (VCM). The compliance market, directed by regulatory frameworks like the European Union's Emissions Trading System (EU ETS), demands that organisations fulfil emission reduction objectives. The voluntary carbon market allows corporations to offset emissions beyond compliance limits. The global voluntary carbon market topped US$1 billion for the first time in 2021, suggesting greater business engagement (Forest Trends, Ecosystem Marketplace 2021). Many industrial companies, including energy, manufacturing, technology, forestry-based industries and aviation, are receiving credits to reach net-zero carbon goals. Recently, massive multinational companies such as Microsoft, Amazon, Tata, Mahendra, Adani Group and Shell have made significant investments in carbon credits to minimise their carbon footprints. Microsoft stated their intention to become carbon negative by 2030, depending mainly on carbon credit buying to offset unavoidable emissions (Microsoft 2020). Similarly, Shell committed to investing 100 million US$ in nature-based solutions (NBS), including plantation, afforestation, reforestation and wetland restoration, to use nature-based solutions to mitigate emissions of around 120 million tonnes of CO_2 per year by 2030 to create carbon offsets (Shell Sustainability Report 2021). In India, the Adani Group is investing US$ 100 billion over the next decade to transition to green energy, green hydrogen, plant 100 million trees by 2030, solar panels and wind turbines. Tata Power is investing up to US$ 9 billion to increase its renewable energy capacity by 2030, and Tata Consultancy Services (TCS) is developing a platform to automate carbon credit calculations. The platform uses a 'sky-and-earth' convergence method that combines satellite data with ground data to measure carbon stock.

The private sector is also focused on places with higher controlling frameworks or buyer demand for sustainability. For example, companies operating in Europe and North America are leading the voluntary market, driven by corporate social responsibility (CSR) programmes and client expectations. The European Union's Emissions Trading System (EU ETS) has also enhanced business interest in green carbon markets and acquired voluntary credits to complement their compliance with the cap-and-trade system (Rogelj et al. 2020). Many companies invest in programmes that create carbon credits, such as plantation forestry, afforestation, reforestation and renewable energy efforts (Kumar et al. 2025). The research exposes that these enterprises are driven not just by the goal to reach carbon reduction objectives but also by the co-benefits like aesthetic value, biodiversity enhancement and water conservation (Blaufelder et al. 2021). Major corporations are progressively investing in carbon capture and storage (CCS) technology, along with natural carbon sinks like forests and wetlands (Meena and Pradhan 2023). Such initiatives enable enterprises to claim carbon credits, which may subsequently be traded on carbon markets (Carton and Andersson 2021). Additionally, some plans may not bring long-term carbon sequestration, as forest-based carbon credits are susceptible to deforestation or forest degradation.

The absence of a reliable method and technique for calculating carbon credits poses a significant challenge. The absence of common global standards makes it difficult for corporations to check the quality of carbon credits and assurance they are genuinely contributing to carbon emission reductions. This lack of homogeneity has led to proposals for tighter rules and regulations and verification methods in both the compliance and voluntary markets (Schneider et al. 2020). Moreover, the cost of carbon sequestration remains too high for many enterprises, particularly those in poor and developing nations, where very limited financial and technological resources. This generates an inequity in the global carbon credit market, with richer corporations and nations able to engage more freely than others (Fajardy et al. 2021). Many companies' pledges to become net-zero carbon are driving the formation of cutting-edge financial products linked to carbon credits, such as green bonds, incentives and loans. In these ways, trades may use carbon credits as a means of financing sustainability initiatives in addition to using them for carbon sequestration (Zhou et al. 2021; Pradhan and Meena 2023b).

Finally, it is predicted that more private sector investment in initiatives like soil carbon, blue carbon, forest carbon sequestration and reforestation would result from the increasing emphasis on nature-based solutions (NBS) as part of global mitigation plans. NBS may supply up to 30% of the global mitigation essential to fulfil the objectives outlined in the Paris Agreement (Griscom et al. 2020). Due to their significant buying of carbon credits and huge investments in initiatives to reduce carbon emissions, the industrial and private sectors have grown to be essential players in the worldwide carbon credit market. Even while they are becoming more involved, there are still large problems with cost, carbon credit price, standardisation, calculation and credit quality. But new developments in corporate finance and market systems are going to boost further involvement, to play a major role in achieving global climate targets.

16.4 Potential of Private and Public Sector Engagement in Carbon Credit Initiatives

In the current scenario of climate change, there are various chances for sustainable development and climate mitigation when the private, public and commercial sectors contribute to carbon credit programmes. The private sector has increasingly used carbon credits as a tool to offset emissions and fund eco-friendly initiatives, driven by government regulations, shareholder pressure, corporate sustainability goals, and market competitiveness. Massive potential has been exhibited by the private sector's participation in REDD++ programmes, as businesses come to value greenery, biodiversity, reduced pollution and community development more than fair carbon offsets (Laing et al. 2016; Pradhan and Meena 2023a). Generating fund plans may improve project viability and boost sustainable energy transitions, and increase forest area and cover. One example is the use of carbon credits to assist local energy communities (Meitern 2024). As a result, there is a market for carbon credits, which encourages the carbon sequestration or reduction of greenhouse gas (GHG) emissions through initiatives like reforestation, afforestation, renewable energy projects, precision farming, climate-smart agriculture and sustainable agriculture. Still, the public and private sectors are also very important since they set emissions create policies, create reduction objectives such as carbon pricing, cap-and-trade systems, and establish regulatory agendas. The agenda and market stability that carbon trading necessitates are provided by public policies that break down difficulties and boost private investment in carbon credit schemes, particularly in developing nations (Averchenkova 2017). The public sector may set up a strong voluntary carbon market mechanism that promotes broad involvement. Such an approach to encourage emission reductions is the two-token economic model (Hou et al. 2022). The way these two industries interact can greatly improve the efficiency of carbon markets. Public–private partnerships (PPPs), for example, facilitate the integration of public sector oversight, regulatory frameworks and incentives with private sector innovation, knowledge and money, resulting in carbon credit systems that are more reliable. By offering tax breaks or loan guarantees, subsidies, government policies etc. can reduce the risk associated with private investments in carbon-reducing technology. This boosts private initiatives to make long-term, large-scale investments in projects like REED+, carbon capture and storage (CCS). Significant issues, including carbon leakage and transparency in carbon credit verification, are also addressed by this partnership. Furthermore, the participation of both businesses in carbon credit programmes inspires investment in low-carbon technology such as renewable energy, solar systems, energy efficiency, agroforestry, sustainable agriculture and integrated farming systems. Huge funding expended by the government for research and development (R&D), along with the private sector's search for affordable solutions, can hasten the commercialisation of new innovative technology. For example, to generate carbon credits, private companies may invest in wind, solar, and bioenergy projects, while governments ensure that the necessary regulations and infrastructure for widespread adoption are in place. The public and commercial sectors are advancing ways for reducing carbon emissions while simultaneously advancing sustainable development through job creation, economic growth and environmental protection. To propel significant climate action and expand the carbon credit markets, the public and private sectors must play complementary roles. When combined, they provide an atmosphere that is favourable to investment, creativity and international collaboration in the shift to a low-carbon, sustainable economy.

16.5 Opportunities for Farmers to Gain Carbon Credit from Private and Industrial Sectors

Farmers have many chances to obtain carbon credits from commercial and industrial sectors, mainly through innovative farming methods and involvement in carbon markets. Adoption of sustainable practices, implementing conservation agricultural practices, may greatly increase soil organic carbon, allowing farmers to produce carbon credits. For example, Italy gave initiatives that have shown potential carbon sequestration of 12.8 t/ha CO_2 through improved soil management (Brenna et al. 2018). Conservation agriculture, like zero tillage, happy seeder and enhanced grazing, can help carbon sequestration, rendering producers' taxable credits in carbon emission trading systems (Weersink et al. 2003; Pradhan et al. 2023). In carbon markets, farmers can participate in volunteer carbon markets, which are increasingly run by commercial businesses. The creation of carbon credits and standards can simplify farmers' entry into these markets, matching their practices with market demands (Phelan et al. 2023). Providing farmers with technical help and education on carbon credit schemes can boost their capacity to participate successfully. This involves comprehending measuring, reporting and verification processes (Mooney and Janoski 2022). Data sharing and privacy, addressing farmers' doubts about data privacy, are vital for boosting participation in carbon markets. Many farmers are ready to provide data for educational purposes, but are cautious to share with private or government organisations (Niles and Han 2022). Economic incentives and carbon credit markets give a new cash stream for farmers, particularly as corporations seek to offset their emissions cost-effectively. The marginal cost of carbon abatement in agriculture is generally lower than in other sectors, making it an appealing alternative for both farmers and companies.

16.6 Role of the Private and Industrial Sectors in Environmental and Social Co-benefits Through Carbon Credit

A carbon credit program relies on the private and industrial sectors to generate co-benefits for society, human health, and the environment. Their involvement improves ecosystem health, sustainable development and community welfare in addition to supporting the reduction in carbon emissions. Industries are becoming more involved in carbon reduction ascendency, but their selfless efforts often have little positive impact on environmental and climate change. Governments need to effectively co-regulate in directives to guarantee more intense disruptive changes in business models (Lister 2018). The feeling of small- and medium-sized enterprises (SMEs) towards local carbon credit programmes, along with their gratitude for social development, environmental development and ecological conservation, recommends an increasing awareness about the wider significances of carbon storage and sequestration (Triana and Ota 2024). Senadheera et al. (2019) suggested that the Hiniduma Bio-link Project in Sri Lanka serves as an example of how carbon credit programmes may improve local communities' livelihood while conserving biodiversity, proving that carbon credits can be used to influence extra social benefits. Business guidelines have significantly reduced energy consumption, climate change, GHG emissions and air pollution in Shanghai's Baoshan District, representing the potential for co-benefits in business sectors through effective policy implementation (Jiang et al. 2016). Though the private and public sectors' involvement in the carbon credits programme is encouraging, difficulties still need to be overcome to guarantee fair access, beliefs and maximise co-benefits for all parties involved. This emphasises the need for continued cooperation and innovation.

16.7 Major Areas for Engagement of the Private Sector in Carbon Credit

The active involvement of both private and public sectors is essential in areas where sustainable land use and environmental stewardship are promoted through carbon credit trading. The production, livestock management, processing and distribution of food, agroforestry, plantation forestry, sustainable fisheries and aquaculture, carbon capture and storage (CCS) and carbon sequestration in agriculture are important areas for private and public sector involvement by sequester carbon of trees, reduction in carbon, enhancing soil carbon and biodiversity conservation. For example, in Zambia, the private companies may take part in an integrated farming system that improves water security, soil quality, air pollution and empowers small and medium-sized enterprises (SMEs) in rural areas (Upla et al. 2022). The involvement of the private sector is essential for the monitoring, reporting and verification of carbon credits in the developing agricultural field and soil carbon market. Farmers must embrace soil quality and health techniques to contribute to the carbon market (Oldfield et al. 2022; Phelan et al. 2023). Businesses involved in food processing and production, tree planting, forest management, and agroforestry initiatives can offer technical and financial support in exchange for carbon credits. Sustainable agricultural farming systems, like zero tillage or no-till farming, crop rotation, cover cropping, climate resilient agriculture, natural farming, precision farming and organic farming, can be encouraged by private enterprises in the agriculture field sector to minimise soil disturbance, scarcity of water and greenhouse gas emissions. The livestock may obtain carbon credits by investment in methane (CH_4) reducing technology, such as biogas systems, vermicomposting, FYM formation, better animal feed and waste management. Commerce that uses carbon credits to implement sustainable innovations will be governed by international agreements such as the Kyoto Protocol, which encourages the reduction in emissions through the Clean Development Mechanism (CDM) method (Anukwonke and Abazu 2022). Benavides (2023) reported that companies have the opportunity to invest in agroecosystems that improve climate change resilience and ensure compliance with international monitoring, reporting and verification requirements through the development of voluntary carbon credit markets. To achieve carbon sequestration, the food supply chain can also invest in methods and technologies for reducing carbon emissions, such as reforestation and afforestation. Carbon credit-based farming system under agroforestry refers to farming systems and practices that incorporate trees, shrubs and other woody vegetation with crop and animal production systems. This combination of trees and agriculture can help atmospheric carbon sequester and store, producing carbon credits that can be traded on carbon markets to offset greenhouse gas emissions. Agroforestry systems provide several benefits like increased biodiversity, improved soil health, reduced soil erosion, improved water quality and differentiated income for farmers. Some major areas for involvement of the private sector in carbon credit sequestration and trading between 2020 and 2022 are shown in Table 16.1.

Table 16.1 Major areas for the engagement of the private sector in carbon credit

Companies	Carbon credit (million)	Major areas for carbon offsets
Shell	9.88	REED+, afforestation/reforestation, forest management, wind power, hydropower, solar power etc.
Volkswagen	9.61	
Chevron	6.02	
Comcast	3.22	
Petro China	1.98	
Microsoft	1.16	
AXA	0.93	
BMW	0.87	
TotalEnergies	0.86	
Apple	0.74	
Allianz	0.54	
BNP Pribas	0.55	
Alfabet	0.18	
Samsung	0.15	

Source: https://alliedoffsets.com/

16.8 Promoting Private and Public Investment

Working together with various investors and forming cross-sector alliances are essential to encourage private and public funding in carbon credit programmes. To establish settings that are favourable to governments, non-governmental organisations, investment, stakeholders, companies and financiers must cooperate. Private investments are made less risky by the public sector support, which includes clear policies, carbon pricing systems, methods and technical support. This may involve making promises, tax breaks and subsidy offers. To encourage companies to contribute to carbon credits while lowering monetary risks and enhancing sustainability over time, facilitating funds also needs to promote clarity and streamline the regulatory environment.

16.9 Assessment of Risk and Business Opportunities

In the current scenario, carbon markets are one of the biggest problems facing farmers worldwide, especially in poor countries. Governments and financial organisations must generate initiatives that support farmers in navigating the intricacies of carbon trading and assurance just payment for their labours. According to the United Nations Framework Convention on Climate Change (UNFCCC), the governments should ratify policies that offer subsidies, low-interest loans or tax breaks to farmers who agree to engage in carbon sequestration techniques.

Besides, the growth of carbon credit collectors can aid in lowering transaction costs and enhancing market accessibility by alliance of the carbon credits of small and marginal farmers into larger, marketable units. But putting these mechanisms in place will cost a lot of money upfront, and both the private and public sectors must contribute.

16.10 Challenges to Realise 'Agri-carbon' Market Potential

The technologies of carbon credit for monitoring, reporting and verification of carbon sequestration can be difficult, expensive and time-consuming, mainly for small and marginal farming. Further, market access and the circumstances nearby carbon credit authorisation are problems troubling farmers in developing countries. Another big challenge is equitable contribution and payment, particularly for smallholders.

But the opportunities are significant. If commercial buyers seeking carbon offsets are connected to farmers working toward building up soil health, they may be a significant step forward in climate mitigation while promoting economic development for rural areas. This heightened demand for carbon credits by industry is expected to increase the need for verified, high-quality carbon sequestration projects and place farmers in a critical role in combating climate change on a global scale.

One of the key mechanisms to achieve world climate targets is that the corporate and industrial sectors participate in carbon credit projects, mainly by buying credits from farmers. Such projects are a pathway to reduce emissions, sustainability and economic resilience by collaborating with companies and rural communities. In the future, farmers might have a crucial role to play in shaping more equitable and sustainable societies, especially if the carbon market expands.

16.11 Investment Barriers, Challenges, Risks and Support Needs

The upcoming perspective of carbon credits faces various risks and challenges. Important challenges include ensuring the honesty and transparency of carbon credit trading, addressing concerns everywhere double counting and developing healthy monitoring, reporting and verification systems. Risks contain potential market instability, regulatory uncertainty and the potential for greenwashing if carbon credits are not appropriately validated and accounted for. India's carbon market is a vital step towards achieving net-zero carbon emissions, but it faces various challenges. Guaranteeing market honesty, transparency and standardisation is vital for its success. Properly positioning the carbon credit trading scheme with India's intended nationally determined contribution goals and global climate standards is essential to gain international recognition.

- *Market instability:* Carbon credit trading can involve significant price variations due to factors like changes in demand, supply and regulatory policies. This trading instability poses risks for stockholders and organisations contributing to carbon trading, as the value of their carbon credits may rise or fall unpredictably, affecting their emissions reduction and financial planning strategies.
- *Greenwashing anxieties:* The upcoming perspective of carbon credits faces risks and challenges, including concerns over greenwashing. Greenwashing refers to the practice of creating false or illusory claims about the environmental benefits of a service, in order to seem more environmentally responsible. This is an important risk in the carbon credit trading, as there are anxieties that some carbon credits may not represent honest emissions reductions or may be used for carbon sequestration without truly mitigating climate change.
- *Measurement uncertainties:* Exactly measuring, verifying and reporting the amount of carbon sequestered or reduced carbon emissions through carbon credit schemes is a main challenge. Different factors like changing environmental conditions, climate change, complex carbon cycles and limitations in monitoring technologies can introduce measurement uncertainties, making it hard to exactly calculate the climate impact of carbon credit activities.
- *Social and moral risks:* The social and moral risks of the upcoming carbon credit trading include concerns around reasonable access, potential exploitation of marginalised communities and questions of individual against collective accountability for carbon emissions reductions. There are challenges in guaranteeing the fair distribution of carbon credit benefits and preventing carbon trading from exacerbating existing socioeconomic inequalities. Cautious policy making and investor engagement will be vital to addressing these complex social and moral considerations as the carbon credit scheme evolves.
- *Scam and double counting:* The upcoming perspective of carbon credits faces risks and challenges, mainly around scam and double counting. A scam can occur when credits are traded multiple times and do not represent honest reductions in carbon emissions. Double counting is when the same reduction in carbon emissions is claimed by multiple stakeholders, leading to an inaccurate accounting of overall carbon emissions. Talking these matters through vigorous monitoring, reporting and verification processes will be vital for the truth and growth of carbon credit trading going forward.

16.12 Policy Framework on Carbon Credit System Through Private and Industrial Sectors in India

Globally, India issued till 2023 the highest 1586 projects and 321 $MtCO_2e$ in terms of carbon credits, accounting world's 18% of total projects and 16% of total carbon credits. Currently, in India there are almost 78 voluntary carbon market projects in the agriculture sector through enhanced irrigation management, precision farming, manure methane digestion, integrated farming system, rice emission reductions and sustainable agriculture.

The policy outline on the carbon credit system in India mentions the government initiatives, rules and regulations to create a market-based mechanism for trading of carbon credits. This system incentivises trades and industries to decrease their greenhouse gas emissions by permitting them to earn carbon credits that can be traded to other entities that require sequestration and reduce their emissions. The aim is to encourage sustainable practices and help India's fulfilment of climate change mitigation targets. The chief policy framework on the carbon credit system is shown in Fig. 16.1.

- *National policy:* India recently passed two bills: the Energy Conservation (Amendment) Bill, 2022 and the Carbon Credit Trading Scheme (CCTS), 2023. This will permit the generation of carbon credits by public and private sector entities in India to reduce emissions. The carbon credit certificates will be domestically tradable between companies and individuals, subsequently setting up a compliance market.
- *Area-specific policies:* The policy agenda on the carbon credit system in India includes area-specific policies that aim to facilitate the generation, incentivise and trading of carbon credits. These area-specific policy frameworks define the eligibility criteria, guidelines, and mechanisms for various industries and sectors to participate in carbon credit trading, thereby promoting sustainable practices and reducing emissions across the Indian economy.
- *Industrial area:* The policy outline on the carbon credit system in India under the industrial area aims to incentivise industries to reduce their greenhouse gas

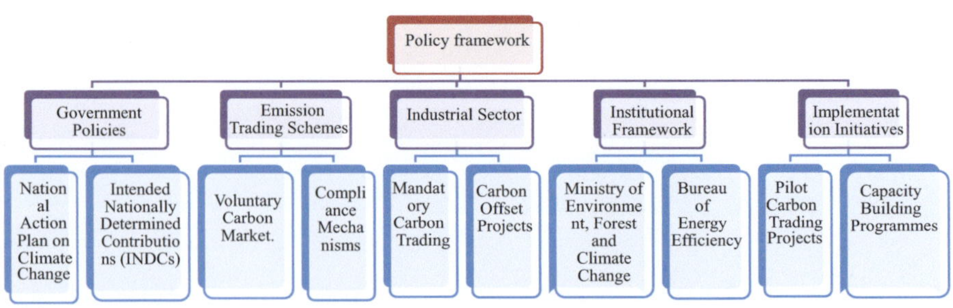

Fig. 16.1 Flowchart of policy framework on carbon credit system through private and industrial sectors in India

emissions. It creates a market-based mechanism where industries can earn carbon credits by applying energy-efficient technologies or adopting renewable energy sources. These carbon credits can then be traded, providing financial benefits to the participating industries and contributing to overall climate change mitigation efforts in India.

- *Forestry sector:* The policy outline on the carbon credit system in India includes necessities for the forestry sector. The forestry sector is outlined as a key area for generating carbon credits through activities like reforestation, afforestation, REDD+ and sustainable forest management. This allows for the generation and carbon credits, which can be used by individuals or companies to offset their greenhouse gas emissions.
- *Expansion of carbon market:* The policy outline on the carbon credit system in India aims to expand a domestic carbon market. It outlines mechanisms to facilitate the carbon credits market generated from emissions reduction schemes. This provides financial incentives for organisations and businesses to invest in green technologies and accept sustainable practices that lower their carbon storage and sequestration.
- *Organisational structure:* The policy outline on the carbon credit system in India frameworks the organisational structure, which involves the key organisations and their roles in handling and supervising the trading of the carbon credit mechanism. This contains certification agencies, controlling bodies and trading platforms that work together to establish the infrastructure, processes and rules essential for the effective monitoring, verification and implementation of the carbon credit system in India.
- *Support and incentives:* The policy outline on the carbon credit system in India offers support and incentives to encourage investment and participation in the carbon emission reduction programme. Key elements contain financial incentives, like tax benefits and the price of carbon credits, as well as technical support and capacity-building programmes to help organisations grow and implement carbon mitigation initiatives. The overall goal is to excite the growth of the domestic carbon market and drive progress towards climate action targets in India.
- *Capacity building:* The policy outline on the carbon credit system in India offers support and incentives for capacity building. This includes initiatives to grow the essential knowledge, skills and infrastructure to successfully contribute to carbon trading, storage and sequestration programmes. The capacity-building efforts aim to empower individuals, communities, businesses and farmers to understand, implement and benefit from the carbon credit system in India.

Various other policy outlines related to India's carbon credit system include the carbon capture trading scheme, renewable energy targets, carbon pricing instruments, energy efficiency mandates, emission intensity reduction goals, trading platforms, price discovery mechanisms, international linkages, financial instruments, tax incentives, carbon finance tools and subsidies, as well as initiatives for skill development and knowledge dissemination.

16.13 Conclusion

The Carbon Credit Trading Scheme in India marks a historic step in the fight against climate change. By inspiring industries to accept eco-friendly practices through market incentives, the scheme enables a move towards cleaner, net-zero zero and sustainable goals. Even though challenges and concerns remain, transparency, proactive measures and collaboration between government and investors promise a greener earth, eco-friendliness and the nation towards a sustainable future. Carbon credit-based farming system contains agricultural practices that aim to increase carbon storage and sequestration or reduce greenhouse gas emissions, for which farmers can earn carbon credits on their farm, which can be traded on carbon credit markets. This farming system may contain practices like no tillage or zero tillage, conventional agriculture, precision agriculture, cover cropping, AI-based agriculture, agroforestry, farm forestry and livestock management strategies that enhance soil carbon, blue carbon and reduced air carbon. A carbon credit-based farming system involves public, private, and commercial entities purchasing carbon credits from farmers who adopt practices that sequester carbon in the soil and reduce greenhouse gas emissions. This generates an economic incentive for farmers to accept more sustainable and integrated farming systems, which can help them adopt and mitigate climate change. The industrial and private business sectors play an important role in the heavy demand for these carbon credits, as long as a carbon credit trading-based mechanism to support carbon-reducing agricultural practices like agroforestry, farm forestry, IFS and conventional agriculture.

References

Anukwonke CC, Abazu CI (2022) Green business through carbon credits. In: Dervash MA, Wani AA (eds) Climate change alleviation for sustainable progression. CRC Press, Boca Raton, pp 1–18. https://doi.org/10.1201/9781003106982-16

Averchenkova A (2017) Engaging private sector and mobilizing private finance through mitigation actions in developing countries. In: Climate finance: theory and practice, pp 55–78. https://doi.org/10.1142/9789814641814_0004

Benavides MA (2023) Building carbon credits in the voluntary market: a viewpoint. Revista Agro 1(1):1–19. https://doi.org/10.47633/agro.v1.i1.4

Blaufelder C, Levy C, Mannion P and Pinner D (2021) A blueprint for scaling voluntary carbon markets to meet the climate challenge. Retrieved from McKinsey Sustainability: www.mckinsey.com/business-functions/sustainability/ourinsights/a-blueprint-for-scaling-voluntary-carbon-markets-tomeet-the-climate-challenge

Brenna S, Acutis M, Tabaglio V and Grandi M (2018) Conservation agriculture as a driver for carbon credit market. https://staging-unicatt.elsevierpure.com/en/publications/conservation-agriculture-as-a-driver-for-carbon-credit-market-9

Carton W, Andersson E (2021) The growing market for voluntary carbon offsets: what's behind the boom? Clim Pol 21(7):1011–1024

Crippa M, Guizzardi D, Banja M, Solazzo E, Muntean M, Schaaf E, Pagani F, Monforti-Ferrario F, Olivier J, Quadrelli R, Risquez Martin A, Taghavi-Moharamli P, Grassi G, Rossi S, Jacome

Felix Oom D, Branco A, San-Miguel-Ayanz J, Vignati E (2022) CO2 emissions of all world countries – 2022 report, Eur 31182 En. Publications Office of the European Union, Luxembourg. https://doi.org/10.2760/730164, JRC130363

Fajardy M, MacDowell N, Fantuzzi A (2021) The challenges of scaling up carbon capture and storage (CCS) technologies. Energy Environ Sci 14(3):1323–1331

Forest Trends, Ecosystem Marketplace (2021) 'Market in motion', state of voluntary carbon markets 2021, installment 1. Forest Trends Association, Washington, DC, p 3

Griscom BW et al (2020) Natural climate solutions for the United States. Sci Adv 4(11):1167

Hamilton K, Sjardin M, Shapiro A, Marcello T (2009) Fortifying the foundation: state of the voluntary carbon markets 2009. New York. Ecosystem Marketplace & New Carbon Finance

Hou D, Hao Y, Du Y (2022) Design of public voluntary carbon market mechanism based on cross chain and token economics. https://doi.org/10.1145/3494106.3528685

Jiang P, Xu B, Geng Y, Dong W, Chen Y, Xue B (2016) Assessing the environmental sustainability with a co-benefits approach: a study of industrial sector in Baoshan District in Shanghai. J Clean Prod 114:114–123. https://doi.org/10.1016/J.JCLEPRO.2015.07.159

Kumar S, Meena RS, Kumar S, Pradhan G, Jangir CK, Ghosh S, Punia H, Sheoran P, Meena R, Ahmad MA, Goyal SK (2025) Designing a diversified Indian mustard production system for energy-carbon-cum-heat use efficiency and sowing dates assessment. GCB Bioenergy 17(6): e70044

Laing T, Taschini L, Palmer C (2016) Understanding the demand for REDD+ credits. Environ Conserv 43(4):389–396. https://doi.org/10.1017/S0376892916000187

Lister J (2018) The policy role of corporate carbon management: co-regulating ecological effectiveness. Global Pol 9(4):538–548. https://doi.org/10.1111/1758-5899.12618

Meena RS, Pradhan G (2023) Industrial garbage-derived biocompost enhances soil organic carbon fractions, CO_2 biosequestration, potential carbon credits, and sustainability index in a rice-wheat ecosystem. Environ Res 116525. https://doi.org/10.1016/j.envres.2023.116525

Meena RS, Pradhan G, Kumar S, Lal R (2023) Using industrial wastes for rice-wheat cropping and food-energy-carbon-water-economic nexus to the sustainable food system. Renew Sust Energ Rev 187:113756. https://doi.org/10.1016/j.rser.2023.113756

Meena RS, Pradhan G, Singh K, Kumar S, Singh AK, Shashidhar KS, Mina KK, Rao CS (2024a) Agriculture models for restoring degraded land to enhance CO_2 biosequestration and carbon credits in the Vindhyan region of India. Sci Total Environ 929:172661. https://doi.org/10.1016/j.scitotenv.2024.172661

Meena RS, Lal R, Kumar S, Pradhan G, Rao CS, Singh AK, Pathak H, Abhilash PC, Kumar A, Sharma SK, Jat ML, Singh S (2024b) Potential of Indian agriculture for capturing atmospheric CO_2 and monetizing carbon credits to the farmers: an overview and policy framework. Adv Agron 188:101–206. https://doi.org/10.1016/bs.agron.2024.06.001

Meena RS, Pradhan G, Nalani Singh K, Singh AK, Verma SK, Mina KK, Kumar S, Chaturvedi RK (2024c) Diversified cropping modules designed for soil restoration, CO_2 sequestration, and generating carbon credits. Land Degrad Dev 35. https://doi.org/10.1002/ldr.5224

Meena RS, Pradhan G, Kumar S (2025) Energy flow, eco-efficiency, and economic circulation with recycled industrial waste compost application in wheat and subsequent rice farming. Sci Total Environ 967:178779. https://doi.org/10.1016/j.scitotenv.2025.178779

Meitern M (2024) Unlocking carbon finance: empowering energy communities for mutual benefit. Renew Sustain Energy Rev 199:114499. https://doi.org/10.1016/j.rser.2024.114499

Microsoft (2020) Microsoft announces it will be carbon negative by 2030. Microsoft News Center, 15 Jan 2020

Mooney S, Janoski K (2022) Economic considerations for the development of a carbon farming scheme. In: Rumpel C (ed) Understanding and fostering soil carbon sequestration. Buddleigh Doods Science Publishers, Cambridge, pp 809–827. https://doi.org/10.19103/AS.2022.0106.26

Niles M, Han G (2022) Interested but Uncertain: Carbon markets and data sharing among US row crop farmers. https://doi.org/10.31235/osf.io/mhv2w

Oldfield E, Eagle AJ, Rubin R, Rudek J, Sanderman J, Gordon DR (2022) Crediting agricultural soil carbon sequestration. Science 375(6586):1222–1225. https://doi.org/10.1126/science.abl7991

Phelan L, Chapman PJ, Ziv G (2023) The emerging global agricultural soil carbon market: the case for reconciling farmers' expectations with the demands of the market. Environ Dev 49:100941. https://doi.org/10.1016/j.envdev.2023.100941

Pradhan G, Meena RS (2023a) Utilizing waste compost to improve the atmospheric CO_2 capturing in the rice-wheat cropping system and energy-cum-carbon credit audit for a circular economy. Sci Total Environ 164572. https://doi.org/10.1016/j.scitotenv.2023.164572

Pradhan G, Meena RS (2023b) Interaction impacts of biocompost on nutrient dynamics and relations with soil biota, carbon fractions index, societal value of CO_2 equivalent, and ecosystem services in the wheat-rice farming. Chemosphere 139695. https://doi.org/10.1016/j.chemosphere.2023.139695

Pradhan G, Meena RS, Kumar S, Lal R (2023) Utilizing industrial wastes as compost in wheat-rice production to improve the above and below-ground ecosystem services. Agric Ecosyst Environ 358:108704. https://doi.org/10.1016/j.agee.2023.108704

Rogelj J et al (2020) Net-zero emissions targets and the role of carbon credits in corporate climate strategies. Nat Clim Chang 10(3):238–244

Schneider L et al (2020) Carbon credit integrity: ensuring high-quality carbon offsets in global markets. J Environ Econ Manag 101(3):101512

Senadheera DL, Wahala WMPSB, Weragoda S (2019) Livelihood and ecosystem benefits of carbon credits through rainforests: A case study of Hiniduma Bio-link, Sri Lanka. Ecosystem Services 37:100933

Shell Sustainability Report (2021) Managing greenhouse gas emissions, p 32. https://reports.shell.com/sustainability-report/2021/_assets/downloads/shell-sustainability-report-2021.pdf

State of Global Climate (2022) WMO-No. 1316, Chair, Publications Board, World Meteorological Organization (WMO)-2023, Geneva, Switzerland

Triana N, Ota T (2024) Assessing preferences for forest carbon credit and co-benefits: a choice experiment case study in Japan. Environ Chall 15:100936. https://doi.org/10.1016/j.envc.2024.100936

Upla P, Reed J, Moombe KB, Kazule BJ, Mulenga B, Ros-Tonen M, Sunderland T (2022) Assessing the potential for private sector engagement in integrated landscape approaches: insights from value-chain analyses in southern Zambia. Land 11(9):1549. https://doi.org/10.3390/land11091549

Weersink A, Joseph S, Beverly DK, Turvey CG (2003) An economic analysis of the potential influence of carbon credits on farm management practices. CAFRI: current agriculture, food and resource issues, Canadian Agricultural Economics Society, issue 4, pp 1–11, September. https://doi.org/10.22004/AG.ECON.45728

Zhou C et al (2021) Financial instruments for accelerating the global carbon credit market: carbon-linked loans and beyond. J Sustain Financ Invest 11(5):673–689

Chapter 17
Carbon-Neutral Alternative in the Rice–Wheat Cropping System

Krishna Kant Mina, Gourisankar Pradhan, Sanskriti Rai, Sandeep Kumar, Manoj Kumar Jhariya, Suchismita Satapathy, Sanjeev Kumar, Amit Kumar Yadav, Ajad Singh Verma, Shambhunath Ghosh, and P. Ashoka

Abstract Globally, controlling greenhouse gas (GHG) emissions is of the highest priority in the modern world, where environmental consciousness is critical and climate issues are becoming more severe. The alarming effects of climate change on agricultural output highlight how urgently feasible emission reduction efforts are needed. The practice of carbon farming (CF), which includes farming methods intended to improve soil and vegetation carbon (C) sequestration, provides a potential answer to the twin problems of reducing climate change and ensuring food security. The most prominent agricultural cultivation system in the world, the rice (*Oryza sativa* L.)–wheat (*Triticum aestivum* L.) cropping system, is used on 24 million hectares in Asia. The deterioration of soil health, water scarcity, labor/energy crises, nutrient deficiencies, low soil organic matter concentration, the release of GHG, and other variables have all put the long-term viability of traditional rice–wheat cropping systems in danger. Numerous studies have been carried out to enhance C-sequestration in terrestrial agro-ecosystems, elucidating the optimal management strategies for optimizing C storage and reducing emissions. It is now essential to encourage the broad use of these efficient methods to lower C footprints

K. K. Mina (✉) · G. Pradhan · S. Rai · S. Satapathy · S. Kumar · A. K. Yadav · A. S. Verma
Institute of Agricultural Sciences, Banaras Hindu University, Varanasi, Uttar Pradesh, India
e-mail: suchismitasatapathy@bhu.ac.in

S. Kumar
ICAR-Indian Agricultural Research Institute, Regional Station, Karnal, Haryana, India
e-mail: sandeepsihag1992@gmail.com

M. K. Jhariya
Department of Farm Forestry, Sant Gahira Guru Vishwavidyalaya, Ambikapur, Chhattisgarh, India

S. Ghosh
ICAR-Indian Institute of Wheat and Barley Research, IARI Karnal Hub, Karnal, Haryana, India

P. Ashoka
Department of Agronomy, College of Agriculture (University of Agricultural Sciences, Dharwad), Haveri, Karnataka, India

R. S. Meena et al. (eds.), *Carbon-Negative Agriculture*, Sustainability Sciences in Asia and Africa, https://doi.org/10.1007/978-981-95-1483-0_17

and improve C sequestration. In this regard, attaining India's target of net zero emissions by 2070 requires incorporating agricultural practices like agroforestry, grassland management, conservation agriculture, cover crops, crop diversification, sustainable handling of crop residue to prevent burning, restoration of forests and wetlands, and control measures. These methods support other sustainable development goals (SDGs) in addition to enhancing crop output and soil health. Enhancing access to markets for smallholder farmers (C credits), raising awareness and technical expertise, and overcoming upfront costs are all necessary to address the difficulties with executing CF in India. A designed policy framework is also necessary for the successful implementation of CF. The system should create supportive facilities, offer financial rewards, and incorporate CF techniques into national agricultural strategies. The implementation of CF practices can also be greatly aided by programs for social responsibility and collaboration between the public and private sectors. The development of scalable and affordable CF technologies, comprehension of the long-term effects of different practices using effective emission evaluation technologies, and the development of integrated models that combine C-sequestration with other ecosystem services. India can improve its agricultural sustainability and meet its climate goals by tackling these issues and utilizing strategic measures.

Keywords Agroforestry · Carbon credit · Organic carbon · Rice

17.1 Introduction

The most effective way to provide food for the enormous population, eradicate poverty, and promote shared prosperity is through the farming industry. However, climate change threatens agriculture-driven development, particularly in many food-insecure areas. To keep the temperature increase below 1.5 °C, the Intergovernmental Panel on Climate Change (IPCC) predicted that 570 Gt CO_2 of global greenhouse gases (GHG) would be needed (Singh et al. 2023). One component of the approach to climate change is agriculture-related mitigation. Agriculture has become a more energy-intensive industry in the past 10 years, and growing environmental concerns have led to a 25% increase in GHG emissions, which has more than doubled in the last 55-years (Sinha et al. 2020; Meena et al. 2025). Agriculture in India is responsible for 16% of the nation's total release of GHG (Panchasara et al. 2021). With 13.5 million hectares of land under rice–wheat rotation, India is the globe's second-largest producer of both rice and wheat (Meena et al. 2023; Sharma et al. 2012; Jangir et al. 2024). High releases of GHG and inadequate usage of resources are further characteristics of the rice–wheat cropping sequence (RWCS). About 76% of the Indo-Gangetic Plain (IGP) is in India, with additional areas in West Bengal, Punjab, Haryana, Uttar Pradesh, and Bihar. Transplanted puddled environments, which generate 84.2 Tg CO_2-Ce yr-1 of methane (CH_4), cover almost all of India's 43.86 million hectares of land (Bhatia et al. 2013). Indian soils used for agriculture are estimated to release 64.6 Tg CO_2-Ce worth of N_2O-N annually, with a global warming potential (GWP) of 148 Tg CO_2-Ce. Approximately one-third of climate change is caused by CH_4 (IPCC 2021). To achieve the 1.5 or 2 °C warming target of the Paris Agreement, CH_4 emissions would need to be significantly reduced after increasing by roughly 25% over the previous 20 years (Nisbet et al. 2019). Between 40% and 50% of global CH_4 emissions are produced by millions of farms with a wide range of sizes and farming methods. It is possible to reduce CH_4 emissions by about 12% by 2030 and 30% by 2050 by implementing suitable soil, nitrogen, and water management strategies (Meena and Pradhan 2023; Saunois et al. 2019). By altering land use and other development-related activities, human actions have impacted the biosphere, which has changed the natural equilibrium of GHG in the atmosphere. Climate change, which is now one of the most significant global issues confronting humanity with effects on farming, biodiversity, drinking and irrigation water supplies, human, animal, and plant health, natural refuse, economic growth, etc., has never been at the forefront of human society before due to environmental concerns. This has led to a global increase in awareness of the detrimental effects of the release of GHG on the climate and the development of appropriate corrective actions using various GHG mitigation processes. One of the possible advantages of the clean development mechanism (CDM) is that investors can obtain costs that are lower than in their own countries, while creators can benefit from financing, efficient technological devices, and environmentally friendly benefits. Managing carbon through "C sequestration" is a much more practical and cost-effective option. Combining perennials and annuals, agroforestry (AF) offers a variety of tangible benefits like fuel, fodder, and small timber, while intangible benefits include improved biological diversity, C sequestration (CS), and ecological enhancement. Despite their immense benefits, the intangible effects, which have no associated cost, have not been fully realized (Hymavathi et al. 2010). Because AF systems are more intricate than monoculture tree/crop scenarios, they balance increased productivity with sustainable ecological services (soil health, hydrology, biodiversity, CS, etc.), enhancing the multifunctionality of agricultural landscapes. Since the inclusion of trees leads to greater CO_2 sequestration from the air through the method of photosynthesis and enhances carbon storage in tree sections for a long time, trees on agricultural landscapes are frequently thought of as carbon sinks (Chauhan et al. 2012; Murthy et al. 2013; Pradhan and Meena 2023a; Swamy and Puri 2005). Therefore, in this study, we focus on generating higher CO_2 sequestration and carbon credits by including suitable tree species in the RWCS region of IGP, India. This study is anticipated to provide a state-of-the-art understanding of the potential for mitigating climate change under RWCS.

17.2 Innovation and Technological Advancement to Generate Carbon Credit

The pursuit of C credits encourages research and development in clean energy technologies and innovative solutions to reduce emissions (Kumar et al. 2025). This can lead to breakthroughs in sustainable practices and technologies. Agricultural sociologists have historically placed a high value on the agricultural community's willingness to embrace novel ideas and advances in recent decades. They have worked together extensively with other agrarian disciplines to better comprehend farmers' processes for making decisions and encourage the adoption of new ideas, technologies, and procedures. Following the lead of economists, agricultural sociologists attempt to explain the factors driving adoption that transcend the framework of economic rationality. Rural sociologists have diligently worked since the late 1940s to develop a structure to govern the adoption of innovations in agriculture, hoping to gain a better understanding of how farmers adopt new technologies and processes. A typical dissemination study covers the unique characteristics of consumers, the ways that innovations in technology flow within social frameworks, and how the links among people and various channels of communication influence their rate of acceptance (Rogers and Singhal 2003).

17.2.1 Promoting Sustainable Practices

The C credit system encourages the adoption of sustainable practices across various sectors, including agriculture, forestry, transportation, and industry. This promotes responsible resource management and reduces environmental degradation. RE techniques that mitigate warming temperatures are currently extensively acknowledged and promoted as environmentally conscious farm management practices, or sustainable agricultural practices (SAPs) (FAO 2018). Numerous socio-economic and environmental advantages are shown when contrasting environmentally friendly farming practices with conventional techniques (Hazra et al. 2020; Kumara et al. 2019). The most common SAPs to enhance the long-term preservation of C in farmed areas include integrated nutrient management, organic amendment, no-tillage, crop rotation, residue retention, intercropping, biochar, and agroforestry. The use of such methods has grown recently as a result of increasing awareness of the importance of field-level proof and SAPs (Chen et al. 2009; Clark et al. 2017; Kumar et al. 2022a, 2023).

17.3 Carbon Credit in Rice–Wheat Cropping System

Carbon credits in rice–wheat cropping systems typically involve practices aimed at reducing GHG emissions, improving soil carbon sequestration, and enhancing the overall sustainability of this intensive agricultural system (Pradhan and Meena 2023b; Meena et al. 2024c). Here are some ways in which carbon credits can be associated with rice–wheat cropping systems.

17.3.1 No-Till Farming

Implementing no-till or reduced tillage practices in rice–wheat systems can reduce carbon emissions from soil disturbance and increase soil carbon sequestration. Carbon credits can be earned based on the calculated emissions reductions and carbon gains resulting from reduced tillage (Mkhabela et al. 2008).

17.3.2 Crop Residue Management

Proper management of crop residues, such as straw and stubble, can significantly impact carbon storage in the soil. Practices like retaining crop residues as mulch or incorporating them into the soil can lead to carbon credit generation (Leharwan et al. 2023). Retaining residues in the field after harvest improves soil health, carbon storage, and soil moisture (Pradhan et al. 2023; Yadav et al. 2021).

17.3.3 Cover Cropping

Introducing cover crops during the fallow period between rice and wheat seasons can improve soil health and sequester carbon. Carbon credits can be earned for the additional carbon stored in the soil due to cover cropping. Cover crops can additionally influence soil carbon sequestration by reducing erosion rates. The impact of erosion on carbon sequestration hinges on the interplay between accelerated decomposition during the transportation process, decelerated decomposition of buried sediment, and alterations in the rates of fresh carbon input and its stabilization at the erosion site (Berhe et al. 2007; Kumar et al. 2020).

17.3.4 *Efficient Fertilizer Use*

Implementing precision agriculture techniques to optimize fertilizer application can reduce nitrous oxide emissions, a potent greenhouse gas associated with nitrogen fertilizers (Punia et al. 2020). Carbon credits can be earned based on emissions reductions. Furthermore, our research findings demonstrate that soil organic carbon (SOC) stocks were significantly higher under organic amendments and integrated nutrient management than under chemical fertilizer application (Kumar et al. 2023; Jhariya et al. 2022). This aligns with previous studies that have consistently shown an increase in SOC stocks using organic amendments. For instance, Wu et al. (2021) observed a substantial 34% elevation in SOC content when employing organic amendments compared to chemical fertilizers. Similar trends of augmented SOC due to organic amendments were also corroborated by Bai et al. (2020) and Li et al. (2021). A comprehensive global meta-analysis conducted by Han et al. (2016) revealed a substantial 36% increase in topsoil organic carbon under integrated nutrient management practices. Furthermore, Padbhushan et al. (2021) conducted a meta-analysis that highlighted how the incorporation of organic inputs led to a remarkable 16–23% augmentation in SOC compared to the use of nitrogen, phosphorous, potassium (NPK) fertilizers.

17.3.5 *Methane Capture*

In rice, CH_4 emissions can be a significant concern. CH_4 capture systems, such as biogas digesters or alternative wetting and drying (AWD) rice cultivation, can reduce CH_4 emissions and potentially generate C credits. The hydrogen peroxide (OH) radical oxidizes most of the atmospheric CH_4 in the troposphere. Through atmospheric chemical cycles, CH_4 has a feedback loop that impacts its lifetime even in the absence of climate change effects. Specifically, higher concentrations of CH_4 result in lower amounts of OH (as it is consumed in reaction with CH_4), which in turn leads to lower amounts of CH_4 destruction (O'Connor et al. 2010; Voulgarakis et al. 2013) (Fig. 17.1).

17.4 Participation in Carbon Markets

To trade C credits associated with rice–wheat cropping systems, farmers or project developers can participate in C markets. Compliance markets or voluntary markets can provide opportunities to sell these credits to entities looking to offset their emissions. The initiatives have made it possible to purchase offsets to comply.

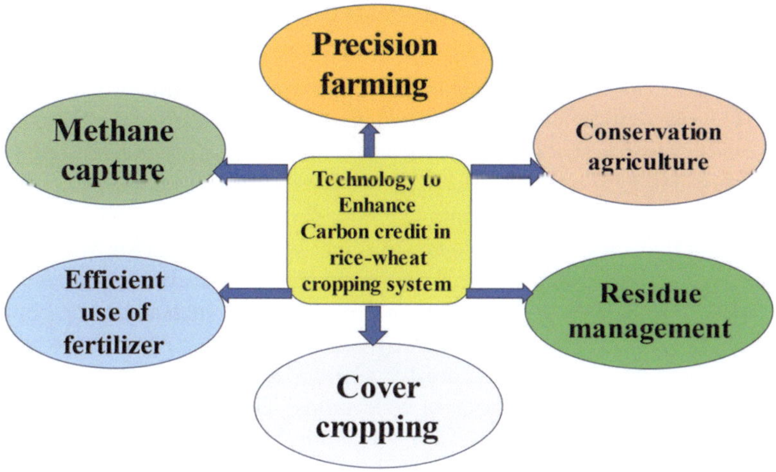

Fig. 17.1 Technology to enhance carbon credit in rice–wheat cropping system

Thus, C offsets are meant to fulfill many functions under an emissions control policy (Gillenwater and Seres 2011). They give more freedom to people looking for the cheapest way to report their emissions of CO_2, especially those who have to comply with regulations. Their inclusion is beneficial to the compliance schemes as well since the offset's component can be changed to make up for issues encountered with the other forms of compliance, including changes in the pricing of the allowances to emit or the costs of reductions at source.

It is important to note that the generation and trading of C credit in agriculture, including rice–wheat systems, can be complex and require rigorous measurement, reporting, and verification. Additionally, the availability and terms of C credit programs can vary by region and market. Farmers interested in participating in C credit initiatives should work with experts or organizations experienced in C offset projects to navigate the process effectively and ensure compliance with relevant standards.

Researchers are developing optimum approaches to management, such as conservation agriculture (CA) approaches and new crop combinations that are more efficient, economical, and thought to be ecologically friendly, as a way to address those challenges (Gathala et al. 2013; Laik et al. 2014). Many CA component technologies, notably rice residue retention, dry drill seeding, and zero or reduced tillage (ZT), have been investigated and tested in cereal systems (Gathala et al. 2011; Ladha et al. 2009). The maize–wheat rotation was initially evaluated as a replacement in place of the traditional rice–wheat rotation due to labor and water shortages, in addition to promoting diversification, notably in the northern IGP (Gathala et al. 2013; Sheoran et al. 2022).

17.5 Carbon Credit in Agroforestry

Integrating trees or woody perennials into the cropping system (agroforestry) can enhance C sequestration and biodiversity. C credits can be earned for the C sequestered by these agroforestry practices (Meena et al. 2024a). C sequestration involves the net removal of CO_2 from the atmosphere and its storage in enduring C reservoirs. These reservoirs encompass elements like aboveground plant biomass, and belowground components including roots and soil microorganisms, along with relatively stable organic and inorganic C forms present in soils and deeper subsurface layers. Additionally, long-lasting products derived from biomass, such as timber, contribute to these C pools. Numerous studies (Kirby and Potvin 2007; Roshetko et al. 2002; Sanchez 2000) suggest that agroforestry systems hold a greater potential for C sequestration compared to pastures or field crops. This hypothesis is founded on the idea that incorporating trees into croplands and pastures can increase net sequestration of aboveground and belowground C, as observed in studies by Haile et al. (2008) and Palm et al. (2004).

Carbon credits in agroforestry refer to the practice of generating and selling C offset credits by implementing agroforestry projects. Agroforestry is a land use system that combines the cultivation of crops or livestock with the cultivation of trees or shrubs on the same piece of land. These trees and shrubs can sequester CO_2 from the atmosphere, thereby helping to mitigate climate change. As a result, agroforestry can be a valuable tool for C sequestration and can generate C credits that can be sold on the C market (Fig. 17.2).

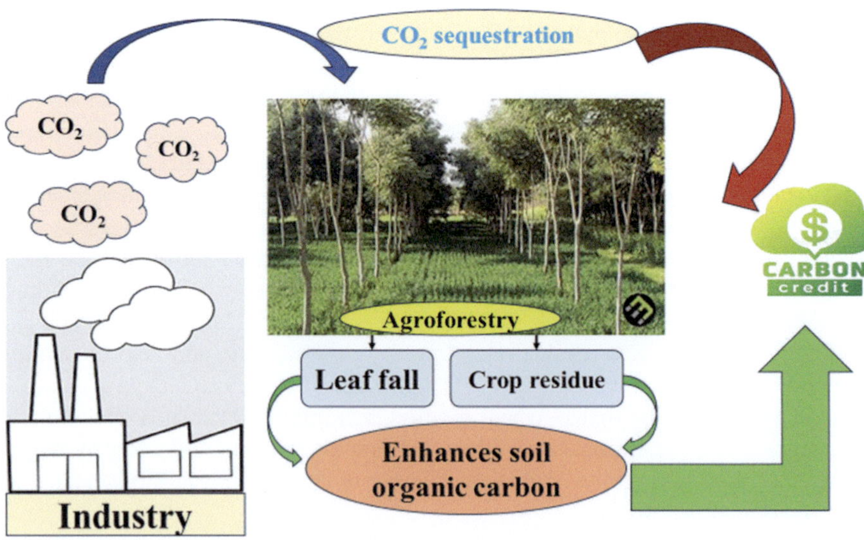

Fig. 17.2 Process of carbon credit generation in an agroforestry system

17.5.1 Suitable Tree Species in the Rice–Wheat Cropping Systems Area

Choosing crops that produce a lot of biomass, reusing residue, applying organic materials, converting annual to perennial crops, moving from monoculture to rotation cropping, and implementing agroforestry systems are all ways to increase the amount of C input in rice–wheat cropping systems. (Nieder and Benbi 2008). An essential choice for C sequestration and generating C credit is agroforestry, a land-use system that intentionally maintains or adds trees or other woody perennials alongside crops for agriculture (Benbi et al. 2012). A survey of the literature on soil C sequestration in agroforestry systems revealed estimates that varied greatly (1.3–173 t C ha^{-1}) depending on the type of tree and age, the climate (Ramachandran Nair et al. 2009). Poplars can be obtained after an extended 6- to 7-year rotation and are one of the tree species that grow the fastest in a suitable climate in rice–wheat cropping systems regions (Mohit Gera et al. 2006). The research work reported that the poplar-based agroecosystem has a higher potential for aboveground CO_2 bio-sequestration, enhanced SOC, and soil organic matter (SOM) compared to other systems in rice growing areas (Chauhan et al. 2011; Mohit Gera et al. 2006). Another study reported that the tree species viz., Acacia (*Acacia nilotica*), poplar (*Populus deltoides*), and Eucalyptus (*Eucalyptus tereticornis*) have a higher potential to generate C credit in rice–wheat cropping systems of the north-west region (Singh et al. 1997).

17.5.2 Here's How Carbon Credits Work in the Context of Agroforestry

17.5.2.1 Carbon Sequestration

Trees and shrubs in agroforestry systems capture CO_2 from the atmosphere through photosynthesis and store it in their biomass and the soil. This process helps to reduce the concentration of GHGs in the atmosphere, which is essential for mitigating climate change. Alterations in land use dynamics play a pivotal role in shaping soil C stocks and influencing the land's capacity to sequester C (Bhat et al. 2012; Roy et al. 2022; Meena et al. 2024b). Human-driven changes in land use possess the potential to mitigate CO_2 emissions by augmenting SOC content (Guo and Gifford 2002; Hutchinson et al. 2007). Additionally, Feliciano et al. (2018) underscored the capacity of Agroforestry to reduce GHG emissions through C sequestration, especially in tropical regions. Hence, increased C sequestration within agricultural landscapes is attainable through strategies such as crop diversification and the adoption of sustainable land management (SLM) practices (Tamburini et al. 2020). Consequently, tree-based land use systems, such as agroforestry, were integrated as GHG offset activities under the Kyoto Protocol in 2001 (Ramachandran Nair et al.

2009). Nevertheless, it is recommended that efforts to mitigate the impacts of climate change by enhancing SOC should be coupled with simultaneous efforts to address various socio-economic challenges (Batjes 2004; Raj et al. 2022). Agroforestry systems have been recognized as ecologically sound approaches to food production when compared to traditional farming and forestry practices, and offer a wide array of socio-economic and environmental advantages (Torralba et al. 2016; Kumar et al. 2022b).

17.6 How to Generate Carbon Credits from Rice Fields

Trading C credits involves buying and selling these credits on C markets. C markets operate in different regions and can be categorized into two primary types: compliance markets and voluntary markets. Here's how you can trade C credits in each of these markets.

17.6.1 Compliance Markets

Compliance markets are typically established by governments or international agreements to help meet emissions reduction targets and regulatory requirements. Trading C credits in these markets is often mandatory for entities that are subject to emissions caps or limits. Here are the general steps involved. To register and verify emission reduction projects effectively, the emissions reduction methods must be farmer-friendly, conceptually straightforward, minimally affecting crop yield, and scientifically sound. This ensures both the adoption by producers and the environmental effectiveness of any agricultural offset protocol.

The Environmental Defense Fund collaborated with agronomic stakeholders and rice producers to adopt three CH_4 emissions reduction methods that meet these criteria.

(1) First, entities or project developers initiate emission reduction projects, such as renewable energy installations, energy efficiency improvements, or afforestation projects. (2) These projects must adhere to specific methodologies and standards established by the regulatory body overseeing the compliance market. (3) Emission reductions are measured, verified, and registered with the regulatory body. Independent third-party auditors often verify the emissions reductions to ensure accuracy.

Alternate wetting and drying (AWD) (eligible for offset crediting in the Mid-South Rice Growing Region) involves cyclically draining and re-wetting rice fields that would otherwise be continuously flooded. This practice has been proven to reduce CH_4 emissions during rice cultivation by 30.5% (Minamikawa et al. 2014). Research indicates that implementing AWD can maintain or even increase yields while offering additional benefits, such as a 30% reduction in water usage (Ye et al.

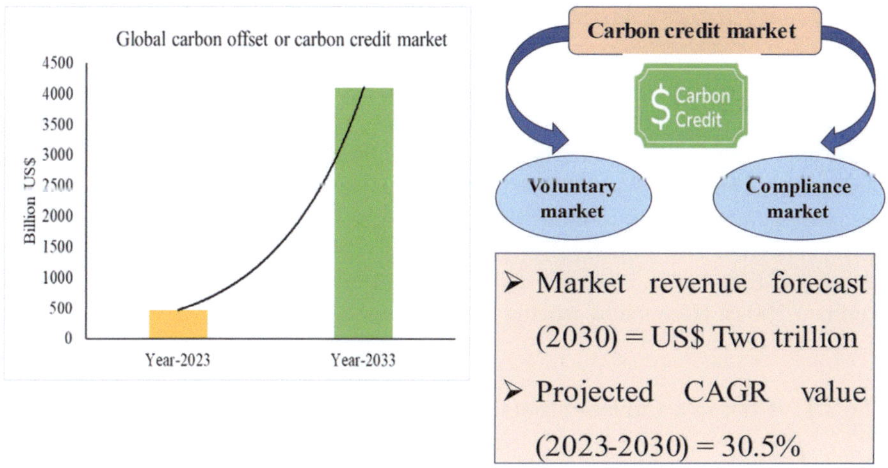

Fig. 17.3 Predicted global carbon credit market. (Modified from Fairfield Market Research 2023)

2013) decreased arsenic levels in rice (Carrijo et al. 2017; Yang et al. 2017) and reduced instances of root rot (Itoh et al. 2011).

Dry seeding (eligible for offset crediting in the California Rice Growing Region) involves sowing rice seeds into dry or moist, but not flooded, soil. This practice is already common in the Mid-South. Dry seeding not only reduces CH_4 emissions (Stallworth 2021). But has also been shown to deliver comparable yields and decrease the need for irrigation, labor, and production costs (Kumar and Ladha 2011).

Receive Carbon Credits
Upon successful verification, the emission reduction projects receive C credits (often called "emission reduction units" or "C allowances") based on the quantity of emissions reduced or sequestered.

Compliance and Trading
(1) Entities subject to emissions limits can use these C credits to comply with emission reduction targets. If they exceed their targets, they can sell excess credits to other entities that need them. (2) Trades typically occur on a regulated exchange or through direct transactions. Regulatory oversight ensures transparency and compliance with market rules. (3) Prices for C credit in compliance markets can fluctuate based on supply and demand, regulatory changes, and the overall market conditions (Fig. 17.3).

17.6.2 Voluntary Markets

Voluntary C markets are not governed by regulatory requirements, and participation is entirely voluntary. Individuals, organizations, and businesses buy C credits to offset their emissions voluntarily. In essence, the development of voluntary standards can be viewed as an industry-led form of 'self-regulation.' In this context, companies collaborate to establish quality standards in response to consumer and media concerns. This initiative serves to proactively address and influence government regulations to stabilize the neoliberal environmental markets (McCarthy and Prudham 2004). Here's how trading works in voluntary markets.

17.6.2.1 Choose a Carbon Offset Project

Rice cultivation is a clear target for the C offset policy given the magnitude of its emissions and capacity for emissions reductions. Worldwide, rice cultivation is the second-largest emitter of CH_4 behind enteric fermentation (FAO 2018), contributing approximately 1% of total global anthropogenic GHG emissions (Smith et al. 2014), and the United States Environmental Protection Agency (EPA) estimates that the technologically feasible reduction potential of rice cultivation worldwide will be 200 million metric tons of CO_2e in 2030 (Aggarwal et al. 2018).

(1) Buyers in voluntary markets select C offset projects they wish to support. These projects can include renewable energy projects, reforestation efforts, CH_4 capture, and more. (2) Ensure the project follows recognized standards and methodologies to guarantee the legitimacy of the C offsets.

17.6.2.2 Purchase Carbon Credits

(1) Buyers purchase C credits directly from the project developers, brokers, or specialized C offset platforms. (2) Prices for voluntary C credits can vary, and buyers can choose the number of credits they wish to purchase based on their C footprint or sustainability goals.

17.6.2.3 Retirement or Use of Credits

(1) Once purchased, the C credits can be "retired," meaning they are permanently removed from circulation, symbolizing the buyer's commitment to offsetting their emissions. (2) Alternatively, some buyers may use these credits to claim C neutrality or reduce their reported emissions. It is important to note that the process of trading C credits can vary depending on the specific market, region, and project type. Additionally, the C credit market is evolving, and new standards and platforms are continually emerging to facilitate trading. When engaging in C credit trading, it is

advisable to work with reputable brokers or platforms and to ensure transparency and verification of the credits being purchased or sold. Systems for recording and monitoring C credit project data, such as credits created, owned, sold, and retired, are known as C offset certifiers (Michaelowa et al. 2019).

India, which was decided wherein it is projected that C-neutral farming is going to grow increasingly popular, may provide several commercial options. The establish ment of technological innovations required for credit certification, among which are those associated with the measurement, obtaining, and processing of data from the field, including the quantity of C found in the soil, is a good example of this. The manufacturing of agricultural supplies needed for C-neutral farming, such as biological pesticides and crops that are capable of preserving C, is a different field offering opportunities for business. Genetic Engineering & Biotechnology News (GEN) Given such possibilities, it is not unanticipated that Indian businesses involved in seeds and agricultural products are expanding into this field of business. Another conceivable fresh opportunity for growers would be offering them financial incentives instead of C credits. Improvement of the entire value chain, connecting farmers who conduct C farming with consumers who respect and want to pay more for their hard work, is one illustration in this field. According to the 2022 World Inequality Report, an environmentally friendly financing plan that offers farmers participating in such projects a beneficial rate of interest may also be considered.

17.7 Measurement, Reporting, and Verification

Accurate measures for lowering GHGs have to involve measurement and measurement, reporting, and verification (MRV). Participating organizations are able to track progress, publicly disclose the results to those who are interested (including donors), and ensure that their information is authentic or of outstanding quality with the support of an efficient MRV system. A high-quality MRV system will generate results that can be analyzed among GHG reduction efforts, shared with scientists, and used to encourage learning. A robust MRV infrastructure can also draw in additional money (Hongo 2018).

Measurement, reporting, and verification (MRV) are critical components of the C credit process. They ensure that C credits are accurately quantified, reported, and verified, thus maintaining the integrity of emissions reductions or C sequestration claims. MRV processes are designed to provide transparency, credibility, and accountability in C credit programs. Here's an overview of each component:

Methodology

Methodologies are typically created by individual project participants and consultants who propose specific methodological approaches tailored to their projects. These proposed methodologies are then evaluated and approved by the United Nations Framework Convention on Climate Change (UNFCCC). The Clean Development Mechanism (CDM) has, thus far, produced over 200 such methodologies, and many of them serve as the foundation for methods in various other C crediting programs. These methodologies are either directly adopted (Joint Implementation, Gold Standard, and Verified C Standard) or adapted to meet specific program requirements (Clean Air Act, California C Offset, Gold Standard, Verified C Standard, Switzerland). The Verified C Standard (VCS) not only accepts CDM and Clean Air Act methodologies but also develops new methodologies through a process that includes public consultations and two rounds of reviews conducted by external validation/verification bodies (Espenan 2023).

Monitoring

Monitoring involves the process of collecting data directly or through the use of proxies, which is essential for calculating emission reductions within a specific scope and time frame (Shishlov and Cochran 2015). In the context of CDM, a monitoring plan is essential for identifying and regularly measuring (or estimating) anthropogenic GHG emissions from sources within the project boundaries. This monitoring plan should adhere to the methodologies, guidance documents, and standards provided by the CDM executive board. For California and Switzerland, their monitoring requirements are tied to their respective local jurisdictions. California has specified these requirements in its local cap-and-trade regulations and compliance offset protocols. The Gold Standard has introduced an innovative approach by not only monitoring GHG reductions but also assessing positive sustainable development goals (SDG) impacts, engaging stakeholders, and incorporating climate-related monitoring parameters.

Reporting

Following the measurement of emissions reductions or C sequestration, project developers are required to report this data to the relevant authorities or C credit standard bodies. Reporting encompasses the processes of aggregating, documenting, and communicating this data to the appropriate authorities (Shishlov and Cochran 2015). Within any program, multiple reporting periods are typically included in the project's crediting period.

In the context of the CDM, the duration of monitoring periods, that is, the frequency of submitting monitoring reports, is not fixed, and larger projects often submit reports more frequently. Australia's Emission Reduction Fund and Carbon Farming Initiative (CFI) participants, on the other hand, are mandated to complete and submit project reports to the Clean Energy Regulator after each reporting period. The initial reporting period commences at the beginning of a project's crediting period, with subsequent reporting periods following immediately after the conclusion of the previous one. The reporting periods under the CFI may vary, ranging from 6 months to 5 years. Notably, credits can only be issued to a project once a report has been received and assessed (Miller and Cox 2014).

Typically, reporting encompasses the following key elements:

(a) Documenting the methods and data sources used for measurement
(b) Providing transparent reporting of results, including the documentation of emissions reductions or C stock changes over time

(c) Submitting reports at regular intervals, such as annually or as required by the specific C credit program. For instance, in British Columbia, project proponents are obligated to report on emission reductions or removals on an annual basis.

Verification
Independent verification is a crucial step in ensuring the accuracy of reported emissions reductions or C sequestration. This verification process may entail the involvement of third-party auditors or verification bodies and is designed to detect errors or potential fraudulent reporting. Typically, verification is carried out by an independent third party, as highlighted by Shishlov and Cochran (2015). The majority of C crediting mechanisms mandate some form of verification by an independent third party, usually accredited by the governing body. For instance, in the case of CDM, the consistency between the project description and the relevant methodology used to calculate emissions reductions, the monitoring plan, and the proper implementation of the project must be periodically verified by an independent auditor accredited by the United Nations Framework Convention on Climate Change (UNFCCC). In contrast, some programs, like the Green Investment Scheme (GIS), do not require any verification. Furthermore, programs like VCS and the Gold Standard (GS) allow for the simultaneous validation of a project and the verification of emissions reductions. The broad acceptance and implementation of low-C agriculture face a range of institutional and technological challenges. Specifically, there is a need to develop new low-C farming practices that consider both technological advancements and economic viability. Additionally, to incentivize farmers to embrace these practices, it will be essential to have robust public policies and supportive initiatives from private organizations.

The central focus of the present research lies in the third challenge, which revolves around assessing the outcomes of implementing low-C farming techniques and the indispensable necessity to establish streamlined monitoring and MRV systems to facilitate these endeavors. Overcoming institutional and technological obstacles is pivotal to driving the widespread adoption of low-C agriculture.

For example, this necessitates the development of innovative and sustainable low-C farming practices while simultaneously evaluating their economic viability. Furthermore, it calls for the presence of robust government initiatives and complementary actions from private organizations to incentivize farmers to embrace these sustainable approaches. This third challenge, which constitutes the primary area of investigation in our ongoing research, focuses on the assessment of the positive impacts derived from the implementation of low-C farming practices and the critical need to institute effective MRV systems to support and authenticate these endeavors. Before low-C agriculture can gain widespread adoption, several institutional and technological challenges must be effectively addressed. For instance, it is essential to devise innovative low-C farming methods while simultaneously assessing their economic viability. Encouraging producers to adopt these strategies will require the implementation of effective government programs and community support for private initiatives.

The third challenge, which is also the primary focus of the present research, entails the evaluation of the benefits of sustainable farming methods after their implementation. This also involves laying the groundwork for the development of robust MRV systems. Previous experiences have shown that a hybrid governance model, involving collaboration between private and public stakeholders, is the most effective approach for managing large-scale surveillance systems that require property-level information. For instance, the use of third-party verification to access privately held information can be costly for governmental entities (Cashore et al. 2003). Therefore, a hybrid governance model in which farmers and other commercial actors, including certifying bodies, cooperate to provide information, may reduce costs for government agencies (Endres 2010).

A hybrid governance system could enhance individual farmers' access to markets. In return for the information they provide, they are granted access to secondary databases made available by public authorities, substantially reducing the cost of private certification (Perosa and Azevedo 2019). Smallholders may face certification expenses that jeopardize their participation in supply chains (Bartley 2011). Hence, there is a need for approaches that result in agreements benefiting both the public and private sectors.

Ideally, MRV procedures should be established to ensure that each step of the process complements the others. For instance, MRV systems have the potential to reduce audit costs by utilizing satellite imagery of cultivated land to validate data self-reported by farmers (Grecchi et al. 2014). To scale up MRV systems, similar approaches are essential, especially in extensive areas where the cost of on-site verification may render expansion unfeasible.

MRV systems have found applications in various industries, including transportation, manufacturing, and agriculture, and have become integral to public policy. They play a crucial role in initiatives like Nationally Appropriate Mitigation Actions (NAMA), which are tailored to specific sectors. For instance, Costa Rica has established a NAMA for its cattle and coffee industries, two of its major exports. To monitor and account for emissions reductions in these programs, specific MRV systems have been developed.

Similarly, Chile has implemented a NAMA for the timber industry, which includes a dedicated MRV framework (FAO 2018). Furthermore, MRV systems are being designed and implemented for specific sectors. For example, the forestry sector has created monitoring systems to link projects and policies (Loveridge 2021). The development and execution of comprehensive processes for measuring forest C stocks and mitigating emissions have been instrumental in the establishment of initiatives like reducing emissions from deforestation and degradation (REDD+) and other C markets centered on forests.

17.8 Policies and Framework for Carbon Crediting

Particularly interesting that the usage of C imprint in the present moment aims to directly access and take into account C pricing measurement along with the potential for follow-up credits. As mentioned, there exist multiple approaches for the price of C. Raising taxes on C at the point of production usually ends in reduced expenses whenever connected to the complete supply chain, as well as minimizing the price of commodities at the point of consumption. The implementation of CO_2 prices is reflected in the formation of financing based on market sustainability. In simple terms, the price could give businesses with a small environmental impact an edge in the marketplace, encouraging the supply of financial resources to these businesses and promoting a transition to a low-C economy. Right now, organizations allocate C pricing through any of the three internal methods: C taxes, the C shadow price, or explicit costing. A C tax, or the price paid per unit of GHG emissions, is frequently adopted by enterprises. This expenditure is added to the regular expenses of the examination. By adopting this management approach, the corporation creates incentives to reduce temporary emissions while fostering long-term innovation; nevertheless, challenges may remain in ensuring sustained impact. As a result, the tax on C grades can be set slightly lower to encourage buy-in from individual company units. As an outcome, the tax on C grades can be a little low to promote individual company unit buy-in.

According to the previous example, documenting all GHG emissions in monetary terms offers an excellent substitute for promoting energy efficiency through immediate C cuts. To attempt to modify the way that things are usually done, each operating unit's effect on the industrial follow-up through taxes on C is reflected within its decision-making, resulting in improvements in the immediate environment. Furthermore, by implementing those changes into behavior, it could be possible to reduce expenditures by building something for climate mitigation. Taxes may be used to boost C reduction objectives that would otherwise call for other funding sources. The instrument might position the organization as an expert in climate strategy. This necessitates that a company has the expertise and equipment essential to measure its GHG emissions consistently. For a business considering implementing such an approach, its lack of a C reference rate in regions without C pricing might serve as a deterrent. The company-wide costs of C are thus believed to be an emerging company strategy.

The hidden costing tackle is the second strategy employed by industrial sectors to regulate the selling price of C. To be able to figure out how the emissions of GHG influence the business's revenue on each investment, the company uses this method by integrating an internal cost of C within every single one of its funding plans. An organization could employ its C shadows price to forecast the effects of possible legislative eventualities on the company's operations and supply mix, alongside applying it for directing investment projects in capital. Phantom pricing of C may therefore help companies in managing hazards and the execution of strategies.

The real-world or forecast prices of C's underlying current emissions trading applications may serve as the foundation for the shadow price of C. In addition to other government policies that implicitly price C, indirect costs can also be calculated using the expense of renewable energy sources or charges on certain products. In a marketplace with strict regulations, businesses can also establish a pricing variety that depends on the CO_2 emissions of an undertaking. Company-used "shadow" pricing often goes above what is subject to the emissions tax.

According to inferred prices, the price of C fluctuates depending on the amount of money a company spends on lowering its GHG emissions and complying with regulations. In this instance, the marginal cost of decreasing these GHG is the expense for a single ton of C. Contrary to the two other C pricing strategies mentioned above, this tool is backwards because it concentrates on the business's previous choices as opposed to what the business can do in the future. Delivering a company further details about both the environmental impact and the cost of adhering to regulations is beneficial. As a result, a company may implement many C pricing approaches at once and decide whether to impose one cost on all of its activities or pursue a differentiated pricing plan.

Based on the initial studies that have been disclosed, both of these main possibilities are not, nevertheless, incompatible with each other. Nevertheless, the main strategy seeks to bring about energy reforms by regulating GHG emissions and reducing global warming by dispersing every commodity as a pot of money to compensate individuals who encounter electricity shortages. Consequently, the national government's initiative has to implement general C verification related to C tax conversion alongside C credits to mitigate these adverse consequences by raising the selling price of C and allocating money for the community at large. In comparison, the second means of payment through emissions credit comes in the form of fees and payouts. In summary, this option is not significantly different from the first; however, revenues are given away to all citizens by dividends that are distributed to all homes. With those with a lower income having the chance to earn more, many households will be regarded as net beneficiaries. Depending on what is agreed upon for both the near and long term, there continues to be a lot of studies and condensing to be done before government initiatives can be put into practice.

17.9 Conclusion

The need to promote sustainable growth (production + sustainability) with fair human rights and duties has gained worldwide recognition due to the impact of climate change. This difficulty has created new chances for underdeveloped and developing nations to trade C and draw spending in GHG mitigation initiatives. With an anticipated annual business of 50 million tons C, land use, land-use change, and forestry (LULUCF) was an important factor in CDM (A/R) for the first commitment period. However, in practice, this could not be realized because of the nonrepetitive achievement of projected tasks due to complex procedures, agroforestry best practices, low stakeholder consciousness, high transaction costs, related multiple risks, etc. In the developing world, there are significant gaps between scientific and operational capacity to make ecological improvements, so to utilize academic research, the limitations mentioned in this paper must be appropriately addressed. Additionally, since the interventions vary at the farmer level in a variety of agroforestry systems and designs, it is crucial to produce system-specific figures rather than general projections. Even though the extra C payments are not realized anytime soon, providing AF adopters with the right technical knowledge and policy support can have a significant impact on ecological services and job creation. With further observable advantages, rising demand for biomass C can contribute to emission reduction through the natural process of photosynthesis. Farmers would have new opportunities if India implemented the AF Policy, which views the tree–crop interaction as a key option for helping climate-smart agriculture and scaling up the practical application of C benefits with the right market and technological assistance. As a confidence-boosting step to improve long-lived C pools in this intricate land use system, related uncertainty and risk must be addressed.

References

Aggarwal A, Arostegui D, DeLyser K, Hewett B, Johnson E, Rudee A (2018) Achieving the mid-century strategy goals for deep decarbonization in agriculture and forestry. Bass Connections, Durham

Bai Y, Zhou Y, He H (2020) Effects of rehabilitation through afforestation on soil aggregate stability and aggregate-associated carbon after forest fires in subtropical China. Geoderma 376: 114548

Bartley T (2011) Transnational governance as the layering of rules: intersections of public and private standards. Theor Inq L 12:517–542

Batjes NH (2004) Estimation of soil carbon gains upon improved management within croplands and grasslands of Africa. Environ Dev Sustain 6:133–143

Benbi DK, Brar K, Toor AS, Singh P, Singh H (2012) Soil carbon pools under poplar-based agroforestry, rice-wheat, and maize-wheat cropping systems in semi-arid India. Nutr Cycl Agroecosyst 92:107–118

Berhe AA, Harte J, Harden JW, Torn MS (2007) The significance of the erosion-induced terrestrial carbon sink. Bioscience 57:337–346

Bhat JA, Kumar M, Negi AK, Pala NA, Todaria NP (2012) Soil organic carbon stock and sink potential in high mountain temperate Himalayan forests of India. Int J Curr Res 4:206–209

Bhatia A, Jain N, Pathak H (2013) Methane and nitrous oxide emissions from Indian rice paddies, agricultural soils and crop residue burning. Greenhouse Gases Sci Technol 3:196–211

Carrijo DR, Lundy ME, Linquist BA (2017) Rice yields and water use under alternate wetting and drying irrigation: a meta-analysis. Field Crop Res 203:173–180

Cashore B, Auld G, Newsom D (2003) Forest certification (eco-labeling) programs and their policy-making authority: explaining divergence among North American and European case studies. Forest Policy Econ 5:225–247

Chauhan SK, Gupta N, Walia R, Yadav S, Chauhan R, Mangat PS (2011) Biomass and carbon sequestration potential of poplar-wheat inter-cropping system in irrigated agro-ecosystem in India. J Agric Sci Technol A 1:575–586

Chauhan BS, Mahajan G, Sardana V, Timsina J, Jat ML (2012) Productivity and sustainability of the rice–wheat cropping system in the Indo-Gangetic Plains of the Indian subcontinent: problems, opportunities, and strategies. Adv Agron 117:315–369

Chen H, Marhan S, Billen N, Stahr K (2009) Soil organic-carbon and total nitrogen stocks as affected by different land uses in Baden-Württemberg (Southwest Germany). J Plant Nutr Soil Sci 172:32–42

Clark KM, Boardman DL, Staples JS, Easterby S, Reinbott TM, Kremer RJ, Kitchen NR, Veum KS (2017) Crop yield and soil organic carbon in conventional and no-till organic systems on a claypan soil. Agron J 109:588–599

Endres JM (2010) Clearing the air: the meta-standard approach to ensuring biofuels environmental and social sustainability. Va Environ L J 28:73

Espenan NP (2023) Improving voluntary carbon markets through standardization and blockchain technology. Wyo L Rev 23:141

Fairfield Market Research (2023) Carbon credit market. https://www.fairfieldmarketresearch.com/report/carbon-credit-market [WWW Document]. FAIRFIELD

FAO (2018) Food and Agriculture Organization of the United Nations, Rome. http://faostat.fao.org 403

Feliciano D, Ledo A, Hillier J, Nayak DR (2018) Which agroforestry options give the greatest soil and above ground carbon benefits in different world regions? Agric Ecosyst Environ 254:117–129

Gathala MK, Ladha JK, Kumar V, Saharawat YS, Kumar V, Sharma PK, Sharma S, Pathak H (2011) Tillage and crop establishment affects sustainability of South Asian rice–wheat system. Agron J 103:961–971

Gathala MK, Kumar V, Sharma PC, Saharawat YS, Jat HS, Singh M, Kumar A, Jat ML, Humphreys E, Sharma DK (2013) Optimizing intensive cereal-based cropping systems addressing current and future drivers of agricultural change in the northwestern Indo-Gangetic Plains of India. Agric Ecosyst Environ 177:85–97

Gillenwater M, Seres S (2011) The clean development mechanism: a review of the first international offset programme. GHG Measure Manage 1:179–203

Grecchi RC, Gwyn QHJ, Bénié GB, Formaggio AR, Fahl FC (2014) Land use and land cover changes in the Brazilian Cerrado: a multidisciplinary approach to assess the impacts of agricultural expansion. Appl Geogr 55:300–312

Guo LB, Gifford RM (2002) Soil carbon stocks and land use change: a meta analysis. Glob Chang Biol 8:345–360

Haile SG, Nair PKR, Nair VD (2008) Carbon storage of different soil-size fractions in Florida silvopastoral systems. J Environ Qual 37:1789–1797

Han P, Zhang W, Wang G, Sun W, Huang Y (2016) Changes in soil organic carbon in croplands subjected to fertilizer management: a global meta-analysis. Sci Rep 6:27199

Hazra KK, Nath CP, Ghosh PK, Swain DK (2020) Inclusion of legumes in rice–wheat cropping system for enhancing carbon sequestration. In: Carbon management in tropical and sub-tropical terrestrial systems. Springer, Singapore, pp 23–36

Hongo T (2018) Prospects of quality infrastructure program and private sector MRV for accelerating the transition towards low-carbon energy system. In: Financing for low-carbon energy transition: unlocking the potential of private capital. Springer, Singapore, pp 359–396

Hutchinson JJ, Campbell CA, Desjardins RL (2007) Some perspectives on carbon sequestration in agriculture. Agric For Meteorol 142:288–302

Hymavathi HN, Kandya AK, Patel LP (2010) Beneficial effects of multiple plantation patterns in agroforestry systems. Indian Forester 136(4):465–475

IPCC (2021) Climate Change 2021: The Physical Science Basis. Contribution of Working Group I to the Sixth Assessment Report of the Intergovernmental Panel on Climate Change. [Masson-Delmotte, V., P. Zhai, A. Pirani, S.L. Connors, C. Péan, S. Berger, N. Caud, Y. Chen, L. Goldfarb, M.I. Gomis, M. Huang, K. Leitzell, E. Lonnoy, J.B.R. Matthews, T.K. Maycock, T. Waterfield, O. Yelekçi, R. Yu, and B. Zhou (eds.)]. Cambridge University

Press, Cambridge, United Kingdom and New York, NY, USA, 2391. https://doi.org/10.1017/9781009157896

Itoh M, Sudo S, Mori S, Saito H, Yoshida T, Shiratori Y, Suga S, Yoshikawa N, Suzue Y, Mizukami H (2011) Mitigation of methane emissions from paddy fields by prolonging midseason drainage. Agric Ecosyst Environ 141:359–372

Jangir CK, Sangwan PS, Panghaal D, Kumar S, Meena RS, Bharti, Jat RD, Singh N (2024) Spatial variability and statistical analysis of soil properties in the rice wheat-based systems of north-west India. Commun Soil Sci Plant Anal 55(8):1205–1223

Jhariya MK, Raj A, Banerjee A, Meena RS, Bargali SS, Kumar S, Nema S, Poonam, Oraon PR (2022) Plan and policies for soil organic carbon management under agroforestry system. In: Plans and policies for soil organic carbon management in agriculture. Singapore, Springer, pp 191–219

Kirby KR, Potvin C (2007) Variation in carbon storage among tree species: implications for the management of a small-scale carbon sink project. For Ecol Manag 246:208–221

Kumar V, Ladha JK (2011) Direct seeding of rice: recent developments and future research needs. Adv Agron 111:297–413

Kumar R, Punia H, Kumar S (2020) Nanomaterials for agriculture input use efficiency. In: Resources use efficiency in agriculture. Springer, Singapore. https://doi.org/10.1007/978-981-15-6953-1_5

Kumar S, Sharma SK, Thakral SK, Bhardwaj KK, Jhariya MK, Meena RS, Jangir CK, Bedwal S, Jat RD, Gaber A, Atta AA (2022a) Integrated nutrient management improves the productivity and nutrient use efficiency of *Lens culinaris* Medik. Sustainability 14(3):1284

Kumar S, Bamboriya SD, Rani K, Meena RS, Sheoran S, Loyal A, Kumawat A, Jhariya MK (2022b) Grain legumes: a diversified diet for sustainable livelihood, food, and nutritional security. In: Advances in legumes for sustainable intensification. Academic Press, London, pp 157–178

Kumar S, Sharma SK, Dhaka AK, Bedwal S, Sheoran S, Meena RS, Jangir CK, Kumar D, Kumar R, Jat RD, Meena AK (2023) Efficient nutrient management for enhancing crop productivity, quality and nutrient dynamics in lentil (*Lens culinaris* Medik.) in the semi-arid region of northern India. PLoS One 18(2):e0280636

Kumar S, Meena RS, Kumar S, Pradhan G, Jangir CK, Ghosh S, Punia H, Sheoran P, Meena R, Ahmad MA, Goyal SK (2025) Designing a diversified Indian mustard production system for energy-carbon-cum-heat use efficiency and sowing dates assessment. GCB Bioenergy 17(6): e70044

Kumara TM, Kandpal A, Pal S (2019) Determinants and impacts of conservation agriculture in South Asia: a meta-analysis of the evidences. Indian J Agric Econ 74:311–320

Ladha JK, Kumar V, Alam MM, Sharma S, Gathala M, Chandna P, Saharawat YS, Balasubramanian V (2009) Integrating crop and resource management technologies for enhanced productivity, profitability, and sustainability of the rice-wheat system in South Asia. In: Integrated crop and resource management in the rice–wheat system of South Asia. ADB, Manila, pp 69–108

Laik R, Sharma S, Idris M, Singh AK, Singh SS, Bhatt BP, Saharawat Y, Humphreys E, Ladha JK (2014) Integration of conservation agriculture with best management practices for improving system performance of the rice–wheat rotation in the eastern Indo-Gangetic Plains of India. Agric Ecosyst Environ 195:68–82

Leharwan M, Kumar Y, Kumar R, Kumar Saraswat P, Kumar R, Kumar Thaliyil Veetil A, Bhattacharjee S, Kumar A, Kumar S (2023) Assessing the effects of conservation tillage and in-situ crop residue management on crop yield and soil properties in rice–wheat cropping system. Sustainability 15(17):12736

Li X, Zhu W, Xu F, Du J, Tian X, Shi J, Wei G (2021) Organic amendments affect soil organic carbon sequestration and fractions in fields with long-term contrasting nitrogen applications. Agric Ecosyst Environ 322:107643

Loveridge R (2021) Beyond win-wins: understanding diverse impacts of complex protected area governance arrangements on human wellbeing and conservation in tropical forests. PhD thesis, University of York, York

McCarthy J, Prudham S (2004) Neoliberal nature and the nature of neoliberalism. Geoforum 35: 275–283

Meena RS, Pradhan G (2023) Industrial garbage-derived biocompost enhances soil organic carbon fractions, CO2 biosequestration, potential carbon credits and sustainability index in a rice-wheat ecosystem. Environ Res 235:116525

Meena RS, Pradhan G, Kumar S, Lal R (2023) Using industrial wastes for rice-wheat cropping and food-energy-carbon-water-economic nexus to the sustainable food system. Renew Sust Energ Rev 187:113756

Meena RS, Pradhan G, Singh K, Kumar S, Singh AK, Shashidhar KS, Mina KK, Rao CS (2024a) Agriculture models for restoring degraded land to enhance CO_2 biosequestration and carbon credits in the Vindhyan region of India. Sci Total Environ 929:172661. https://doi.org/10.1016/j.scitotenv.2024.172661

Meena RS, Lal R, Kumar S, Pradhana G, Rao CS, Singh AK, Pathak H, Abhilash PC, Kumar A, Sharma SK, Jat ML, Singh S (2024b) Potential of Indian agriculture for capturing atmospheric CO2 and monetizing carbon credits to the farmers: an overview and policy framework. Adv Agron 188:101–206. https://doi.org/10.1016/bs.agron.2024.06.001

Meena RS, Pradhan G, Nalani SK, Singh AK, Verma SK, Mina KK, Kumar S, Chaturvedi RK (2024c) Diversified cropping modules designed for soil restoration, CO_2 sequestration, and generating carbon credits. Land Degrad Dev 35. https://doi.org/10.1002/ldr.5224

Meena RS, Pradhan G, Kumar S (2025) Energy flow, eco-efficiency, and economic circulation with recycled industrial waste compost application in wheat and subsequent rice farming. Sci Total Environ 967:178779. https://doi.org/10.1016/j.scitotenv.2025.178779

Michaelowa A, Shishlov I, Hoch S, Bofill P, Espelage A (2019) Overview and comparison of existing carbon crediting schemes. NEFCO, Helsinki

Miller M, Cox S (2014) Overview of variable renewable energy regulatory issues: a clean energy regulators initiative report. National Renewable Energy Lab (NREL), Golden

Minamikawa K, Fumoto T, Itoh M, Hayano M, Sudo S, Yagi K (2014) Potential of prolonged midseason drainage for reducing methane emission from rice paddies in Japan: a long-term simulation using the DNDC-Rice model. Biol Fertil Soils 50:879–889

Mkhabela MS, Madani A, Gordon R, Burton D, Cudmore D, Elmi A, Hart W (2008) Gaseous and leaching nitrogen losses from no-tillage and conventional tillage systems following surface application of cattle manure. Soil Tillage Res 98:187–199

Mohit Gera MG, Giresh Mohan GM, Bisht NS, Neelu Gera NG (2006) Carbon sequestration potential under agroforestry in Rupnagar District of Punjab. Indian Forester. 132(5):543–555

Murthy IK, Gupta M, Tomar S, Munsi M, Tiwari R, Hegde GT, Ravindranath NH (2013) Carbon sequestration potential of agroforestry systems in India. J Earth Sci Clim Change 4:1–7

Nieder R, Benbi DK (2008) Carbon and nitrogen in the terrestrial environment. Springer, Dordrecht

Nisbet EG, Manning MR, Dlugokencky EJ, Fisher RE, Lowry D, Michel SE, Myhre CL, Platt SM, Allen G, Bousquet P (2019) Very strong atmospheric methane growth in the 4 years 2014–2017: implications for the Paris agreement. Glob Biogeochem Cycles 33:318–342

O'Connor FM, Boucher O, Gedney N, Jones CD, Folberth GA, Coppell R, Friedlingstein P, Collins WJ, Chappellaz J, Ridley J (2010) Possible role of wetlands, permafrost, and methane hydrates in the methane cycle under future climate change: a review. Rev Geophys 48:RG4005

Padbhushan R, Sharma S, Kumar U, Rana DS, Kohli A, Kaviraj M, Parmar B, Kumar R, Annapurna K, Sinha AK (2021) Meta-analysis approach to measure the effect of integrated nutrient management on crop performance, microbial activity, and carbon stocks in Indian soils. Front Environ Sci 9:724702

Palm C, Tomich T, Van Noordwijk M, Vosti S, Gockowski J, Alegre J, Verchot L (2004) Mitigating GHG emissions in the humid tropics: case studies from the Alternatives to Slash-and-Burn Program (ASB). Environ Dev Sustain 6:145–162

Panchasara H, Samrat NH, Islam N (2021) Greenhouse gas emissions trends and mitigation measures in Australian agriculture sector—a review. Agriculture 11:85

Perosa BB, Azevedo PFDE (2019) A evolução de mecanismos de governança ambiental: Uma Ferramenta analítica institucional aplicado ao setor de biocombustíveis. Ambiente Soc 22: e02602

Pradhan G, Meena RS (2023a) Interaction impacts of biocompost on nutrient dynamics and relations with soil biota, carbon fractions index, societal value of CO2 equivalent and ecosystem services in the wheat-rice farming. Chemosphere 139695:139695

Pradhan G, Meena RS (2023b) Utilizing waste compost to improve the atmospheric CO2 capturing in the rice-wheat cropping system and energy-cum-carbon credit audit for a circular economy. Sci Total Environ:164572. https://doi.org/10.1016/j.scitotenv.2023.164572

Pradhan G, Meena RS, Kumar S, Lal R (2023) Utilizing industrial wastes as compost in wheat-rice production to improve the above and below-ground ecosystem services. Agric Ecosyst Environ 358:108704. https://doi.org/10.1016/j.agee.2023.108704

Punia H, Tokas J, Malik A, Satpal, Rani A, Gupta P, Kumari A, Mor VS, Bhuker A, Kumar S (2020) Solar radiation and nitrogen use efficiency for sustainable agriculture. In: Resources use efficiency in agriculture. Springer, Singapore, pp 177–212

Raj A, Jhariya MK, Banerjee A, Meena RS, Nema S, Khan N, Yadav SK, Pradhan G (2022) Agroforestry a model for ecological sustainability. In: Natural resources conservation and advances for sustainability. Elsevier, San Diego, pp 289–307

Ramachandran Nair PK, Mohan Kumar B, Nair VD (2009) Agroforestry as a strategy for carbon sequestration. J Plant Nutr Soil Sci 172:10–23

Rogers EM, Singhal A (2003) Empowerment and communication: lessons learned from organizing for social change. Ann Int Commun Assoc 27:67–85

Roshetko JM, Delaney M, Hairiah K, Purnomosidhi P (2002) Carbon stocks in Indonesian homegarden systems: can smallholder systems be targeted for increased carbon storage? Am J Altern Agric 17:138–148

Roy O, Meena RS, Kumar S, Jhariya MK, Pradhan G (2022) Assessment of land use systems for CO_2 sequestration, carbon credit potential, and income security in Vindhyan Region, India. Land Degrad Dev 33:670–682

Sanchez PA (2000) Linking climate change research with food security and poverty reduction in the tropics. Agric Ecosyst Environ 82:371–383

Saunois M, Stavert AR, Poulter B, Bousquet P, Canadell JG, Jackson RB, Raymond PA, Dlugokencky EJ, Houweling S, Patra PK (2019) The global methane budget 2000–2017. Earth Syst Sci Data Discuss 2019:1–136

Sharma AR, Jat ML, Saharawat YS, Singh VP, Singh R (2012) Conservation agriculture for improving productivity and resource-use efficiency: prospects and research needs in Indian context. Indian J Agron 57:131–140

Sheoran S, Gupta M, Kumari S, Kumar S, Rakshit S (2022) Meta-QTL analysis and candidate genes identification for various abiotic stresses in maize (*Zea mays* L.) and their implications in breeding programs. Mol Breed 42(5):26

Shishlov I, Cochran I (2015) Tendances Carbone no. 105. Joint implementation: the good, the bad and how to avoid the ugly. Institute for Climate Economics-I4CE, Paris

Singh G, Singh NT, Dagar JC, Singh H, Sharma VP (1997) An evaluation of agriculture, forestry and agroforestry practices in a moderately alkali soil in northwestern India. Agrofor Syst 37: 279–295

Singh S, Chaturvedi S, Nayak P, Dhyani VC, Nandipamu TMK, Singh DK, Gudapaty P, Mathyam P, Srinivasrao K, Govindaraju K (2023) Carbon offset potential of biochar based straw management under rice-wheat system along Indo-Gangetic Plains of India. Sci Total Environ 897:165176

Sinha R, Soni P, Perret SR (2020) Environmental and economic assessment of paddy based cropping systems in Middle Indo-Gangetic plains, India. Enviro Sustain Indic 8:100067

Smith P, Bustamante M, Ahammad H, Clark H, Dong H, Elsiddig EA, Haberl H, Harper R, House J, Jafari M (2014) Agriculture, forestry and other land use (AFOLU). In: Climate change 2014: mitigation of climate change. Contribution of working group III to the fifth assessment report of the intergovernmental panel on climate change. Cambridge University Press, Cambridge, pp 811–922
Stallworth SD (2021) Weedy rice (Oryza sativa ssp.): an untapped genetic resource for abiotic stress tolerant traits for rice improvement. Mississippi State University, Mississippi State
Swamy SL, Puri S (2005) Biomass production and C-sequestration of Gmelina arborea in plantation and agroforestry system in India. Agrofor Syst 64:181–195
Tamburini G, Bommarco R, Wanger TC, Kremen C, Van Der Heijden MGA, Liebman M, Hallin S (2020) Agricultural diversification promotes multiple ecosystem services without compromising yield. Science. Advances 6:eaba1715
Torralba M, Fagerholm N, Burgess PJ, Moreno G, Plieninger T (2016) Do European agroforestry systems enhance biodiversity and ecosystem services? A meta-analysis. Agric Ecosyst Environ 230:150–161
Voulgarakis A, Naik V, Lamarque J-F, Shindell DT, Young PJ, Prather MJ, Wild O, Field RD, Bergmann D, Cameron-Smith P (2013) Analysis of present day and future OH and methane lifetime in the ACCMIP simulations. Atmos Chem Phys 13:2563–2587
Wu L, Zhang S, Ma R, Chen M, Wei W, Ding X (2021) Carbon sequestration under different organic amendments in saline-alkaline soils. Catena 196:104882
Yadav GS, Babu S, Das A, Datta M, Mohapatra KP, Singh R, Singh VK, Rathore SS, Chakraborty M (2021) Productivity, soil health, and carbon management index of Indian Himalayan intensified maize-based cropping systems under live mulch based conservation tillage practices. Field Crop Res 264:108080
Yang J, Zhou Q, Zhang J (2017) Moderate wetting and drying increases rice yield and reduces water use, grain arsenic level, and methane emission. Crop J 5:151–158
Ye Y, Liang X, Chen Y, Liu J, Gu J, Guo R, Li L (2013) Alternate wetting and drying irrigation and controlled-release nitrogen fertilizer in late-season rice. Effects on dry matter accumulation, yield, water and nitrogen use. Field Crop Res 144:212–224

MIX
Papier aus verantwortungsvollen Quellen
Paper from responsible sources
FSC® C105338

If you have any concerns about our products,
you can contact us on
ProductSafety@springernature.com

In case Publisher is established outside the EU,
the EU authorized representative is:
Springer Nature Customer Service Center GmbH
Europaplatz 3, 69115 Heidelberg, Germany

Printed by Libri Plureos GmbH
in Hamburg, Germany